Springer Proceedings in Advanced Robotics 8

The Springer Proceedings in Advanced Robotics (SPAR) publishes new developments and advances in the fields of robotics research, rapidly and informally but with a high quality.
The intent is to cover all the technical contents, applications, and multidisciplinary aspects of robotics, embedded in the fields of Mechanical Engineering, Computer Science, Electrical Engineering, Mechatronics, Control, and Life Sciences, as well as the methodologies behind them.

The publications within the "Springer Proceedings in Advanced Robotics" are primarily proceedings and post-proceedings of important conferences, symposia and congresses. They cover significant recent developments in the field, both of a foundational and applicable character. Also considered for publication are edited monographs, contributed volumes and lecture notes of exceptionally high quality and interest.

An important characteristic feature of the series is the short publication time and world-wide distribution. This permits a rapid and broad dissemination of research results.

More information about this series at http://www.springer.com/series/15556

Jadran Lenarcic · Vincenzo Parenti-Castelli
Editors

Advances in Robot Kinematics 2018

Springer

Editors
Jadran Lenarcic
J. Stefan Institute
Ljubljana, Slovenia

Vincenzo Parenti-Castelli
Department of Industrial Engineering
University of Bologna
Bologna, Italy

ISSN 2511-1256 ISSN 2511-1264 (electronic)
Springer Proceedings in Advanced Robotics
ISBN 978-3-030-06608-6 ISBN 978-3-319-93188-3 (eBook)
https://doi.org/10.1007/978-3-319-93188-3

Printed on acid-free paper

This Springer imprint is published by the registered company Springer International Publishing AG part of Springer Nature
The registered company address is: Gewerbestrasse 11, 6330 Cham, Switzerland

Preface

The series of international conferences Advances in Robot Kinematics (ARK) was organized for the first time in 1988 in Ljubljana (Slovenia); therefore, in 2018, we celebrate its 30th anniversary. ARK symposia are regularly organized, every two years, in various European cities. Since 1992, ARK is held under the patronage of the International Federation for the Promotion of Machine Sciences (IFToMM). Few people would have thought, in the early years of the symposium, that it would last for so long and, after thirty years, it would continue to stir the interest of so many researchers around the world. Kinematics, the motion of mechanisms, remains one of the most fundamental aspects of robot design, analysis, and control. It is also relevant to other scientific domains, such as biomechanics, sports, molecular biology. In 2018, the symposium, which is the sixteenth in a row, is held in Bologna (Italy), the city of the oldest European university. It is organized by the Group of Robotics, Automation and Biomechanics (GRAB), at the Department of Industrial Engineering of the University of Bologna.

The first book on Advances in Robot Kinematics was published by Springer in 1991. Since 1994, a new issue has been published every two years. Each edited book is linked to the corresponding symposium, in which the participants exchange their results and opinions in a meeting that brings together the world's best researchers and scientists in the field of robot kinematics, as well as young students. The current book is the 14th in the series of Springer (and Kluwer). It is a result of a peer-review process intended to select the newest and the most original achievements in the field. It contains 53 contributions. A large team of reviewers contributed their critical and insightful recommendations. The articles in this book show once again that robot kinematics is an exciting domain, with a wide number of research challenges that go well beyond the field of robotics.

First and foremost, we are grateful to the authors who participated in the project with their enthusiasm and strong commitment. We are grateful to Springer, but especially to Nathalie Jacobs, Cynthia Feenstra, and Anneke Pot, who have made this publication possible. We are indebted to the members of the GRAB team for organizing the symposium. Above all, we are grateful to our younger colleagues Marco Carricato and Rocco Vertechy, who performed a magnificent job in preparing the book for publication. Without them, things would not have taken place as efficiently and rapidly. We finally gratefully acknowledge the generous support provided by the sponsors of ARK 2018: Marchesini Group, L3 Calzoni, and Coesia GD.

We hope that this new ARK book will again reach the shelves of scholars, researchers, and students who are attracted to the unique field of robot kinematics around the world.

Jadran Lenarčič
Vincenzo Parenti-Castelli

Contents

Design

Cable Driven Parallel Manipulators

Biorobotics

Theoretical Kinematics

Kinematic Interpretation of the Study Quadric's Ambient Space

Georg Nawratil[✉]

Institute of Discrete Mathematics and Geometry,
Vienna University of Technology, Vienna, Austria
nawratil@geometrie.tuwien.ac.at

Abstract. It is well known that real points of the Study quadric (sliced along a 3-dimensional generator space) correspond to displacements of the Euclidean 3-space. But we still lack of a kinematic meaning for the points of the ambient 7-dimensional projective space P^7. This paper gives one possible interpretation in terms of displacements of the Euclidean 4-space. From this point of view we also discuss the extended inverse kinematic map and motions corresponding to straight lines in P^7. Moreover we present an application of this interpretation in the context of interactive motion design.

Keywords: Kinematic map · Study quadric · SE(4)
Circular Darboux 2-Motion

1 Introduction

$\mathfrak{Q} := q_0 + q_1\mathbf{i} + q_2\mathbf{j} + q_3\mathbf{k}$ with $q_0, \ldots, q_3 \in \mathbb{R}$ is an element of the skew field of quaternions $\mathbb{H}$, where $\mathbf{i}, \mathbf{j}, \mathbf{k}$ are the quaternion units. The scalar part is q_0 and the pure part equals $q_1\mathbf{i} + q_2\mathbf{j} + q_3\mathbf{k}$, which is also denoted by $\mathfrak{q}$. The conjugated quaternion to $\mathfrak{Q} = q_0 + \mathfrak{q}$ is given by $\widetilde{\mathfrak{Q}} := q_0 - \mathfrak{q}$ and $\mathfrak{Q}$ is called a unit-quaternion for $\mathfrak{Q} \circ \widetilde{\mathfrak{Q}} = 1$, where $\circ$ denotes the quaternion multiplication. Finally we can embed points P of the Euclidean 3-space E^3 with coordinates (p_1, p_2, p_3) with respect to the Cartesian frame $(\mathsf{O}; x_1, x_2, x_3)$ into the set of pure quaternions by:

$$\iota_3 : \mathbb{R}^3 \to \mathbb{H} \quad \text{with} \quad (p_1, p_2, p_3) \mapsto \mathfrak{p} := p_1\mathbf{i} + p_2\mathbf{j} + p_3\mathbf{k}. \tag{1}$$

Combining two quaternions by the dual unit ε with the property $\varepsilon^2 = 0$ yields the set of dual quaternions $\mathbb{H} + \varepsilon\mathbb{H}$. An element $\mathfrak{E} + \varepsilon\mathfrak{T}$ of $\mathbb{H} + \varepsilon\mathbb{H}$ is called dual unit-quaternion if $\mathfrak{E}$ is a unit-quaternion and the following condition holds:

$$\mathfrak{T} \circ \widetilde{\mathfrak{Q}} + \mathfrak{Q} \circ \widetilde{\mathfrak{T}} = 0 \quad \Longleftrightarrow \quad e_0t_0 + e_1t_1 + e_2t_2 + e_3t_3 = 0. \tag{2}$$

We denote the set of dual unit-quaternions by $\mathbb{E}$. By skipping the so-called Study condition of Eq. (2) we get a superset of $\mathbb{E}$, which we call $\mathbb{F}$. Note that both sets $\mathbb{E}$ and $\mathbb{F}$ build a group with respect to the quaternion multiplication.

J. Lenarcic and V. Parenti-Castelli (Eds.): ARK 2018, SPAR 8, pp. 3–11, 2019.
https://doi.org/10.1007/978-3-319-93188-3_1

Based on the usage of $\mathfrak{E} + \varepsilon\mathfrak{T} \in \mathbb{E}$ it can be shown (e.g. [4, Sect. 3.3.2.2]) that the mapping of points $\mathsf{P} \in E^3$ induced by any element of SE(3), can be written as:

$$\delta_3 : \mathbb{H} \to \mathbb{H} \quad \text{with} \quad \mathfrak{p} \mapsto \mathfrak{E} \circ \mathfrak{p} \circ \widetilde{\mathfrak{E}} + (\mathfrak{T} \circ \widetilde{\mathfrak{E}} - \mathfrak{E} \circ \widetilde{\mathfrak{T}}). \tag{3}$$

Moreover the mapping of Eq. (3) is an element of SE(3) for any $\mathfrak{E}, \mathfrak{T}$ implying a dual quaternion $\mathfrak{E} + \varepsilon\mathfrak{T} \in \mathbb{E}$. Note that $\delta_3(\mathfrak{p})$ is again a pure quaternion, where the first summand $\mathfrak{E} \circ \mathfrak{p} \circ \widetilde{\mathfrak{E}}$ is the rotational component and the term in the parentheses corresponds to the translational part.

As both dual quaternions $\pm(\mathfrak{E} + \varepsilon\mathfrak{T}) \in \mathbb{E}$ imply the same Euclidean motion of E^3, we consider the homogeneous 8-tuple $(e_0 : \ldots : e_3 : t_0 : \ldots : t_3)$. These so-called Study parameters can be interpreted as points of a projective 7-dimensional space P^7. Therefore there is a bijection between SE(3) and all real points of P^7 located on the Study quadric $\Phi \subset P^7$ given by Eq. (2), which is sliced along the 3-dimensional generator-space $G: \ e_0 = e_1 = e_2 = e_3 = 0$. Points of this generator-space are called "Pseudosomen" by Study [20].

1.1 Motivation

It is well known (e.g. [14]) that there exists a bijection between the set $\mathscr{L}$ of lines of the projective 3-space and all real points of the so-called Plücker quadric

$$\Psi: \quad l_{01}l_{23} + l_{02}l_{31} + l_{03}l_{12} = 0 \tag{4}$$

of P^5, where the homogeneous 6-tuple $(l_{01} : l_{02} : l_{03} : l_{23} : l_{31} : l_{12})$ are the Plücker coordinates of the lines. This bijection $\mathscr{L} \to \Psi$ is also known as Klein mapping.

Remark 1. Clearly there is an analogy between the Study quadric Φ and the Plücker quadric Ψ. In contrast to SE(3) the set $\mathscr{L}$ is compact, i.e. the quadric Ψ has not to be sliced. But the 2-dimensional generator space $l_{01} = l_{02} = l_{03} = 0$ corresponds to the set of ideal lines, which therefore are the analog of the "Pseudosomen". ◇

Moreover one can define an *extended Klein mapping* which identifies each point of P^5 with a linear complex of lines in P^3. As the latter can always be seen as a path-normal complex of instantaneous screws (different from the zero screw), which only vary in speed (i.e. they are real proportional), one ends up with a bijection between points of P^5 and *homogeneous screw coordinates* $\$:= (s_{01} : s_{02} : s_{03} : s_{23} : s_{31} : s_{12})$. The corresponding path-normals fulfill the condition

$$s_{01}l_{23} + s_{02}l_{31} + s_{03}l_{12} + s_{23}l_{01} + s_{31}l_{02} + s_{12}l_{03} = 0, \tag{5}$$

which geometrically means that they are located in the intersection of Ψ and the polar plane of $\$$ with respect to Ψ. This line/screw-geometric explanation of the complete P^5 raises the question for a kinematic interpretation of the points of the Study quadric's ambient space P^7, which is still missing.

Moreover this study is motivated by the *extended inverse kinematic map* $\kappa^{-1}: P^7 \setminus G \to SE(3)$ discussed by Pfurner, Schröcker and Husty (PSH) in [13]. Under this map any point of $P^7 \setminus G$ is identified with a displacement of SE(3), which corresponds to a point on $\Phi \setminus G$. These two points are linked by the so-called (cf. [19]) PSH-map: $P^7 \setminus G \to \Phi \setminus G$, which can be written in terms of dual quaternions as:

$$\varphi : \mathbb{F} \to \mathbb{E} \quad \text{with} \quad \mathfrak{E} + \varepsilon\mathfrak{T} \mapsto \mathfrak{E} + \varepsilon\left[\mathfrak{T} - \tfrac{1}{2}\left(\mathfrak{T} \circ \widetilde{\mathfrak{E}} + \mathfrak{E} \circ \widetilde{\mathfrak{T}}\right) \circ \mathfrak{E}\right]. \quad (6)$$

In this context it should also be mentioned that the PSH-map is the analogue of the map in P^5 sending a screw \$ to its axis (cf. [19] and [9, Remark 10]).

Remark 2. Note that the basic principle of κ^{-1} was already mentioned by Ge and Purwar [3, Section 4], who used it inter alia in [15,16]. ◇

We do not want to interpret points of the ambient space P^7 of Φ by means of κ^{-1} as this map is not bijective (in contrast to the *extended Klein mapping*). We overcome this problem by interpreting the points of P^7 as displacements of a motion group of the Euclidean 4-space E^4 (cf. Sect. 2). From this point of view we discuss the PSH-map (cf. Sect. 2) and straight lines in P^7 (cf. Sect. 3). We conclude the paper (cf. Sect. 4) by noting an application for the interactive design of rational motions.

2 Kinematic Interpretation as a Subgroup of SE(4)

We embed points P of E^4 with coordinates (p_0, p_1, p_2, p_3) with respect to the Cartesian frame $(\mathsf{O}; x_0, x_1, x_2, x_3)$ into the set of quaternions by the mapping:

$$\iota_4 : \mathbb{R}^4 \to \mathbb{H} \quad \text{with} \quad (p_0, p_1, p_2, p_3) \mapsto \mathfrak{P} := p_0 + p_1\mathbf{i} + p_2\mathbf{j} + p_3\mathbf{k} = p_0 + \mathfrak{p}. \quad (7)$$

Let us identify E^3 with any hyperplane $x_0 = k$ with $k \in \mathbb{R}$. Moreover we consider the set X_4 of displacements of SE(4), which fixes the direction of the x_0-axis. It is an easy task to show that X_4 is a subgroup[1] of SE(4).

Theorem 1. *The mapping of points $\mathsf{P} \in E^4$ induced by any element of X_4, can be written as:*

$$\delta_4 : \mathbb{H} \to \mathbb{H} \quad \textit{with} \quad \mathfrak{P} \mapsto \mathfrak{E} \circ \mathfrak{P} \circ \widetilde{\mathfrak{E}} - 2\mathfrak{E} \circ \widetilde{\mathfrak{T}}. \quad (8)$$

Moreover the mapping of Eq. (8) is an element of X_4 for any $\mathfrak{E}, \mathfrak{T}$ implying a dual quaternion $\mathfrak{E} + \varepsilon\mathfrak{T} \in \mathbb{F}$.

[1] The nomenclature is due to the fact that X_4 can be seen as the 4-dimensional analogue of a Schoenflies group of SE(3), which is denoted by X.

Proof. Direct computation shows that the direction of the x_0-axis remains fixed under the mapping of Eq. (8). Therefore any displacement of this form is an element of X_4. It remains to show that any element of X_4 can be written in this way. Therefore we use the following corollary which follows immediately from [10, Theorem 3.1]:

Corollary 1. *The mapping* $\mathfrak{P} \mapsto \mathfrak{E} \circ \mathfrak{P} \circ \widetilde{\mathfrak{E}}$ *is a rotation about the plane* Λ *through the origin* O *spanned by* $\mathsf{I} := (1, 0, 0, 0)$ *and* $\mathsf{E} := (e_0, e_1, e_2, e_3)$*, where the rotation angle* λ *is two times the enclosed angle of these two vectors; i.e.* $0° \leq \frac{\lambda}{2} \leq 180°$. *The angle of rotation* $\lambda \geq 0$ *is given with respect to the orientation of* Λ*, which is implied by the oriented triangle* $\mathsf{O}, \mathsf{I}, \mathsf{E}$.

The last sentence also implies that $\mathfrak{E}$ and $-\mathfrak{E}$ yield the same rotation. Moreover from this geometric interpretation it is clear that any rotation about a plane through O containing I can be written in the form $\mathfrak{P} \mapsto \mathfrak{E} \circ \mathfrak{P} \circ \widetilde{\mathfrak{E}}$.

We proceed with the translation part $-2\mathfrak{E} \circ \widetilde{\mathfrak{T}}$ of Eq. (8): As shown in [10, Proof of Theorem 2.6] any translation vector of $\mathbb{R}^4$ can be generated in this way. Moreover there is a bijection between $-2\mathfrak{E} \circ \widetilde{\mathfrak{T}}$ and $\mathfrak{T}$ for a given unit-quaternion $\mathfrak{E}$.

The final ingredient for our proof (cf. [1, Chapter 9] and [2, Chapter I]) is that any displacement in E^4 (beside the identity transform) fixing the x_0-direction is either

(a) a translation or
(b) a pure rotation about a plane Γ parallel to the x_0-direction or
(c) a composition of a pure rotation about a plane Γ parallel to the x_0-direction and a translation parallel to Γ.

Clearly, all pure translations (including the identity) can be written as in Eq. (8) by setting $\mathfrak{E} = 1$. Moreover any displacements of item (b) and (c) can be archived by the composition of a rotation about a plane Λ parallel to Γ, which contains the origin O, and a translation. If the translation vector is orthogonal to Λ then we obtain case (b) otherwise we end up with case (c). This closes the proof of Theorem 1. □

As both dual quaternions $\pm(\mathfrak{E} + \varepsilon\mathfrak{T}) \in \mathbb{F}$ correspond to the same X_4-motion of E^4, we consider again the homogeneous 8-tuple $(e_0 : \ldots : e_3 : t_0 : \ldots : t_3)$. Therefore there is a bijection between X_4 and all real points of $P^7 \setminus G$. Due to

$$\delta_4(\mathfrak{P}) = p_0 - 2(e_0t_0 + e_1t_1 + e_2t_2 + e_3t_3) + \delta_3(\mathfrak{p}) \tag{9}$$

the mapping of Eq. (8) restricted to the *pure part* equals the mapping of Eq. (3). Moreover displacements of X_4 fixing the hyperplanes $x_0 = k$ correspond to points on $\Phi \setminus G$. These points imply SE(3)-displacements in the hyperplanes $x_0 = k$, which we identify with E^3. This completes the kinematic interpretation of all points of $P^7 \setminus G$; i.e. the Study parameter space with exception of the "Pseudosomen".

Remark 3. The normalized Grassmann plane coordinates of the plane Γ, which has been mentioned in item (b) and (c) of the proof above, are given in [11]. Note that the intersection of the hyperplane $x_0 = k$ and the plane Γ yields the axis of the corresponding displacement (pure rotation or screw displacement) in E^3. ◇

Now we want to clarify the meaning of the extended inverse kinematic mapping κ^{-1} in terms of our given interpretation. Therefore we compute the action of $\varphi(\mathfrak{E} + \varepsilon\mathfrak{T}) \in \mathbb{E}$ on a point $\mathsf{P} \in E^4$ according to Eq. (8), which yields:

$$\mathfrak{P} \mapsto \mathfrak{E} \circ \mathfrak{P} \circ \widetilde{\mathfrak{E}} - 2\mathfrak{E} \circ \widetilde{\mathfrak{T}} + (\mathfrak{T} \circ \widetilde{\mathfrak{E}} + \mathfrak{E} \circ \widetilde{\mathfrak{T}}). \tag{10}$$

The term in the parentheses is a pure scalar which equals the scalar part of $2\mathfrak{E}\circ\widetilde{\mathfrak{T}}$. Therefore the inverse kinematic mapping is nothing else than an orthogonal projection onto the hyperplane $x_0 = k$, which we have identified with E^3.

Remark 4. In order to make the things more descriptive we stress the following lower-dimensional analogue: Consider the Schoenflies motion group X with respect to the direction x_3 and its subgroup of planar motions SE(2) parallel to the x_1x_2-plane. If we apply an orthogonal projection along the x_3-direction (analogue of κ^{-1}) to a X-motion we obtain a planar one. For an illustrating figure and further comments please see [11, Fig. 1 and Remark 4]. ◇

3 Straight Lines in the Study Parameter Space

It is known (cf. [16, Section 3.2], [19, Section 4], [18, Section 3]) that straight lines in P^7 are in general send by the PSH-map to vertical Darboux motions (see e.g. [2, pp. 321–322] and [8]). In special cases they correspond to rotations about fixed axes or translations along fixed directions.

As preparatory work for the study of straight lines of P^7 in terms of X_4-motions, we recall some results on so-called Darboux 2-motions, which were studied by Karger [6] in E^n. In [6, Definition 1] he defined a Darboux 2-motion as a motion, where all trajectories are (1) planar and (2) affine equivalent in the kinematical sense. The second condition is redundant for E^4 due to a result given by Karger in [7]. Therefore all motions of SE(4) with planar trajectories, which are known [6, Theorem 4] to be either ellipses or line segments, are given in [6]. We focus on Darboux 2-motions in E^4, where all points have circular trajectories.

Theorem 2. *A circular Darboux 2-motion, which is neither spherical nor a pure translation, has to be composed of a rotation about a fixed plane and a circular translation parallel to this plane.*

Proof. In the last remark of [6], Karger mentioned that there are two types of Darboux 2-motions in E^4. In the following we study them separately:

1. **Type 1** *(characterized by Karger's parameters $k_1 = 1$, $\delta_1 = \delta_2 = 1$; cf. [6]):* According to [6, Theorem 2] this motion can be written in matrix form (adapted in notation) for a proper choice of coordinate frames as:

$$\begin{pmatrix} p_0' \\ p_1' \\ p_2' \\ p_3' \end{pmatrix} = \begin{pmatrix} 1 & 0 & 0 & 0 \\ 0 & 1 & 0 & 0 \\ 0 & 0 & \cos(\tau) & -\sin(\tau) \\ 0 & 0 & \sin(\tau) & \cos(\tau) \end{pmatrix} \begin{pmatrix} p_0 \\ p_1 \\ p_2 \\ p_3 \end{pmatrix} + \begin{pmatrix} \frac{\beta}{2\sqrt{\gamma}} \left(\gamma \sin(\tau) + 1 - \cos(\tau)\right) \\ \frac{\beta}{2\sqrt{\gamma}} \sqrt{\gamma^2 - 1}\,(1 - \cos(\tau)) \\ 0 \\ \frac{\nu}{2} (\cos(\tau) - 1) \end{pmatrix}, \tag{11}$$

where $\beta, \gamma, \nu \in \mathbb{R}$ with $\beta \geq 0$, $\gamma \geq 1$, $\nu \geq 0$ and τ denoting the motion parameter. We apply the half-angle substitution $\cos(\tau) = \frac{1-t^2}{1+t^2}$ and $\sin(\tau) = \frac{2t}{1+t^2}$ and introduce homogenous coordinates $p_i' := \frac{h_0'}{h'}$ for $i = 0, \ldots, 3$. Now we can intersect the trajectories with the hyperplane at infinity $h' = 0$ which implies $t = \pm I$ where I denotes the complex unit. Then it can easily be seen that $\nu = 0$ is a necessary condition for these two intersection points to be located on the absolute sphere $h_0'^2 + h_1'^2 + h_2'^2 + h_3'^2 = 0$ independent of the point (p_0, p_1, p_2, p_3). As a consequence the rotation plane (x_0x_1-plane) is fixed. Then points of this plane can only have circular trajectories if the translation is circular.
2. **Type 2** *(characterized by Karger's parameters $k_1 = 2$, $\delta_1 = \delta_2 = 0$; cf. [6]):* According to [6, Theorem 2] this motion can be composed of a spherical motion about the origin O, where the orthogonal matrix is given in [6, Eq. (26)], and a translational vector of the form $(0, 0, \star(\tau), 0)^T$ (cf. [6, Eq. (27)]). Therefore the origin can only have a linear trajectory or it is fixed ($\Rightarrow$ spherical motion) if the expression $\star(\tau)$ vanishes independently of τ. This finishes the proof. □

Remark 5. On the first sight it is not clear how the remaining parameters $\beta \geq 0$ and $\gamma \geq 1$ of Type 1 have to be chosen such that we get a circular translation. A short computation shows that we get the following expression:

$$c\,(\sin(\tau), 1 - \cos(\tau), 0, 0)^T \tag{12}$$

if we set $\beta := c/\sqrt{\gamma}$ with $c \geq 0$ and then applying the limit $\gamma \to \infty$ to the translation vector of Eq. (11). Now it can easily be verified that the trajectory of each point $\mathsf{P} \in E^4$ is circular for $c > 0$. Up to the best knowledge of the author, the existence of these circular Darboux 2-motions has not been mentioned in the literature until now. Moreover they belong to the class of Borel-Bricard motions of E^4 (i.e. all points have hyperspherical trajectories). Further examples of this class are given in [21,22]. ◇

Theorem 3. *A straight line $\in P^7 \setminus G$ corresponds to one of the following X_4-motions and vice versa:*

(i) a translation along a fixed direction,
(ii) a rotation about a fixed plane,

(iii) a circular Darboux 2-motion, which is neither spherical nor a pure translation.

Proof. Given are two different displacements of X_4 by $\mathfrak{E}_1 + \varepsilon\mathfrak{T}_1 \in \mathbb{F}$ and $\mathfrak{E}_2 + \varepsilon\mathfrak{T}_2 \in \mathbb{F}$ which span a straight line in $P^7 \setminus G$. The corresponding quaternion $\mathfrak{E}(s) + \varepsilon\mathfrak{T}(s) \in \mathbb{F}$ of the motion can be written in dependence of the parameter s as:

$$\mathfrak{E}(s) + \varepsilon\mathfrak{T}(s) := \frac{\mathfrak{E}_s}{\sqrt{\mathfrak{E}_s \circ \widetilde{\mathfrak{E}_s}}} + \varepsilon\frac{\mathfrak{T}_s}{\sqrt{\mathfrak{E}_s \circ \widetilde{\mathfrak{E}_s}}} \quad \text{with} \quad \begin{cases} \mathfrak{E}_s := s\mathfrak{E}_1 + (1-s)\mathfrak{E}_2 \\ \mathfrak{T}_s := s\mathfrak{T}_1 + (1-s)\mathfrak{T}_2 \end{cases} \tag{13}$$

If the given two displacements have the same orientation ($\mathfrak{E}_1 = \pm\mathfrak{E}_2$) then the motion is trivially a pure translation along a fixed direction; i.e. case (i). Therefore we can assume $\mathfrak{E}_1 \neq \pm\mathfrak{E}_2$ for the remainder of the proof.

Moreover if the expression $\mathfrak{T}(s) \circ \widetilde{\mathfrak{E}}(s) + \mathfrak{E}(s) \circ \widetilde{\mathfrak{T}}(s)$ is constant $-k$ for all $s \in \mathbb{R}$, then the relative motion between each two poses of this motion is located on the Study quadric. Therefore the SE(3)-motion in the hyperplane $x_0 = k$ has to be a pure rotation about a fixed line. As a consequence we get case (ii).

If we also exclude this scenario we are left with the general case. We can compute the trajectories by means of Eq. (8). As the obtained coordinates are rational quadratic polynomials in s, the trajectories are conic sections. It remains to show that they are circles, which can be done analogously to Theorem 2; i.e. both intersection points with the hyperplane at infinity have to be located on the absolute sphere. This can be verified by e.g. direct computations with MAPLE.

Now we tackle the reverse direction, that every X_4-motion of type (i), (ii) or (iii) corresponds to a straight line in $P^7 \setminus G$. For (i) and (ii) this is trivial and left to the reader. For (iii) we proceed as follows: Without loss of generality we can choose a reference frame in a way that the rotation plane is the x_0x_1-plane. This corresponds to the rotation matrix of Eq. (11). But as we cannot chose the x_0-direction freely (it is predetermined) the translation vector of Eq. (12) has to be rotated about the x_2x_3-plane by the angle ρ, which yields:

$$c\left(\cos(\rho)\sin(\tau) - \sin(\rho)(1-\cos(\tau)), \sin(\rho)\sin(\tau) + \cos(\rho)(1-\cos(\tau)), 0, 0\right)^T.$$

It can be checked that the corresponding straight line in $P^7 \setminus G$ is obtained by:

$$\mathfrak{E}_1 = 1, \quad \mathfrak{T}_1 = 0, \quad \mathfrak{E}_2 = \mathbf{i}, \quad \mathfrak{T}_2 = -c\cos(\rho) + c\sin(\rho)\mathbf{i} \quad \text{and} \quad s = \frac{1}{1+t}.$$

As a change of the reference frame within the group X_4 implies a linear transformation of the corresponding motion parameters in P^7 ($\Rightarrow$ a straight line is mapped to a straight line), the proof is completed. □

4 Conclusion

The given kinematic interpretation of all points of the Study quadric's ambient space P^7 in terms of X_4-displacements can e.g. be used for the interactive design of rational motions in the following way:

We start with a *projective de Casteljau construction* in P^7. Due to the given interpretation the control points and Farin points of this construction in P^7 correspond to poses in E^4, which can be projected orthogonally along the x_0-direction onto E^3. In addition we label the obtained poses of E^3 by the x_0-coordinate[2]. As the resulting Farin poses and control poses in E^3 are not affinely distorted (as done by some other known methods [5,17]), the user can modify very intuitively the control structure (Farin and control poses can be changed by *mouse action* and their x_0-heights by the *scroll wheel*). This is illustrated in [11, Fig. 2]. Moreover, as a meaningful optimization criterion of the resulting rational motion one can use the minimization of the maximal translation along the x_0-direction during the motion.
Finally the interested reader is referred to

- ⋆ [12], where the line-symmetry of the circular Darboux 2-motions in E^4 and their corresponding basic surfaces are discussed,
- ⋆ [11, Section 4], where in analogy to linear complexes of lines (cf. Sect. 1.1) some basics on linear complexes of SE(3)-displacements are given.

Acknowledgements. The author is supported by Grant No. P 24927-N25 of the Austrian Science Fund FWF within the project "Stewart Gough platforms with self-motions".

References

1. Berger, M.: Geometry I. Springer, Heidelberg (1987)
2. Bottema, O., Roth, B.: Theoretical Kinematics. North-Holland, Amsterdam (1979)
3. Ge, Q.J., Purwar, A.: Spears, oriented screw displacements and their image space. In: Proceedings of the 11th World Congress in Mechanism and Machine Science, Tianjin, China (2004)
4. Husty, M., Karger, A., Sachs, H., Steinhilper, W.: Kinematik und Robotik. Springer, Heidelberg (1997)
5. Jüttler, B., Wagner, M.: Kinematics and animation. In: Farin, G., et al. (eds.) Handbook of Computer Aided Geometric Design, pp. 723–748. North Holland (2001)
6. Karger, A.: Darboux motions in E_n. Czech. Math. J. **29**(2), 303–317 (1979)
7. Karger, A.: 3D-Darboux motions in 4-dimensional Euclidean space. Math. Pannon. **6**(1), 133–143 (1995)
8. Krames, J.: Zur aufrechten Ellipsenbewegung des Raumes (Über symmetrische Schrotungen III). Monatsh. Math. Phys. **46**(1), 38–50 (1937)
9. Nawratil, G.: Point-models for the set of oriented line-elements - a survey. Mech. Mach. Theory **111**, 118–134 (2017)
10. Nawratil, G.: Fundamentals of quaternionic kinematics in Euclidean 4-space. Adv. Appl. Clifford Algebras **26**(2), 693–717 (2016)
11. Nawratil, G.: Kinematic interpretation of the Study quadric's ambient space. arXiv:1708.02622 (2017)

[2] In German such a map is known as "kotierte Projektion".

12. Nawratil, G.: Alternative interpretation of the Plücker quadric's ambient space and its application. In: Proceedings of the 18th International Conference on Geometry and Graphics. Springer (accepted)
13. Pfurner, M., Schröcker, H.-P., Husty, M.: Path planning in kinematic image space without the study condition. In: Lenarcic, J., Merlet, J.-P. (eds.) Advances in Robot Kinematics 2016, pp. 285–292. Springer, Cham (2018)
14. Pottmann, H., Wallner, J.: Computational Line Geometry. Springer, Heidelberg (2001)
15. Purwar, A., Ge, Q.J.: On the effect of dual weights in computer aided design of rational motions. J. Mech. Des. **127**, 967–972 (2005)
16. Purwar, A., Ge, Q.J.: Kinematic convexity of rigid body displacements. In: Proceedings of the ASME International Design Engineering Technical Conference and Computer and Information in Engineering Conference, Montreal, Canada (2010)
17. Röschel, O.: Rational motion design – a survey. Comput. Aided Des. **30**(3), 169–178 (1998)
18. Schröcker, H.-P.: From A to B: new methods to interpolate two poses. J. Geom. Graph. arXiv:1706.04539 (accepted)
19. Selig, J.M., Wu, Y., Carricato, M.: Motion interpolation in lie subgroups and symmetric subspaces. In: Zeghloul, S., et al. (eds.) Computational Kinematics, pp. 467–474. Springer, Cham (2017)
20. Study, E.: Geometrie der Dynamen. Teubner, Leipzig (1903)
21. Vogler, H.: Zwangläufe im höchstens vierdimensionalen euklidischen Raum $\mathbb{E}_\nu$ mit affin durchlaufenen Bahnpunktsmengen. Grazer Math. Ber. **304**, 128–143 (1989)
22. Wunderlich, W.: Zur Schraubung im vierdimensionalen euklidischen Raum. J. reine angew. Math. **285**, 79–99 (1976)

Input-Output Equation for Planar Four-Bar Linkages

M. John D. Hayes[1(✉)], Manfred L. Husty[2], and Martin Pfurner[2]

[1] Carleton University, Ottawa, ON, Canada
john.hayes@carleton.ca
[2] University of Innsbruck, Unit Geometry and CAD, Innsbruck, Austria
https://carleton.ca/mae/people/m-john-d-hayes

Abstract. In this paper the generalised input-output (I-O) equation for planar $4R$ function generators is derived in a new way, leading to the algebraic form of the well known Freudenstein equation. The long term goal is to develop a generalised method to derive constraint based algebraic I-O equations that can be used for continuous approximate synthesis, where the synthesis equations are integrated between minimum and maximum input angle values resulting in a linkage whose objective function has been optimised over every output angle. In this paper we use a planar projection of Study's soma and the Cartesian displacement constraints for the dyads. These are mapped to the image space leading to four constraint equations in terms of the image space coordinates and the sines and cosines of the input and output angles. Using the tangent of the half angle substitution the trigonometric equations are converted to algebraic ones. Algebraic methods are used to eliminate the image space coordinates, then the polynomial resultants are found to obtain common roots leading to the desired equations.

Keywords: Function generators · Continuous approximate synthesis Kinematic mapping · Polynomial resultants

1 Introduction

A planar $4R$ function generator correlates driver and follower angles in a functional relationship. The mechanism essentially generates the function $\varphi = f(\psi)$, or vice versa, see Fig. 1. Design methods typically employ the Freudenstein synthesis equations to identify link lengths required to generate the function [2,4]. For over-determined sets of prescribed input-output (I-O) angle pairs, these equations are linear in the three unknown Freudenstein parameters, which are ratios

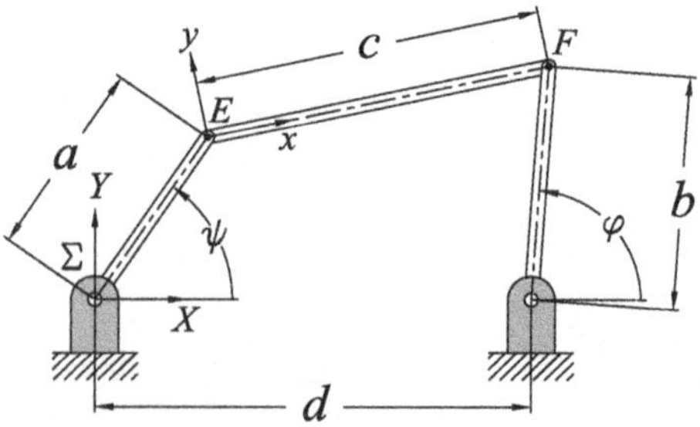

Fig. 1. $4R$ function generator.

J. Lenarcic and V. Parenti-Castelli (Eds.): ARK 2018, SPAR 8, pp. 12–19, 2019.
https://doi.org/10.1007/978-3-319-93188-3_2

of the link lengths, and can be solved for using any least squares method to minimise a specified performance error. To the best of the authors knowledge, there are no alternative algebraic models of the function generator displacement equations in the accessible literature.

It has been observed [5,7] that as the cardinality of the prescribed discrete I-O data set increases, the corresponding four-bar linkages that minimise the Euclidean norm of the design and structural errors tend to converge to the same linkage. The important implication is that minimising the Euclidean norm, or any p-norm, of the structural error can be accomplished indirectly by minimising the same norm of the design error. In [5] the trigonometric Freudenstein synthesis equations are integrated in the range between minimum and maximum input values, thereby reposing the discrete approximate synthesis problem as a continuous one whereby the objective function is optimised over the entire I-O range. Hence, our long-term goal is to determine a general method to derive motion constraint based algebraic I-O equations that may be used together with the method of continuous approximate synthesis [5] to obtain the very best linkage that can generate an arbitrary function. The goal of this paper is to develop one in the hope of providing new insight into the continuous approximate synthesis of function generators, while mitigating numerical integration error. Of course, the same equation will be obtained by making the tangent half-angle substitutions directly into the Freudenstein equation then collecting terms after factoring, normalising, and eliminating non-zero factors. But that must be the case since the geometric relations require that outcome, however this is irrelevant because the goal is to generalise a method to develop constraint based algebraic I-O equations for continuous approximate synthesis of planar, spherical, and spatial linkages. This paper represents the first step in that long journey.

2 Geometric and Algebraic Approach

The Freudenstein equation relating the input to the output angles of a planar $4R$ four-bar mechanism, with link lengths as in Fig. 1, was first put forward in [3]. In the equation the angle ψ is traditionally selected to be the input while φ is the output angle:

$$k_1 + k_2 \cos(\varphi_i) - k_3 \cos(\psi_i) = \cos(\psi_i - \varphi_i). \tag{1}$$

Equation (1) is linear in the k_i Freudenstein parameters, which are defined in terms of the link length ratios as

$$\left.\begin{array}{ll} k_1 \equiv & \frac{(a^2+b^2+d^2-c^2)}{2ab}, \\ k_2 \equiv & \frac{d}{a}, \\ k_3 \equiv & \frac{d}{b}. \end{array}\right\} \Leftrightarrow \begin{cases} d = & 1, \\ a = & \frac{1}{k_2}, \\ b = & \frac{1}{k_3}, \\ c = & (a^2+b^2+d^2-2abk_1)^{1/2}. \end{cases}$$

The new idea starts the same as with the Freudenstein method, writing the displacement constraints in terms of the I-O angles. Continuing with tradition, we select ψ to be the input angle and φ to be the output angle. Let Σ be a non moving Cartesian coordinate system with coordinates X and Y whose origin is located at the centre of the ground fixed link R-pair with length a. Let E be a coordinate system that moves with the coupler of length c whose origin is at the centre of the distal R-pair of link a, having basis directions x and y.

The displacement constraints for the origin of E can be expressed as

$$\begin{aligned} X - a\cos\psi &= 0, \\ Y - a\sin\psi &= 0, \end{aligned} \tag{2}$$

while those for point F, located at the centre of the distal R-pair on the output link with length b are

$$\begin{aligned} X - d - b\cos\varphi &= 0, \\ Y - b\sin\varphi \quad &= 0. \end{aligned} \tag{3}$$

Next, we use a planar projection of Study's soma coordinates [8] to establish the I-O equation. Any displacement in Euclidean space, E_3, can be mapped in terms of the coordinates of a 7-dimensional projective image space using the transformation [1]

$$\mathbf{T} = \begin{bmatrix} x_0^2 + x_1^2 + x_2^2 + x_3^2 & 0 & 0 & 0 \\ 2(-x_0y_1 + x_1y_0 - x_2y_3 - x_3y_2) & x_0^2 + x_1^2 - x_2^2 - x_3^2 & 2(x_1x_2 - x_0x_3) & 2(x_1x_3 + x_0x_2) \\ 2(-x_0y_2 + x_1y_3 + x_2y_0 - x_3y_1) & 2(x_1x_2 + x_0x_3) & x_0^2 - x_1^2 + x_2^2 - x_3^2 & 2(x_2x_3 - x_0x_1) \\ 2(-x_0y_3 - x_1y_2 + x_2y_1 + x_3y_0) & 2(x_1x_3 - x_0x_2) & 2(x_2x_3 + x_0x_1) & x_0^2 - x_1^2 - x_2^2 + x_3^2 \end{bmatrix} \tag{4}$$

This transforms the coordinates of any point described in a moving $3D$ coordinate system E to the coordinates of the same point in a relatively fixed $3D$ coordinate system Σ (assuming that the two frames are initially coincident) after a given displacement of E relative to Σ in terms of the coordinates of a point on the Study quadric, S_6^2. In order for a point in the image space to represent a real displacement, and therefore to be located on S_6^2, the non-zero condition of $x_1^2 + x_2^2 + x_3^2 + x_4^2 \neq 0$ must be satisfied.

The transformation matrix $\mathbf{T}$ simplifies considerably when we consider displacements that are restricted to the plane. Three degrees of freedom are lost and hence four Study parameters vanish. The displacements may be restricted to any plane. Without loss in generality, we may select one of the principal planes in Σ. Thus, we arbitrarily select the plane $Z = 0$. Since E and Σ are assumed to be initially coincident, this means

$$\begin{bmatrix} W \\ X \\ Y \\ 0 \end{bmatrix} = \mathbf{T} \begin{bmatrix} w \\ x \\ y \\ 0 \end{bmatrix}, \tag{5}$$

leaving us with the four soma coordinates

$$(x_0 : x_3 : y_1 : y_2). \tag{6}$$

The non-zero condition is now $x_0^2+x_3^2 \neq 0$, and the fourth row and column of the reduced $\mathbf{T}$ contains only this condition as the last element with zeros elsewhere leading to the trivial equation $Z = z = 0$. We can therefore eliminate the fourth row and column and normalise the coordinates with the nonzero condition giving the planar mapping equation

$$\mathbf{T} = \frac{1}{x_0^2+x_3^2}\begin{bmatrix} x_0^2+x_3^2 & 0 & 0 \\ 2(-x_0y_1+x_3y_2) & x_0^2-x_3^2 & -2x_0x_3 \\ -2(x_0y_2+x_3y_1) & 2x_0x_3 & x_0^2-x_3^2 \end{bmatrix}. \tag{7}$$

We can now express a point in Σ in terms of the soma coordinates and the corresponding point coordinates in E as

$$\begin{bmatrix} 1 \\ X \\ Y \end{bmatrix} = \mathbf{T}\begin{bmatrix} 1 \\ x \\ y \end{bmatrix} = \frac{1}{x_0^2+x_3^2}\begin{bmatrix} x_0^2+x_3^2 \\ 2(-x_0y_1+x_3y_2)+(x_0^2-x_3^2)x-(2x_0x_3)y \\ -2(x_0y_2+x_3y_1)+(2x_0x_3)x+(x_0^2-x_3^2)y \end{bmatrix}. \tag{8}$$

The novelty of the approach begins with creating two Cartesian vector constraint equations containing the nonhomogeneous coordinates in Eqs. (2) and (3), but substituting the values in Eq. (8) for (X, Y). These two vectors equations are $\mathbf{F}_1 = \mathbf{0}$ and $\mathbf{F}_2 = \mathbf{0}$:

$$\mathbf{F}_1 = \frac{1}{x_0^2+x_3^2}\begin{bmatrix} 2(-x_0y_1+x_3y_2)+(x_0^2-x_3^2)x-2x_0x_3y-(a\cos\psi)(x_0^2+x_3^2) \\ -2(x_0y_2+x_3y_1)+2x_0x_3x+(x_0^2-x_3^2)y-(a\sin\psi)(x_0^2+x_3^2) \end{bmatrix} = \mathbf{0};$$

$$\mathbf{F}_2 = \frac{1}{x_0^2+x_3^2}\begin{bmatrix} 2(-x_0y_1+x_3y_2)+(x_0^2-x_3^2)x-2x_0x_3y-(b\cos\varphi+d)(x_0^2+x_3^2) \\ -2(x_0y_2+x_3y_1)+2x_0x_3x+(x_0^2-x_3^2)y-(b\sin\varphi)(x_0^2+x_3^2) \end{bmatrix} = \mathbf{0}.$$

Now we determine equations for the coupler. The coordinate system that moves with the coupler has its origin, point E, on the centre of the R-pair, as in Fig. 1, having coordinates $(x, y) = (0, 0)$, while point F is on the R-pair centre on the other end having coordinates $(x, y) = (c, 0)$. One more vector equation, $\mathbf{H}_1$ is obtained by substituting $(x, y) = (0, 0)$ in $\mathbf{F}_1$, and another, $\mathbf{H}_2$ is obtained by substituting $(x, y) = (c, 0)$ in $\mathbf{F}_2$. Next $\mathbf{H}_1$ and $\mathbf{H}_2$, two rational expressions, are converted to factored normal form. This is the form where the numerator and denominator are relatively prime polynomials with integer coefficients. The denominators for both $\mathbf{H}_1$ and $\mathbf{H}_2$ are the nonzero condition $x_0^2+x_3^2$, which can safely be factored out of each equation leaving the following two vector equations with polynomial elements:

$$\mathbf{H}_1 = \begin{bmatrix} -a\cos\psi(x_0^2+x_3^2)+2(-x_0y_1+x_3y_2) \\ -a\sin\psi(x_0^2+x_3^2)-2(x_0y_1+x_3y_2) \end{bmatrix} = \mathbf{0}; \tag{9}$$

$$\mathbf{H}_2 = \begin{bmatrix} -(b\cos\varphi+d)(x_0^2+x_3^2)+c(x_0^2-x_3^2)+2(-x_0y_1+x_3y_2) \\ -b\sin\varphi(x_0^2+x_3^2)+2c(x_0x_3)-2(x_0y_2+x_3y_1) \end{bmatrix} = \mathbf{0}. \tag{10}$$

The system of four displacement constraints on the I-O equations are $\mathbf{H}_1 = \mathbf{0}$ and $\mathbf{H}_2 = \mathbf{0}$. However, these are trigonometric equations. We convert them to algebraic ones using the tangent of the half-angle substitutions

$$u = \tan\frac{\psi}{2}, \quad v = \tan\frac{\varphi}{2},$$

and

$$\cos\psi = \frac{1-u^2}{1+u^2}, \quad \sin\psi = \frac{2u}{1+u^2},$$
$$\cos\varphi = \frac{1-v^2}{1+v^2}, \quad \sin\varphi = \frac{2v}{1+v^2}.$$

The usual constraint equations in the kinematic mapping image space are obtained by considering $\mathbf{H}_1$ and $\mathbf{H}_2$ with the tangent of the half-angles, giving four new algebraic polynomials when considering the individual elements converted to factored normal form. The denominators are u^2+1 and v^2+1 which can safely be factored out because they are always non-vanishing. The resulting four algebraic equations are expressed in terms of the elements of $\mathbf{K}_1 = \mathbf{0}$ and $\mathbf{K}_2 = \mathbf{0}$:

$$\mathbf{K}_1 = \begin{bmatrix} (au^2 - a)(x_0^2 + x_3^2) + 2u^2(-x_0y_1 + x_3y_2) + 2(-x_0y_2 + x_3y_1) \\ -2au(x_0^2 + x_3^2) - 2(1+u^2)(-x_0y_2 + x_3y_1) \end{bmatrix} = \mathbf{0}; \quad (11)$$

$$\mathbf{K}_2 = \begin{bmatrix} (v^2(b-d) + b - d)(x_0^2 + x_3^2) + (cv^2 + c)(x_0^2 - x_3^2) + \\ 2(1+v^2)(-x_0y_1 + x_3y_2) \\ 2cv^2x_0x_3 - 2(v^2 + bv + 1)(x_0^2 + x_3^2) + 2cx_0x_3 \end{bmatrix} = \mathbf{0}. \quad (12)$$

Factoring the resultant of the first and second elements of $\mathbf{K}_1 = \mathbf{0}$ with respect to u, as well as the first and second elements of $\mathbf{K}_2 = \mathbf{0}$ with respect to v yields the two displacement constraint equations in the image space:

$$a^2(x_0^2 + x_3^2) - 4(y_1^2 + y_2^2) = 0,$$

$$(b^2 - c^2 - d^2)(x_0^2 + x_3^2) + 2cd(x_0^2 - x_3^2) + 4c(x_0y_1 + x_3y_2) + 4d(-x_0y_1 + x_3y_2) - 4(y_1^2 + y_2^2) = 0.$$

Inspection of the quadratic forms of these two equations reveals that they are two hyperboloids of one sheet, which is exactly what is expected for two *RR* dyads [6]. But these are not the constraints we are looking for. We want to eliminate the image space coordinates using $\mathbf{K}_1 = \mathbf{0}$ and $\mathbf{K}_2 = \mathbf{0}$ to obtain an algebraic polynomial with the tangent half angles u and v as variables and link lengths as coefficients.

To obtain this algebraic polynomial we start by setting the homogenising coordinate $x_0 = 1$, which can safely be done since we are only concerned with real finite displacements. Next, observe that the two equations represented by the elements of $\mathbf{K}_1 = \mathbf{0}$ have a simpler form than those of $\mathbf{K}_2 = \mathbf{0}$, and are linear in y_1 and y_2. Solving these two equations for y_1 and y_2 reveals that

$$y_1 = \frac{1}{2}\frac{a(u^2 - 2ux_3 - 1)}{u^2+1}, \quad (13)$$
$$y_2 = \frac{1}{2}\frac{a(u^2x_3 + 2u - x_3)}{u^2+1}. \quad (14)$$

Equations (13) and (14) reveal the common denominator of u^2+1, which can never be less than 1, and hence may be factored out. Now we back-substitute these expressions for y_1 and y_2 into Eqs. (11) and (12), thereby eliminating these image space coordinates, and factor the resultant with respect to x_3 which yields four factors, the first three being

$$4c^2, \quad (u^2+1)^3, \quad (v^2+1)^3.$$

None of these three factors can ever be zero and at the same time represent a real displacement constraint, hence they are eliminated. The remaining factor is a polynomial with only u and v as variables and link lengths a, b, c, and d, as coefficients. This is exactly the constraint equation we desire. It is factored, and the terms collected then distributed over u and v revealing

$$\begin{aligned}(a-b+c+d)(a-b-c+d)u^2v^2+(a+b-c+d)(a+b+c+d)u^2+\\(a+b-c-d)(a+b+c-d)v^2-8abuv+(a-b+c-d)(a-b-c-d)=0.\end{aligned} \quad (15)$$

Equation (15) is an algebraic polynomial of degree four which represents the I-O equation for any planar $4R$ mechanism. It has two singular points at infinity, namely those of the X- and Y-axes. These two singular points are either double points, or acnodes, i.e. isolated, or hermit points in the solution set of a polynomial equation in two real variables. When both are double points the mechanism is a double crank, when both are acnodes the mechanism is a double rocker. In the event the mechanism is a folding four-bar then the degree of Eq. (15) is less than four.

Freudenstein's equation [4] is linear in the ratios of the link lengths and therefore is ideally suited to identifying link lengths that minimise some mechanism performance error in a least squares sense for approximate synthesis. The corresponding algebraic form of Freudenstein's equation is Eq. (15), which is quadratic in the link lengths a, b, c, and d, but still lends itself to linear least squares error minimisation subject to quadratic constraints, and the method presented in [5]. However, in the following example we shall use exact synthesis, using only three of the prescribed sets on I-O pairs, and leave the approximate case to future work.

3 Example

This example serves to demonstrate that Eq. (15) can be used to identify link lengths to create a $4R$ mechanism to generate an arbitrary function. Here, the function is specified in terms of the tangent of the half angle parameters $v = f(u)$ as

$$v = 2 + \tan\left(\frac{u}{1+u^2}\right). \quad (16)$$

Eight I-O pairs $[u, v]$ were specified as, using Fig. 1 for reference,

$$[0,2], \left[\frac{1}{4}, \frac{30055}{13419}\right], \left[\frac{1}{2}, \frac{49597}{20471}\right], \left[\frac{3}{4}, \frac{48857}{19383}\right], \left[1, \frac{64699}{25409}\right], \left[\frac{5}{4}, \frac{25536}{10091}\right], \left[\frac{3}{2}, \frac{110471}{44235}\right], \left[2, \frac{49597}{20471}\right].$$

Since a function generator is scalable, and hence only the ratios of the link lengths are needed, we set $d = 1$ and solve for the remaining three using the first, fourth, and eighth I-O pairs, giving, in generic units of length, $a = -0.23$, $b = 1.43$, and $c = 1.20$. Another linkage was identified using the second, fourth, and seventh I-O pairs giving nearly identical link lengths. Note that it is not uncommon in computational methods to obtain negative link lengths. These lengths are directed distances, and $a = -0.23$ means that the distance is directed from the distal R-pair in link a to the origin, instead of the other way around, as in [9].

In Fig. 2 the prescribed I-O function is plotted as the dashed curve, and the I-O function generated by the two identified linkages are plotted as the solid curves. The generated function shows good fidelity relative to the prescribed function over a reasonable range of I-O angles.

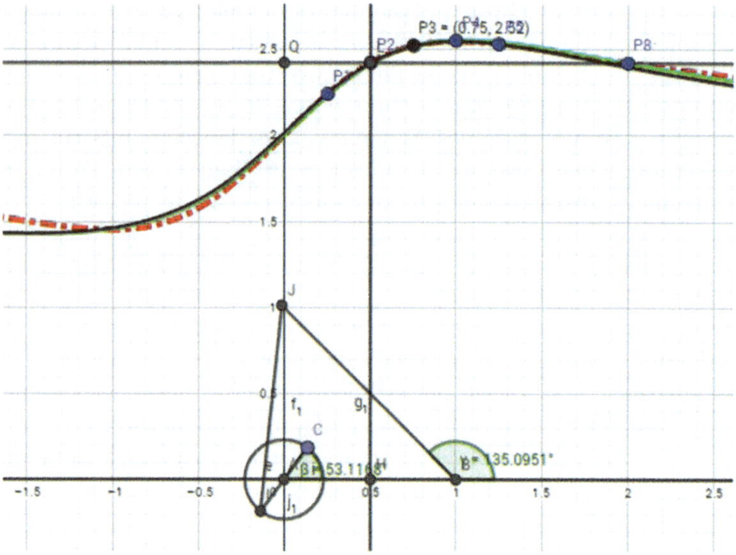

Fig. 2. Graphical representation of results.

4 Conclusions

In this paper a new method for deriving the I-O equations of planar $4R$ function generators was presented. The Cartesian displacement constraints of the two dyads comprising a planar $4R$ mechanism are expressed in terms of lengths and angles. This set of general constraint equations is mapped to a planar projection of Study's soma coordinates. The reason for using this unconventional form of planar kinematic mapping is to be able to apply these methods to spherical and spatial function generators in future work, where all eight soma coordinates will be needed. The result of this step is two arrays in terms of the I-O angles, link lengths, and soma coordinates. The equations are converted from trigonometric expressions to algebraic ones with the tangent of the half angle substitutions. These equations are used to eliminate the soma coordinates. What remains is factored using resultants, and all non-zero factors are eliminated, ultimately leaving an algebraic polynomial that is of degree four in the tangent of the half angle parameters, and quadratic in the link lengths.

The I-O equation derived in this paper is an algebraic, however nonlinear, polynomial in terms of the link lengths. Regardless, this algebraic formulation will significantly mitigate the effect of round-off errors observed in the numerical integration of the trigonometric Freudenstein synthesis equations [5]. While the very same I-O equation is, necessarily, obtained starting from the Freudenstein equation, the point of the presented material is to generalise the derivation of function generator I-O equations. The ultimate goal is to use continuous I-O data sets to synthesise the very best linkage to generate an arbitrary planar, spherical, or spatial function. Derivation of the planar algebraic I-O equation is one of the first steps towards this goal.

References

1. Bottema, O., Roth, B.: Theoretical Kinematics. Dover Publications Inc., New York (1990)
2. Freudenstein, F.: An analytical approach to the design of fourlink mechanisms. Trans. ASME **77**, 483–492 (1954)
3. Freudenstein, F.: Design of four-link mechanisms. Ph.D. thesis, Columbia University, New York (1954)
4. Freudenstein, F.: Approximate synthesis of four-bar linkages. Trans. ASME **77**, 853–861 (1955)
5. Guigue, A., Hayes, M.: Continuous approximate synthesis of planar function-generators minimising the design error. Mech. Mach. Theor. **101**, 158–167 (2016). https://doi.org/10.1016/j.mechmachtheory.2016.03.012
6. Hayes, M., Husty, M.: On the kinematic constraint surfaces of general three-legged planar robot platforms. Mech. Mach. Theor. **38**(5), 379–394 (2003)
7. Hayes, M.J.D., Parsa, K., Angeles, J.: The effect of data-set cardinality on the design and structural errors of four-bar function-generators. In: Proceedings of the Tenth World Congress on the Theory of Machines and Mechanisms, Oulu, pp. 437–442 (1999)
8. Study, E.: Geometrie der Dynamen. Teubner Verlag, Leipzig (1903)
9. Todorov, T.: Synthesis of four bar mechanisms as function generators by Freudenstein-Chebyshev. J. Robot. Mech. Eng. Res. **1**(1), 1–6 (2015)

Principal Kinematic Inequalities

Gregory S. Chirikjian(✉)

Johns Hopkins University, Baltimore, MD 21218, USA
gchirik1@jhu.edu
http://rpk.lcsr.jhu.edu

Abstract. One hundred years ago, Blaschke and Poincaré derived exact closed-form formulas for integrals over $SE(n)$ of Euler characteristics of intersections of n-dimensional bodies for $n = 2, 3$. Three decades thereafter Chern extended these formulas to the n-dimensional case. These results form a core tool in the fields of convex geometry, integral geometry, geometric probability, and stochastic geometry. These results, often referred to as "Principal Kinematic Formulae," are extended here to the case of integrals of set-indicator functions, resulting in inequalities for the case of non-convex bodies. These results are relevant to assessing the frequency of occurrence of collisions that occur in sample-based robot motion planning.

Keywords: Integral geometry · Brunn-Minkowski
Kinematic formula · Convex bodies · Inequalities

1 Introduction and Literature Review

As is well known in Kinematics, $SE(n)$ is the group of rigid body motions with group law [1,4,14,18,20,27]

$$(R_1, \mathbf{t}_1) \circ (R_2, \mathbf{t}_2) = (R_1 R_2, R_1 \mathbf{t}_2 + \mathbf{t}_1)$$

where $R_i \in SO(n)$, the special orthogonal group consisting of $n \times n$ rotation matrices. The above group law makes $SE(n)$ a semi-direct product,

$$SE(n) = \mathbb{R}^n \rtimes SO(n)$$

with a natural bi-invariant volume element with which to integrate [7,8], despite the fact that $SE(n)$ does not have a bi-invariant metric. This paper makes extensive use of the bi-invariant integration measure for $SE(n)$ to compute bounds on the volume of the part of $SE(n)$ that puts moving bodies in collision. The derived results hold even when these bodies are not convex.

Let B_i be a finite solid body in $\mathbb{R}^n$ with closed surface ∂B_i for $i \in \{0, 1\}$. This paper is concerned with properties of the intersection of two such bodies. In particular, suppose that B_0 is held fixed, and for an arbitrary rigid-body motion $g = (R, \mathbf{t}) \in SE(n)$,

$$g \cdot B_1 \doteq \{R\mathbf{x} + \mathbf{t} \,|\, \mathbf{x} \in B_1\}$$

is a rigidly moved copy of B_1.

J. Lenarcic and V. Parenti-Castelli (Eds.): ARK 2018, SPAR 8, pp. 20–27, 2019.
https://doi.org/10.1007/978-3-319-93188-3_3

It is then possible to define functions of motion such as

$$p(g) \doteq \phi(B_0 \cap g \cdot B_1) \text{ and } s(g) \doteq \psi(\partial B_0 \cap g \cdot \partial B_1).$$

For example, for each fixed value of $g \in SE(n)$, ϕ might be the Euler characteristic, volume, or indicator function of the intersection body, and ψ might be the Euler characteristic, area, or mean curvature of the intersection surface.

A major part of the field of integral geometry/geometric probability is concerned with computing closed-form expressions for integrals of functions such as $p(g)$ and $s(g)$. One of the primary results from this field is that if ϕ is the Euler characteristic, then a closed-form expression results for the integral

$$\Xi(B_0, B_1) \doteq \int_{SE(n)} \chi(B_0 \cap g \cdot B_1)\, dg$$

where χ is the Euler characteristic and dg is the Haar measure for the group $SE(n)$. This dg is the unique measure (up to scaling) that has the property that for all $f \in L^1(SE(3))$

$$\int_{SE(3)} f(g)\, dg = \int_{SE(3)} f(g^{-1})\, dg = \int_{SE(3)} f(h \circ g)\, dg = \int_{SE(3)} f(g \circ h)\, dg \quad (1)$$

for arbitrary $h \in SE(3)$. In particular, for rigid-body motions in the plane parameterized in terms of translations expressed in Cartesian coordinates (x, y) and rotation angle θ, $dg = dxdyd\theta$. For motions in 3D with translations expressed in Cartesian coordinates (x, y, z) and rotations expressed in terms of ZXZ Euler angles (α, β, γ), $dg = \sin\beta\, d\alpha d\beta d\gamma dx dy dz$. More generally, the procedure for constructing the expression for dg in any dimension and in any parameterization is outlined in [7,8].

Remarkably, the closed form integral for $\Xi(B_0, B_1)$ can be obtained for arbitrary compact bodies (not only convex ones) with continuous piecewise differentiable boundaries, and can be expressed in the form [2,3,5,16,21,25,26]

$$\Xi(B_0, B_1) = \sum_{i=0}^{n} c_i P_i(B_0) P_{n-i}(B_1) \quad (2)$$

where each c_i is a known constant and each $P_i(B_k)$ is an integral of the body (or its bounding surface), such as the Euler characteristic, volume, surface area, integral of mean curvature, etc. Equation (2) is called the Principal Kinematic Formula.

Due to Hadwiger's characterization theorem [12,13], it is also possible to compute closed form integrals when ϕ is taken to be the volume of intersection, or ψ is taken to be any of the other P_i functions (which are called mixed-volumes in Brunn-Minkowski theory, or quermassintegrals in Geometric Probability).

The following subsections review this formula in the cases of planar and spatial bodies.

1.1 The 2D Case for Convex Bodies

Let C_0 and C_1 be two compact convex planar bodies with corresponding continuous piecewise differentiable boundaries ∂C_0 and ∂C_1. Let $A(C_i)$ denote the area of C_i and let $L(C_i)$ denote its perimeter, i.e., the length of the curve ∂C_i. For arbitrary bodies B_0 and B_1 use the same definitions where L counts in addition the total length of any internal boundaries for a non-simply connected body. Then the following theorem holds.

Theorem 1. *(Blaschke, [2, 3]): Given arbitrary compact planar bodies B_0 and B_1 with continuous piecewise differentiable boundaries, then (2) evaluates as*

$$\Xi(B_0, B_1) = 2\pi[A(B_0)\chi(B_1) + A(B_1)\chi(B_0)] + L(B_0)L(B_1). \tag{3}$$

and for planar compact convex bodies C_0 and C_1, then (2) evaluates as

$$\Xi(C_0, C_1) = 2\pi[A(C_0) + A(C_1)] + L(C_0)L(C_1). \tag{4}$$

For the proof see [2,3,16,25].

1.2 The 3D Case for Convex Bodies

Let C_i for $i \in \{0, 1\}$ be compact convex bodies in $\mathbb{R}^3$. If the spatial body C_i has a continuous piecewise differentiable surface, ∂C_i, that we can compute the total surface area

$$\int_{\partial C_i} dS = F(C_i)$$

Furthermore, if κ denotes the Gaussian curvature at each point on the surface, we can compute (via the Gauss-Bonnet Theorem):

$$\int_{\partial C_i} \kappa \, dS = 2\pi \, \chi(\partial C_i)$$

where $\chi(\partial C_i)$ is the Euler characteristic of the bounding surface. The Euler characteristic of a body and the surface that bounds it are related as

$$\chi(\partial C_i) = 2 \cdot \chi(C_i).$$

Moreover, for simply connected planar and spatial bodies $\chi(C_i) = \iota(C_i) = 1$ where the *indicator function* on any measurable body, B, (not necessarily convex and perhaps not even connected) is defined by:

$$\iota(B) \doteq \begin{cases} 1 \text{ if } B \neq \varnothing \\ 0 \text{ for } B = \varnothing \end{cases}$$

In differential geometry, κ is not the only kind of curvature. A second kind of curvature is defined at every point on a surface. This is the *mean curvature*, m. The total mean sectional curvature is defined as

$$M(C_i) = \int_{\partial C_i} m \, dS.$$

Let $V(C_i)$ denote the volume of C_i. Then we have the following theorem.

The evaluation of M and F for nonconvex simply connected bodies follows in the same way, and for non-simply-connected bodies it follows by extending the computations to the surfaces of any internal voids.

Theorem 2. *(Blaschke/Poincaré, [2, 3, 21]): Given arbitrary compact 3D bodies B_0 and B_1 with continuous piecewise differentiable boundaries, then (2) evaluates as*

$$\Xi(B_0, B_1) = 8\pi^2[V(B_0)\chi(B_1)+V(B_1)\chi(B_0)]+2\pi[F(B_0)M(B_1)+F(B_1)M(B_0)]. \tag{5}$$

Given compact convex bodies C_0 and C_1 in $\mathbb{R}^3$, then (2) evaluates as

$$\Xi(C_0, C_1) = 8\pi^2[V(C_0) + V(C_1)] + 2\pi[F(C_0)M(C_1) + F(C_1)M(C_0)]. \tag{6}$$

For the proof see [2,3,16,25].

1.3 Main Goal of This Paper

In the context of robot motion planning algorithms, one would like to assess the amount of free space that a body has to move without colliding with other bodies. As such, it becomes useful to evaluate

$$\mathcal{I}(B_0, B_1) \doteq \int_{SE(n)} \iota(B_0 \cap g \cdot B_1)\, dg.$$

Moreover, if the intersection body $B_0 \cap g \cdot B_1$ is convex, then $\iota(B_0 \cap g \cdot B_1) = \chi(B_0 \cap g \cdot B_1)$. If this holds for all motions $g \in SE(n)$ which contribute to the integrals, then the integrals themselves will be equal. In particular, since intersections of convex bodies are always either convex or empty, given compact convex bodies C_0 and C_1 we can say

$$\mathcal{I}(C_0, C_1) = \Xi(C_0, C_1),$$

which can be evaluated using the Principal Kinematic Formula (2).

However, in general since for nonconvex bodies in $\mathbb{R}^n$, it can happen that intersections can either have multiple components (which have $\chi > 1$) and toroidal regions (which have $\chi < 1$), these both differ from the value of $\iota = 1$ on nonempty intersections. Hence, for nonconvex bodies in general

$$\mathcal{I}(B_0, B_1) \neq \Xi(B_0, B_1).$$

The main goal of this paper is therefore to establish bounds of the form

$$\Xi(B_0^L, B_1^L) \leq \mathcal{I}(B_0, B_1) \leq \Xi(B_0^U, B_1^U) \tag{7}$$

which justifies the title of this paper.

1.4 Literature Review

The topics of convex and integral geometry have a long and distinguished history, often intertwining with kinematics. Classical works in this area include [6,11–13,15,21,22]. Comprehensive yet accessible monographs introducing this topic include [23,25,26]. Relatively recent expositions of these topics include [16,17]. Extensions of the principal kinematic formula include [19,24].

Ideas related to this formula have been explored previously by the author in [9,10,28], which provide motivations for the current presentation.

2 Integral Geometry of the Indicator Function

In this section, the properties of the Haar measure for $SE(3)$ are combined with properties of the indicator function to produce the desired inequalities.

In combination with the invariance of the integral in (1), we find that

$$\mathcal{I}(C_0, C_1) = \int_{SE(n)} \iota(C_0 \cap g^{-1} \cdot C_1)\, dg = \int_{SE(n)} \iota(C_1 \cap g \cdot C_0)\, dg = \mathcal{I}(C_1, C_0) \tag{8}$$

Due to the semi-direct-product structure of $SE(n)$, the integral of any integrable function $f : SE(n) \to \mathbb{R}$ can be decomposed as

$$\int_{SE(n)} f(g)\, dg = \int_{\mathbb{R}^n} \int_{SO(n)} f(R, \mathbf{t})\, dR\, d\mathbf{t} = \int_{SO(n)} \int_{\mathbb{R}^n} f(R, \mathbf{t})\, d\mathbf{t}\, dR.$$

Applying this principle, we find that

$$\mathcal{I}(B_0, B_1) = \int_{SO(n)} \int_{\mathbb{R}^n} \iota(B_0 \cap (R \cdot B_1 + \mathbf{t}))\, d\mathbf{t}\, dR.$$

If $R \cdot B_1$ is denoted as B_2, note that for each fixed R,

$$\int_{\mathbb{R}^n} \iota(B_0 \cap (B_2 + \mathbf{t}))\, d\mathbf{t} = V(B_0 \oplus (-B_2))$$

where $\oplus$ denotes the Minkowski sum, and $-B_i$ denotes the centrally inverted version of B_i which maps each point $\mathbf{x} \in B_i$ to $-\mathbf{x}$. Since $-I$ and $R \in SO(3)$ commute,

$$\mathcal{I}(B_0, B_1) = \int_{SO(n)} V(B_0 \oplus (-R \cdot B_1))\, dR. \tag{9}$$

3 Principal Kinematic Inequalities

Other than for the case of convex bodies, general equalities for $\mathcal{I}(B_0, B_1)$ have not been derived over the past 100 years of integral geometry. Here several inequalities are derived.

3.1 Inscribing and Circumscribing Convex Bodies

Suppose that we have two not-necessarily-convex bodies, B_0 and B_1. Let B_0 be stationary, and let B_1 be mobile. If it is possible to inscribe a convex body $C_i^L \subseteq B_i$ for $i \in \{0,1\}$, and to enclose B_i with a convex body C_i^U (e.g., the convex hull), then obviously (7) can be achieved as

$$\Xi(C_0^L, C_1^L) \leq \mathcal{I}(B_0, B_1) \leq \Xi(C_0^U, C_1^U).$$

3.2 Using the Brunn-Minkowski Inequality

Building on (9), and raising the Brunn-Minkowski inequality

$$V(B_0 \oplus B_2)^{1/n} \geq V(B_0)^{1/n} + V(B_2)^{1/n}$$

to the nth power, and observing that $V(-R \cdot B_1) = V(B_1)$ gives

$$\mathcal{I}(B_0, B_1) \geq Vol(SO(n)) \sum_{k=0}^{n} \binom{n}{k} |V(B_0)|^{k/n} |V(B_1)|^{(n-k)/n}. \qquad (10)$$

Note that $Vol(SO(2)) = 2\pi$ and $Vol(SO(3)) = 8\pi^2$. Similarly, an upper-bound can be obtained from the reverse Brunn-Minkowski inequality.

3.3 Special Non-convex Cases When $\mathcal{I}(B_0, B_1) \leq \Xi(B_0, B_1)$

In the plane, if B_0 and B_1 are simply connected (but not necessarily convex), then there is no way to generate an annular region by performing intersections of the form $B_0 \cap g \cdot B_1$, and hence

$$\iota(B_0 \cap g \cdot B_1) \leq \chi(B_0 \cap g \cdot B_1).$$

This is not always true in 3D because, for example, two cup-like objects can intersect at their lips producing a solid torus with $\chi = 0$ rather than $\iota = 1$. And the same can happen when intersecting a cup-shaped object and a convex object. However, two 3D non-convex spire-shaped objects (such as the volume of revolution generated from a nonconvex planar curve) will never produce such tori, nor will the intersection of a spire and a convex body. And so, in these cases $\mathcal{I}(B_0, B_1) \leq \Xi(B_0, B_1)$ as well.

3.4 Special Non-convex Cases When $\mathcal{I}(B_0, B_1) = \Xi(B_0, B_1)$

In every dimension, if all nonempty intersections of B_0 and $g \cdot B_1$ always have a single component, then $\iota(B_0 \cap g \cdot B_1) = \chi(B_0 \cap g \cdot B_1)$, regardless of whether or not B_0 and B_1 are convex. Consequently, in such cases $\mathcal{I}(B_0, B_1) = \Xi(B_0, B_1)$.

Such examples can be constructed. Let D be a solid disk/ball and let C_0 be a convex body. Choose a translation $\mathbf{t} \in \partial C_0$ and define $B_0 \doteq C_0 \cap (\mathbf{t} + D)$. If D is chosen to be small enough, then B_0 can be guaranteed to be simply connected. In this case, given a second disk/ball $D' \subset D$, we find that $B_0 \cap g \cdot D'$ has a single component, and so $\mathcal{I}(B_0, D') = \Xi(B_0, D')$. Moreover, if C_1 is a convex body such that $C_1 \subset D'$ and the curvatures at every point on ∂C_1 are greater than those of D', then it is guaranteed that $\iota((C_0 \cap (\mathbf{t} + D)) \cap g \cdot C_1) = \chi((C_0 \cap (\mathbf{t} + D)) \cap g \cdot C_1)$, and hence

$$\mathcal{I}(C_0 \cap (\mathbf{t} + D), C_1) = \Xi(C_0 \cap (\mathbf{t} + D), C_1). \tag{11}$$

Relaxing the curvature conditions results in the upper bound $\mathcal{I} \leq \Xi$.

4 Conclusions

In the fields of integral and convex geometry, the principal kinematic formula plays a central role. Two versions of this formula typically appear: (1) closed-form expressions for the integral over motion of the Euler characteristic of nonconvex bodies; (2) closed-form expressions for the integral over motion of any quermassintegral/mixed-volume of convex bodies. In both cases, Hadwiger's characterization theorem applies. However, in robot motion planning one is concerned with determining which regions of the configuration space of a rigid body correspond to collision. Such collisions are detected by evaluating the set-indicator function of potentially non-convex moving bodies with stationary obstacles. Integral and convex geometry do not address this directly, and so inequalities are derived and presented to assess the integral over motion of the indicator function of intersections of bodies.

References

1. Angeles, J.: Rational Kinematics. Springer, New York (1988)
2. Blaschke, W.: "Einige Bemerkungen über Kurven und Flächen konstanter Breite," Ber. Kgl. Sächs. Akad. d. Wiss. Leipzig **67**, 290–297 (1915)
3. Blaschke, W.: Vorlesungen über Integralgeometrie. Deutscher Verlag der Wissenschaften, Berlin (1955)
4. Bottema, O., Roth, B.: Theoretical Kinematics. Dover Publications, Inc., New York (1990). Reprinted
5. Chern, S.-S.: On the kinematic formula in the Euclidean space of N dimensions. Am. J. Math. **74**(1), 227–236 (1952)
6. Chern, S.-S.: On the kinematic formula in integral geometry. J. Math. Mech. **16**(1), 101–118 (1966)
7. Chirikjian, G.S., Kyatkin, A.B.: Harmonic Analysis for Engineers and Applied Scientists. Dover, Mineola (2016)
8. Chirikjian, G.S.: Stochastic Models, Information Theory, and Lie Groups: Volume 2 - Analytic Methods and Modern Applications. Birkhäuser, Boston (2011)
9. Chirikjian, G.S.: Parts entropy and the principal kinematic formula. In: Proceedings of IEEE Conference on Automation Science and Engineering, Washington D.C., pp. 864–869, 23–26 August 2008
10. Chirikjian, G.S., Yan, Y.: The kinematics of containment. In: Proceedings of the Advances in Robot Kinematics, pp. 355–364. Springer, July 2014

11. Fu, J.H.G.: Kinematic formulas in integral geometry. Indiana Univ. Math. J. **39**(4), 1115–1154 (1990)
12. Hadwiger, H.: Vorlesungen über Inhalt, Oberfläche und Isoperimetrie. Springer, Berlin (1957)
13. Hadwiger, H.: Altes und Neues über Konvexe Körper. Birkhäuser Verlag, Basel (1955)
14. Karger, A., Novák, J.: Space Kinematics and Lie Groups. Gordon and Breach Science Publishers, New York (1985)
15. Kendall, M.G., Moran, P.A.P.: Geometrical Probability. Griffin's Statistical Monographs, London (1963)
16. Klain, D.A., Rota, G.-C.: Introduction to Geometric Probability. Cambridge University Press, Cambridge (1997)
17. Mani-Levitska, P.: A simple proof of the kinematic formula. Monatschefte für Mathematik **105**, 279–285 (1988)
18. McCarthy, J.M.: Introduction to Theoretical Kinematics. MIT Press, Cambridge (1990). (Revised and available as an iBook as of 2013)
19. Miles, R.E.: The fundamental formula of Blaschke in integral geometry and geometrical probability, and its iteration, for domains with fixed orientations. Austral. J. Statist. **16**, 111–118 (1974)
20. Park, F.C.: The optimal kinematic design of mechanisms. Ph.D. Thesis, Division of Engineering and Applied Sciences, Harvard University, Cambridge, MA (1991)
21. Poincaré, H.: Calcul de Probabilités, 2nd edn., Paris (1912). (Reprinted by BiblioLife in 2009)
22. Pólya, G.: Über geometrische Wahrscheinlichkeiten. S.-B. Akad. Wiss. Wien **126**, 319–328 (1917)
23. Ren, D.-L.: Topics in Integral Geometry. World Scientific Publishing, Singapore (1994)
24. Rother, W., Zähle, M.: A short proof of the principal kinematic formula and extensions. Trans. Am. Math. Soc. **321**, 547–558 (1990)
25. Santaló, L.: Integral Geometry and Geometric Probability. Cambridge University Press, Cambridge (2004). (Originally published in 1976 by Addison-Wesley)
26. Schneider, R., Weil, W.: Stochastic and Integral Geometry. Springer, Berlin (2008)
27. Selig, J.M.: Geometrical Fundamentals of Robotics, 2nd edn. Springer, New York (2005)
28. Yan, Y., Chirikjian, G.S.: Closed-form characterization of the Minkowski Sum and difference of two ellipsoids. Geometricae Dedicata **177**, 103–128 (2015)

Displacement Varieties for Some PUP Linkages

Jonathan M. Selig(✉)

London South Bank University, London SE1 0AA, U.K.
seligjm@lsbu.ac.uk

Abstract. In this work sub-varieties of the Study quadric of rigid-body displacements are investigated. The particular sub-varieties considered are the possible displacements which can be performed by the end-effector of certain PUP linkages. In general these varieties are linear projections of a 4-dimensional Segre variety in 15-dimensional projective space. The different linkages correspond to projections with different centres. For one particular linkage the intersection of three such sub-varieties is studied, this intersection corresponds to the assembly configurations of a parallel robot with 3 such PUP legs.

Keywords: Study quadric · Segre varieties · 3PUP manipulator

1 Introduction

In a series of papers Dai and co-workers introduced and studied several novel parallel manipulators consisting of 3 PUP legs, [1,4,9]. In this work the geometry of the individual legs are considered. That is, the linkages considered here are comprised of a serial chain with a prismatic joint, two revolute joints with their axes meeting orthogonally and a final prismatic joint. There are several types of such a linkage depending on the angles between directions of the prismatic joints and the revolute joints. Only a few examples are studied in this work, the types chosen are those actually used in the work of Dai *et al.*

The main object of study is the displacement variety of the linkage. This is the algebraic variety of possible rigid-body displacements that the end-effector of the linkage can perform. The model used for the rigid-body displacements will be the Study quadric, a non-singular quadric in $\mathbb{P}^7$. With homogeneous coordinate $(a_0 : a_1 : a_2 : a_3 : c_0 : c_1 : c_2 : c_3)$ for the 7-dimensional projective space, the equation of the Study quadric is,

$$a_0c_0 + a_1c_1 + a_2c_2 + a_3c_3 = 0.$$

Points in this 6-dimensional quadric variety are in 1-to-1 correspondence with the set of rigid-body displacements of space, with the exception of the 3-dimensional plane $a_0 = a_1 = a_2 = a_3 = 0$. This 3-plane will be denoted A_∞ in the following and consists of a set of "ideal elements" of the group of rigid-body displacements.

J. Lenarcic and V. Parenti-Castelli (Eds.): ARK 2018, SPAR 8, pp. 28–36, 2019.
https://doi.org/10.1007/978-3-319-93188-3_4

The displacement varieties considered will be sub-varieties of the Study quadric defined by homogeneous algebraic equations in the homogeneous coordinates of $\mathbb{P}^7$. For more details see [6].

2 Generalities

Expressing elements of the group of rigid displacements as dual quaternions, the displacements generated by prismatic or revolute joints are lines in the Study quadric. For example, rotations about the z-axis are represented as dual quaternions of the form $g = (\alpha + \beta k)$. Since the rotation axis passes through the origin here, there is no dual part to the element. The parameters α and β are arbitrary but cannot be zero simultaneously. Only the ratio of the parameters is important, so α and β can be considered as homogeneous parameters of a projective line $\mathbb{P}^1$. Dividing both α and β by a common factor does not change the group element, so they may be divided by $\sqrt{\alpha^2 + \beta^2}$ and hence these parameter can be considered as the cosine and sine of a rotation angle.

For a prismatic joint, say in the z direction, the corresponding $\mathbb{P}^1$ of group elements is given by, $g = \lambda + \varepsilon k \mu$. In this case the ratio μ/λ gives the distance translated along the joint.

The algebraic variety formed as a product of projective spaces is known as a Segre variety. A product of 4 lines, $\mathbb{P}^1 \times \mathbb{P}^1 \times \mathbb{P}^1 \times \mathbb{P}^1$ gives a 4-dimensional variety in $\mathbb{P}^{15}$. The standard map from the Segre variety to $\mathbb{P}^{15}$ is given by identifying the 16 homogeneous coordinates of $\mathbb{P}^{15}$ with the 16 possible monomials formed from the coordinates of the 4 $\mathbb{P}^1$s, one coordinate from each line. As a simple example consider $\mathbb{P}^1 \times \mathbb{P}^1$ where the coordinates of the lines are $(\alpha_1 : \beta_1)$ and $(\alpha_2 : \beta_2)$. The Segre mapping to $\mathbb{P}^3$ is then given by,

$$(\alpha_1 : \beta_1) \times (\alpha_2 : \beta_2) \longmapsto (\alpha_1\alpha_2 : \alpha_1\beta_2 : \beta_1\alpha_2 : \beta_1\beta_2).$$

The points in the image of the map, the points in the Segre variety, satisfy the single quadratic equation, $X_1X_4 - X_2X_3 = 0$, where X_i are the homogeneous coordinates of the $\mathbb{P}^3$. The case of $\mathbb{P}^1 \times \mathbb{P}^1 \times \mathbb{P}^1$ is treated in some detail in [5].

There are several ways to find the degree of the Segre variety $\mathbb{P}^1 \times \mathbb{P}^1 \times \mathbb{P}^1 \times \mathbb{P}^1$, extending the methods used in [2], using homology or cohomology theory. A simple method that shows that the degree of an n-fold product of lines, $(\mathbb{P}^1)^n$ has degree $n!$ is given in [3]. Hence, the degree of $\mathbb{P}^1 \times \mathbb{P}^1 \times \mathbb{P}^1 \times \mathbb{P}^1$ is 24. The variety also satisfies a large number of quadratic equations.

In this work the varieties are all linear projections of this Segre variety to $\mathbb{P}^7$, the different types of linkage are given by different projections, that is projections from different $\mathbb{P}^7$s in the $\mathbb{P}^{15}$. Note that this would also apply to general 4R serial linkages.

3 First Type

Figure 1 shows four different types of PUP linkage, each in a standard configuration. In this section the leftmost of these mechanisms will be considered.

Fig. 1. Four different types of PUP linkage

By choosing the origin of coordinate to be at the centre of the U-joint and with the two revolute joints of the U joint aligned with the x and y axes, the possible displacements that the linkage can perform may be parameterised as,

$$g = (\lambda_1 + \mu_1\varepsilon k)(\alpha_2 + \beta_2 i)(\alpha_3 + \beta_3 j)(\lambda_4 + \mu_4\varepsilon k)$$

Multiplying out this product gives,

$$g = a_0 + a_1 i + a_2 j + a_3 k + c_0\varepsilon + c_1\varepsilon i + c_2\varepsilon j + c_3\varepsilon k$$

where,

$$\begin{aligned} a_0 &= \lambda_1\lambda_4\alpha_2\alpha_3, & c_0 &= -(\lambda_1\mu_4 + \mu_1\lambda_4)\beta_2\beta_3, \\ a_1 &= \lambda_1\lambda_4\beta_2\alpha_3, & c_1 &= (\lambda_1\mu_4 - \mu_1\lambda_4)\alpha_2\beta_3, \\ a_2 &= \lambda_1\lambda_4\alpha_2\beta_3, & c_2 &= -(\lambda_1\mu_4 - \mu_1\lambda_4)\beta_2\alpha_3, \\ a_3 &= \lambda_1\lambda_4\beta_2\beta_3, & c_3 &= (\lambda_1\mu_4 + \mu_1\lambda_4)\alpha_2\alpha_3. \end{aligned}$$

By inspection these coordinates satisfy 3 linearly independent homogeneous quadratic equations,

$$\begin{aligned} 0 &= a_0a_3 - a_1a_2, \\ 0 &= a_0c_0 + a_3c_3, \\ 0 &= a_0c_0 + a_1c_1 + a_2c_2 + a_3c_3. \end{aligned}$$

The last equation here is just the equation of the Study quadric. Using a symbolic algebra computer package (Singular) it can be determined that the three equations generate a prime ideal. Hence the intersection of the three quadrics determined by the equations form an irreducible projective variety. Thus the degree of the variety is simply $2 \times 2 \times 2 = 8$.

Notice that the variety determined by the 3 quadric contains the 3-plane A_∞ of ideal displacements. This make it difficult to study structures composed of two or three such legs combined in parallel—the intersection of two or three such varieties will not form a complete intersection and hence Bézout's theorem cannot be applied. To investigate this we need to look at the general case for this type of linkage. Consider the PUP linkage illustrated in Fig. 2. To keep the computations as simple as possible the origin is still placed at the centre of the U joint with the x and y axes aligned with the first and second revolute joints.

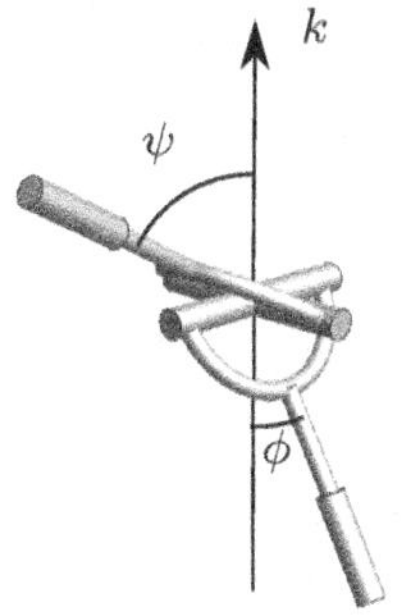

Fig. 2. General PUP linkage of the first type.

The most general starting position for such a linkage can me found from this configuration by a rigid change of coordinates.

With the starting position illustrated the possible rigid displacements of the end-effector relative to the base can be parameterised as,

$$g = (\lambda_1 + \mu_1\varepsilon(\sin\phi j + \cos\phi k))(\alpha_2 + \beta_2 i)(\alpha_3 + \beta_3 j)(\lambda_4 + \mu_4\varepsilon(\sin\psi i + \cos\psi k))$$

multiplying this out gives the same results as above for $a_0, \ldots, a_3$ but for the c_i coordinates the results are,

$$\begin{aligned}
c_0 &= -(\lambda_1\mu_4\cos\psi + \mu_1\lambda_4\cos\phi)\beta_2\beta_3 - (\mu_1\lambda_4\sin\phi)\alpha_2\beta_3 - (\lambda_1\mu_4\sin\psi)\beta_2\alpha_3,\\
c_1 &= (\lambda_1\mu_4\cos\psi - \mu_1\lambda_4\cos\phi)\alpha_2\beta_3 + (\mu_1\lambda_4\sin\phi)\beta_2\beta_3 + (\lambda_1\mu_4\sin\psi)\alpha_2\alpha_3,\\
c_2 &= -(\lambda_1\mu_4\cos\psi - \mu_1\lambda_4\cos\phi)\beta_2\alpha_3 + (\mu_1\lambda_4\sin\phi)\alpha_2\alpha_3 + (\lambda_1\mu_4\sin\psi)\beta_2\beta_3,\\
c_3 &= (\lambda_1\mu_4\cos\psi + \mu_1\lambda_4\cos\phi)\alpha_2\alpha_3 - (\mu_1\lambda_4\sin\phi)\beta_2\alpha_3 - (\lambda_1\mu_4\sin\psi)\alpha_2\beta_3.
\end{aligned}$$

Since the results for the a_is are the same, the variety will still satisfy, $a_0a_3 - a_1a_2 = 0$. Also since these are just group multiplications the equation for the Study quadric will also be satisfied. The parametrisation also satisfies the quadratic equation,

$$\begin{aligned}
&(-2a_0\cos\phi\cos\psi + a_1\cos\psi\sin\phi + a_2\cos\phi\sin\psi - a_3\sin\phi\sin\psi)c_0\\
&\qquad + (a_0\cos\psi\sin\phi - a_2\sin\phi\sin\psi - a_3\cos\phi\sin\psi)c_1\\
&\qquad + (a_0\cos\phi\sin\psi - a_1\sin\phi\sin\psi - a_3\cos\psi\sin\phi)c_2\\
&\quad - (a_0\sin\phi\sin\psi + a_1\cos\phi\sin\psi + a_2\cos\psi\sin\phi + 2a_3\cos\phi\cos\psi)c_3 = 0.
\end{aligned}$$

This quadratic equation, and the other quadratic equations in the Study coordinates, can each be written as an 8×8 symmetric matrix. Partitioned into 4×4 sub-matrices the quadratic has the shape,

$$\mathbf{g}^T Q\mathbf{g} = (\bar{a}^T, \bar{c}^T)\begin{pmatrix} 0 & \bar{M} \\ \bar{M} & 0 \end{pmatrix}\begin{pmatrix} \bar{a} \\ \bar{c} \end{pmatrix},$$

where $\mathbf{g}$ is the vector of Study coordinates $\mathbf{g} = (a_0, a_1, a_2, a_3, c_0, c_1, c_2, c_3)^T$ which can be written in partitioned form with $\bar{a} = (a_0, a_1, a_2, a_3)^T$ and

$\bar{c} = (c_0, c_1, c_2, c_3)^T$. The symmetric matrix $\bar{M}$ is given by,

$$\bar{M} = \begin{pmatrix} -2\cos\phi\cos\psi & \cos\psi\sin\phi & \cos\phi\sin\psi & -\sin\phi\sin\psi \\ \cos\psi\sin\phi & 0 & -\sin\phi\sin\psi & -\cos\phi\sin\psi \\ \cos\phi\sin\psi & -\sin\phi\sin\psi & 0 & -\cos\psi\sin\phi \\ -\sin\phi\sin\psi & -\cos\phi\sin\psi & -\cos\psi\sin\phi & -2\cos\phi\cos\psi \end{pmatrix}.$$

Now suppose the PUP leg is subject to a rigid displacement. What will happen to the three quadratic equations that specify the displacement sub-variety of the linkage? Under such a displacement the group element of Study coordinates transforms according to,

$$\begin{pmatrix} a_0 \\ \mathbf{a} \\ c_0 \\ \mathbf{c} \end{pmatrix} \longmapsto \begin{pmatrix} 1 & 0 & 0 & 0 \\ 0 & R & 0 & 0 \\ 0 & 0 & 1 & 0 \\ 0 & TR & 0 & R \end{pmatrix} \begin{pmatrix} a_0 \\ \mathbf{a} \\ c_0 \\ \mathbf{c} \end{pmatrix}.$$

Here $\mathbf{a} = (a_1, a_2, a_3)^T$, $\mathbf{c} = (c_1, c_2, c_3)^T$, R is the 3×3 rotation matrix of the displacement and T is the $\times 3$ antisymmetric matrix corresponding to the translation vector. This means that the symmetric matrix of the quadric $\mathcal{Q}$ transforms under such a displacement to,

$$\mathcal{Q} \longmapsto \begin{pmatrix} \bar{R} & 0 \\ \bar{B} & \bar{R} \end{pmatrix}^{-T} \begin{pmatrix} 0 & \bar{M} \\ \bar{M} & 0 \end{pmatrix} \begin{pmatrix} \bar{R} & 0 \\ \bar{B} & \bar{R} \end{pmatrix}^{-1} = \begin{pmatrix} \bar{W} & \bar{U} \\ \bar{U} & 0 \end{pmatrix}$$

where

$$\bar{R} = \begin{pmatrix} 1 & 0 \\ 0 & R \end{pmatrix} \qquad \text{and} \qquad \bar{B} = \begin{pmatrix} 0 & 0 \\ 0 & TR \end{pmatrix}$$

and thus,

$$\bar{W} = \bar{R}(\bar{B}\bar{M} + \bar{M}\bar{B}^T)\bar{R}^T \qquad \text{and} \qquad \bar{U} = \bar{R}\bar{M}\bar{R}^T.$$

This final form is really all that will be used in the following.

Returning to the first quadric the displacement variety lies on; $a_0a_3 - a_1a_2 = 0$, this can also be written as a partitioned 8×8 matrix,

$$\mathcal{K} = \begin{pmatrix} \bar{N} & 0 \\ 0 & 0 \end{pmatrix}, \qquad \text{where} \qquad \bar{N} = \begin{pmatrix} 0 & 0 & 0 & 1 \\ 0 & 0 & -1 & 0 \\ 0 & -1 & 0 & 0 \\ 1 & 0 & 0 & 0 \end{pmatrix}.$$

Hence after a rigid displacement this will be transformed to,

$$\mathcal{K} \longmapsto \begin{pmatrix} \bar{R}\bar{N}\bar{R}^T & 0 \\ 0 & 0 \end{pmatrix}.$$

Now consider a parallel structure consisting of base and platform connected by 3 PUP legs of this type. The possible assembly configurations of such a structure must lie on the intersection of seven quadrics in $\mathbb{P}^7$; the Study quadric and three quadrics of each of the forms,

$$(\bar{a}^T, \bar{c}^T)\mathcal{Q}_i \begin{pmatrix} \bar{a} \\ \bar{c} \end{pmatrix} = 0 \quad \text{and} \quad (\bar{a}^T, \bar{c}^T)\mathcal{K}_i \begin{pmatrix} \bar{a} \\ \bar{c} \end{pmatrix} = 0, \qquad i = 1, 2, 3.$$

These seven quadrics all contain the 3-plane of ideal points A_∞, so the intersection is not a complete intersection. However, only the residual variety, the one remaining after A_∞ has been deleted, is of interest since the points on the ideal 3-plane do not correspond to physical assemblies of the mechanism.

So consider a linear projection with centre A_∞ onto the 3-plane given by $c_0 = c_1 = c_2 = c_3 = 0$. This second plane will be called A_0, the image of the map is usually known as the spherical indicatrix of the mechanism. Using the 4 quadrics, $\mathcal{Q}_1$, $\mathcal{Q}_2$, $\mathcal{Q}_3$ and the Study quadric $\mathcal{Q}_s$ the projection can be inverted. Notice that the 4 quadrics can be thought of as linear equations in the c_i coordinates, that is,

$$2\bar{a}^T\bar{c} = 0 \quad \text{and} \quad 2\bar{a}^T\bar{U}_i\bar{c} = -\bar{a}^T\bar{W}_i\bar{a}, \qquad i = 1,\ 2,\ 3,$$

where the first equation is the Study quadric. Let $\tilde{a}$ be a point in A_0 and form the 4×4 matrix $\bar{\Lambda}$ which has rows $2\tilde{a}^T\bar{U}_1$, $2\tilde{a}^T\bar{U}_2$, $2\tilde{a}^T\bar{U}_3$ and $2\tilde{a}^T$. Also write $\bar{\eta}$ as the 4-vector with entries, $-\tilde{a}^T\bar{W}_1\tilde{a}$, $-\tilde{a}^T\bar{W}_2\tilde{a}$, $-\tilde{a}^T\bar{W}_3\tilde{a}$ and 0. The 4 quadrics can be written in the matrix form,

$$\bar{\Lambda}\bar{c} = \bar{\eta}.$$

Using linear algebra we can invert the equation to produce a point in $\mathbb{P}^7$,

$$\bar{a} = \det(\bar{\Lambda})\tilde{a} \qquad \text{and} \qquad \bar{c} = \mathrm{Adj}(\bar{\Lambda})\bar{\eta}$$

where Adj denotes the transposed matrix of cofactors usually called the adjugate matrix. So given a point $\tilde{a}$, in A_0 this gives a rational map to the intersection of the 4 quadrics in $\mathbb{P}^7$. The map clearly inverts the projection to A_0 and hence shows that the intersection of the 4 quadrics (with the component A_∞ deleted) is birationally equivalent to a $\mathbb{P}^3$, namely A_0. Notice that the image of a point $\tilde{a}$ has degree 5 in the coordinates of $\tilde{a}$.

To project the other three quadrics $\mathcal{K}_1$, $\mathcal{K}_2$ and $\mathcal{K}_3$ to A_0 the coordinates c_0, c_1, c_2 and c_3 must be eliminated, but the quadrics do not contain these variables. Hence the quadrics project to three 2-dimensional quadrics in A_0. In general the intersection of three such quadrics will give $2^3 = 8$ discrete points and, after mapping back to $\mathbb{P}^7$, this gives 8 possible assembly modes for the structure.

There are however, a number of possible special cases. These are cases where the displacement sub-variety of the 3 PUP parallel mechanism has a component of one or two dimensions and hence the machine acquires one or two degrees of freedom. Two particular cases are particularly simple. Suppose the U joints of the three legs are all parallel. That is, in the home position, the U joints are simply translates of each other with no rotation. In this case the three quadrics

$\mathcal{K}_1$, $\mathcal{K}_2$ and $\mathcal{K}_3$ are identical and hence the mechanism will be 2-dimensional. The displacement sub-variety will be birationally equivalent to a 2-dimensional quadric in $\mathbb{P}^3$. In the second special case, suppose that only two of the U joints are parallel in the home position. Now the spherical indicatrix is the intersection of a pair of 2-dimensional quadric and so is, in general, an elliptic quartic curve.

It is also possible that the three 2-dimensional quadrics intersect in a curve, necessarily not a complete intersection. Only a few possibilities can arise, for the intersection:

- A twisted cubic curve.
- A conic and two discrete points.
- Two lines and a two discrete points.
- A single line and four discrete points.

See for example, [7].

The question arises, which of these possibilities can occur as the displacement variety of a 3 UPU mechanism? Remember that each of the 3 quadrics are related to each other by rigid rotations. One possibility that can occur is the line and 4 points. This is because the three legs can be arranged in such a way that the U joints in two of the legs are rotated around a line in the quadric determined by the third leg. This line will thus lie in all three quadrics and hence their intersection will consist of the line and four discrete points, some or all of these discrete points might bc complex points however. Note that since the rotational motion of the platform will be a rotation about a fixed direction here, the complete motion of the platform will be a Schönflies motion. Moreover, it can be seen that the image of a line in A_0 will generally be a curve in the Study quadric with degree 5.

4 Second and Third Types

The second and third types of PUP linkage as shown in Fig. 1, are distinguished by the fact that direction of one of the P joints is parallel to the axis of its neighbouring R joint. For the second type the two joints with parallel axes are the last two while in the third type they are the first two. This type of linkage could be analysed in a similar fashion to the work in the previous section. But in these cases there is a simpler method because here the displacement sub-variety can be identified with a purely geometric problem. Consider for definiteness the second type of linkage here. Call the axis of the final R joint ℓ, the direction of this line is parallel to the direction of the final P joint. Also let π be the plane through the centre of the U joint and perpendicular to the first R joint of the universal joint. Now it is clear that the line ℓ lies in the plane π and any motion of the linkage will preserve this incidence.

The variety of displacements which move a line so that it remains in a fixed plane can be see to be a Segre variety $\mathbb{P}^3 \times \mathbb{P}^1$. This is a product of subgroups: a rotation about the line ℓ, in its initial configuration can be followed by a planar displacement taking π to itself. According to [2], the degree of a Segre variety $\mathbb{P}^3 \times \mathbb{P}^1$ is $\binom{3+1}{1} = 4$. The space of displacements that this second type of PUP can achieve will be an open set in this Segre variety. This is because the linkage cannot move the axis of the final R joint parallel to the axis of the first R joint. These impossible displacements comprise translations in the plane but parallel to the first R joint combined with any rotation about the second R joint. These impossible displacements form a $\mathbb{P}^1 \times \mathbb{P}^1$ Segre variety and so the displacements that the second type of PUP linkage can achieve are a $\mathbb{P}^3 \times \mathbb{P}^1$ with a $\mathbb{P}^1 \times \mathbb{P}^1$ deleted.

Clearly the third type of PUP linkage is the inversion of the second type so the methods of [8] could be used to study its displacement variety. However, it is clear that the possible displacements of the end-effector can be viewed as the displacements of a plane such that the plane always contains a fixed line. This is simply a planar motion followed by a rotation about the line and hence is also be a Segre variety $\mathbb{P}^1 \times \mathbb{P}^3$. Hence the displacements that this third type of PUP linkage can achieve also comprise an open set in a $\mathbb{P}^1 \times \mathbb{P}^3$ Segre variety.

It is well know, see [2] for example, that a $\mathbb{P}^3 \times \mathbb{P}^1$ Segre variety lies in 6 linearly independent quadrics in $\mathbb{P}^7$. For the second type of PUP with home position illustrated in Fig. 1 above the 6 quadrics can be found as follows. The dual quaternions representing the rigid displacement of the end-effector can be written,

$$g = (\lambda_1 + \mu_1 \varepsilon k)(\alpha_2 + \beta_2 i)(\alpha_3 + \beta_3 k)(\lambda_4 + \mu_4 \varepsilon k),$$

where, as above, the α_i, λ_j and so forth are the homogeneous parameters of the joints. Expanding this equation and collecting terms gives,

$$\begin{aligned} a_0 &= \lambda_1\lambda_4\alpha_2\alpha_3, & c_0 &= -(\lambda_1\mu_4 + \mu_1\lambda_4)\alpha_2\beta_3, \\ a_1 &= \lambda_1\lambda_4\beta_2\alpha_3, & c_1 &= (\lambda_1\mu_4 - \mu_1\lambda_4)\beta_2\beta_3, \\ a_2 &= -\lambda_1\lambda_4\beta_2\beta_3, & c_2 &= (\lambda_1\mu_4 - \mu_1\lambda_4)\beta_2\alpha_3, \\ a_3 &= \lambda_1\lambda_4\alpha_2\beta_3, & c_3 &= (\lambda_1\mu_4 + \mu_1\lambda_4)\alpha_2\alpha_3. \end{aligned}$$

The six quadratic equations satisfied by these group elements can be expressed as,

$$\text{Rank}\begin{pmatrix} a_0 & a_1 & c_2 & c_3 \\ a_3 & -a_2 & c_1 & -c_0 \end{pmatrix} = 1.$$

That is the six possible 2×2 sub-determinants of this matrix must vanish.

To find the quadrics for the general case the geometric problem of displacing a general line ℓ so that it remains on a fixed plane π could be considered. There are several other ways to express this geometry, suppose ℓ_1, ℓ_2 and ℓ_3 are three linearly independent lines lying in the plane π, then any rigid displacement which keeps ℓ in the plane π will also preserve the reciprocity between ℓ and the three other lines ℓ_1, ℓ_2, ℓ_3. This will only produce 3 quadrics however.

5 Conclusion

The results here suggest the novel possibility that a mobile over-constrained parallel manipulator can have assembly configurations of different dimensions. This phenomenon does not seem to have attracted much attention in the parallel robotics literature to date. It is however, well known for closed loop mechanisms such as the cyclohexane molecule.

Associating a linkage with a purely geometrical problem provides a powerful technique for understanding the linkage's displacement variety. There are now many examples in the literature, probably beginning with the identification of the spherical-spherical linkage with the problem of moving a point in such a manner that it remains on a fixed sphere. The first PUP linkage considered above can also be associated with a geometrical problem. Consider a pair of lined planes. A lined plane in a plane with a distinguished line lying in the plane. In terms of the linkage the two planes are the planes normal to the two R joints and passing through the point where the axes of the R joints cross. The distinguished lines are formed by translating the centre point of the U joint in the direction of the first or last P joints. Now assume that the first lined plane is fixed and consider the possible rigid displacements of the second which move it in such a way that the planes remain perpendicular and the lines remain in contact. It is not too difficult to see that these are precisely the possible displacements of the end-effector of this PUP linkage relative to its home position.

Limitations of space has precluded the study of a fourth type of PUP linkage where the direction of the final P joint lies in the same plane as the axes of the two R joints. Preliminary results suggest that the displacement variety of such a linkage is not too different from the that of the first type considered here.

References

1. Gan, D., Dai, J.S.: Geometry constraint and branch motion evolution of 3-PUP parallel mechanisms with bifurcated motion. Mech. Mach. Theor. **61**, 168–183 (2013). https://doi.org/10.1016/j.mechmachtheory.2012.09.011
2. Harris, J.: Algebraic Geometry a First Course. Springer Verlag, New York (1992)
3. Polizzi, F.: Degree of the segre product of lines (2014). (https://mathoverflow.net/users/7460/francesco-polizzi). (version: 2014-06-11)
4. Rodriguez-Leal, E., Dai, J.S., Pennock, G.R.: Inverse kinematics and motion simulation of a 2-DOF parallel manipulator with 3-PUP legs. In: Kecskeméthy, A., Müller, A. (eds.) Computational Kinematics, Proceedings of the 5th International Workshop on Computational Kinematics, pp. pp. 85–92. Springer, Berlin, Heidelberg (2009). https://doi.org/10.1007/978-3-642-01947-0_11
5. Selig, J.M.: Some remarks on the *RRR* linkage. In: Lenarčič, J., Khatib, O. (eds.) Advances in Robot Kinematics, pp. 77–85. Springer International Publishing, Cham (2014). https://doi.org/10.1007/978-3-319-06698-1_9
6. Selig, J.M.: Geometric Fundamentals of Robotics. Springer Verlag, New York (2005)
7. Semple, J.G., Roth, L.: Introduction to Algebraic Geometry. Clarendon Press, Oxford (1949). Reprinted in 1985 with index
8. Stigger, T., Husty, M.L.: Constraint equations of inverted kinematic chains. In: Zeghloul, S., Romdhane, L., Laribi, M. (eds.) Computational Kinematics. Mechanisms and Machine Science, vol. 50. Springer, Cham (2018). https://doi.org/10.1007/978-3-319-60867-9_56
9. Zhang, K., Dai, J.S., Fang, Y.: Constraint analysis and bifurcated motion of the 3PUP parallel mechanism. Mechanism and Mach. Theor. **49**, 256–269 (2012). https://doi.org/10.1016/j.mechmachtheory.2011.10.004

Computational Kinematics

Accurate Computation of Quaternions from Rotation Matrices

Soheil Sarabandi and Federico Thomas(✉)

Institut de Robòtica i Informàtica Industrial (CSIC-UPC), Llorens Artigas 4-6, 08028 Barcelona, Spain
soheil.sarabandi@gmail.com, fthomas@iri.upc.edu

Abstract. The main non-singular alternative to 3×3 proper orthogonal matrices, for representing rotations in $\mathbb{R}^3$, is quaternions. Thus, it is important to have reliable methods to pass from one representation to the other. While passing from a quaternion to the corresponding rotation matrix is given by Euler-Rodrigues formula, the other way round can be performed in many different ways. Although all of them are algebraically equivalent, their numerical behavior can be quite different. In 1978, Shepperd proposed a method for computing the quaternion corresponding to a rotation matrix which is considered the most reliable method to date. Shepperd's method, thanks to a voting scheme between four possible solutions, always works far from formulation singularities. In this paper, we propose a new method which outperforms Shepperd's method without increasing the computational cost.

Keywords: Quaternions · Rotation matrices

1 Introduction

Arbitrary rotations in $\mathbb{R}^3$ can be represented using proper orthogonal 3×3 matrices (*rotation matrices* for short) of the form:

$$\mathbf{R} = \begin{pmatrix} r_{11} & r_{12} & r_{13} \\ r_{21} & r_{22} & r_{23} \\ r_{31} & r_{32} & r_{33} \end{pmatrix}. \tag{1}$$

Nevertheless, according to Euler rotation theorem, every rotation in three dimensions is defined by its axis, given by a unit vector $\mathbf{n} = (n_x, n_y, n_z)$, and its angle, the amount of rotation about that axis, given by θ. Clearly, $\mathbf{n}$ and θ provide a much more compact and meaningful information about a rotation than the 9 entries of a rotation matrix.

The information about the axis and angle of rotation is usually organized as a quaternion so that the product of two quaternions gives us the quaternion corresponding to the composition of the rotations that the two matrices represent, in the same way as the product of two rotation matrices gives us the matrix

J. Lenarcic and V. Parenti-Castelli (Eds.): ARK 2018, SPAR 8, pp. 39–46, 2019.
https://doi.org/10.1007/978-3-319-93188-3_5

representing the rotation corresponding to the composition of the two rotations the matrices represent. The four elements of the quaternion are given by

$$q_1 = \cos\frac{\theta}{2}, \qquad q_2 = n_x \sin\frac{\theta}{2}, \qquad q_3 = n_y \sin\frac{\theta}{2}, \qquad q_4 = n_z \sin\frac{\theta}{2}, \tag{2}$$

which are commonly known as Euler parameters. These quantities are not independent, they are related through the following equation:

$$q_1^2 + q_2^2 + q_3^2 + q_4^2 = 1. \tag{3}$$

As a consequence, when operating with quaternions, we need to be sure that (3) is always satisfied.

Because both, quaternions and rotation matrices, are useful for certain kinematics calculations, the need arises to convert between these representations. It is well known [1] that a 3×3 proper orthogonal rotation matrix can be expressed in terms of the quaternion components as:

$$\mathbf{R} = \begin{pmatrix} q_1^2 + q_2^2 - q_3^2 - q_4^2 & 2(q_2 q_3 - q_1 q_4) & 2(q_2 q_4 + q_1 q_3) \\ 2(q_2 q_3 + q_1 q_4) & q_1^2 - q_2^2 + q_3^2 - q_4^2 & 2(q_2 q_4 - q_1 q_2) \\ 2(q_2 q_4 - q_1 q_3) & 2(q_3 q_4 + q_1 q_2) & q_1^2 - q_2^2 - q_3^2 + q_4^2 \end{pmatrix}. \tag{4}$$

In contrast, there are many methods to obtain the quaternion corresponding to a given rotation matrix. They include algebraic, trigonometric, and numerical methods (see [2] for a recent survey). If we limit our analysis to algebraic methods, that is, to those methods consisting in solving the system of nonlinear equations resulting from equating the matrices in (1) and (4), Shepperd's algorithm is considered as the best one from the numerical point of view. This paper is devoted to the derivation of an alternative method which is shown to be numerically better conditioned than Shepperd's method.

By observing (4), it is easy to conclude that (q_1, q_2, q_3, q_4) and $(-q_1, -q_2, -q_3, -q_4)$ lead to the same rotation matrix. In mathematical terms, it is said that quaternions provide a double covering of the space of rotations. Since this is a 2-to-1 map, it cannot be smoothly inverted. This has important consequences in all methods that compute the quaternion corresponding to a rotation matrix: they all have to deal with singularities, and they all give the same solution within an undetermined but physically unimportant overall sign.

This paper is organized as follows. In the next section, we briefly summarize Shepperd's method. In Sect. 3 we present an alternative method which, in Sect. 4, is shown to be numerically more accurate than Shepperd's method. Finally, Sect. 5 summarizes the main points.

2 Shepperds's Method

Since it was first proposed in [3], Shepperd's method remains as one of the most popular methods for computing the quaternion corresponding to a rotation matrix. It improves on Hughes' method [5] via a voting scheme in which the quaternion is computed without numerical instabilities.

It is worth to mention that, although the method is associated with Shepperd, its origins can be traced back at least to a book by E. Study published in 1910 (see [4] for details).

The equations resulting from equating the matrices in (1) and (4) can be expressed as:

$$4q_1^2 = 1 + r_{11} + r_{22} + r_{33} \tag{5}$$

$$4q_2^2 = 1 + r_{11} - r_{22} - r_{33} \tag{6}$$

$$4q_3^2 = 1 - r_{11} + r_{22} - r_{33} \tag{7}$$

$$4q_3^2 = 1 - r_{11} - r_{22} + r_{33} \tag{8}$$

$$4q_3q_4 = r_{23} + r_{32} \tag{9}$$

$$4q_2q_4 = r_{31} + r_{13} \tag{10}$$

$$4q_2q_3 = r_{12} + r_{21} \tag{11}$$

$$4q_1q_2 = r_{32} - r_{23} \tag{12}$$

$$4q_1q_3 = r_{13} - r_{31} \tag{13}$$

$$4q_1q_4 = r_{21} - r_{12} \tag{14}$$

In Hughes' method, q_1 is calculated first and then it is treated very differently from the remaining three parameters. Since we can solve the system of Eqs. (5)–(14) for any of the four Euler parameters first, there are four different formulas for computing the quaternion as a function of the entries of the rotation matrix, all of them formally equivalent. Numerically, however, these four formulas are not identical and, depending on the rotation matrix, one of them is numerically better conditioned than the others. From the system of Eqs. (5)–(14), we arrive at these four different solutions:

$$\mathbf{u}_1 = \frac{1}{2}\begin{pmatrix} (1+r_{11}+r_{22}+r_{33})^{\frac{1}{2}} \\ (r_{32}-r_{23})/(1+r_{11}+r_{22}+r_{33})^{\frac{1}{2}} \\ (r_{13}-r_{31})/(1+r_{11}+r_{22}+r_{33})^{\frac{1}{2}} \\ (r_{21}-r_{12})/(1+r_{11}+r_{22}+r_{33})^{\frac{1}{2}} \end{pmatrix}, \tag{15}$$

$$\mathbf{u}_2 = \frac{1}{2}\begin{pmatrix} (r_{32}-r_{23})/(1+r_{11}-r_{22}-r_{33})^{\frac{1}{2}} \\ (1+r_{11}-r_{22}-r_{33})^{\frac{1}{2}} \\ (r_{12}+r_{21})/(1+r_{11}-r_{22}-r_{33})^{\frac{1}{2}} \\ (r_{31}+r_{13})/(1+r_{11}-r_{22}-r_{33})^{\frac{1}{2}} \end{pmatrix}, \tag{16}$$

$$\mathbf{u}_3 = \frac{1}{2}\begin{pmatrix} (r_{13}-r_{31})/(1-r_{11}+r_{22}-r_{33})^{\frac{1}{2}} \\ (r_{12}+r_{21})/(1-r_{11}+r_{22}-r_{33})^{\frac{1}{2}} \\ (1-r_{11}+r_{22}-r_{33})^{\frac{1}{2}} \\ (r_{23}+r_{32})/(1-r_{11}+r_{22}-r_{33})^{\frac{1}{2}} \end{pmatrix}, \tag{17}$$

$$\mathbf{u}_4 = \frac{1}{2}\begin{pmatrix} (r_{21}-r_{12})/(1-r_{11}-r_{22}+r_{33})^{\frac{1}{2}} \\ (r_{31}+r_{13})/(1-r_{11}-r_{22}+r_{33})^{\frac{1}{2}} \\ (r_{32}+r_{23})/(1-r_{11}-r_{22}+r_{33})^{\frac{1}{2}} \\ (1-r_{11}-r_{22}+r_{33})^{\frac{1}{2}} \end{pmatrix}. \tag{18}$$

When computing any of the above solutions, numerical issues arise when square rooting, or when dividing by, very small numbers [6]. To obtain the better conditioned solution for each case, the ordinal number i of the largest element in the following vector is determined:

$$\begin{pmatrix} r_{11}+r_{22}+r_{33} \\ r_{11} \\ r_{22} \\ r_{33} \end{pmatrix}. \tag{19}$$

Then, the best solution, from the numerical point of view, is considered to be $\mathbf{u}_i$.

This four-fold multiplicity of the solution arises in other methods. For example, the one presented in [7], based on geometric arguments, was shown to be equivalent to this method.

3 Our Method

Let us suppose we only want to compute q_1. Then, from (5), we have that

$$q_1 = \frac{1}{2}\sqrt{1 + r_{11} + r_{22} + r_{33}}. \tag{20}$$

The term inside the above square root lies in the interval $[0, 4]$. Indeed, observe that $\text{Trace}(\mathbf{R}) = r_{11} + r_{22} + r_{33} = 2\cos\theta + 1$ [8, Sect. 2.3]. Unfortunately, numerical problems arise when this term gets close to zero. In practice, it can even become negative due to rounding errors [2]. Since this term coincides with $2 + 2\cos\theta$, it becomes ill-conditioned when $\theta \to \pi$. Observe that (20) only takes into account the diagonal entries of $\mathbf{R}$. To obtain an alternative formula involving all the elements of the rotation matrix, let us substitute in (3) the values of q_1^2, q_2^2, q_3^2, and q_4^2 obtained from (5), (12), (13), and (14), respectively. The result is:

$$\frac{1+r_{11}+r_{22}+r_{33}}{4} + \left(\frac{r_{32}-r_{23}}{4q_1}\right)^2 + \left(\frac{r_{13}-r_{31}}{4q_1}\right)^2 + \left(\frac{r_{21}-r_{12}}{4q_1}\right)^2 = 1. \tag{21}$$

Solving the above equation for q_1, we obtain

$$q_1 = \frac{1}{2}\sqrt{\frac{(r_{32}-r_{23})^2+(r_{13}-r_{31})^2+(r_{21}-r_{12})^2}{3-r_{11}-r_{22}-r_{33}}}. \tag{22}$$

Now, the term in the denominator of (22) also lies in the interval $[0, 4]$. Since this denominator coincides with $2 - 2\cos\theta$, (22) is ill-conditioned for $\theta \to 0$.

When this happens, the diagonal of $\mathbf{R}$ is dominant and, as a consequence, the numerator in (22) tends also to be small. Thus, (20) and (22) can be seen as complementary. Therefore, it is reasonable to establish a threshold for the trace of $\mathbf{R}$ —to be determined for optimal results— above which it is preferable to use (22) instead of (20). In other words, we have that

$$q_1 = \begin{cases} \frac{1}{2}\sqrt{1+r_{11}+r_{22}+r_{33}}, & \text{if } r_{11}+r_{22}+r_{33} > \eta \\ \frac{1}{2}\sqrt{\frac{(r_{32}-r_{23})^2+(r_{13}-r_{31})^2+(r_{21}-r_{12})^2}{3-r_{11}-r_{22}-r_{33}}}, & \text{otherwise} \end{cases} \tag{23}$$

where η has to be determined. Extending this reasoning to the computation of the other elements of the quaternion, the result is:

$$q_2 = \begin{cases} \frac{1}{2}\sqrt{1+r_{11}-r_{22}-r_{33}}, & \text{if } r_{11}-r_{22}-r_{33} > \eta \\ \frac{1}{2}\sqrt{\frac{(r_{32}-r_{23})^2+(r_{12}+r_{21})^2+(r_{31}+r_{13})^2}{3-r_{11}+r_{22}+r_{33}}}, & \text{otherwise} \end{cases} \tag{24}$$

$$q_3 = \begin{cases} \frac{1}{2}\sqrt{1-r_{11}+r_{22}-r_{33}}, & \text{if } -r_{11}+r_{22}-r_{33} > \eta \\ \frac{1}{2}\sqrt{\frac{(r_{13}-r_{31})^2+(r_{12}+r_{21})^2+(r_{23}+r_{32})^2}{3+r_{11}-r_{22}+r_{33}}}, & \text{otherwise} \end{cases} \tag{25}$$

$$q_4 = \begin{cases} \frac{1}{2}\sqrt{1-r_{11}-r_{22}+r_{33}}, & \text{if } -r_{11}-r_{22}+r_{33} > \eta \\ \frac{1}{2}\sqrt{\frac{(r_{21}-r_{12})^2+(r_{31}+r_{13})^2+(r_{32}+r_{23})^2}{3+r_{11}+r_{22}-r_{33}}}, & \text{otherwise} \end{cases} \tag{26}$$

Due to the presence of square roots, the signs of q_i, $i = 1, \ldots, 4$ are undefined. As in other methods where these signs are undefined [2], if we assume that q_1 is positive, then, according to (12), (13), and (14), we have to assign the signs of $r_{32}-r_{23}$, $r_{13}-r_{31}$, and $r_{21}-r_{12}$, to q_2, q_3, and q_4, respectively.

4 Computation of the Optimal Threshold and Comparison with Shepperd's Method

To find the optimal threshold in our formulation, we here perform a statistical analysis of the proposed method using single-precision floating-point arithmetics in MATLAB®. To this end, we generate 10^6 random unit quaternions using the algorithm described in [9] (it actually generates uniformly distributed points in $\mathbb{S}^4$). For each generated quaternion, we obtain the corresponding rotation matrix using (4), and then recover the original quaternion using the proposed method with different thresholds. The error committed is evaluated as the norm of the vector difference between the original and recovered quaternions. In general, this is not a good way to compute the distance between two quaternions. Nevertheless, since in our case the error is assumed to be very small, the length of the vector connecting both orientations in $\mathbb{S}^4$ is going to coincide with the value of the angle formed by them if seen from the center of $\mathbb{S}^4$ (and this angle can be taken as a distance between any two elements of the 3D rotation group $SO(3)$ [10]).

The results of this statistical analysis, for different values of the threshold, are compiled in Table 1. In this table, the first four columns refer to the error performance. The first one shows the percentage of cases in which the original quaternion is recovered without error. The other three correspond to the error committed in the worst-case, the average error, and the standard deviation of the error, respectively. Finally, we have two columns with the time performance. The first column gives the average time required to compute a quaternion from a rotation matrix; and the second column, the time required in the best of the cases. These results have been obtained for a MATLAB® implementation running on an Intel® Core™i7 with 32 GB of RAM.

Table 1. Performance of the proposed method, for different thresholds, and that of Shepperd's method

Threshold (η)	Quaternions recovered without error	Worst-case error $\times 10^{-6}$	Average error $\times 10^{-6}$	Standard deviation $\times 10^{-6}$	Average time (μs)	Best-case time (μs)
−0.75	19.00%	16.066	0.0332	2.4190	6.93	5.85
−0.50	25.43%	0.136	0.0248	0.0346	7.27	5.36
−0.25	27.61%	0.119	0.0230	0.0329	6.56	5.36
0.00	28.00%	0.123	0.0227	0.0325	6.52	5.36
0.25	27.67%	0.146	0.0254	0.0370	6.36	5.36
0.50	26.63%	0.149	0.0243	0.0342	6.36	5.36
1.00	24.05%	0.189	0.0279	0.0384	6.33	5.36
1.50	21.72%	0.248	0.0323	0.0434	6.35	5.36
2.00	19.20%	0.304	0.0387	0.0510	6.41	5.36
3.00	15.56%	11.164	0.0715	0.6980	6.97	5.56
Shepperd's method	24.40%	0.170	0.0304	0.0410	7.27	6.10

The last row in Table 1 contains the performance figures for Shepperd's method. From this analysis, we conclude that, although the choice of the threshold is not critical, the best results are obtained at about $\eta = 0$ for which the proposed method clearly outperforms Shepperd's method. Indeed, for $\eta = 0$, the proposed method gives the lowest value for the worse-case error, average error,

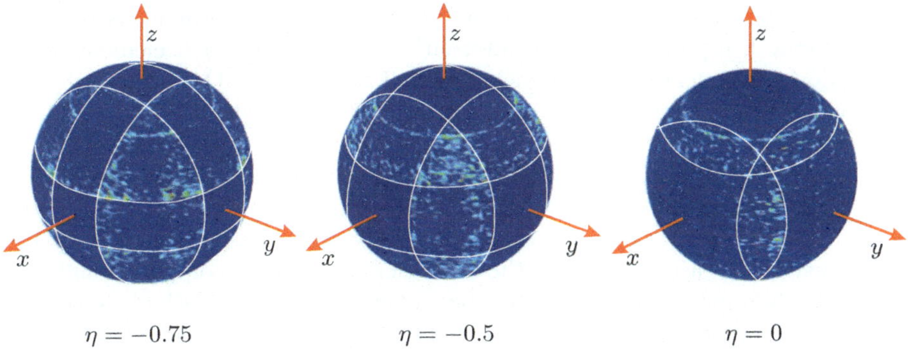

Fig. 1. A discretization of $\mathbb{S}^3$ permits to visualize which orientations can lead to the highest errors. As the threshold in increased from $\eta = -0.75$ to $\eta = 0$, the spherical triangle region centered at each octant is reduced till it vanishes.

and standard deviation. Moreover, its computational cost is not higher than the celebrated Shepperd's method.

To get a deeper insight into the behavior of the proposed method as the threshold varies, let us now explore a visual analysis. To this end, let us discretize the unit sphere $\mathbb{S}^3$. Each sample gives us an orientation for the rotation axis (n_x, n_y, n_z). Then, we discretize the rotated angle in the interval $[0, \pi]$. For each of these samples, we have a quaternion which is converted into a rotation matrix and recovered back using the proposed method. For all the samples of the rotated angle, we only keep the obtained maximum error. Thus, for each orientation in $\mathbb{R}^3$, we have an associated maximum error to which we can assign a color. The result of this experiment is shown in Fig. 1 for the three different thresholds, where brighter colors correspond to higher errors. For $\eta = -0.75$, 25 regions are clearly visible. The lowest errors are obtained for rotation axes corresponding to points contained in the spherical quadrangles centered around the coordinate axes. The highest errors are obtained for rotation axes corresponding to points contained in the spherical triangles centered in each octant. Now, if we increase the threshold, these spherical triangles, where the highest errors are obtained, are reduced. For $\eta = 0$, they have almost completely disappeared. This gives a visual justification of why the optimal threshold is located at $\eta = 0$.

The origin of the pattern of circles centered around the $z-$axis, which do not appear around the other two axes, despite all formulas are symmetric with respect to the three coordinate axes, seem to be an artifact consequence of discretizing the sphere following circles of constant longitude.

5 Conclusion

A new method for computing the quaternion corresponding to a given rotation matrix has been presented. The key idea of this new method is that, instead of generating four alternative solutions for the whole quaternion, as in Shepperd's method, it works with two alternative solutions for each element of the quaternion. This means that the method implicitly works with up to 16 alternative solutions for the whole quaternion, which accounts for its improved global numerical behavior.

If we take a quaternion at random and we compute the corresponding rotation matrix, the probability of recovering exactly the original quaternion from this matrix using Shepperd's method is about 24%, while using our method this probability is increased to 28%, without an increment in the computational cost. It is certainly a modest improvement, but a large number of methods have been proposed to solve the problem [2] and Shepherd's method has remained the best one till now.

Acknowledgments. This work has been partially supported by the Spanish Ministry of Economy and Competitiveness through projects DPI2014-57220-C2-2-P, DPI2017-88282-P, and MDM-2016-0656.
We also want to express our gratitude to Antonio B. Martínez for the financial support of the first author, and to the anonymous reviewers whose comments have led to a number of valuable improvements in the text.

References

1. Goldstein, H.: Classical Mechanics. Addison-Wesley, Cambridge (1951)
2. Sarabandi, S., Thomas, F.: A survey on the computation of Euler parameters from rotation matrices (submitted for publication)
3. Shepperd, S.W.: Quaternion from rotation matrix. J. Guidance Control **1**(3), 223–224 (1978)
4. Husty, M.L., Pfurner, M., Schröcker, H.-P., Brunnthaler, K.: Algebraic methods in mechanism analysis and synthesis. Robotica **25**(6), 661–675 (2007)
5. Hughes, P.C.: Spacecraft Attitude Dynamics. Wiley, Hoboken (1986)
6. Goldberg, D.: What every computer scientist should know about floating-point arithmetic. ACM Comput. Surv. **23**(1), 5–48 (1991)
7. Shuster, M.D., Natanson, G.A.: Quaternion computation from a geometric point of view. J. Astronaut. Sci. **41**(4), 545–556 (1993)
8. Angeles, J.: Fundamentals of Robotic Mechanical Systems. Theory, Methods, and Algorithms. Springer, New York (1997)
9. Marsaglia, G.: Choosing a point from the surface of a sphere. Ann. Math. Stat. **43**, 645–646 (1972)
10. Huynh, D.Q.: Metrics for 3D rotations: comparison and analysis. J. Math. Imaging Vis. **35**(2), 155–164 (2009)

Redundancy Parametrization in Globally-Optimal Inverse Kinematics

Enrico Ferrentino(✉) and Pasquale Chiacchio

Università degli Studi di Salerno, 84084 Fisciano, SA, Italy
{eferrentino,pchiacchio}@unisa.it
http://www.automatica.unisa.it/all_people.php

Abstract. Redundancy resolution schemes based on the optimization of integral performance indices through calculus of variations do not provide a minimum parametrization of redundancy, i.e. the one with the minimum number of variables and differential equations, which is rather implicitly used in techniques based on dynamic programming. This communication investigates the possible minimum parametrizations and the conditions by which they become unrepresentative of the redundancy. Techniques are proposed to choose the right parametrization, depending on the manipulator kinematic characteristics and trajectory. While reducing the number of differential equations to be solved, they pave the way to the design of more effective redundancy resolution algorithms based on dynamic programming.

Keywords: Inverse kinematics · Redundant robots
Redundancy parametrization

1 Introduction

Because their kinematic chain provides more degrees of freedom than those required to execute a given task, redundant manipulators are characterized by an higher level of dexterity. This feature can be exploited to achieve several objectives, which are usually desirable for a multitude of real applications, such as local or global optimization of performance indices, improvement of manipulability, singularity avoidance, obstacle avoidance and constrained motions [9]. This communication focuses on redundancy resolution via global optimization of performance indices, i.e. performance indices of an integral form, which mainly makes use of calculus of variations and dynamic programming.

In [8], said n the number of joints and $m < n$ the number of variables describing the task, by using the *Pontryagin's Maximum Principle* (PMP), it is demonstrated that finding the globally-optimal joint space trajectory means solving a *Two-Point Boundary Value Problem* (TPBVP) made of $2n$ differential equations and $2n$ boundary conditions. In [6], an equivalent TPBVP is derived from the Euler-Lagrange necessary conditions. In [7], for a specific class of performance indices, a reduced-order form of the TPBVP is found, made of only $2n - m$ differential equations and boundary conditions.

J. Lenarcic and V. Parenti-Castelli (Eds.): ARK 2018, SPAR 8, pp. 47–55, 2019.
https://doi.org/10.1007/978-3-319-93188-3_6

In [4,5], to different purposes, a parameter, termed *redundancy parameter*, is introduced in the kinematic equations. It is made of as many elements as $r = n - m$, i.e. the degree of redundancy, and is such that, if given, for each point on the end-effector path, the set of all possible inverse kinematics solutions becomes finite. On the basis of this simple principle, the configuration space is parametrized with respect to the redundancy parameter and the globally-optimal (or Pareto-optimal) solution is found through a dynamic-programming-inspired algorithm.

In this communication, we first recall the generic formulation of a TPBVP and its reduced form as first theorized in [7], with some important extensions. These aspects are addressed in Sect. 2. In Sect. 3, on top of the state-of-the-art, we demonstrate that, if the dynamic programming problem can be parametrized with respect to only one r-dimensional parameter, the same has to be possible for the TPBVP. We come to a new formulation of the problem which allows to solve as many differential equations as the degree of redundancy, leading to a drastic simplification of the problem. In Sect. 4, we investigate the conditions by which the minimum parametrization becomes unrepresentative of the redundancy and provide important guidelines necessary to the selection of the right parametrization, paving the way to the design of more effective redundancy resolution schemes based on dynamic programming. In Sect. 5, we draw the conclusions of our work.

2 Classic Formulation of a TPBVP with Calculus of Variations

Let us consider the case of a generic redundant manipulator such as $\mathbf{q}$ and $\dot{\mathbf{q}} \in \Re^n$ represent the vectors of joint positions and velocities, respectively, and $\mathbf{k} : \Re^n \rightarrow \Re^m$ is the direct kinematic mapping function from the joint space to the configuration space, i.e.

$$\mathbf{x}(t) = \mathbf{k}(\mathbf{q}(t)) \tag{1}$$

with $\mathbf{x}(t) \in \Re^m$ representing the task variable trajectory.

The problem of globally optimizing a performance index G, subject to the kinematic constraints (1) can be formalized as in [7]:

$$\begin{aligned} \min_{\mathbf{q}(t)} \; & \int_{t_i}^{t_f} G(\mathbf{q}, \dot{\mathbf{q}}, t) \, dt \\ \text{s.t.} \; & \mathbf{x}(t) = \mathbf{k}(\mathbf{q}(t)) \end{aligned} \tag{2}$$

where t_i and t_f are the initial and final time instants, respectively.

Said $\boldsymbol{\mu} \in \Re^m$ the vector of Lagrange multipliers, the usage of calculus of variations applied to the Lagrangian function

$$L = G(\mathbf{q}, \dot{\mathbf{q}}, t) + \boldsymbol{\mu}^T[\mathbf{x} - \mathbf{k}(\mathbf{q})] \tag{3}$$

leads to the following set of n second-order differential equations, equivalent to $2n$ first-order differential equations, and m algebraic equations:

$$\frac{\partial L}{\partial \mathbf{q}} - \frac{d}{dt}\left(\frac{\partial L}{\partial \dot{\mathbf{q}}}\right) = 0 \qquad (4)$$
$$\mathbf{x}(t) = \mathbf{k}(\mathbf{q}(t))$$

In addition to (4), the following $2n$ two-point boundary conditions have to be satisfied [6]:

$$\left(\frac{\partial L}{\partial \dot{\mathbf{q}}}\right)^T \delta\mathbf{q}(t_B) = 0 \quad \text{for } t_B = t_i, t_f \qquad (5)$$

The problem given by (4) and (5) is the TPBVP to be solved to obtain the solution minimizing the integral cost.

Typical performance indices can be found in [3,6,7]. A peculiarity of such indices is that the necessary conditions (4) can all be written in the general form

$$\ddot{\mathbf{q}} = \mathbf{J}_W^{\dagger}(\ddot{\mathbf{x}} - \dot{\mathbf{J}}\dot{\mathbf{q}}) + \mathbf{P}_W \mathbf{A}(\mathbf{q}, \dot{\mathbf{q}}, t) \qquad (6)$$

where $\mathbf{J}$ is the manipulator's Jacobian, $\mathbf{W}$ is a generic non-singular $n \times n$ symmetric weights matrix, $\mathbf{J}_W^{\dagger} = \mathbf{W}^{-1}\mathbf{J}^T\left[\mathbf{J}\mathbf{W}^{-1}\mathbf{J}^T\right]^{-1}$ is the weighted pseudoinverse of the Jacobian for non-singular $\mathbf{J}\mathbf{W}^{-1}\mathbf{J}^T$, $\mathbf{P}_W = (\mathbf{I} - \mathbf{J}_W^{\dagger}\mathbf{J})$ is the $\mathbf{W}$-dependent null-space projector of the Jacobian, $\mathbf{I}$ is the identity matrix of proper dimension and $\mathbf{A}$ is a column vector whose expression depends on the specific performance index. Expression (6), such as (4), corresponds to $2n$ first-order differential equations.

The reader may verify that this is indeed the form obtained for the most common performance indices mentioned in the literature and certainly includes a broader category of solutions.

A reduced-order form of (4) was first found by *Martin et al.* in [7]. Although they demonstrated it for a specific performance index, we extend their results to all the solutions that can be expressed as (6).

Said $\boldsymbol{\gamma}$ the vector of null-space velocities leading to internal motion, $\mathbf{N}$ a null space basis of the Jacobian and $\mathbf{N}_W = \mathbf{W}\mathbf{N}$ the weighted null-space basis, the following holds:

$$\boldsymbol{\gamma} = \mathbf{N}_W^T \dot{\mathbf{q}} \qquad (7)$$

The generic first-order solution is

$$\dot{\mathbf{q}} = \mathbf{J}_W^{\dagger}\dot{\mathbf{x}} + \mathbf{P}_W \dot{\boldsymbol{\phi}} \qquad (8)$$

Considering that both $\mathbf{N}_W^T \mathbf{J}_W^{\dagger} = \mathbf{0}$ and $\mathbf{N}_W^T \mathbf{P}_W = \mathbf{N}_W^T$ hold, by folding (8) in (7), $\boldsymbol{\gamma}$ can also be written as

$$\boldsymbol{\gamma} = \mathbf{N}_W^T \dot{\boldsymbol{\phi}} \qquad (9)$$

Now, by substituting $\mathbf{P}_W = \mathbf{N}(\mathbf{N}_W^T\mathbf{N})^{-1}\mathbf{N}_W^T$ in (8) and using (9), we get to

$$\dot{\mathbf{q}} = \mathbf{J}_W^{\dagger}\dot{\mathbf{x}} + \mathbf{N}(\mathbf{N}_W^T\mathbf{N})^{-1}\boldsymbol{\gamma} \qquad (10)$$

On the other hand, by differentiating (7), we have:

$$\dot{\gamma} = \dot{\mathbf{N}}_W^T \dot{\mathbf{q}} + \mathbf{N}_W^T \ddot{\mathbf{q}} \tag{11}$$

and by folding (6) in the expression above, we finally obtain:

$$\dot{\gamma} = \dot{\mathbf{N}}_W^T \dot{\mathbf{q}} + \mathbf{N}_W^T \mathbf{A}(\mathbf{q}, \dot{\mathbf{q}}, t) \tag{12}$$

Equations (10) and (12) together represent a set of $n+r = 2n-m$ equations in $\mathbf{q}$ and γ, which constitutes the reduced order form of (6). As instance, for a 7-DOF manipulator and a 6-DOF task, such a technique would allow to integrate only 8 equations instead of 14.

In [2], it is stated that the reduced order form given by (10) and (12) is not applicable to a general performance index such as the kinetic energy. However, since its optimization leads to differential equations in the form of (6), we demonstrate in this paper that the reduced form applies to kinetic energy as well. In addition, it applies to the variation of the weighted manipulability index [3] and to any other quantity that might be expressed through a weight matrix.

3 Minimum Number of Equations

Minimum parametrizations of redundancy can be obtained from the idea that the m algebraic equations in (4) and their m derivatives can be used in (6), leading to a system of $2(n-m) = 2r$ differential equations. For a 7-DOF manipulator and a 6-DOF task, it is equivalent to integrating only 2 equations instead of 14.

3.1 Parametrization Through Joint(s) Selection

For the sake of simplicity, let us assume that the performance index to minimize is the square norm of joint velocities, but the dissertation can be easily generalized to any other performance index that can be managed with the Euler-Lagrange method, leading to n second-order differential equations. For the square norm of velocities, the Euler-Lagrange conditions lead to the following differential equations:

$$\ddot{\mathbf{q}} = \mathbf{J}^{\dagger}(\mathbf{q}) \left[\ddot{\mathbf{x}} - \dot{\mathbf{J}}(\mathbf{q}, \dot{\mathbf{q}})\dot{\mathbf{q}}\right] \tag{13}$$

Without loss of generality, and to the only purpose of simplifying the notation, assume that $n - m = 1$, i.e. one degree of redundancy, and that we want to parametrize the redundancy with one of the joints, generically called q_i. We then pick the i-th equation from the set above:

$$\ddot{q}_i = \mathbf{J}^{\dagger}_{i,all}(\mathbf{q}) \left[\ddot{\mathbf{x}} - \dot{\mathbf{J}}(\mathbf{q}, \dot{\mathbf{q}})\dot{\mathbf{q}}\right] \tag{14}$$

where with $\mathbf{J}^{\dagger}_{i,all}(\mathbf{q})$ we refer to the i-th raw of $\mathbf{J}^{\dagger}(\mathbf{q})$.

The terms on the right-hand side of (14) are dependent on q_i, as well as on the other joints. Define as $\mathbf{q}_r \in \mathcal{R}^m$ the position vector of the remaining joints:

$$\mathbf{q}_r = \left[q_1 \dots q_{i-1}\ q_{i+1} \dots q_n\right]^T \tag{15}$$

Equation (14) can then be rewritten as:

$$\ddot{q}_i = \mathbf{J}^{\dagger}_{i,all}(q_i, \mathbf{q}_r)\left[\ddot{\mathbf{x}} - \dot{\mathbf{J}}(q_i, \mathbf{q}_r, \dot{q}_i, \dot{\mathbf{q}}_r)\dot{\mathbf{q}}\right] \tag{16}$$

Now consider the direct kinematics equations in (1) in which q_i and $\mathbf{q}_r$ are highlighted:

$$\mathbf{x} = \mathbf{k}(q_i, \mathbf{q}_r) \tag{17}$$

For a redundant manipulator, far from singularities, the inverse kinematics admits an infinite number of solutions. If q_i is driven by the differential equations above, the other joints can be found by inverting Eq. (17), which returns a finite set of solutions. We select $\mathbf{q}_r$ from such a finite set:

$$\mathbf{q}_r = \mathbf{k}^{-1}(\mathbf{x}, q_i) \tag{18}$$

Also, if q_i is given, $\mathbf{k}(q_i, \mathbf{q}_r) = \mathbf{k}(\mathbf{q}_r)$, which implies that the Jacobian of $\mathbf{k}$ is a square matrix. Differentiating Eq. (17) with respect to time, we obtain:

$$\dot{\mathbf{x}} = \mathbf{J}(q_i, \mathbf{q}_r)\dot{\mathbf{q}} \tag{19}$$

Define $\mathbf{J}_{all,i}$ as the i-th column of $\mathbf{J}$ and $\mathbf{J}_r$ as the square matrix obtained by removing $\mathbf{J}_{all,i}$ from $\mathbf{J}$, getting to the following equation:

$$\dot{\mathbf{x}} = \mathbf{J}_{all,i}(q_i, \mathbf{q}_r)\dot{q}_i + \mathbf{J}_r(q_i, \mathbf{q}_r)\dot{\mathbf{q}}_r \tag{20}$$

Assuming that the kinematic substructure (i.e. the kinematic chain obtained by removing q_i) is not in a singularity, we can solve for $\dot{\mathbf{q}}_r$, obtaining:

$$\dot{\mathbf{q}}_r = \mathbf{J}_r^{-1}(q_i, \mathbf{q}_r)\left(\dot{\mathbf{x}} - \mathbf{J}_{all,i}(q_i, \mathbf{q}_r)\dot{q}_i\right) \tag{21}$$

Putting all the results together, we obtain the following differential equations:

$$\begin{aligned} \ddot{q}_i &= \mathbf{J}^{\dagger}_{i,all}(q_i, \mathbf{q}_r)\left[\ddot{\mathbf{x}} - \dot{\mathbf{J}}(q_i, \mathbf{q}_r, \dot{q}_i, \dot{\mathbf{q}}_r)\dot{\mathbf{q}}\right] \\ \dot{q}_i &= \frac{dq_i}{dt} \end{aligned} \tag{22}$$

which, together with Eqs. (18) and (21), is demonstrated to be the reduced form of (13), with the minimum number (two) of differential equations.

3.2 Parametrization Through Joints Combination

Alternatively to joint selection, another possibility is to parametrize the redundancy with respect to some joint combination $\mathbf{v} = \mathbf{k}_v(\mathbf{q})$, with $\mathbf{k}_v : \Re^n \rightarrow \Re^r$. In this perspective, once the function $\mathbf{k}_v$ has been defined, the augmented direct kinematics can be defined from (1) as:

$$\begin{bmatrix} \mathbf{x} \\ \mathbf{v} \end{bmatrix} = \begin{bmatrix} \mathbf{k}(\mathbf{q}) \\ \mathbf{k}_v(\mathbf{q}) \end{bmatrix} = \mathbf{k}_a(\mathbf{q}) \tag{23}$$

By differentiating Eq. (23), we obtain:

$$\begin{bmatrix} \dot{\mathbf{x}} \\ \dot{\mathbf{v}} \end{bmatrix} = \begin{bmatrix} \mathbf{J}(\mathbf{q}) \\ \mathbf{J}_v(\mathbf{q}) \end{bmatrix} \dot{\mathbf{q}} = \mathbf{J}_a(\mathbf{q})\dot{\mathbf{q}} \tag{24}$$

The reader may recognize that such a formulation is equivalent to that of the augmented Jacobian technique [9] used in the frame of redundancy resolution via task augmentation. Here, although the same augmented Jacobian is obtained, $\mathbf{v}$ (i.e. the additional task) is not intended to be given along the trajectory, but has to be computed as part of the solution to the global optimization problem.

The objective is now to reorganize Eq. (13) in order to reduce them to a system of $2(n-m)$ equations in $\mathbf{v}$ and $\dot{\mathbf{v}}$. To do so, differentiate the second set of equations in (24), getting to:

$$\ddot{\mathbf{v}} = \mathbf{J}_v(\mathbf{q})\ddot{\mathbf{q}} + \dot{\mathbf{J}}_v(\mathbf{q})\dot{\mathbf{q}} \tag{25}$$

By substituting Eq. (13) and adding the differential equation linking the parameter to its derivative, we obtain

$$\begin{aligned} \ddot{\mathbf{v}} &= \mathbf{J}_v(\mathbf{q})\mathbf{J}^{\dagger}(\mathbf{q}) \left[\ddot{\mathbf{x}} - \dot{\mathbf{J}}(\mathbf{q}, \dot{\mathbf{q}})\dot{\mathbf{q}}\right] + \dot{\mathbf{J}}_v(\mathbf{q})\dot{\mathbf{q}} \\ \dot{\mathbf{v}} &= \frac{d\mathbf{v}}{dt} \end{aligned} \tag{26}$$

which is the minimum number of differential equations if a combination of the joints is used. Similarly to the joint selection case, if $\mathbf{v}$ is driven by the differential equations above, $\mathbf{q}$ and $\dot{\mathbf{q}}$ can be obtained by inverting Eqs. (23) and (24).

Both the joints selection and joints combinations methods reduce the complexity of the TPBVP and are especially important for dynamic programming, where, because of the optimization burden, the complexity has to be reduced to the very minimum.

4 Selecting the Right Parametrization

In the previous section we have shown that different minimum parametrizations are possible. The problem now is to determine which the most suitable parametrization is for a given manipulator and an assigned task trajectory. In this section, we show that some redundancy parameters could be unrepresentative of the redundancy. Thus, after concluding that the selection cannot be randomly made, we provide a procedure to choose an effective parametrization.

4.1 Joint(s) Selection

We have seen above that the minimum number of differential equations obtained by "fixing" one or more of the joints makes sense only when $\mathbf{J}_r$ is full-rank and, thus, can be inverted. Hence, in order to ensure that Eq. (21) can be used and that, as a consequence, the number of solutions of (18) is finite, singularities for the reduced Jacobians need to be identified. It is worth remarking that a singular Jacobian is a necessary condition for Eq. (18) to admit an infinite set of solutions, but not sufficient, i.e. singularities might be spotted for the reduced Jacobian not leading to an infinite number of solutions.

The following procedure to select the "redundant" joint(s) can be used:

1. Extract the possible square matrices $\mathbf{J}_r$ obtained from $\mathbf{J}$ by removing as many columns as the number of redundancy degrees. To the purpose of simplifying the dissertation, let us assume that $n - m = 1$ and call such matrices $\mathbf{J}_r^{(i)}$, with $i = 1, 2, \ldots, n$. The superscript in parentheses indicates the column removed from $\mathbf{J}$ to obtain the reduced Jacobian.
2. Identify conditions for which $\mathbf{J}_r^{(i)}$ becomes singular. For some manipulators, such as the ones of Fig. 1, it is possible to obtain such conditions analytically and some a-priori considerations could be made. Otherwise a trajectory shall be provided and the distance from singularities shall be computed numerically in the same way as for non-redundant manipulators along the specified trajectory.
3. Once a joint is selected, which also implies a choice of a $\mathbf{J}_r^{(i)}$, trajectories that make it singular cannot be executed. On the other hand, once a trajectory is given, none of the joints whose reduced Jacobian is singular for at least one trajectory point can be chosen to parametrize the redundancy.

In principle, it is possible that a trajectory is such that each of the reduced Jacobians gets rank-deficient for at least one point on the trajectory. If this is the case, one parametrization is not enough to represent the redundancy on the whole trajectory. Assume, as instance, that q_2 is being used for $t \in [t_0, t_k[$ and that $\mathbf{x}(t_k)$ forces the joints to make $\mathbf{J}_r^{(2)}$ singular. At time t_k, one could take the states computed up until then and use them as initial condition for the same differential equations written for a different joint. In other words, Eq. (22) can be switched from one joint to another anytime $\mathbf{J}_r$ is about to be singular.

Figure 1 provides two examples of a 3D (left) and a planar (right) manipulator to which a 2D and a 1D task are assigned respectively.

In both cases, if the manipulators were asked to track the depicted trajectories, only one value of q_2 would allow the end-effector to stay on the path. Note that, with q_2 fixed, both manipulators are still redundant with respect to the task they have to fulfill, which means that q_2 is not a suitable choice to represent redundancy.

If the manipulator is still redundant, even though $n - m$ variables are fixed, it means that two or more of the rows of $\mathbf{J}_r$ are linearly dependent, which, in turn, implies that $\mathbf{J}_r$ is not full rank for at least one point on the trajectory.

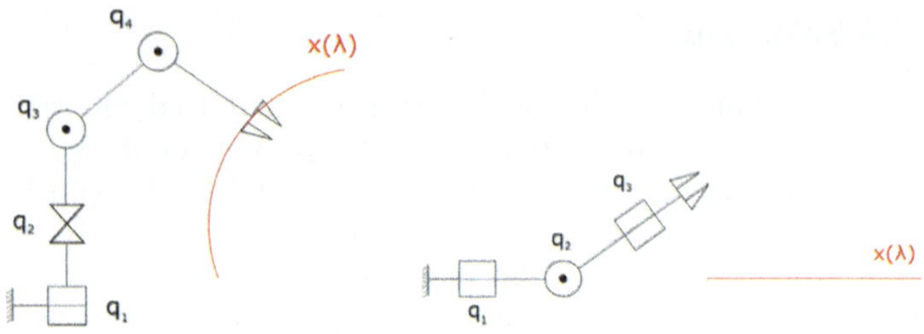

Fig. 1. Kinematic structures for which the assigned task makes the joint q_2 meaningless in terms of redundancy parametrization

Thus, if we attempted to solve

$$\mathbf{q}_r = \mathbf{k}^{-1}(\mathbf{x}, q_2) \tag{27}$$

we would still get an infinite number of solutions.

It is also worth noticing that this issue is not strictly related to the fact that, said the selected joint i, $\dot{q}_i(t) = 0\ \forall t$, but to the condition that the manipulator is still redundant in achieving the task even when q_i is fixed. In fact, it is possible, in principle, to assign tasks, especially in constrained environments, for which one joint has to be still along the whole trajectory, but only a finite number of configurations is admissible for the other joints (i.e. the manipulator is no longer redundant with respect to the assigned task). In this case, joint q_i (just like any other joint) is still a suitable choice to parametrize the redundancy.

4.2 Joints Combination

As noted in [9], the augmented Jacobian suffers from the so-called *algorithmic* singularities, which are conditions in which $\mathbf{J}_a$ is rank-deficient, even though the manipulator is not singular. To the purpose of parametrizing the redundancy, one could choose the suitable joints combination which allows keeping the singularity away from the assigned task.

Also, it can be demonstrated that:

$$rk(\mathbf{J}_a) = n \iff \mathcal{R}(\mathbf{J}^T) \cap \mathcal{R}(\mathbf{J}_v^T) = \varnothing \tag{28}$$

where $\mathcal{R}(\mathbf{M})$ corresponds to the range space of a generic matrix $\mathbf{M}$.

Unlike the augmented Jacobian redundancy resolution method, here we are not interested in assigning a second task to the manipulator, but to ensure that the selected parametrization is always meaningful along the trajectory and, possibly, regardless of the trajectory. This implies that $\mathbf{J}_v$ can be suitably selected to make condition (28) always verified. As instance this can be done by computing a basis of $\mathcal{R}(\mathbf{J}_v^T)$ that completes the basis of $\mathcal{R}(\mathbf{J}^T)$ to n, i.e.

$$\mathcal{R}(\mathbf{J}_v^T) = \mathcal{R}^{\perp}(\mathbf{J}^T) \tag{29}$$

If this condition is satisfied, the two mappings are orthogonal [9] and $\mathbf{J}_a$ is always full-rank.

This parametrization chosen, $\mathbf{v}$ is not attributable, in general, to any physical quantity that one may easily visualize, even for very simple 3-DOF planar manipulators. Also, its analytic form is not trivial to find. As a consequence, the system of equations results to be more suitable for numerical CLIK-based implementations [1].

Lastly, it is important to notice that the formulation presented here represents a more general case of the joint selection method presented in Sect. 3.1, whose outcomes could be easily re-found by imposing $n-m=1$ and $\mathbf{v}=v=q_i$.

5 Conclusions

We demonstrated that TPBVPs obtained from calculus of variations can be drastically simplified to integrate as many differential equations as twice the redundancy degree, regardless the total number of joints. We showed that such a simplification could lead to a lack of representativeness in parametrizing redundancy. While this can be overcome when dealing with differential equations by choosing a non-minimum parametrization, in dynamic programming, choosing the right parametrization is central to resolve the redundancy. We believe that such results pave the way to the design of more effective redundancy resolution schemes based on dynamic programming, which we will address elsewhere.

References

1. Chiacchio, P., Chiaverini, S., Sciavicco, L., Siciliano, B.: Closed-loop inverse kinematics schemes for constrained redundant manipulators with task space augmentation and task priority strategy. Int. J. Robot. Res. **10**(4), 410–425 (1991)
2. Choi, B.W., Won, J.H., Chung, M.J.: Optimal redundancy resolution of a kinematically redundant manipulator for a cyclic task. J. Robot. Syst. **9**(4), 481–503 (1992)
3. Gotlih, K., Troch, I., Jezernik, K.: Global optimal control of redundant robot. Robotica **14**(2), 131–140 (1996)
4. Guigue, A., Ahmadi, M., Hayes, M.J.D., Langlois, R., Tang, F.C.: A dynamic programming approach to redundancy resolution with multiple criteria. In: IEEE International Conference on Robotics and Automation, pp. 1375-1380 (2007)
5. Guigue, A., Ahmadi, M., Langlois, R., Hayes, M.J.: Pareto optimality and multiobjective trajectory planning for a 7-DOF redundant manipulator. IEEE Trans. Robot. **26**(6), 1094–1099 (2010)
6. Kazerounian, K., Wang, Z.: Global versus local optimization in redundancy resolution of robotic manipulators. Int. J. Robot. Res. **7**(5), 3–12 (1988)
7. Martin, D.P., Baillieul, J., Hollerbach, J.M.: Resolution of kinematic redundancy using optimization techniques. IEEE Trans. Robot. Autom. **5**(4), 529–533 (1989)
8. Nakamura, Y., Hanafusa, H.: Optimal redundancy control of robot manipulators. Int. J. Robot. Res. **6**(1), 32–42 (1987)
9. Siciliano, B., Khatib, O.: Springer Handbook of Robotics. Springer, New York (2016)

Solution for the Direct Kinematics Problem of the General Stewart-Gough Platform by Using only Linear Actuators' Orientations

Stefan Schulz(✉), Arthur Seibel, and Josef Schlattmann

Workgroup on System Technologies and Engineering Design Methodology, Hamburg University of Technology, 21073 Hamburg, Germany
{st.schulz,arthur.seibel,j.schlattmann}@tuhh.de

Abstract. In this paper, we present a solution for the direct kinematics problem of the general Stewart-Gough platform where only the orientations of the linear actuators are measured with accelerometer sensors. By using an extended Kalman filter with the inverse kinematics equations, we can uniquely determine the manipulator platform's pose. In this context, we investigate the solution's accuracy and derive the minimum number of iterations under static conditions and validate our concept with experiments on a state-of-the-art Stewart-Gough platform.

Keywords: Direct kinematics problem · Parallel kinematics
Stewart-Gough platform · Linear actuators' orientations

1 Introduction

The determination of the direct kinematics is one of the major problems of parallel mechanisms and one of the most researched topics. Unlike the inverse kinematics problem, which can be uniquely solved for most parallel mechanisms, the direct kinematics problem is analytically not solvable due to non-linear closure constraints. The direct kinematics problem means finding the actual position and orientation, known as pose, of the moveable manipulator platform from the active joints' coordinates. In order to realize a high-precision position and motion control, it has to be solved accurately, robustly, and as quickly as possible.

In general, the direct kinematics problem can be written as a system of non-linear kinematic constraint equations resulting from inverse kinematics. The resultant univariate polynomial is a function of the closure conditions and, therewith, of the parallel mechanism's architecture. Its degree or rank directly corresponds to the number of possible solutions of the direct kinematics problem, i. e., the number of possible assembly modes or poses of the manipulator platform.

For the most general case of the Stewart-Gough platform that consists of six linear actuators connecting the fixed base platform with the moveable manipulator platform, the number of possible assembly modes has an upper bound of

J. Lenarcic and V. Parenti-Castelli (Eds.): ARK 2018, SPAR 8, pp. 56–64, 2019.
https://doi.org/10.1007/978-3-319-93188-3_7

40 [4,8,9]. The corresponding univariate polynomial therewith enables finding all the solutions of the direct kinematics problem but does not help identifying the actual pose of the manipulator platform.

In order to distinguish the actual pose from all the possible assembly modes, numerous numerical methods were proposed, see, e. g., [3,6,7]. The most common ones are genetic algorithms, interval analysis, and iterative techniques, such as Newton-Raphson algorithms. On the one hand, iterative methods need a good initial estimation to calculate the error between the measured linear actuators' lengths and the lengths that correspond to the pose estimate. On the other hand, they require a determinable pose that is sufficiently far away from a singular configuration. In conclusion, the initial pose estimate influences both the pose the algorithms converge to (that is not necessarily the actual one) and the computation time that corresponds to the number of iterations.

As an alternative to additional numerical procedures, we can use additional sensor information to transform the system of non-linear kinematic constraint equations into an explicit or linear problem where a unique solution can be found. This enables determining the actual manipulator platform's pose faster, more accurately, and independent from initial pose estimates. In this context, every proposed sensor concept uses at least one linear actuator's length, and most of them require numerical methods to guarantee a unique solution, see, e. g., [1,2,5,12,14]. In fact, additional sensor concepts with a minimum number of sensors (a) are only applicable to parallel kinematics with special architectures, or (b) have a reduced accuracy due to the increased equations' complexity, or (c) have the problem of selecting the actual pose from a set of candidates, see [2,14].

In a previous work, we proposed a comprehensive sensor concept for uniquely determining the manipulator platform's pose by measuring only two of the linear actuators' orientations and the orientation of the manipulator platform [10]. In this concept, we completely avoid length measurements because (a) length sensors are expensive, (b) they generally require initial length measurements or reference drives (except for length sensors with absolute encoders, which, however, are considerably more expensive), (c) the lengths are generally determined indirectly so that deformations and/or backlashes in the linear actuators and/or joints cannot be detected, and (d) at least nine length sensors are required [5] (three of them on additional legs, which leads to additional costs and reduced working range) to enable a unique solution. However, measuring the manipulator platform's orientation is always conflicted with the yaw-angle drift and needs numerical procedures [13]. In order to avoid measuring the manipulator platform's orientation, in this paper, we investigate using the orientations of all six linear actuators to determine the manipulator platform's pose because the minimal set of three orientations does not guarantee a unique solution [11,15]. In particular, we use an extended Kalman filter, which evaluates new measurements to predict the system's state and correct previous predictions.

The remainder of this paper is as follows. In Sect. 2, our approach and the essential mathematical background are presented. In Sect. 3, we evaluate our concept with experiments on a common Stewart-Gough platform. Finally, in Sect. 4, our results are summarized and discussed.

Throughout the paper, we use the following notation. The body-fixed frames of the base and the manipulator platform are denoted as {1} and {2}, respectively. The position vector of the kth joint $J_{k,1}$ is given by $\boldsymbol{p}_{1J_{k,1}}$ and the vector between the joints $J_{k,1}$ and $J_{k,2}$ of the platforms {1} and {2} by $\boldsymbol{p}_{J_{k,1}J_{k,2}}$ with $k \in \{1, \ldots, 6\}$. Using inverse kinematics, this vector can be written as

$$ {}^{1}\boldsymbol{p}_{J_{k,1}J_{k,2}} = {}^{1}\boldsymbol{p}_{12} + {}^{1}\boldsymbol{R}_{2} \cdot {}^{2}\boldsymbol{p}_{2J_{k,2}} - {}^{1}\boldsymbol{p}_{1J_{k,1}} \tag{1} $$

with respect to platform {1}. Here, $\boldsymbol{p}_{12}$ is the vector connecting the origins of the platforms {1} and {2} and ${}^{1}\boldsymbol{R}_{2}$ is the rotation matrix from platform {2} into platform {1}, where

$$ {}^{1}\boldsymbol{R}_{2} = {}^{1}\boldsymbol{R}_{2,\alpha} \cdot {}^{1}\boldsymbol{R}_{2,\beta} \cdot {}^{1}\boldsymbol{R}_{2,\gamma} \tag{2} $$

with

$$ {}^{1}\boldsymbol{R}_{2,\alpha} = \begin{bmatrix} 1 & 0 & 0 \\ 0 & \cos\alpha & \sin\alpha \\ 0 & -\sin\alpha & \cos\alpha \end{bmatrix}, {}^{1}\boldsymbol{R}_{2,\beta} = \begin{bmatrix} \cos\beta & 0 & -\sin\beta \\ 0 & 1 & 0 \\ \sin\beta & 0 & \cos\beta \end{bmatrix}, {}^{1}\boldsymbol{R}_{2,\gamma} = \begin{bmatrix} \cos\gamma & \sin\gamma & 0 \\ -\sin\gamma & \cos\gamma & 0 \\ 0 & 0 & 1 \end{bmatrix}, \tag{3} $$

and α, β, and γ are the rotation angles around the axis x, y, and z, respectively. The orientation of $\boldsymbol{p}_{J_{k,1}J_{k,2}}$ is referred to as $\boldsymbol{r}_{J_{k,1}J_{k,2}}$, which has unit length.

2 Approach

We determine the pose

$$ \boldsymbol{p} = \left[{}^{1}\boldsymbol{p}_{12}^{\top}\ \alpha\ \beta\ \gamma\right]^{\top} = \left[x\ y\ z\ \alpha\ \beta\ \gamma\right]^{\top} \tag{4} $$

from the six linear actuators' orientations that are combined in the measurement vector $\boldsymbol{z}$:

$$ \boldsymbol{z} = \left[\boldsymbol{r}_{J_{1,1}J_{1,2}} \cdots \boldsymbol{r}_{J_{6,1}J_{6,2}}\right]^{\top}. \tag{5} $$

Using Eq. (1), we can describe the relation between the measurements $\boldsymbol{z}$ and the pose $\boldsymbol{p}$ by the (overconstrained) measurement model $\boldsymbol{h}(\boldsymbol{p})$:

$$ \boldsymbol{z} = \left[\frac{{}^{1}\boldsymbol{p}_{12}+{}^{1}\boldsymbol{R}_{2}\cdot{}^{2}\boldsymbol{p}_{2J_{1,2}}-{}^{1}\boldsymbol{p}_{1J_{1,1}}}{\left\|{}^{1}\boldsymbol{p}_{12}+{}^{1}\boldsymbol{R}_{2}\cdot{}^{2}\boldsymbol{p}_{2J_{1,2}}-{}^{1}\boldsymbol{p}_{1J_{1,1}}\right\|} \cdots \frac{{}^{1}\boldsymbol{p}_{12}+{}^{1}\boldsymbol{R}_{2}\cdot{}^{2}\boldsymbol{p}_{2J_{6,2}}-{}^{1}\boldsymbol{p}_{1J_{6,1}}}{\left\|{}^{1}\boldsymbol{p}_{12}+{}^{1}\boldsymbol{R}_{2}\cdot{}^{2}\boldsymbol{p}_{2J_{6,2}}-{}^{1}\boldsymbol{p}_{1J_{6,1}}\right\|}\right]^{\top} =: \boldsymbol{h}(\boldsymbol{p}). \tag{6} $$

In the following, we derive an extended Kalman filter to estimate the state vector $\boldsymbol{p}$ based on the measurements $\boldsymbol{z}$. We set the pose's dynamics to be a random walk, while the measurements are given by Eq. (6):

$$ \boldsymbol{p}(i) = \boldsymbol{f}\big(\boldsymbol{p}(i-1)\big) = \boldsymbol{p}(i-1) + \boldsymbol{w}(i-1), \qquad \boldsymbol{z}(i) = \boldsymbol{h}\big(\boldsymbol{p}(i)\big) + \boldsymbol{v}(i), \tag{7} $$

where $\boldsymbol{w}(i) \sim \mathcal{N}(\boldsymbol{0}, \boldsymbol{Q})$ and $\boldsymbol{v}(i) \sim \mathcal{N}(\boldsymbol{0}, \boldsymbol{C})$ with $\boldsymbol{Q}$ and $\boldsymbol{C}$ being the covariance matrices of the process noise and the observation noise, respectively. Here, i is the iteration step with $i = 1, \ldots, n$, and $n \in \mathbb{N}$. We assume the measurement vector

$\boldsymbol{z}$ to be Gaussian distributed with its individual, constant variances $\sigma(\boldsymbol{z}_j)$, $j \in \{x, y, z, \alpha, \beta, \gamma\}$. The variances are stored as diagonal elements in the covariance matrix $\boldsymbol{C}$. The extended Kalman filter requires the Jacobians $\boldsymbol{J}_f(i)$ of $\boldsymbol{f}(\boldsymbol{p})$, which is $\boldsymbol{I}_6$ (an identity matrix of size six), and $\boldsymbol{J}_h(i)$ of $\boldsymbol{h}(\boldsymbol{p})$, which is

$$\boldsymbol{J}_h(i) = \frac{\mathrm{d}\boldsymbol{h}(i)}{\mathrm{d}\boldsymbol{p}(i)} = \left[\frac{\partial \boldsymbol{h}^\top(i)}{\partial x(i)} \cdots \frac{\partial \boldsymbol{h}^\top(i)}{\partial \gamma(i)}\right]^\top . \tag{8}$$

The predicted pose and covariance estimates $\boldsymbol{p}^-$ and $\boldsymbol{P}^-$ are

$$\boldsymbol{p}^-(i) = \boldsymbol{p}^+(i-1) + \boldsymbol{w}(i), \qquad \boldsymbol{P}^-(i) = \boldsymbol{P}^+(i-1) + \boldsymbol{Q}, \tag{9}$$

where $^-$ indicates 'a priori' and $^+$ 'a posteriori' estimates. For step $i = 1$, initial pose and covariance estimates $\boldsymbol{p}^+(0)$ and $\boldsymbol{P}^+(0)$ are necessary. Innovation residual $\boldsymbol{\kappa}$, residual covariance $\boldsymbol{S}$, and Kalman gain $\boldsymbol{K}$ can then be obtained by

$$\boldsymbol{\kappa}(i) = \boldsymbol{z}(i) - \boldsymbol{h}\big(\boldsymbol{p}(i)\big), \qquad \boldsymbol{S}(i) = \boldsymbol{J}_h(i) \cdot \boldsymbol{P}^-(i) \cdot \boldsymbol{J}_h^\top(i) + \boldsymbol{C}(i), \tag{10}$$

$$\boldsymbol{K}(i) = \boldsymbol{P}^-(i) \cdot \boldsymbol{J}_h(i) \cdot \boldsymbol{S}^{-1}(i). \tag{11}$$

Finally, the updated pose and covariance estimates $\boldsymbol{p}^+$ and $\boldsymbol{P}^+$ are

$$\boldsymbol{p}^+(i) = \boldsymbol{p}^-(i) + \boldsymbol{K}(i) \cdot \boldsymbol{\kappa}(i), \qquad \boldsymbol{P}^+(i) = \big(\boldsymbol{I}_6 - \boldsymbol{K}(i) \cdot \boldsymbol{J}_h(i)\big) \cdot \boldsymbol{P}^-(i). \tag{12}$$

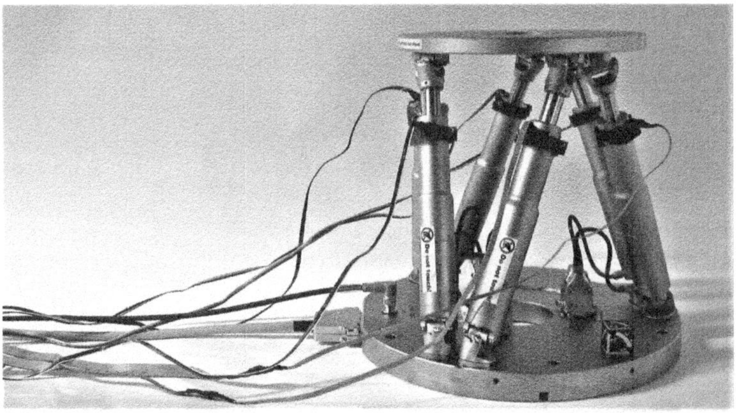

Fig. 1. Photo of the H-840 6-axis Stewart-Gough platform by Physik Instrumente.

3 Experiments

In order to validate our sensor concept, we perform several experiments using the H-840 6-axis Stewart-Gough platform by Physik Instrumente, as shown in Fig. 1. The joint coordinates are given in Table 1. The linear actuators' orientations are calculated from the accelerometer values of the inertial measurement units InvenSense MPU-6050. The distance error δ and the orientation error ϵ are calculated as follows:

$$\delta = \left\| {}^{1}\boldsymbol{p}_{12} - {}^{1}\hat{\boldsymbol{p}}_{12} \right\|, \qquad \epsilon = \tan^{-1}\left(\frac{2\left\| {}^{1}\hat{\boldsymbol{R}}_{2}^{\top} \cdot {}^{1}\boldsymbol{R}_{2} \right\|}{\operatorname{tr}\left({}^{1}\hat{\boldsymbol{R}}_{2}^{\top} \cdot {}^{1}\boldsymbol{R}_{2} \right) - 1} \right), \tag{13}$$

where ${}^{1}\boldsymbol{p}_{12}$ and ${}^{1}\boldsymbol{R}_{2}$ are the actual and ${}^{1}\hat{\boldsymbol{p}}_{12}$ and ${}^{1}\hat{\boldsymbol{R}}_{2}$ the calculated position and orientation of the manipulator platform, respectively, c. f. [1].

Table 1. Coordinates of the base and the manipulator platform joint's fulcrums with respect to the platform's origins. All units are in mm.

No.	Base platform joints	Manipulator platform joints
1	$\begin{bmatrix} 145.4 & 21.0 & 0.0 \end{bmatrix}^{\top}$	$\begin{bmatrix} 64.2 & 69.2 & 0.0 \end{bmatrix}^{\top}$
2	$\begin{bmatrix} -54.5 & 136.4 & 0.0 \end{bmatrix}^{\top}$	$\begin{bmatrix} 27.8 & 90.2 & 0.0 \end{bmatrix}^{\top}$
3	$\begin{bmatrix} -90.9 & 115.4 & 0.0 \end{bmatrix}^{\top}$	$\begin{bmatrix} -92.0 & 21.0 & 0.0 \end{bmatrix}^{\top}$
4	$\begin{bmatrix} -90.9 & -115.4 & 0.0 \end{bmatrix}^{\top}$	$\begin{bmatrix} -92.0 & -21.0 & 0.0 \end{bmatrix}^{\top}$
5	$\begin{bmatrix} -54.5 & -136.4 & 0.0 \end{bmatrix}^{\top}$	$\begin{bmatrix} 27.8 & -90.2 & 0.0 \end{bmatrix}^{\top}$
6	$\begin{bmatrix} 145.4 & -21.0 & 0.0 \end{bmatrix}^{\top}$	$\begin{bmatrix} 64.2 & -69.2 & 0.0 \end{bmatrix}^{\top}$

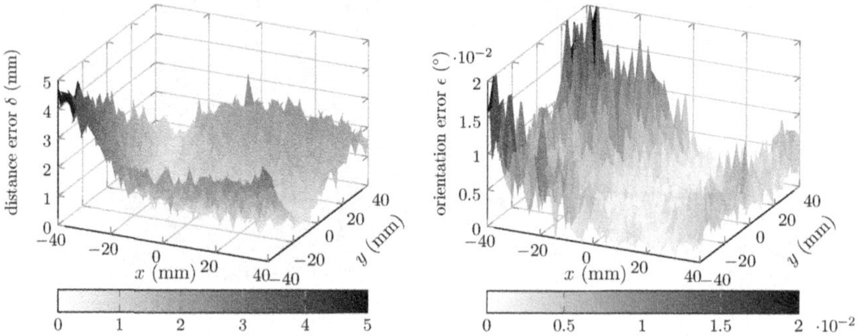

Fig. 2. Surface plot of the distance error δ and the orientation error ϵ for the 1,681 manipulator platform poses of the plane $[x\, y\, 0\, 0\, 0\, 0]^{\top}$, where $\{x, y\} \in [-40, 40]$. For each pose, 100 measurements are taken.

In the first experiment, we measure 1,681 predefined, planar poses, where $\{x, y\} \in [-40, 40]$ with a step size of 2 mm and z, α, β, and γ are zero. Here, we take 100 measurements per pose and use the measurement vectors to calculate the actual pose with the extended Kalman filter. For the experiments, we use the following parameters:

$$\boldsymbol{p}^{+}(0) = \boldsymbol{0}_{6,1}, \qquad \boldsymbol{P}^{+}(0) = 0.1\,\boldsymbol{I}_6, \qquad \boldsymbol{C} = 10^{-4}\,\boldsymbol{I}_{18}, \qquad \boldsymbol{Q} = 0.1\,\boldsymbol{I}_6. \tag{14}$$

Figure 2 shows two surface plots of the remaining distance error δ and the remaining orientation error ϵ after 100 iterations. The distance error spreads between 0 mm and 4.5 mm and the orientation error between 0° and 0.2°. The distance errors have their maxima at $(x, y) = (-40\,\text{mm}, -40\,\text{mm})$ and $(x, y) = (40\,\text{mm}, -40\,\text{mm})$ while the orientation errors have their maxima at $(x, y) = (-40\,\text{mm}, -40\,\text{mm})$ and $(x, y) = (-40\,\text{mm}, 40\,\text{mm})$. The minimal distance error is reached at $x > -10\,\text{mm}$ and $y = 0\,\text{mm}$ and at a diagonal towards $(x, y) = (-40\,\text{mm}, 40\,\text{mm})$. The minimal orientation error is reached at $x > 0\,\text{mm}$.

In the second experiment, we measure the linear actuators' orientations for six randomly chosen poses, see Table 2. We again use the initial pose and covariance estimates from Eq. (14). Figure 3 shows the distance error δ and the orientation error ϵ for the six poses $\boldsymbol{p}^1$–$\boldsymbol{p}^6$ in dependence of the iteration number i. For each pose, the distance error δ and the orientation error ϵ decrease drastically with increasing iteration number. Starting from a relatively high distance error between 21 mm and 33 mm, after ten iterations, the error lies between 2.7 mm and 7 mm. Finally, the error remains stable between 0.7 mm and 2.3 mm. The same applies for the orientation error. Starting between 0.03° and 0.46°, the error decreases to a value between 0.003° and 0.007°. In conclusion, above $i = 20$, neither the distance error nor the orientation error improves significantly.

Table 2. Investigated static poses $\boldsymbol{p}^1$–$\boldsymbol{p}^6$. All units are in mm for x, y, and z and in ° for α, β, and γ.

	$\boldsymbol{p}^1$	$\boldsymbol{p}^2$	$\boldsymbol{p}^3$	$\boldsymbol{p}^4$	$\boldsymbol{p}^5$	$\boldsymbol{p}^6$
x	30.084	−10.408	18.041	2.146	−34.190	−23.483
y	−10.511	23.179	17.755	19.544	−11.408	7.081
z	−9.112	2.565	−3.597	13.033	−10.147	−7.125
α	−2.492	0.498	0.679	−3.706	2.947	1.545
β	1.161	4.173	−4.251	0.691	−1.888	1.890
γ	−0.802	−6.424	−13.394	−0.906	0.857	7.437

For the third experiment, we define a circular trajectory with a radius of 20 mm and 50 target poses of the manipulator platform in the x–z plane. The trajectory is described by the following equation:

$$\begin{bmatrix} x\ y\ z\ \alpha\ \beta\ \gamma \end{bmatrix}^{\mathsf{T}} = \begin{bmatrix} 20\cos\left(\frac{2\pi(t-1)}{50}\right)\ 0\ 20\sin\left(\frac{2\pi(t-1)}{50}\right)\ 0\ 0\ 0 \end{bmatrix}^{\mathsf{T}}, \tag{15}$$

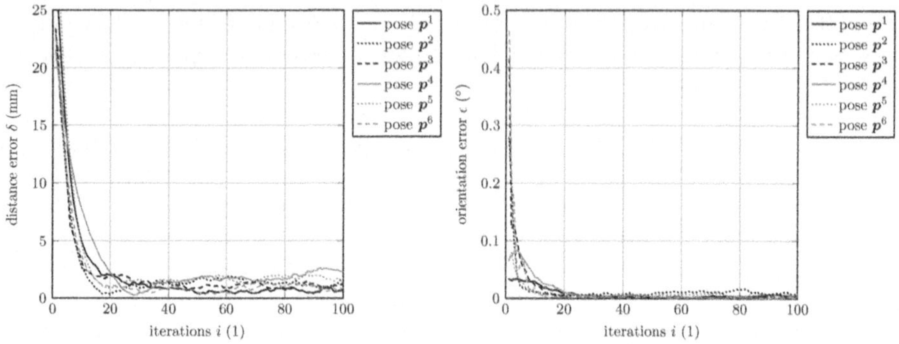

Fig. 3. Distance error δ and orientation error ϵ for the six poses $\boldsymbol{p}^1$–$\boldsymbol{p}^6$. All poses are calculated with the same initial pose and covariance estimates.

and $t = 1 \ldots 50$. For each pose of the trajectory, we take several measurements of the linear actuators' orientations and calculate the pose using the introduced extended Kalman filter. The results for four different numbers of iterations are shown in Fig. 4. The plots in the left graph are obtained with $\boldsymbol{p}^+(0) = \mathbf{0}_{6,1}$ as initial pose estimate for each pose. It can be seen that a minimum of 30 iterations is necessary to achieve accurate results, and the trajectory can be identified as a circle. For the results in the right graph, we use the calculated values from the previous pose as initial pose estimates for the new pose, i. e., we realize a continuous pose detection. Both plots are shifted by a constant offset of 2.5 mm. If we consider the previous pose as initial pose estimate, one iteration is sufficient. Note that, here, the distance between two consecutive poses has to be small enough and the first pose needs more iterations.

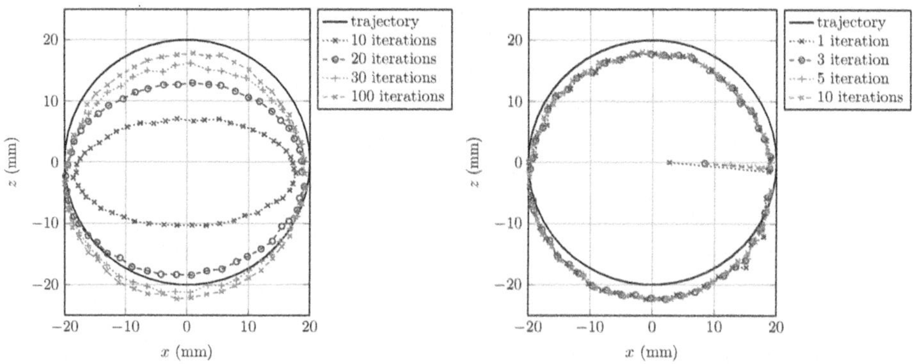

Fig. 4. Calculated poses in the x–z plane for a circular trajectory with 50 target poses for different numbers of iterations per pose. Displayed are the results with $\mathbf{0}_{6,1}$ (left) and the previous pose (right) as initial pose estimate.

4 Conclusions

Instead of measuring the linear actuators' lengths and still requiring numerical procedures, in this paper, we used the linear actuators' orientations and an extended Kalman filter to solve the direct kinematics problem of the general Stewart-Gough platform and derive the manipulator platform's pose. We then evaluated the approach with three static experiments to derive the maximum possible accuracy, the minimum number of iterations, and the effect of continuous pose detections. For the planar measurements, we derived accurate results with a maximum error of 5 mm at the edges. This error occurs mainly in the z-axis and probably results from the low-cost inertial measurement units which have a high variance and a non-linearity that could not be completely removed by the calibration. We furthermore discovered that 30 iterations are sufficient for an accurate pose determination. If the previous pose is considered, then only one iteration is necessary. Our results can be further improved by using more accurate measurement units, optimized covariance matrices $\boldsymbol{C}$ and $\boldsymbol{Q}$ to enhance the extended Kalman filter's accuracy and performance, and/or a more sophisticated Bayesian filter. Future investigations include reducing the linear actuator's orientations to a minimal set, comparing the extended Kalman filter with the Newton-Raphson algorithm in terms of accuracy, iterations, and computational costs, as well as enabling a real-time pose control.

Acknowledgements. This work was partly supported by the German Research Foundation (DFG) under grant SCHL 275/15-1 and the German Federal Ministry for Economic Affairs and Energy under grant 03THW02G04.

References

1. Baron, L., Angeles, J.: The decoupling of the direct kinematics of parallel manipulators using redundant sensors. In: IEEE International Conference on Robotics and Automation, San Diego, CA, USA, pp. 974–979 (1994)
2. Chiu, Y.J.: Forward kinematics of a general fully parallel manipulator with auxiliary sensors. Int. J. Robot. Res. **20**(5), 401–414 (2001)
3. Gosselin, C.M.: Parallel computational algorithms for the kinematics and dynamics of planar and spatial parallel manipulators. ASME J. Dyn. Syst. Meas. Control **118**(1), 22–28 (1996)
4. Hunt, K.H., Primrose, E.J.F.: Assembly configurations of some in-parallel-actuated manipulators. Mech. Mach. Theory **28**(1), 31–42 (1993)
5. Merlet, J.P.: Closed-form resolution of the direct kinematics of parallel manipulators using extra sensors data. In: IEEE International Conference on Robotics and Automation, Atlanta, GA, USA, pp. 200–204 (1993)
6. Merlet, J.P.: Solving the forward kinematics of a Gough-type parallel manipulator with interval analysis. Int. J. Robot. Res. **23**(3), 221–235 (2004)
7. Mimura, N., Funahashi, Y.: A new analytical system applying 6 DOF parallel link manipulator for evaluating motion sensation. In: IEEE International Conference on Robotics and Automation, Nagoya, Japan, pp. 227–233 (1995)

8. Mourrain, B.: The 40 'generic' positions of a parallel robot. In: International Symposium on Symbolic and Algebraic Computation, Kiev, Ukraine, pp. 173–182 (1993)
9. Raghavan, M.: The Stewart platform of general geometry has 40 configurations. J. Mech. Des. **115**(2), 277–282 (1993)
10. Schulz, S., Seibel, A., Schreiber, D., Schlattmann, J.: Sensor concept for solving the direct kinematics problem of the Stewart-Gough platform. In: IEEE/RSJ International Conference on Intelligent Robots and Systems, Vancouver, BC, Canada, pp. 1959–1964 (2017)
11. Seibel, A., Schulz, S., Schlattmann, J.: On the direct kinematics problem of parallel mechanisms. J. Robot. **2018**, 9 p. (2018)
12. Stoughton, R., Arai, T.: Optimal sensor placement for forward kinematics evaluation of a 6-DOF parallel link manipulator. In: IEEE/RSJ International Workshop on Intelligent Robots and Systems, Osaka, Japan, pp. 785–790 (1991)
13. Valenti, R.G., Dryanovski, I., Xiao, J.: Keeping a good attitude: a quaternion-based orientation filter for IMUs and MARGs. Sensors **15**(9), 19302–19330 (2015)
14. Vertechy, R., Parenti-Castelli, V.: Accurate and fast body pose estimation by three point position data. Mech. Mach. Theory **42**(9), 1170–1183 (2007)
15. Zhou, X.S., Roumeliotis, S.I.: Determining the robot-to-robot 3D relative pose using combinations of range and bearing measurements: 14 minimal problems and closed-form solutions to three of them. In: IEEE/RSJ International Conference on Intelligent Robots and Systems, Taipei, Taiwan, pp. 2983–2990 (2010)

Novel Plücker Operators and a Dual Rodrigues Formula Applied to the IKP of General 3R Chains

Bertold Bongardt(✉)

TU Braunschweig, Institut für Robotik und Prozessinformatik,
38106 Braunschweig, Germany
b.bongardt@tu-bs.de

Abstract. Based on new concepts for three (6×6)-matrix operators for Plücker vectors, the paper presents a novel adjoint formulation of the Rodrigues equation for spatial displacements in the trigonometric form. Due to the structural similarity of this dualized equation to its spherical pendant, the principle of transference is applicable to lift the analytic solution to the inverse kinematics problem (IKP) of generic spherical 3R chains yielding a simple analytic solution for generic spatial 3R chains based on the geometry of lines and screws.

Keywords: Rodrigues' formula · Plücker vectors · Screw theory
Line geometry · Dual numbers · Principle of transference
Inverse kinematics

1 Introduction

Plücker vectors provide homogeneous coordinates of lines in space and represent a crucial concept for studying kinematics in the perspective of screw theory. In this document, three matrices associated to a Pücker vector are formally defined which allow to derive a novel adjoint representation of Rodrigues' formula [7] in trigonometric form. In that specific form, an adjoint matrix of a displacement is represented as an 'affine-trigonometric' combination of the orthogonal basis provided by the three matrices. The class of generic spatial 3R chains, also called 'regional manipulators', is an important class for the (modular) design of novel mechanisms. Thus, the systematic study of their kinematics and specific properties is of high interest [4,5,9]. As an example application of the novel Rodrigues equation, a screw-based analytic solution to the inverse kinematics problem (IKP) of regional manipulators is presented applying the principle of transference to the solution approach for the corresponding, spherical problem.

Prerequisites. A Plücker vector is a six-dimensional vector $\binom{n}{m}$, consisting firstly of a direction $\boldsymbol{n}$, and secondly of a moment $\boldsymbol{m}$, in the so-called ray-coordinate order. A line in 3D-space in form of a Plücker vector is briefly denoted here as $\boldsymbol{\Lambda} = \binom{n}{m}$; its moment $\boldsymbol{m}$ is defined by the cross product of an arbitrary, fixed

J. Lenarcic and V. Parenti-Castelli (Eds.): ARK 2018, SPAR 8, pp. 65–73, 2019.
https://doi.org/10.1007/978-3-319-93188-3_8

point of the line, an *anchor* $\boldsymbol{a}$, with the direction of the line, as $\boldsymbol{m} = \boldsymbol{a} \times \boldsymbol{n}$. A unit Plücker vector is a Plücker vector that features either $\|\boldsymbol{n}\| = 1$, or $\|\boldsymbol{m}\| = 1$ and $\|\boldsymbol{n}\| = 0$; here indicated by a hat, as $\hat{\boldsymbol{\Lambda}} = \left(\begin{smallmatrix} \hat{\boldsymbol{n}} \\ \hat{\boldsymbol{m}} \end{smallmatrix}\right)$. The corresponding unit Plücker vector $\left(\begin{smallmatrix} \hat{\boldsymbol{n}} \\ \hat{\boldsymbol{m}} \end{smallmatrix}\right)$ associated to a Plücker vector $\left(\begin{smallmatrix} \boldsymbol{n} \\ \boldsymbol{m} \end{smallmatrix}\right)$ is obtained by means of the 'normalization' function $\eta : \left(\begin{smallmatrix} \boldsymbol{n} \\ \boldsymbol{m} \end{smallmatrix}\right) \mapsto \left(\begin{smallmatrix} \hat{\boldsymbol{n}} \\ \hat{\boldsymbol{m}} \end{smallmatrix}\right)$ given by $\eta\left(\begin{smallmatrix} \boldsymbol{n} \\ \boldsymbol{m} \end{smallmatrix}\right) := \frac{1}{l} \cdot \left(\begin{smallmatrix} \boldsymbol{n} \\ \boldsymbol{m} \end{smallmatrix}\right)$ via the Plücker vector length $l := \left\{ \|\boldsymbol{n}\| \text{ for } \|\boldsymbol{n}\| \neq 0,\ \|\boldsymbol{m}\| \text{ for } \|\boldsymbol{n}\| = 0 \right\}$. A screw in 3D-space in form of a unit Plücker vector is briefly denoted as $\hat{\$} = \left(\begin{smallmatrix} \hat{\boldsymbol{n}} \\ \hat{\boldsymbol{m}} \end{smallmatrix}\right)$; its moment is determined by $\hat{\boldsymbol{m}} = \boldsymbol{a} \times \hat{\boldsymbol{n}} + h \cdot \hat{\boldsymbol{n}}$ with a pitch h. A finite twist, representing a spatial displacement, is obtained as product of an angle with a unit screw, as $\$ = \phi \cdot \hat{\$}$. Denoting a dual number by $\tilde{z} = x + \epsilon \cdot y$, with $\epsilon^2 = 0$, a Plücker vector is expressed in form of a dual 3-vector $\tilde{\$} = \boldsymbol{n} + \epsilon \cdot \boldsymbol{m}$, instead of the form as a real 6-vector. The (6×6)-left adjoint matrix form of a (4×4)-homogeneous displacement matrix $\boldsymbol{D} = \left(\begin{smallmatrix} \boldsymbol{R} & \boldsymbol{t} \\ \boldsymbol{0} & 1 \end{smallmatrix}\right) \in SE(3)$ is denoted in the sequel as $\boldsymbol{D}^{\mathrm{Adl}} = \left(\begin{smallmatrix} \boldsymbol{R} & \boldsymbol{0} \\ \boldsymbol{t}^{\otimes}\boldsymbol{R} & \boldsymbol{R} \end{smallmatrix}\right)$, where $\boldsymbol{t}^{\otimes}$ expresses the skew-symmetric matrix associated to the $\boldsymbol{t}$-vector, see definition in (1). The operator $*$ expresses the combination of transposition and multiplication, for example, as $\boldsymbol{A} * \boldsymbol{B} = \boldsymbol{A}^T \cdot \boldsymbol{B}$, determining a standard inner product.

Structure. In Sect. 2, the spherical case is treated: first, the matrix operators are introduced; next, the trigonometric Rodrigues formula is stated using these operators; at last, the IKP is solved based on the Rodrigues equation. In Sect. 3, the novel, generalized concepts are presented for lines and screws following an identical structure. In Sect. 4, brief conclusions are drawn.

2 Spherical Geometry

2.1 Vector Matrix Operators

In accordance to [1], three (3×3)-matrices, the *cross-matrix* $\boldsymbol{a}^{\otimes}$, the *unit-matrix* $\boldsymbol{a}^{\circledcirc}$, and the *square-matrix* $\boldsymbol{a}^{\oslash}$, associated to a vector $\boldsymbol{a} \in \mathbb{R}^3$ are defined:

$$\boldsymbol{a}^{\otimes} := \begin{pmatrix} 0 & -a_3 & a_2 \\ a_3 & 0 & -a_1 \\ -a_2 & a_1 & 0 \end{pmatrix} \qquad \boldsymbol{a}^{\circledcirc} := -\,\boldsymbol{a}^{\otimes} \cdot \boldsymbol{a}^{\otimes} \qquad \boldsymbol{a}^{\oslash} := \boldsymbol{a} \cdot \boldsymbol{a}^T \,. \tag{1}$$

The summation of the unit-matrix and the square-matrix of a unit vector $\hat{\boldsymbol{a}}$ equals the identity matrix:

$$\hat{\boldsymbol{a}}^{\circledcirc} + \hat{\boldsymbol{a}}^{\oslash} = \boldsymbol{I} \,. \tag{2}$$

In the next paragraph, the geometric relevance of the three introduced matrices, $\boldsymbol{a}^{\otimes}$, $\boldsymbol{a}^{\circledcirc}$, and $\boldsymbol{a}^{\oslash}$, is illustrated with an example.

Projections and Decomposition. With the square-matrix $\hat{\boldsymbol{n}}^{\oslash}$, the projection of a vector $\boldsymbol{b}$ onto (an axis given by) a unit vector $\hat{\boldsymbol{n}}$, the *projection vector*, is computed as the matrix-vector product

$$\pi(\boldsymbol{b}; \hat{\boldsymbol{n}}) = \hat{\boldsymbol{n}}^{\oslash} \cdot \boldsymbol{b} \,. \tag{3}$$

The unit-matrix $\hat{\boldsymbol{n}}^{\circledcirc}$ is used to determine the projection of a vector into the planar orthogonal complement of a unit vector, the *rejection vector*, as the product

$$\tau(\boldsymbol{b}; \hat{\boldsymbol{n}}) = \hat{\boldsymbol{n}}^{\circledcirc} \cdot \boldsymbol{b} \,. \tag{4}$$

The matrix-vector product of the cross-matrix $\hat{\boldsymbol{n}}^{\otimes}$ with a vector $\boldsymbol{b}$ determines the *orthogonal vector* to $\pi(\,\boldsymbol{b};\hat{\boldsymbol{n}}\,)$ and $\tau(\,\boldsymbol{b};\hat{\boldsymbol{n}}\,)$, that is $\tau(\,\boldsymbol{b};\hat{\boldsymbol{n}}\,)$ rotated by $+\frac{\pi}{2}$ along $\hat{\boldsymbol{n}}$, denoted here as

$$\chi(\,\boldsymbol{b};\hat{\boldsymbol{n}}\,) = \hat{\boldsymbol{n}}^{\otimes}\cdot\boldsymbol{b}\,. \tag{5}$$

The projection vector and the rejection vector together provide an orthogonal decomposition of a vector $\boldsymbol{b}$ with respect to an axis $\hat{\boldsymbol{n}}$ as

$$\boldsymbol{b} = \pi(\,\boldsymbol{b};\hat{\boldsymbol{n}}\,) + \tau(\,\boldsymbol{b};\hat{\boldsymbol{n}}\,)\,. \tag{6}$$

The introduced concepts are illustrated by an example in Fig. 1.

2.2 Matrix Exponential and Rodrigues Equation

For a 'vector' of the Lie algebra $so(3)$, a product of a rotation angle ϕ and the cross-matrix form of a rotation axis, $\hat{\boldsymbol{n}^{\otimes}}$, the matrix exponential generates an 'element' of the Lie group $SO(3)$, an orientation matrix $\boldsymbol{R}$, as

$$\boldsymbol{R} = \mathrm{rot}(\,\phi;\hat{\boldsymbol{n}}\,) = \exp(\,\phi\cdot\hat{\boldsymbol{n}^{\otimes}}\,)\,.$$

The trigonometric form [3] of the Rodrigues equation is given as

$$\exp(\,\phi\cdot\hat{\boldsymbol{n}^{\otimes}}\,) = \cos\phi\cdot\hat{\boldsymbol{n}}^{\circledcirc} + \sin\phi\cdot\hat{\boldsymbol{n}^{\otimes}} + \hat{\boldsymbol{n}^{\oslash}}\,, \tag{7}$$

providing a finite sum expression for the matrix exponential of $so(3)$ in terms of the three (matrix-shaped) 'basis vectors' defined in (1).

2.3 Inverse Kinematics of Generic Spherical 3R Chains

The finite inverse kinematics problem of generic spherical 3R chains is to find joint configurations $\boldsymbol{q} = (\phi_{12}, \phi_{23}, \phi_{34})$ that satisfy

$$\boldsymbol{R}_{14} = \mathrm{rot}(\,\phi_{12};\hat{\boldsymbol{n}}_{12}\,)\cdot\mathrm{rot}(\,\phi_{23};\hat{\boldsymbol{n}}_{23}\,)\cdot\mathrm{rot}(\,\phi_{34};\hat{\boldsymbol{n}}_{34}\,)\,,$$

for a given constellation of joint axes $\hat{\boldsymbol{n}}_{12}, \hat{\boldsymbol{n}}_{23}$, and $\hat{\boldsymbol{n}}_{34}$ (given by design), and a reference rotation $\boldsymbol{R}_{14}$ of the distal link with respect to the proximal link. For a fixed axes constellation, and with $\boldsymbol{R}_{\mathrm{ij}} = \mathrm{rot}(\,\phi_{\mathrm{ij}};\hat{\boldsymbol{n}}_{\mathrm{ij}}\,)$, the problem is briefly expressed as

$$\boldsymbol{R}_{14} = \boldsymbol{R}_{12}\cdot\boldsymbol{R}_{23}\cdot\boldsymbol{R}_{34}\,.$$

For a solution, the ansatz [1,6] is given by the chain of equations

$$\begin{aligned}\hat{\boldsymbol{n}}_{12} * \boldsymbol{R}_{14}\cdot\hat{\boldsymbol{n}}_{34} &= \hat{\boldsymbol{n}}_{12} * \boldsymbol{R}_{12}\cdot\boldsymbol{R}_{23}\cdot\boldsymbol{R}_{34}\cdot\hat{\boldsymbol{n}}_{34} = \hat{\boldsymbol{n}}_{12} * \boldsymbol{R}_{23}\cdot\hat{\boldsymbol{n}}_{34}\\ &= \hat{\boldsymbol{n}}_{12} * \left(\cos\phi_{23}\cdot\hat{\boldsymbol{n}}_{23}^{\circledcirc} + \sin\phi_{23}\cdot\hat{\boldsymbol{n}^{\otimes}}_{23} + \hat{\boldsymbol{n}^{\oslash}}_{23}\right)\cdot\hat{\boldsymbol{n}}_{34}\,,\end{aligned} \tag{8}$$

using the Rodrigues Eq. (7). By means of the four parameters

$$a := \hat{\boldsymbol{n}}_{12} * \hat{\boldsymbol{n}}_{23}^{\circledcirc} \cdot \hat{\boldsymbol{n}}_{34} \qquad c := \hat{\boldsymbol{n}}_{12} * \left(\boldsymbol{R}_{14} - \hat{\boldsymbol{n}}_{23}^{\oslash}\right) \cdot \hat{\boldsymbol{n}}_{34}$$

$$b := \hat{\boldsymbol{n}}_{12} * \hat{\boldsymbol{n}}_{23}^{\otimes} \cdot \hat{\boldsymbol{n}}_{34} \qquad d := \sqrt{a^2 + b^2 - c^2}\,,$$

Equation (8) is transformed into the scalar constraint equation

$$a \cdot \cos\phi_{23} + b \cdot \sin\phi_{23} = c \tag{9}$$

that features two solutions $\phi_{23}^{(+)}, \phi_{23}^{(-)} := \phi_{23}^{(0)} \pm \phi_{23}^{(\Delta)}$ for the second joint ϕ_{23} expressed via the two auxiliary terms, $\phi_{23}^{(0)} := \operatorname{atan2}(b, a)$ and $\phi_{23}^{(\Delta)} := \operatorname{atan2}(d, c)$.

For each of the two values, $\phi_{23}^{(+)}$ and $\phi_{23}^{(-)}$, corresponding values for the first and for the third joint are obtained, with $\boldsymbol{R}_{23} = \operatorname{rot}(\,\phi_{23}; \hat{\boldsymbol{n}}_{23}\,)$, via

$$\phi_{12} = \operatorname{acos3}\left(\,\boldsymbol{R}_{14} \cdot \hat{\boldsymbol{n}}_{34}, \boldsymbol{R}_{23} \cdot \hat{\boldsymbol{n}}_{34}; -\hat{\boldsymbol{n}}_{12}\,\right)$$

$$\phi_{34} = \operatorname{acos3}\left(\,\boldsymbol{R}_{14}^{T} \cdot \hat{\boldsymbol{n}}_{12}, \boldsymbol{R}_{23}^{T} \cdot \hat{\boldsymbol{n}}_{12}; +\hat{\boldsymbol{n}}_{34}\,\right)\,,$$

where the trivariate inverse trigonometric function 'acos3' [1] is given as

$$\operatorname{acos3}(\,\boldsymbol{a}, \boldsymbol{b}; \hat{\boldsymbol{n}}\,) = \operatorname{ornt3}(\,\boldsymbol{a}, \boldsymbol{b}, \hat{\boldsymbol{n}}\,) \cdot \operatorname{acos}(\,\hat{\mathcal{T}}(\,\boldsymbol{b}; \hat{\boldsymbol{n}}\,) * \hat{\mathcal{T}}(\,\boldsymbol{b}; \hat{\boldsymbol{n}}\,)\,)\,,$$

using the normalized planar projection $\hat{\mathcal{T}}(\,\boldsymbol{b}; \hat{\boldsymbol{n}}\,) = \frac{\mathcal{T}(\,\boldsymbol{b};\hat{\boldsymbol{n}}\,)}{\|\mathcal{T}(\,\boldsymbol{b};\hat{\boldsymbol{n}}\,)\|}$ and the orientation

$$\operatorname{ornt3}(\,\boldsymbol{a}, \boldsymbol{b}, \boldsymbol{c}\,) = \operatorname{sgn}(\,\det(\boldsymbol{a}\ \boldsymbol{b}\ \boldsymbol{c})\,)\,. \tag{10}$$

Details about singular and infeasible reference orientations, degenerate axes constellations, and the treatment of the signs of the involved angles are given in [1].

3 Spatial Geometry

3.1 Plücker Matrix Operators

Three novel (6×6)-matrices are introduced as line-generalizations to the three (3×3) vector-matrices defined in Sect. 2.1. The *cross-matrix* associated to a six-dimensional Plücker vector $\boldsymbol{\$} = \left(\begin{smallmatrix} \boldsymbol{n} \\ \boldsymbol{m} \end{smallmatrix}\right)$ is defined as

$$\begin{pmatrix} \boldsymbol{n} \\ \boldsymbol{m} \end{pmatrix}^{\boxtimes} := \begin{pmatrix} \boldsymbol{n}^{\otimes} & \boldsymbol{0} \\ \boldsymbol{m}^{\otimes} & \boldsymbol{n}^{\otimes} \end{pmatrix}\,. \tag{11}$$

The line-geometric *unit-matrix* is defined as

$$\begin{pmatrix} \boldsymbol{n} \\ \boldsymbol{m} \end{pmatrix}^{\boxdot} := -\begin{pmatrix} \boldsymbol{n} \\ \boldsymbol{m} \end{pmatrix}^{\boxtimes} \cdot \begin{pmatrix} \boldsymbol{n} \\ \boldsymbol{m} \end{pmatrix}^{\boxtimes}\,. \tag{12}$$

The affine *square-matrix* is introduced[1] as

$$\begin{pmatrix} \boldsymbol{n} \\ \boldsymbol{m} \end{pmatrix}^{\boxslash} := \boldsymbol{I} - \begin{pmatrix} \boldsymbol{n} \\ \boldsymbol{m} \end{pmatrix}^{\boxdot}\,. \tag{13}$$

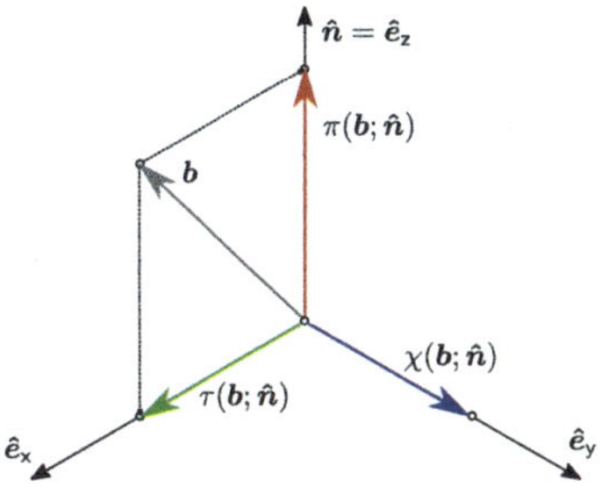

Fig. 1. The projection vector $\pi(\boldsymbol{b}; \hat{\boldsymbol{n}})$ (red), the rejection vector $\tau(\boldsymbol{b}; \hat{\boldsymbol{n}})$ (green), and the orthogonal vector $\chi(\boldsymbol{b}; \hat{\boldsymbol{n}})$ (blue) of a vector $\boldsymbol{b} = \frac{1}{10} \cdot (6, 0, 8)^T$ (gray) with respect to the axis $\hat{\boldsymbol{n}} = (0, 0, 1)^T$. The numerical values obtained from (4), (5), and (6) are $\pi(\boldsymbol{b}; \hat{\boldsymbol{n}}) = \frac{1}{10} \cdot (0, 0, 8)^T$, $\tau(\boldsymbol{b}; \hat{\boldsymbol{n}}) = \frac{1}{10} \cdot (6, 0, 0)^T$, and $\chi(\boldsymbol{b}; \hat{\boldsymbol{n}}) = \frac{1}{10} \cdot (0, 6, 0)^T$.

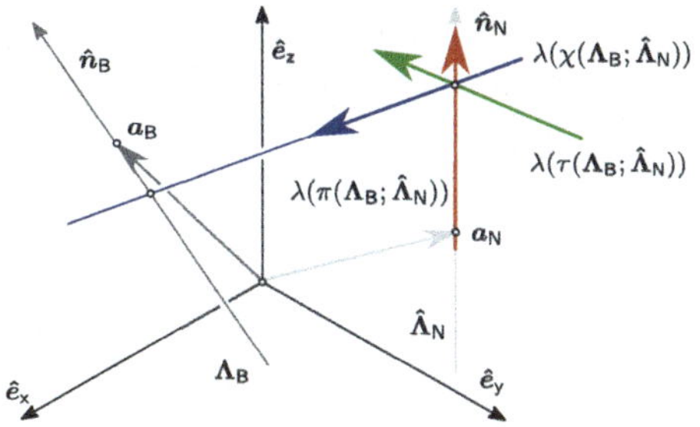

Fig. 2. Line projection $\lambda(\pi(\boldsymbol{\Lambda}_{\mathsf{B}}; \hat{\boldsymbol{\Lambda}}_{\mathsf{N}}))$, line rejection $\lambda(\tau(\boldsymbol{\Lambda}_{\mathsf{B}}; \hat{\boldsymbol{\Lambda}}_{\mathsf{N}}))$, and line orthogonal $\lambda(\chi(\boldsymbol{\Lambda}_{\mathsf{B}}; \hat{\boldsymbol{\Lambda}}_{\mathsf{N}}))$ of a line $\boldsymbol{\Lambda}_{\mathsf{B}}$ with direction $\hat{\boldsymbol{n}}_{\mathsf{B}} = \frac{1}{\sqrt{21}} \cdot (-1, -4, 2)^T$ and anchor $\boldsymbol{a}_{\mathsf{B}} = \frac{1}{10} \cdot (6, 0, 8)^T$ wrt. line $\hat{\boldsymbol{\Lambda}}_{\mathsf{N}}$ with direction $\hat{\boldsymbol{n}}_{\mathsf{N}} = (0, 0, 1)^T$ and anchor $\boldsymbol{a}_{\mathsf{N}} = \frac{1}{10} \cdot (-4, 4, 2)^T$. All lines pass the closest point to $\boldsymbol{\Lambda}_{\mathsf{B}}$ on $\hat{\boldsymbol{\Lambda}}_{\mathsf{N}}$ ('point projection').

In analogy to the spherical case, the relevance of the three novel matrices, for the geometry of lines and screws, is illustrated next.

Projections and Decomposition. The orthogonal projection of a line vector ($\boldsymbol{\Lambda}_{\mathsf{B}}$) onto a unit line ($\hat{\boldsymbol{\Lambda}}_{\mathsf{N}}$), the *line projection*, is determined with the square-matrix $\hat{\boldsymbol{\Lambda}}_{\mathsf{N}}^{\boxslash}$ from (13) via the matrix-vector product

$$\pi\big(\boldsymbol{\Lambda}_{\mathsf{B}}; \hat{\boldsymbol{\Lambda}}_{\mathsf{N}}\big) = \hat{\boldsymbol{\Lambda}}_{\mathsf{N}}^{\boxslash} \cdot \boldsymbol{\Lambda}_{\mathsf{B}} \,,$$

generalizing the projection vector, defined in (3). The projection into the orthogonal complement of a unit line $\hat{\boldsymbol{\Lambda}}_{\mathsf{N}}$, the *line rejection*, is employed by means of the unit matrix $\hat{\boldsymbol{\Lambda}}_{\mathsf{N}}^{\boxdot}$ from (12) via the product

$$\tau\big(\boldsymbol{\Lambda}_{\mathsf{B}}; \hat{\boldsymbol{\Lambda}}_{\mathsf{N}}\big) = \hat{\boldsymbol{\Lambda}}_{\mathsf{N}}^{\boxdot} \cdot \boldsymbol{\Lambda}_{\mathsf{B}} \,,$$

generalizing the rejection vector, defined in (4). The common orthogonal of two lines, the *line orthogonal*, is determined via the cross-matrix $\hat{\boldsymbol{\Lambda}}_{\mathsf{N}}^{\boxtimes}$ from (11) via the matrix-vector multiplication

$$\chi\big(\boldsymbol{\Lambda}_{\mathsf{B}}; \hat{\boldsymbol{\Lambda}}_{\mathsf{N}}\big) = \hat{\boldsymbol{\Lambda}}_{\mathsf{N}}^{\boxtimes} \cdot \boldsymbol{\Lambda}_{\mathsf{B}}$$

generalizing the orthogonal vector, defined in (5). For each of the three line operations, the obtained results generally involve non-zero pitches, indicated in compact notations, $\$_{\pi\mathsf{BN}} = \pi\big(\boldsymbol{\Lambda}_{\mathsf{B}}; \hat{\boldsymbol{\Lambda}}_{\mathsf{N}}\big)$, $\$_{\tau\mathsf{BN}} = \tau\big(\boldsymbol{\Lambda}_{\mathsf{B}}; \hat{\boldsymbol{\Lambda}}_{\mathsf{N}}\big)$, and $\$_{\chi\mathsf{BN}} = \chi\big(\boldsymbol{\Lambda}_{\mathsf{B}}; \hat{\boldsymbol{\Lambda}}_{\mathsf{N}}\big)$,

[1] In contrast to its vector analogue (1), the square-matrix of a Plücker vector (13) is not introduced as an outer product, but via the related identity of (2). The unit and the square matrix feature the term $\mp(\boldsymbol{m}^{\otimes} \cdot \boldsymbol{n}^{\otimes} + \boldsymbol{n}^{\otimes} \cdot \boldsymbol{m}^{\otimes})$ in the lower right block.

as screw vectors. For deriving the corresponding line vectors, satisfying the orthogonality of moment and direction, $\boldsymbol{m} \times \boldsymbol{n} = 0$, the ‘alignment’ function $\lambda : \$ \mapsto \boldsymbol{\Lambda}$, defined as $\lambda(\$) = \lambda\left(\left(\begin{smallmatrix} \boldsymbol{n} \\ \boldsymbol{m} \end{smallmatrix}\right)\right) = (\boldsymbol{n}, \boldsymbol{m} - \frac{\boldsymbol{n} * \boldsymbol{m}}{\boldsymbol{n} * \boldsymbol{n}} \cdot \boldsymbol{n})^T = \boldsymbol{\Lambda}$ is applicable: $\boldsymbol{\Lambda}_{\pi \mathsf{BN}} = \lambda(\,\$_{\pi \mathsf{BN}}\,)$, $\boldsymbol{\Lambda}_{\tau \mathsf{BN}} = \lambda(\,\$_{\tau \mathsf{BN}}\,)$, and $\boldsymbol{\Lambda}_{\chi \mathsf{BN}} = \lambda(\,\$_{\chi \mathsf{BN}}\,)$. The decomposition of a Plücker vector $\$_{\mathsf{B}}$ with respect to a unit line $\hat{\boldsymbol{\Lambda}}_{\mathsf{N}}$ is given by the screws of the line projection and the line complement as

$$\$_{\mathsf{B}} = \pi(\,\$_{\mathsf{B}}; \hat{\boldsymbol{\Lambda}}_{\mathsf{N}}\,) + \tau(\,\$_{\mathsf{B}}; \hat{\boldsymbol{\Lambda}}_{\mathsf{N}}\,)\,,$$

generalizing (6). The lines associated to the introduced Plücker matrix computations are illustrated by an example in Fig. 2.

3.2 Matrix Exponentials and Rodrigues Equations

Homogeneous Form. The (4×4)-homogeneous matrix $\boldsymbol{D} = \left(\begin{smallmatrix} \boldsymbol{R} & \boldsymbol{t} \\ \mathbf{0} & 1 \end{smallmatrix}\right)$ of a spatial displacement, an element of the Lie group $SE(3)$, is generated from a vector-element of the Lie algebra $se(3)$, a product of a rotation angle ϕ and the (4×4)-cross-matrix form of a unit twist $\hat{\$}^{\otimes} = \left(\begin{smallmatrix} \hat{\boldsymbol{n}}^{\otimes} & \boldsymbol{m} \\ \mathbf{0} & 0 \end{smallmatrix}\right)$, by the matrix exponential

$$\boldsymbol{D} = \begin{pmatrix} \boldsymbol{R} & \boldsymbol{t} \\ \mathbf{0} & 1 \end{pmatrix} = \operatorname{dsp}(\,\phi; \hat{\$}\,) := \exp\left(\,\phi \cdot \begin{pmatrix} \hat{\boldsymbol{n}}^{\otimes} & \boldsymbol{m} \\ \mathbf{0} & 0 \end{pmatrix}\right).$$

In contrast to the (trigonometric form of the) Rodrigues formula (7) for $\boldsymbol{R}$, direct, finite-sum expressions for $\boldsymbol{D}$ and $\boldsymbol{t}$ in terms of the screw-geometric parameters involve cubical and fractional expressions [2,7,8].

Adjoint Form. The (6×6)-adjoint matrix $\boldsymbol{D}^{\mathrm{Adl}} = \left(\begin{smallmatrix} \boldsymbol{R} & \mathbf{0} \\ \boldsymbol{t}^{\otimes}\boldsymbol{R} & \boldsymbol{R} \end{smallmatrix}\right)$ of a spatial displacement is generated from a product of an angle ϕ and the (6×6)-cross-matrix form (11) of a unit twist $\hat{\$}^{\boxtimes} = \left(\begin{smallmatrix} \hat{\boldsymbol{n}}^{\otimes} & \mathbf{0} \\ \hat{\boldsymbol{m}}^{\otimes} & \hat{\boldsymbol{n}}^{\otimes} \end{smallmatrix}\right)$, by the matrix exponential

$$\boldsymbol{D}^{\mathrm{Adl}} = \begin{pmatrix} \boldsymbol{R} & \mathbf{0} \\ \boldsymbol{t}^{\otimes}\boldsymbol{R} & \boldsymbol{R} \end{pmatrix} = \operatorname{adsp}(\,\phi; \hat{\$}\,) := \exp\left(\,\phi \cdot \begin{pmatrix} \hat{\boldsymbol{n}}^{\otimes} & \mathbf{0} \\ \hat{\boldsymbol{m}}^{\otimes} & \hat{\boldsymbol{n}}^{\otimes} \end{pmatrix}\right).$$

By means of the Plücker matrix operators introduced in Sect. 3.1, a novel, finite-sum expression of a Lie group element in terms of its screw-geometric parameters, a dual Rodrigues equation, is obtained with

$$\exp\left(\,\phi \cdot \begin{pmatrix} \hat{\boldsymbol{n}}^{\otimes} & \mathbf{0} \\ \hat{\boldsymbol{m}}^{\otimes} & \hat{\boldsymbol{n}}^{\otimes} \end{pmatrix}\right) = \cos\phi \cdot \begin{pmatrix} \hat{\boldsymbol{n}} \\ \hat{\boldsymbol{m}} \end{pmatrix}^{\boxdot} + \sin\phi \cdot \begin{pmatrix} \hat{\boldsymbol{n}} \\ \hat{\boldsymbol{m}} \end{pmatrix}^{\boxtimes} + \begin{pmatrix} \hat{\boldsymbol{n}} \\ \hat{\boldsymbol{m}} \end{pmatrix}^{\boxslash}. \tag{14}$$

This adjoint representation of the spatial Rodrigues equation resembles the trigonometric form of the spherical analogue (7) consistently.[2]

[2] For screws with non-zero shift $s = \phi \cdot h$, pitch $h \neq 0$, identity (14) is extended to $\exp(\,\phi \cdot \hat{\$}^{\boxtimes}\,) = \cos\phi \cdot \hat{\$}^{\boxdot} + \sin\phi \cdot \hat{\$}^{\boxtimes} + \hat{\$}^{\boxslash} + (\phi - \sin\phi) \cdot s \cdot \exp(\phi \cdot \left(\begin{smallmatrix} \mathbf{0} & \mathbf{0} \\ \hat{n}^{\otimes} & \mathbf{0} \end{smallmatrix}\right))..$

3.3 Inverse Kinematics of Generic Spatial 3R Chains

The finite inverse kinematics problem of generic spatial 3R chains is to find joint configurations $\boldsymbol{q} = (\phi_{12}, \phi_{23}, \phi_{34})$ that satisfy

$$\boldsymbol{D}_{14} = \mathrm{dsp}(\,\phi_{12}; \hat{\boldsymbol{\Lambda}}_{12}\,) \cdot \mathrm{dsp}(\,\phi_{23}; \hat{\boldsymbol{\Lambda}}_{23}\,) \cdot \mathrm{dsp}(\,\phi_{34}; \hat{\boldsymbol{\Lambda}}_{34}\,)\ ,$$

for a given constellation of joint axes $\hat{\boldsymbol{\Lambda}}_{12}$, $\hat{\boldsymbol{\Lambda}}_{23}$, and $\hat{\boldsymbol{\Lambda}}_{34}$ (given by design), and a reference displacement $\boldsymbol{D}_{14}$ of the distal link with respect to the proximal link. Equivalently to the homogeneous formulation, the constraint equation is re-stated in the adjoint formulation as

$$\boldsymbol{D}_{14}^{\mathrm{Adl}} = \mathrm{adsp}(\,\phi_{12}; \hat{\boldsymbol{\Lambda}}_{12}\,) \cdot \mathrm{adsp}(\,\phi_{23}; \hat{\boldsymbol{\Lambda}}_{23}\,) \cdot \mathrm{adsp}(\,\phi_{34}; \hat{\boldsymbol{\Lambda}}_{34}\,)\ .$$

For a fixed axes constellation, and with $\boldsymbol{M}_{\mathrm{ij}}^{\mathrm{Adl}} = \mathrm{adsp}(\,\phi_{\mathrm{ij}}; \hat{\boldsymbol{\Lambda}}_{\mathrm{ij}}\,)$, the problem is briefly expressed as

$$\boldsymbol{D}_{14}^{\mathrm{Adl}} = \boldsymbol{M}_{12}^{\mathrm{Adl}} \cdot \boldsymbol{M}_{23}^{\mathrm{Adl}} \cdot \boldsymbol{M}_{34}^{\mathrm{Adl}}\ .$$

By means of the introduced tools, the principle of transference can be readily applied to the ansatz for the spherical problem from Sect. 2.3 yielding

$$\begin{aligned}\hat{\boldsymbol{\Lambda}}_{12} * \boldsymbol{D}_{14}^{\mathrm{Adl}} \cdot \hat{\boldsymbol{\Lambda}}_{34} &= \hat{\boldsymbol{\Lambda}}_{12} * \boldsymbol{M}_{12}^{\mathrm{Adl}} \cdot \boldsymbol{M}_{23}^{\mathrm{Adl}} \cdot \boldsymbol{M}_{34}^{\mathrm{Adl}} \cdot \hat{\boldsymbol{\Lambda}}_{34} = \hat{\boldsymbol{\Lambda}}_{12} * \boldsymbol{M}_{23}^{\mathrm{Adl}} \cdot \hat{\boldsymbol{\Lambda}}_{34} \\ &= \hat{\boldsymbol{\Lambda}}_{12} * \Big(\cos\phi_{23} \cdot \hat{\boldsymbol{\Lambda}}_{23}^{\boxdot} + \sin\phi_{23} \cdot \hat{\boldsymbol{\Lambda}}_{23}^{\boxtimes} + \hat{\boldsymbol{\Lambda}}_{23}^{\boxslash}\Big) \cdot \hat{\boldsymbol{\Lambda}}_{34}\ ,\end{aligned}$$

employing the dual Rodrigues equation in adjoint trigonometric form (14). In analogy to the spherical case, four parameters are introduced as

$$\begin{aligned} a &:= \hat{\boldsymbol{\Lambda}}_{12} * \hat{\boldsymbol{\Lambda}}_{23}^{\boxdot} \cdot \hat{\boldsymbol{\Lambda}}_{34} & c &:= \hat{\boldsymbol{\Lambda}}_{12} * \big(\boldsymbol{D}_{14}^{\mathrm{Adl}} - \hat{\boldsymbol{\Lambda}}_{23}^{\boxslash}\big) \cdot \hat{\boldsymbol{\Lambda}}_{34} \\ b &:= \hat{\boldsymbol{\Lambda}}_{12} * \hat{\boldsymbol{\Lambda}}_{23}^{\boxtimes} \cdot \hat{\boldsymbol{\Lambda}}_{34} & d &:= \sqrt{a^2 + b^2 - c^2}\ ,\end{aligned}$$

yielding a scalar equation for the second joint angle ϕ_{12} of the shape of (9) that has the two solutions $\phi_{23}^{(+)}$, $\phi_{23}^{(-)} = \mathrm{atan2}(b, a) \pm \mathrm{atan2}(d, c)$.

The values for the first and for the third joint, ϕ_{12} and ϕ_{34}, appropriate to a second joint angle ϕ_{23}, are computed with a procedure that is an affine generalization to the trivariate inverse trigonometric function 'acos3' (Sect. 2.3): At first, the unit lines of the line rejections of $\hat{\boldsymbol{\Lambda}}_{\mathsf{A}}$ and $\hat{\boldsymbol{\Lambda}}_{\mathsf{B}}$ with respect to $\hat{\boldsymbol{\Lambda}}_{\mathsf{N}}$ are computed

$$\hat{\boldsymbol{\Lambda}}_{\tau\mathsf{AN}} = \lambda(\,\eta(\,\tau(\,\hat{\boldsymbol{\Lambda}}_{\mathsf{A}}; \hat{\boldsymbol{\Lambda}}_{\mathsf{N}}\,)\,)\,) \qquad \hat{\boldsymbol{\Lambda}}_{\tau\mathsf{BN}} = \lambda(\,\eta(\,\tau(\,\hat{\boldsymbol{\Lambda}}_{\mathsf{B}}; \hat{\boldsymbol{\Lambda}}_{\mathsf{N}}\,)\,)\,)\ ,$$

using the normalization function η (Sect. 1) and the alignment function λ (Sect. 3.1). At second, the dual inner product

$$\left(\begin{smallmatrix}\boldsymbol{n}_{\mathsf{P}}\\ \boldsymbol{m}_{\mathsf{P}}\end{smallmatrix}\right) \circledast \left(\begin{smallmatrix}\boldsymbol{n}_{\mathsf{Q}}\\ \boldsymbol{m}_{\mathsf{Q}}\end{smallmatrix}\right) = \boldsymbol{n}_{\mathsf{P}} * \boldsymbol{n}_{\mathsf{Q}} + \epsilon \cdot \big(\boldsymbol{n}_{\mathsf{P}} * \boldsymbol{m}_{\mathsf{Q}} + \boldsymbol{m}_{\mathsf{P}} * \boldsymbol{n}_{\mathsf{Q}}\big)$$

is determined for $\hat{\boldsymbol{\Lambda}}_{\tau\mathsf{AN}}$ and $\hat{\boldsymbol{\Lambda}}_{\tau\mathsf{BN}}$ as $\tilde{\vartheta} = \vartheta + \epsilon \cdot g = \hat{\boldsymbol{\Lambda}}_{\tau\mathsf{AN}} \circledast \hat{\boldsymbol{\Lambda}}_{\tau\mathsf{BN}}$. As a third step, the dual angle between the lines $\hat{\boldsymbol{\Lambda}}_{\mathsf{A}}$ and $\hat{\boldsymbol{\Lambda}}_{\mathsf{B}}$ with respect to $\hat{\boldsymbol{\Lambda}}_{\mathsf{N}}$ is obtained via the dual acos-function as

$$\text{acos}(\vartheta + \epsilon \cdot g) = \text{acos}(\vartheta) + \epsilon \cdot \frac{-g}{\sin(\text{acos}(\vartheta))} \,.$$

In the assumed context of generic axes constellations, a degenerate situation with $\text{acos}(\vartheta) = 0$ is disregarded. In spite of the considered case of pure rotational joints, a non-vanishing dual part indicates an infeasible solution triplet. As in the spherical case, the sign of the (real part of the dual) angle needs to be adapted to the orientation (10) of the unit directions, $\text{ornt3}(\hat{\boldsymbol{n}}_{\mathsf{A}}, \hat{\boldsymbol{n}}_{\mathsf{B}}, \hat{\boldsymbol{n}}_{\mathsf{N}})$.

The three steps are applied to the triplet of Plücker vectors $(\mathsf{A}, \mathsf{B}, \mathsf{N}) \cong (\boldsymbol{D}_{14}^{\text{Adl}} \cdot \hat{\boldsymbol{\Lambda}}_{34}, \boldsymbol{M}_{23}^{\text{Adl}} \cdot \hat{\boldsymbol{\Lambda}}_{34}, \hat{\boldsymbol{\Lambda}}_{12})$ to obtain an angle ϕ_{12} for the first joint, and to the triplet $(\mathsf{A}, \mathsf{B}, \mathsf{N}) \cong ((\boldsymbol{D}_{14}^{\text{Adl}})^{-1} \cdot \hat{\boldsymbol{\Lambda}}_{12}, (\boldsymbol{M}_{23}^{\text{Adl}})^{-1} \cdot \hat{\boldsymbol{\Lambda}}_{12}, \hat{\boldsymbol{\Lambda}}_{34})$ to obtain an angle ϕ_{34} for the third joint. The computations are executed twice: the adjoint matrix $\boldsymbol{M}_{23}^{\text{Adl}} = \text{adsp}(\phi_{23}; \hat{\boldsymbol{\Lambda}}_{23})$ is once computed for $\phi_{23}^{(+)}$ and once for $\phi_{23}^{(-)}$. For an admissible reference displacement $\boldsymbol{D}_{14}^{\text{Adl}}$, (at least) one of the obtained solution triplets, $(\phi_{12}^{(+)}, \phi_{23}^{(+)}, \phi_{34}^{(+)})$, and $(\phi_{12}^{(-)}, \phi_{23}^{(-)}, \phi_{34}^{(-)})$, provides an analytical solution to the finite inverse kinematics problem.

4 Conclusions

Three (6×6)-matrix operators associated to a 6D unit line vector are introduced that generalize three (3×3)-matrix operators associated to a 3D unit vector. Next to their applicability as projection operators, the matrices provide a means to state a novel, adjoint trigonometric form of the Rodrigues equation for the case of spatial displacements: the usability of the obtained finite-sum expression for the (6×6)-matrix exponential is demonstrated by employing it to obtain a screw-based analytical solution to the inverse kinematics problem of general 3R chains. Due to the structural similarity of the affine Rodrigues equation to its linear counterpart, the solution is obtained by applying the principle of transference to lift the approach for solving the spherical problem to the spatial case. In the future, variants, properties, and further applications of the introduced operators and identities are studied. The working directions include the position analysis and the workspace analysis of generic spatial 3R chains and of related designs.

References

1. Bongardt, B.: Geometric characterization of the workspace of nonorthogonal rotation axes. J. Geom. Mech. **6**(2), 141–166 (2014)
2. Gallier, J., Xu, D.: Computing exponentials of skew symmetric matrices and logarithms of orthogonal matrices. Int. J. Robot. Autom. **18**, 10–20 (2000)
3. Gelman, H.: A note on the time dependence of the effective axis and angle of a rotation. J. Res. Natl Bur. Stand. B. Math. Sci. **75B.3,4**, 165–171 (1971)

4. Husty, M.L., Ottaviano, E., Ceccarelli, M.: A geometrical characterization of workspace singularities in 3R manipulators. In: Advances in Robot Kinematics: Analysis and Design (2008)
5. Husty, M.L., Pfurner, M., Schröcker, H.P.: A new and efficient algorithm for the inverse kinematics of a general serial 6R manipulator. Mech. Mach. Theory **42**(1), 66–81 (2007)
6. Piovan, G., Bullo, F.: On coordinate-free rotation decomposition: Euler angles about arbitrary axes. IEEE Trans. Robot. **28**(3), 728–733 (2012)
7. Selig, J.M.: Geometric Fundamentals of Robotics. Springer, New York (2005)
8. Trainelli, L., Croce, A.: A comprehensive view of rotation parametrization. In: European Congress on Computational Methods in Applied Sciences and Engineering (ECCOMAS 2004) (2004)
9. Wenger, P.: Cuspidal and Noncuspidal Robot Manipulators. Robotica **25**(6), 677–689 (2007)

Analytical Determination of a Sphere Inside Which the Stewart Platform Translates Without Suffering Any Leg Interference

Anirban Nag and Sandipan Bandyopadhyay(✉)

Department of Engineering Design, Indian Institute of Technology Madras, Chennai, India
nag.anirban16@gmail.com, sandipan@iitm.ac.in,
https://ed.iitm.ac.in/~sandipan

Abstract. This paper presents a formulation to identify a sphere in the fixed-orientation workspace of a Stewart platform manipulator, such that no pair of legs of the platform would suffer an interference between them, if the geometric centre of the moving platform remained inside this sphere. Identification of such a sphere is an important step in the design and analysis of such manipulators. The geometric formulation leads to certain algebraic equations, and finally to the solution of a degree 8 univariate equation. The mathematical formulation is applied to two different situations, and the results obtained are verified by comparison with those obtained from an established open-source library designed to detect interference between solids.

Keywords: Leg-interference · Parallel manipulator

1 Introduction

Determination of the safe working zone (SWZ) of a parallel manipulator (PM) is an important step in its analysis and design [4]. To compute the SWZ of a PM, one has to first find a subset of its workspace that is free of the gain-type as well as the loss-type singularities, and further, obtain regions inside this space in which the *end-effector*/moving platform of the manipulator can move in a manner such that it does not encounter the issues of interference among the links of the manipulator as well as physical limits on the motion of the joints [8]. For the semi-regular Stewart platform manipulator (SRSPM), a sphere has been found in its constant-orientation workspace in [7], such that the manipulator is safe from the gain-type singularities as long as the centre of the moving platform (denoted by $\boldsymbol{p}_c$) remains inside the said sphere. A further step is taken towards computing the SWZ in this paper. A new sphere is computed with the same centre as the singularity-free sphere, such that the legs of the manipulator are free of interference while $\boldsymbol{p}_c$ is confined to this sphere.

J. Lenarcic and V. Parenti-Castelli (Eds.): ARK 2018, SPAR 8, pp. 74–82, 2019.
https://doi.org/10.1007/978-3-319-93188-3_9

The problem of leg-interference has been studied by Merlet et al. [6] using the technique of interval analysis. Merlet has also presented a formulation for the detection of leg-interference via the computation of the least distance between the "centrelines" of the legs, which are approximated as cylinders [5]. Various approaches to the detection of interference between two cylinders in space and the computation of the shortest distance between them have been proposed in [1,3,10]. However, none of the reported works identifies such interference-free regions analytically, as done in this paper, to the best of the knowledge of the authors.

The following steps are followed to obtain the said sphere. First, the legs are approximated geometrically as cylinders (of identical diameter), whose axes coincide with the kinematic axes of the respective legs. The condition of interest is one in which the cylinder corresponding to one leg is *tangential* to another. Equivalently, the limiting condition is that the *shortest distance between the axes of the said cylinders equal the sum of the radii of the cylinders.* This condition defines certain *quadratic* surfaces (denoted by $S_3(x, y, z) = 0$, where x, y, z are the coordinates of the point $\boldsymbol{p}_c$) in the translation workspace of the SPM, whose nature varies depending on the occurrence of the shortest distance at the end-points of one or both of the axes, or none. In the next and final step, the smallest sphere is found such that it is tangential to the surface S_3, with its centre at a given point of interest. An 8-degree polynomial needs to be solved at this stage, the solutions of which determine the points of tangency of the said sphere with the surface S_3, thereby determining its radius. It may be noted that the final result pertains to the smallest of the spheres, corresponding to *all* the real roots of the subproblem described above. This exercise needs to be repeated 15 times, to cover all the 15 distinct pairs of legs exhaustively.

The formulation is illustrated with the help of two numerical examples. The steps are documented in detail, so as to explain the different geometric situations that may arise. Also, an independent numerical validation is carried out in the following manner: a large number (about thirty thousand) of locations of $\boldsymbol{p}_c$ are generated in a random manner, and the afore-mentioned 15 combinations of legs are checked for interference, for each of these points, using the open-source collision-detection tool, `Bullet 3` [2]. It is found, in all the examples, that all the locations associated with the leg-interference lie on the surface S_3, and outside the sphere identified to be safe (from the perspective of leg interference), as expected.

The rest of the paper is organised as follows: Sect. 2 presents the formulation of the problem. Numerical examples have been presented in Sect. 3. The paper is concluded in Sect. 4.

2 Problem Formulation

The links of the SPM are assumed to be bounded in cylinders, as shown in Fig. 1, so as to ease the detection of interference between the legs. It is known that the least distance between two cylinders is given by the length of the common

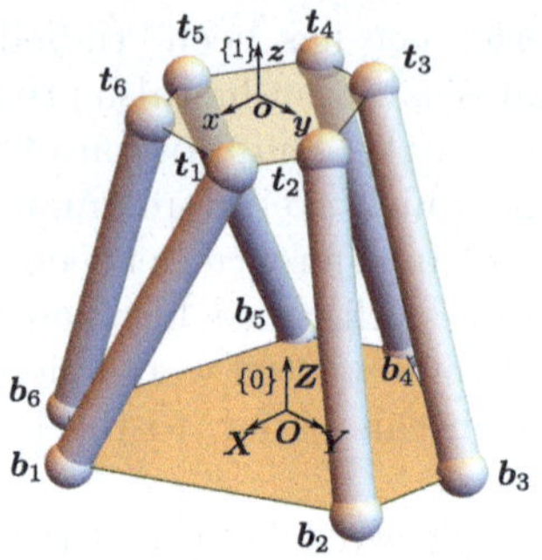

Fig. 1. Schematic representation of the SRSPM

perpendicular of their axes, minus the sum of the radii of the cylinders. However, in this context, the axes must be treated as *line segments* and not lines, as the cylinders are finite in size. Any point ${}^0\boldsymbol{p}_i(u)$ on the axis of the ith leg (denoted by L_i) is given by:

$$ {}^0\boldsymbol{p}_i(u_i) = {}^0\boldsymbol{b}_i + ({}^0\boldsymbol{t}_i - {}^0\boldsymbol{b}_i)u_i. \tag{1}$$

In Eq. (1), ${}^0\boldsymbol{b}_i$ are the vertices of the (fixed) base platform and ${}^0\boldsymbol{t}_i$ the corresponding vertices on the moving platform, expressed in the fixed frame $\{0\}$ as: ${}^0\boldsymbol{t}_i = {}^0\boldsymbol{p}_c + {}^0_1\boldsymbol{R}^1\boldsymbol{t}_i$, $i = 1, \ldots, 6$, where ${}^1\boldsymbol{t}_i$ expresses the vertices of the moving platform in the moving frame, $\{1\}$; ${}^0\boldsymbol{p}_c = [x, y, z]^\top$ denotes the location of the geometric centre of the moving platform in $\{0\}$, and ${}^0_1\boldsymbol{R}$ defines the *rotation* matrix relating the frame $\{1\}$ to the frame $\{0\}$, which is parametrised in terms of *Rodrigues parameters*, denoted by $\{c_1, c_2, c_3\}$.

The distance between a pair of points[1] $\boldsymbol{p}_i(u_i) \in L_i$ and $\boldsymbol{p}_j(u_j) \in L_j$, is defined as: $d^2(u_i, u_j) = (\boldsymbol{p}_i(u_i) - \boldsymbol{p}_j(u_j)) \cdot (\boldsymbol{p}_i(u_i) - \boldsymbol{p}_j(u_j))$. The minimum of d occurs at u_1^* and u_2^*, which satisfy the equations:

$$ \frac{\partial d(u_i^*, u_j^*)}{\partial u_i} = 0, \quad \frac{\partial d(u_i^*, u_j^*)}{\partial u_j} = 0. \tag{2}$$

Geometrically, $\boldsymbol{p}_i^*(u_i^*)$ and $\boldsymbol{p}_j^*(u_j^*)$ locate the *feet of the common perpendicular* on the two lines, and define the desired shortest distance as $d^* = ||\boldsymbol{p}_i^*(u_i^*) - \boldsymbol{p}_j^*(u_j^*)||$. Since the scope of the problem is limited to a fixed orientation of the moving platform, d^* is a function of (x, y, z), and the critical surface is obtained when d^* equals twice the radius of the cylinders bounding the legs (denoted by r):

$$ S_3 := {d^*}^2(x, y, z) - 4r^2 = 0. \tag{3}$$

The surface S_3 turns out to be *quadratic* in x, y, z, which is obtained in the *closed-form* as functions of the architecture and the orientation parameters.

[1] The leading superscript "0" is omitted henceforth as all the points are expressed in $\{0\}$ by default.

The next task is to obtain the *smallest* sphere, S_p, of radius r_0, centred at a given point of interest, $\boldsymbol{p}_0 = [x_0, y_0, z_0]^\top$, which is tangential to this surface:

$$S_p := (x - x_0)^2 + (y - y_0)^2 + (z - z_0)^2 - r_0^2 = 0. \tag{4}$$

The gradients to the surfaces would align at the points of tangency, leading to:

$$\nabla S_p \times \nabla S_3 = \mathbf{0} \Rightarrow g_i(x, y, z) = 0, \quad i = 1, 2, 3. \tag{5}$$

Equation (5) yields three scalar equations, of which only *two* are independent. Selecting any two of the three equations and augmenting these to Eq. (3), a system of three quadratic equations in the three unknowns are obtained. The variables y, z are eliminated from these equations, to obtain a degree 16 univariate equation in x, which contains the spurious component $x^8 = 0$, arising as an artifact of the process of elimination (see Eq. (9)). The details are omitted for the want of space, but some specific examples are described in Sect. 3. Suffice it to note at this point that the smallest *real* value of r_0 (computed from Eq. (4)) defines the sphere, free of link interferences (between the chosen pair of legs).

The above procedure covers only one subset of the possible geometric scenarios, namely, the case where L_i are *infinite lines*, i.e., $u_1^*, u_2^* \in \mathbb{R}$. However, the physically meaningful range for u_1^*, u_2^* is $[0, 1]$, as the cylinders are finite (see Eq. (1)). Therefore, when $\mathbb{R} \ni u_1^*, u_2^* \notin [0, 1]$, the physically meaningful shortest distance occurs at the *end-points* of L_i and/or L_j, i.e, at $\boldsymbol{b}_i$ and/or $\boldsymbol{t}_j$, $i, j = 1, 2$, $i \neq j$. Additional checks need to be incorporated in the computational algorithm, to ensure an exhaustive coverage of all the *physically meaningful* cases. Finally, all the computations done on a pair of legs need to be repeated to cover all the 15 distinct pairs of legs of an SPM, and the smallest real value from all of these computations finally defines S_p. While the details cannot be included in this paper for the want of space, it is attempted to convey the concepts to the reader through the numerical examples in the following section, which cover multiple geometric cases.

3 Numerical Examples

The theory developed is illustrated via two numerical examples. The manipulator used in both the examples is an SRSPM. Radius of the circum-circle of the base platform is scaled to unity, without any loss of generality, thus rendering all the linear dimensions unit-less. Radius of the circum-circle of the moving platform is 0.5, and the radius of the leg-cylinders is $r = 0.1$. For both the examples the centre of the desired sphere is chosen to be at $\boldsymbol{p}_0 = [0, 0, 1]^\top$. The orientation of the manipulator is chosen differently, so as to illustrate two distinct geometric situations. The first example presents the case where two legs have a point of tangency on their curved surfaces, whereas the second example depicts the situation where the *end-cap* of one cylinder touches the cylindrical surface of the other tangentially. The stages of computation are presented in detail to facilitate an understanding of the method. The entire analysis has been performed inside the computer algebra system `Mathematica 10.4` [9], which is governed by its internal variables `MachinePrecision` and `Accuracy`, both of which are set to 16, ensuring that the computed numbers are accurate up to at least 16 digits after the decimal point.

3.1 Example 1: $u_1^*, U_2^* \in [0, 1]$

This example pertains to the orientation of the moving platform given by the Rodrigues parameters $c_1 = 0.124$, $c_2 = 0.187$, $c_3 = 0.556$. Following Sect. 2, the

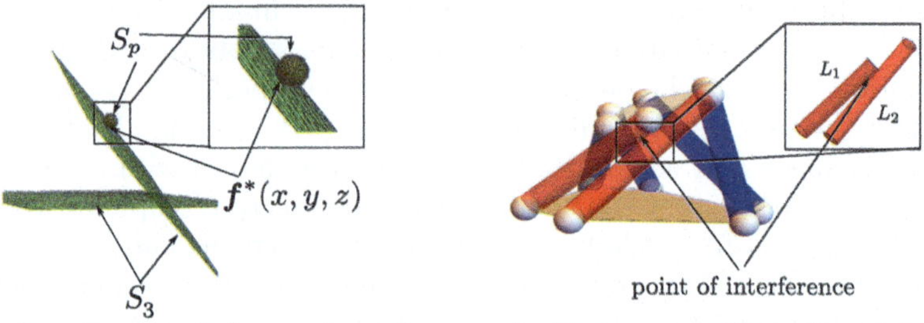

a: Safe sphere and the surface S_3 b: Tangency of legs 1 and 2

Fig. 2. Example 1: tangency of the cylinders occurs inside their finite spans

required equations in three unknowns x, y, z are obtained as:

$$S_3(x, y, z) := \alpha_0 + \alpha_1 x + \alpha_2 x^2 + \alpha_3 y + \alpha_4 xy + \alpha_5 y^2 + \alpha_6 z + \alpha_7 xz \ldots + \alpha_8 yz + \alpha_9 z^2 = 0, \quad (6)$$

$$g_1(x, y, z) := \beta_0 + \beta_1 x + \beta_2 y + \beta_3 xy + \beta_4 y^2 + \beta_5 z + \beta_6 z + \beta_7 xz \ldots + \beta_8 yz + \beta_9 z^2 = 0, \quad (7)$$

$$g_3(x, y, z) := \gamma_0 x + \gamma_1 x^2 + \gamma_2 y + \gamma_3 xy + \gamma_4 y^2 + \gamma_5 xz + \gamma_6 yz = 0. \quad (8)$$

The coefficients $\alpha_i, \beta_i, \gamma_i$, are functions of the architecture and the orientation parameters. It is observed that g_3 is linear in z, which is solved for first, and substituted in the remaining two equations, which become quartics in x, y. Eliminating y between these equations by computing their resultant w.r.t. y, a univariate equation in x is obtained, with the following structure :

$$x^8(\zeta_0 x^8 + \ldots + \zeta_8) = 0, \quad (9)$$

where ζ_i, are functions of the architecture and the orientation parameters. The monic form of the non-trivial factor, when equated to zero, leads to the equation:

$$x^8 - 5.514x^7 + 11.546x^6 - \ldots + 0.009 \times 10^{-5}x - 0.194 \times 10^{-3} = 0. \quad (10)$$

The solutions of Eqs. (6–8) are computed for all the real roots of Eq. (10), leading to the numerical results tabulated in Table 1.

Table 1. Numerical results for Example 1

No.	x	y	z	r_0
1	2.114	1.765	−0.707	3.240
2	0.017	−0.356	0.046	1.017
3	0.608	−0.665	0.177	1.221
4	0.601	−0.671	0.175	1.221
5	0.207	−1.305	0.405	1.449
6	−0.065	0.014	0.928	**0.098**

The radius of the minimal sphere is found to be 0.098 for the tangency of legs 1 and 2. The point of tangency is found as $\boldsymbol{f}^* = [-0.065, 0.014, 0.928]^\top$ (see Fig. 3). The corresponding line parameters are found to be $u_1^* = 0.947$, and $u_2^* = 0.609$, hence the tangency occurs in the spans of the curved surfaces of the cylinders, as can be seen in Fig. 2b. None of the remaining pairs of legs interfere in this case.

a: Link-interference free sphere of the SRSPM

b: Close-up of the tangency between the legs

Fig. 3. Example 2: closest approach of the legs involves the end-point of the axis of one of the cylinders, i.e., $\boldsymbol{t}_1$ in this case.

3.2 Example 2: $u_1^*, u_2^* \notin [0, 1]$

The orientation parameters in this example are: $c_1 = 0.063$, $c_2 = 0.110$, $c_3 = 0.259$. In this case, too, the computation follows the same initial steps as in Example 1. The pair of legs in question are also the same, i.e., legs 1 and 2. However, it is found that the smallest sphere of radius $r_0 = 0.424$ corresponds to $u_1^* = 1.59$ and $u_2^* = 1.02$, i.e., $u_1^*, u_2^* \notin [0, 1]$. Therefore, the tangency of the cylinders does *not* happen within their finite spans, and hence one needs to check the end-points of each cylinder, against the entire span as well as the end-points of the other cylinder, to compute their closest approach[2]. A different

[2] In this situation, the least distance is acutally lesser than the sum of the radii. However, Eq. (3) may still be used, since the result is guaranteed to be conservative.

set of equations are derived for the same, and solved, leading to the numerical results summarised in Table 2. From Table 2, it can be observed that only in cases 1 and 4, the closest approaches happen either inside or at the extremity of the finite spans of the cylinders. Of these, the first case, in which the point t_1 is closest to the line segment L_2 at $u_2^* = 0.561 \in [0, 1]$, leads to the required result, with a r_0 value of 0.541.

Table 2. Results for checks of the end-points against the infinite spans of the cylinder axes

Case	End-point	Line (segment)	r_0	u_1^*/u_2^*
1	$\boldsymbol{t}_1(u_1 = 1)$	L_2	**0.541**	$u_2^* = \mathbf{0.561}$
2	$\boldsymbol{t}_2(u_2 = 1)$	L_1	0.425	$u_1^* = 1.565$
3	$\boldsymbol{b}_1(u_1 = 0)$	L_2	1.255	$u_2^* = -0.827$
4	$\boldsymbol{b}_2(u_2 = 0)$	L_1	0.830	$u_1^* = 0.757$

Validation of the Results. In the following, the numerical results of Example 2 are validated against a popular open-source collision-detection library, namely, `Bullet 3`. A cloud of 30,000 points (i.e., possible locations of $\boldsymbol{p}_c(x, y, z)$) is generated randomly inside a sphere of radius 0.56, which is concentric with S_p, but intensionally larger. It is verified, that for no point *inside* the *safe* sphere S_p, there is any interference/tangency between the cylinders covering the legs. As expected, the points *outside* S_p lead to interference among at least one of the pairs of cylinders, *iff* the point lies on (i.e., $S_3 = 0$) or beneath/inside (i.e., $S_3 < 0$) the surface S_3 (see Fig. 4).

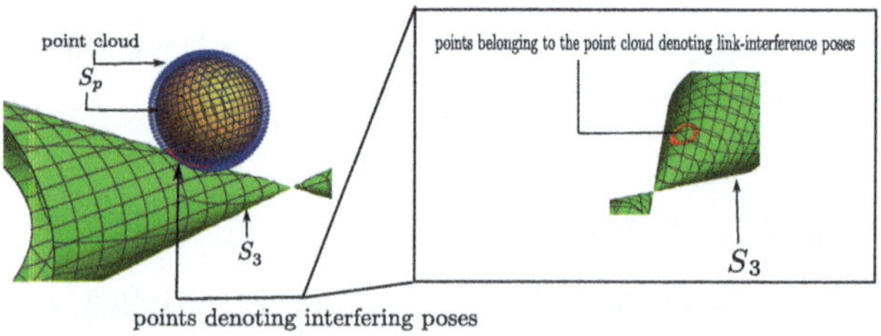

Fig. 4. Example 2: Validation of the results

3.3 Discussions

This work does not cover a number of associated issues, e.g., the possibility of one or more of the legs interfering with the platforms, or with the spherical joints. Also, a number of special cases arise in the mathematical formulation, which need to be studied further. For example, while detecting the interference of the end-cap of one cylinder with the infinite cylindrical surface of the other, the resulting final univariate polynomial equation decomposes into multiple factors, thereby reducing the individual degrees of the polynomial equations to be solved in this case. Further, the method is demonstrated on an SRSPM, but it is equally applicable to SPMs of *any* architecture. One additional step that is required in the algorithm is to check that the sign of S_3 is positive, when it is evaluated at $\boldsymbol{p}_0$. A failure in this check implies that the centre of the desired sphere itself is associated with an interference, and hence the rest of the computation is futile.

4 Conclusion

This paper presents a mathematical formulation to identify a sphere in the constant-orientation workspace of a Stewart platform manipulator, such that the legs of the manipulator are *provably* free of interference, so long as the geometric centre of the moving platform remains inside the said sphere. Using very simple geometric concepts, algebraic equations are derived and solved, to analyse all the possible cases exhaustively. The computation involves the solution of a polynomial of degree 8 and another of degree 6 in some cases (for the end-point against the infinite line case), making it very efficient, and ideally suited for incorporation in a design algorithm. The formulation is explained in detail by application to two different examples. Also, the results are numerically validated against those obtained from an open-source collision-detection library, namely, `Bullet 3`, showing perfect results in 30,000 random test cases.

References

1. Agarwal, S., Srivatsan, R.A., Bandyopadhyay, S.: Analytical determination of the proximity of two right-circular cylinders in space. J. Mech. Robot. **8**(4), 041010 (2016)
2. Coumans, E.: Bullet physics simulation. In: ACM SIGGRAPH 2015 Courses, SIGGRAPH 2015. ACM, New York (2015)
3. Ketchel, J.S., Larochelle, P.M.: Line based collision detection of cylindrical rigid bodies. In: ASME 2004 International Design Engineering Technical Conferences and Computers and Information in Engineering Conference, pp. 1261–1271. American Society of Mechanical Engineers (2004)
4. Kilaru, J., Karnam, M.K., Agarwal, S., Bandyopadhyay, S.: Optimal design of parallel manipulators based on their dynamic performance. In: Proceedings of the 14th IFToMM World Congress, 25–30 October, Taiwan, pp. 406–412 (2015)
5. Merlet, J.P.: Geometrical determination of the workspace of a constrained parallel manipulator. In: Advances in Robot Kinematics, Ferrare, Italy, pp. 7–14, September 1992
6. Merlet, J.P., Daney, D.: Legs interference checking of parallel robots over a given workspace or trajectory. In: Proceedings of the IEEE International Conference on Robotics and Automation, pp. 757–762 (2006)

7. Nag, A., Reddy, V., Agarwal, S., Bandyopadhyay, S.: Identifying singularity-free spheres in the position workspace of semi-regular Stewart platform manipulators. In: Lenarčič, J., Merlet, J.P. (eds.) Advances in Robot Kinematics 2016, pp. 421–430. Springer, Cham (2018)
8. Srivatsan, R.A., Bandyopadhyay, S.: Determination of the safe working zone of a parallel manipulator. In: Thomas, F., Perez Gracia, A. (eds.) Proceedings of the 6th International Workshop on Computational Kinematics (CK2013), pp. 201–208. Springer Netherlands, Dordrecht (2014)
9. Wolfram Research, Inc.: Mathematica, Version 10.4. Champaign, IL (2016)
10. Zsombor-Murray, P.: Intrusion, proximity and stationary distance. In: Zeghloul, S., Romdhane, L., Laribi, M.A. (eds.) Computational Kinematics: Proceedings of the 7th International Workshop on Computational Kinematics, Futuroscope-Poitiers, France, May 2017, pp. 475–482. Springer, Cham (2018)

Higher-Order Relative Kinematics of Rigid Body Motions: A Dual Lie Algebra Approach

Daniel Condurache(✉)

Technical University of Iasi, D. Mangeron Street no. 59, 700050 Iasi, Romania
daniel.condurache@tuiasi.ro

Abstract. In this paper, using the tensor calculus and the dual numbers algebra, a new computing method for studying the higher-order accelerations field properties is proposed in the case of the general rigid body motion. For the spatial kinematic chains, equations that allow the determination of the nth order accelerations field are given, using a Brockett-like formula specific to the dual algebra. The results are coordinate-free and in a closed form. In particular cases, the properties for velocities, accelerations, jerks and jounces fields are given. This approach uses the isomorphism between the Lie group of the rigid displacements $S\mathbb{E}_3$ and the Lie group of the orthogonal dual tensors.

Keywords: Higher-order kinematics · Dual algebra · Lie group

1 Introduction

The kinematic analysis of multibody systems has been traditionally considered as the determination of the positions, velocities, accelerations, jerks and jounces of their constitutive members. This is an old field with a long history, which has attracted the attention of mathematicians and engineers. Michel Chasles discovered (1834) that any rigid body displacement is equivalent to a screw displacement [5]. Screw theory is an efficient mathematical tool for the study of spatial kinematics. The pioneering work of Ball [2], the treatises of Hunt [6], and Phillips [10] and the multitude of contributions appearing in the literature are evidence of this. The isomorphism between screw theory and the Lie algebra, se(3), of the Special Euclidean group, $S\mathbb{E}_3$, provide with a wealth of results and techniques from modern differential geometry and Lie group theory [3, 7, 8, 12, 13].

A kinematic mapping relates the motion of a rigid body to the joint motions of a kinematic chain. Its time derivatives yield the twist, acceleration, jerk and jounce etc. of the body. Time derivatives of the twists of members in a kinematic chain and derivatives of screws are essential operations in kinematics. Recognizing the Lie group nature of rigid body motions, and correspondingly the Lie algebra nature of screws, Karger [7], Rico et al. [11], Lerbet [8] and Müller [9] derived closed form expressions of higher-order time derivatives of twist.

In this paper, using the tensor calculus and the dual numbers algebra, a new computing method for studying the higher-order accelerations field properties is proposed in the case of the general rigid body motion. For the spatial kinematic chains, equations that allow the determination of the nth order accelerations field are given,

J. Lenarcic and V. Parenti-Castelli (Eds.): ARK 2018, SPAR 8, pp. 83–91, 2019.
https://doi.org/10.1007/978-3-319-93188-3_10

using a Brockett-like formula. The crucial observation is that the nth order time derivative of twist of the terminal body in a kinematic chain can be determined by propagating the kth order time derivative of twists of the bodies in the chain, for $k = \overline{0,n}$. The results are coordinate-free and in a closed form.

2 Higher-Order Kinematics of the Rigid Body

The general framework of this section is a rigid body that moves with respect to a fixed reference frame $\{\mathbb{R}^0\}$. Consider another reference frame $\{\mathbb{R}\}$ originated at a point Q and attached to the rigid body. Let us note by $\boldsymbol{\rho}$ and $\mathbf{r}$ the position vectors of an arbitrary point P of the rigid body with respect to the $\{\mathbb{R}^0\}$ and respectively $\{\mathbb{R}\}$. On may write:

$$\boldsymbol{\rho} = \boldsymbol{\rho}_Q + \mathbf{R}\mathbf{r} \tag{1}$$

where the constant vector $\mathbf{r}$ is the relative position vector of point P with respect to the reference frame $\{\mathbb{R}\}$.

The parametric equations of the rigid body motion is:

$$\begin{cases} \boldsymbol{\rho}_Q = \boldsymbol{\rho}_Q(t) \\ \mathbf{R} = \mathbf{R}(t) \end{cases}, t \in \mathbf{I} \in \mathbb{R}, \tag{2}$$

where $\boldsymbol{\rho}_Q = \boldsymbol{\rho}_Q(t) \in \mathbf{V}_3, \mathbf{R} = \mathbf{R}(t) \in S\mathbb{O}_3$. Let $\mathbf{V}_3$ be the set of vectors from the three-dimensional Euclidean space and $S\mathbb{O}_3$ the set of proper-orthogonal tensors.

The nth order acceleration of that point of the rigid body located by the position vector $\boldsymbol{\rho}$ is denoted by $\mathbf{a}_{\boldsymbol{\rho}}^{[n]}, n \in \mathbb{N}^*$:

$$\mathbf{a}_{\boldsymbol{\rho}}^{[n]} \overset{\text{def}}{=} \frac{\mathrm{d}^n \boldsymbol{\rho}}{\mathrm{d}t^n} = \boldsymbol{\rho}^n = \mathbf{a}_Q^{[n]} + \boldsymbol{\phi}_n[\boldsymbol{\rho} - \boldsymbol{\rho}_Q], n \in \mathbb{N}^*, \tag{3}$$

where $\mathbf{a}_Q^{[n]}$ is the nth order acceleration of the bodys' fixed point Q and $\boldsymbol{\phi}_n = \mathbf{R}^{(n)}\mathbf{R}^T$ represents the nth order accelerations tensor. We will denote $\tilde{\boldsymbol{\omega}} = \boldsymbol{\phi}_1 = \dot{\mathbf{R}}\mathbf{R}^T$ the skew-symmetric tensor of instantaneous angular velocity. Let $\mathbf{v} = \dot{\boldsymbol{\rho}}_Q - \boldsymbol{\omega} \times \boldsymbol{\rho}_Q$ be the velocity of a point of the rigid body which at a given moment of time passes through the origin of the reference frame $\{\mathbb{R}^0\}$.

The following theorem holds:

Theorem 1. *The nth order accelerations field of the rigid body motion is given by*

$$\mathbf{a}_{\boldsymbol{\rho}}^{[n]} = \mathbf{a}_n + \boldsymbol{\phi}_n\boldsymbol{\rho}, n \in \mathbb{N}^* \tag{4}$$

where the vectors $\mathbf{a}_n$, called the nth order vector invariant, and the tensors $\boldsymbol{\phi}_n$ satisfy the following relationships of recurrence:

$$\begin{cases} \mathbf{a}_{n+1} = \dot{\mathbf{a}}_n + \boldsymbol{\phi}_n \mathbf{v} \\ \mathbf{a}_1 = \mathbf{v} \end{cases}, n \in \mathbb{N}^* \tag{5}$$

$$\begin{cases} \boldsymbol{\phi}_{n+1} = \dot{\boldsymbol{\phi}}_n + \boldsymbol{\phi}_n \tilde{\boldsymbol{\omega}} \\ \boldsymbol{\phi}_1 = \tilde{\boldsymbol{\omega}} \end{cases}, n \in \mathbb{N}^* \tag{6}$$

Proof: Taking into account that $\mathbf{a}_{\boldsymbol{\rho}}^{[n]} - \boldsymbol{\phi}_n \boldsymbol{\rho} = \mathbf{a}_{\mathbf{Q}}^{[n]} - \boldsymbol{\phi}_n \boldsymbol{\rho}_Q \stackrel{\text{def}}{=} \mathbf{a}_n, \forall n \in \mathbb{N}^*$ and $\forall \boldsymbol{\rho} \in \mathbf{V}_3$, the Eq. (5) results from Eq. (3).

We shall further define the spatial twist of rigid body by a pair of vectors $(\boldsymbol{\omega}, \mathbf{v})$.

Let $\mathbf{L}(\mathbf{V}_3, \mathbf{V}_3)$ be the set of the Euclidean tensors, defined as $\mathbb{R}$-linear applications from $\mathbf{V}_3$ to $\mathbf{V}_3$. With respect to the addition and multiplication of tensors, the set $\mathbf{L}(\mathbf{V}_3, \mathbf{V}_3)$ form a non-commutative unitary ring denoted by $\mathcal{A}$. Let $\mathcal{A}[X]$ be the set of the polynomials with coefficients in the non-commutative ring $\mathcal{A}$. A generic element of $\mathcal{A}[X]$ has the form:

$$\mathbf{P}(X) = \mathbf{A}_0 + \mathbf{A}_1 X + \cdots + \mathbf{A}_m X^m, \mathbf{A}_k \in \mathcal{A}, k = \overline{0, m} \tag{7}$$

The following theorem holds:

Theorem 2. *There is a unique polynomial with the coefficients in the non-commutative ring of Euclidean tensors* $\mathbf{L}(\mathbf{V}_3, \mathbf{V}_3)$ *such that the vector respectively the tensor invariants of the* n^{th} *order accelerations may be written as:*

$$\begin{matrix} \mathbf{a}_n = \mathbf{P}_n \mathbf{v} \\ \boldsymbol{\Phi}_n = \mathbf{P}_n \tilde{\boldsymbol{\omega}} \end{matrix}, n \in \mathbb{N}^* \tag{8}$$

where $\mathbf{P}_n$ *fulfills the relationship of recurrence, with* $\mathbf{D} = \frac{d}{dt}$ *the time derivative operator*:

$$\begin{cases} \mathbf{P}_{n+1} = \mathbf{D}\mathbf{P}_n + \mathbf{P}_n(\tilde{\boldsymbol{\omega}}), n \in \mathbb{N}^* \\ \mathbf{P}_1 = \mathbf{I} \end{cases} \tag{9}$$

It follows that:

$$\begin{gathered} \mathbf{P}_1 = \mathbf{I} \\ \mathbf{P}_2 = \mathbf{D} + \tilde{\boldsymbol{\omega}} \\ \mathbf{P}_3 = \mathbf{D}^2 + \tilde{\boldsymbol{\omega}}\mathbf{D} + 2\dot{\tilde{\boldsymbol{\omega}}} + \tilde{\boldsymbol{\omega}}^2 \\ \mathbf{P}_4 = \mathbf{D}^3 + \tilde{\boldsymbol{\omega}}\mathbf{D}^2 + (3\dot{\tilde{\boldsymbol{\omega}}} + \tilde{\boldsymbol{\omega}}^2)\mathbf{D} + \ddot{\tilde{\boldsymbol{\omega}}} + 3\dot{\tilde{\boldsymbol{\omega}}}\tilde{\boldsymbol{\omega}} + 2\tilde{\boldsymbol{\omega}}\dot{\tilde{\boldsymbol{\omega}}} + \tilde{\boldsymbol{\omega}}^3 \end{gathered} \tag{10}$$

From Eqs. (8) and (10) written for the particular cases $n = \overline{1, 4}$, it follows:

- the velocity field invariants:

$$\begin{cases} \mathbf{a}_1 = \mathbf{v} \\ \boldsymbol{\Phi}_1 = \tilde{\boldsymbol{\omega}} \end{cases} \tag{11}$$

- the acceleration field invariants:

$$\begin{cases} \mathbf{a}_2 = \dot{\mathbf{v}} + \tilde{\boldsymbol{\omega}}\mathbf{v} \\ \boldsymbol{\Phi}_2 = \dot{\tilde{\boldsymbol{\omega}}} + \tilde{\boldsymbol{\omega}}^2 \end{cases} \tag{12}$$

- the jerk field invariants:

$$\begin{cases} \mathbf{a}_3 = \ddot{\mathbf{v}} + \tilde{\boldsymbol{\omega}}\dot{\mathbf{v}} + 2\dot{\tilde{\boldsymbol{\omega}}}\mathbf{v} + \tilde{\boldsymbol{\omega}}^2\mathbf{v} \\ \boldsymbol{\Phi}_3 = \ddot{\tilde{\boldsymbol{\omega}}} + \tilde{\boldsymbol{\omega}}\dot{\tilde{\boldsymbol{\omega}}} + 2\dot{\tilde{\boldsymbol{\omega}}}\tilde{\boldsymbol{\omega}} + \tilde{\boldsymbol{\omega}}^3 \end{cases} \tag{13}$$

- the jounce field invariants:

$$\begin{cases} \mathbf{a}_4 = \dddot{\mathbf{v}} + \tilde{\boldsymbol{\omega}}\ddot{\mathbf{v}} + \left(3\dot{\tilde{\boldsymbol{\omega}}} + \tilde{\boldsymbol{\omega}}^2\right)\dot{\mathbf{v}} + \ddot{\tilde{\boldsymbol{\omega}}}\mathbf{v} + 3\dot{\tilde{\boldsymbol{\omega}}}\tilde{\boldsymbol{\omega}}\mathbf{v} + 2\tilde{\boldsymbol{\omega}}\dot{\tilde{\boldsymbol{\omega}}}\mathbf{v} + \tilde{\boldsymbol{\omega}}^3\mathbf{v} \\ \boldsymbol{\Phi}_4 = \dddot{\tilde{\boldsymbol{\omega}}} + \tilde{\boldsymbol{\omega}}\ddot{\tilde{\boldsymbol{\omega}}} + \left(3\dot{\tilde{\boldsymbol{\omega}}} + \tilde{\boldsymbol{\omega}}^2\right)\dot{\tilde{\boldsymbol{\omega}}} + \ddot{\tilde{\boldsymbol{\omega}}}\tilde{\boldsymbol{\omega}} + 3\dot{\tilde{\boldsymbol{\omega}}}\tilde{\boldsymbol{\omega}}^2 + 2\tilde{\boldsymbol{\omega}}\dot{\tilde{\boldsymbol{\omega}}}\tilde{\boldsymbol{\omega}} + \tilde{\boldsymbol{\omega}}^4 \end{cases} \tag{14}$$

Remark 1. The higher-order time derivative of spatial twist completely solves the problem of determining the field of the nth order accelerations of rigid body motion.

3 Higher–Order Kinematics in Dual Lie Algebra

Let the rigid body motion be given by the following parametric equation in a given reference frame:

$$\begin{cases} \boldsymbol{\rho} = \boldsymbol{\rho}(t) \in V_3 \\ \mathbf{R} = \mathbf{R}(t) \in SO_3 \end{cases}, \tag{15}$$

with $t \in \mathbf{I} \subseteq \mathbb{R}$ is the time variable.

The orthogonal dual tensor that describes the rigid body motion is [1, 4, 12]:

$$\underline{\mathbf{R}} = (\mathbf{I} + \varepsilon\widetilde{\boldsymbol{\rho}})\mathbf{R} \tag{16}$$

In the relation (16), the skew symmetric tensor associated to the vector $\boldsymbol{\rho}$ is denoted by $\widetilde{\boldsymbol{\rho}}$, and ε is the dual quantity, $\varepsilon^2 = 0$, $\varepsilon \neq 0$ [1, 4, 12]. It can be demonstrated that there exists a unit dual vector $\underline{\mathbf{u}}$ and a dual angle $\underline{\alpha}$ such that: $\underline{\mathbf{R}} = \underline{\mathbf{I}} + \sin\underline{\alpha}\widetilde{\underline{\mathbf{u}}} + (1 - \cos\underline{\alpha})\widetilde{\underline{\mathbf{u}}}^2$. We will denote with $\underline{\mathbf{V}}_3$ the set of dual vectors.

The tensors $\underline{\boldsymbol{R}}$ form a Lie group denoted by $\underline{\mathbb{SO}}_3 = \{\underline{\mathbf{R}} \in \mathbf{L}(\underline{\mathbf{V}}_3, \underline{\mathbf{V}}_3) | \underline{\mathbf{R}}\underline{\mathbf{R}}^T = \underline{\mathbf{I}}, \det\underline{\mathbf{R}} = 1.\}$. The Lie group $\underline{\mathbb{SO}}_3$ is isomorphic to the Lie group of the rigid displacement $\mathbb{SE}_3$ [4]. The corresponding Lie algebra of the group $\underline{\mathbb{SO}}_3$ is isomorphic to the dual vector algebra $\underline{\mathbf{V}}_3$, having as internal operation the vector product for dual vectors. The rigid motion given by the Eq. (15) is a parameterized curve: $\underline{\mathbf{R}} = \underline{\mathbf{R}}(t)$ in the $\underline{\mathbb{SO}}_3$ Lie group.

The dual angular velocity for the rigid body motion (15) is given by [4]:

$$\underline{\boldsymbol{\omega}} = \mathrm{vect}\dot{\underline{\mathbf{R}}}\underline{\mathbf{R}}^T \tag{17}$$

It can be demonstrated that:

$$\underline{\boldsymbol{\omega}} = \boldsymbol{\omega} + \varepsilon \mathbf{v} \tag{18}$$

where $\boldsymbol{\omega} = \mathrm{vect}\dot{\mathbf{R}}\mathbf{R}^T$ is the instantaneous angular velocity of the rigid body and $\mathbf{v} = \dot{\boldsymbol{\rho}} - \boldsymbol{\omega} \times \boldsymbol{\rho}$ is the linear velocity of a point of the rigid body that coincides with the origin of the reference frame at that given moment of time.

The dual angular velocity $\underline{\boldsymbol{\omega}}$, also called "dual twist", completely characterizes the distribution of the velocities field of the rigid body.

By taking the $(n-1)$-th derivative with respect to time of the equation $\mathbf{v} = \dot{\boldsymbol{\rho}} - \boldsymbol{\omega} \times \boldsymbol{\rho}$, by considering the Leibniz rule, we will obtain:

$$\mathbf{v}^{(n-1)} = \boldsymbol{\rho}^{(n)} - \sum\nolimits_{k=0}^{n-1} C_{n-1}^{k} \tilde{\boldsymbol{\omega}}^{(n-1-k)} \boldsymbol{\rho}^{(k)} \tag{19}$$

We will denote by $\mathbf{a}_{\boldsymbol{\rho}}^{[n]} = \boldsymbol{\rho}^{(n)}, n \in \mathbb{N}$, the nth order acceleration of the point of rigid body given by the position vector $\boldsymbol{\rho}$ and by:

$$\mathbf{A}_{\boldsymbol{\rho}}^{[n]} = \mathbf{a}_{\boldsymbol{\rho}}^{[n]} - \sum\nolimits_{k=1}^{n-1} C_{n-1}^{k} \tilde{\boldsymbol{\omega}}^{(n-1-k)} \mathbf{a}_{\boldsymbol{\rho}}^{[k]} \tag{20}$$

the nth order reduced acceleration, from Eqs. (19) and (20) it follows:

$$\mathbf{A}_{\boldsymbol{\rho}}^{[n]} = \mathbf{v}^{(n-1)} + \tilde{\boldsymbol{\omega}}^{(n-1)} \boldsymbol{\rho}, n \in \mathbb{N}^*. \tag{21}$$

The Eq. (21) proves the helicoidally character of the nth order reduced accelerations field of the rigid body. By denoting

$$\underline{\mathbf{A}}_{\boldsymbol{\rho}}^{[n]} = \boldsymbol{\omega}^{(n-1)} + \varepsilon \mathbf{A}_{\boldsymbol{\rho}}^{[n]} \tag{22}$$

the nth order dual reduced accelerations, it follows the invariant:

$$e^{\varepsilon \tilde{\boldsymbol{\rho}}} \underline{\mathbf{A}}_{\boldsymbol{\rho}}^{[n]} = \underline{\boldsymbol{\omega}}^{(n-1)}, n \in \mathbb{N}^* \tag{23}$$

where $e^{\varepsilon \tilde{\boldsymbol{\rho}}} = \mathbf{I} + \varepsilon \tilde{\boldsymbol{\rho}}$. For n=1, from (23), on obtain the instantaneous screw axis of velocity field.

The line corresponding to the dual vectors $\underline{\boldsymbol{\omega}}^{(n-1)}, n \in \mathbb{N}^*$, represents the loci where the vectors $\mathbf{A}_{\boldsymbol{\rho}}^{[n]}, n \in \mathbb{N}^*$, have the minimum norm value. Supplementary:

$$\min_{\boldsymbol{\rho} \in V_3} \left\| \mathbf{A}_{\boldsymbol{\rho}}^{[n]} \right\| = \left\| \mathrm{Du} \left| \underline{\boldsymbol{\omega}}^{(n-1)} \right| \right\|, n \in \mathbb{N}^*. \tag{24}$$

From Eq. (21) it follows an interesting interpretation for $\mathbf{v}^{(n-1)}$, $(n-1)^{th}$ derivation on $\mathbf{v}$ is the nth order reduced acceleration of a point of the rigid body which at a given moment of time passes through the origin of the fixed reference frame.

For $\boldsymbol{\rho} = 0$, from Eq. (19), it follows:

$$\mathbf{v}^{(n-1)} = \mathbf{a}_n - \sum_{k=1}^{n-1} C_{n-1}^k \tilde{\boldsymbol{\omega}}^{(n-1-k)} \mathbf{a}_k \tag{25}$$

with $\mathbf{a}_k, k = \overline{1,n}$, the kth order vector invariant of the motion. The inverse of Eq. (25) is Eq. $(8)_1$.

By taking the $(n-1)^{th}$ order derivative of the equation $\dot{\mathbf{R}} = \tilde{\omega}\mathbf{R}$, taking into account the fact that the nth acceleration tensor is $\boldsymbol{\phi}_n = \mathbf{R}^{(n)}\mathbf{R}^T$, it follows:

$$\tilde{\boldsymbol{\omega}}^{(n-1)} = \boldsymbol{\phi}_n - \sum_{k=1}^{n-1} C_{n-1}^k \tilde{\boldsymbol{\omega}}^{(n-1-k)} \boldsymbol{\phi}_k \tag{26}$$

The inverse of Eq. (26) is Eq. $(8)_2$.

If we denote $\mathbf{T} = (\mathbf{v}, \tilde{\boldsymbol{\omega}})$ and by $\boldsymbol{\Psi}_n = (\mathbf{a}_n, \boldsymbol{\Phi}_n)$, $n \in \mathbb{N}^*$, for the case of the velocities, accelerations, jerks and jounces, the Eqs. (25), (26) and (8) are written:

$$\begin{bmatrix} \mathbf{T} \\ \dot{\mathbf{T}} \\ \ddot{\mathbf{T}} \\ \dddot{\mathbf{T}} \end{bmatrix} = \begin{bmatrix} \mathbf{I} & 0 & 0 & 0 \\ -\tilde{\boldsymbol{\omega}} & \mathbf{I} & 0 & 0 \\ -2\dot{\tilde{\boldsymbol{\omega}}} & -\tilde{\boldsymbol{\omega}} & \mathbf{I} & 0 \\ -\ddot{\tilde{\boldsymbol{\omega}}} & -3\dot{\tilde{\boldsymbol{\omega}}} & -\tilde{\boldsymbol{\omega}} & \mathbf{I} \end{bmatrix} \begin{bmatrix} \boldsymbol{\Psi}_1 \\ \boldsymbol{\Psi}_2 \\ \boldsymbol{\Psi}_3 \\ \boldsymbol{\Psi}_4 \end{bmatrix} \tag{27}$$

$$\begin{bmatrix} \boldsymbol{\Psi}_1 \\ \boldsymbol{\Psi}_2 \\ \boldsymbol{\Psi}_3 \\ \boldsymbol{\Psi}_4 \end{bmatrix} = \begin{bmatrix} \mathbf{I} & 0 & 0 & 0 \\ \tilde{\boldsymbol{\omega}} & \mathbf{I} & 0 & 0 \\ 2\dot{\tilde{\boldsymbol{\omega}}} + \tilde{\boldsymbol{\omega}}^2 & \tilde{\boldsymbol{\omega}} & \mathbf{I} & 0 \\ \ddot{\tilde{\boldsymbol{\omega}}} + 2\tilde{\boldsymbol{\omega}}\dot{\tilde{\boldsymbol{\omega}}} + 3\dot{\tilde{\boldsymbol{\omega}}}\tilde{\boldsymbol{\omega}} + \tilde{\boldsymbol{\omega}}^3 & 3\dot{\tilde{\boldsymbol{\omega}}} + \tilde{\boldsymbol{\omega}}^2 & \tilde{\boldsymbol{\omega}} & \mathbf{I} \end{bmatrix} \begin{bmatrix} \mathbf{T} \\ \dot{\mathbf{T}} \\ \ddot{\mathbf{T}} \\ \dddot{\mathbf{T}} \end{bmatrix} \tag{28}$$

Theorem 3. *The nth order accelerations field of a rigid body in a general motion is uniquely determined by the kth order time derivative of a dual twist $\underline{\boldsymbol{\omega}}$, $k = \overline{0, n-1}$.*

4 Higher–Order Kinematics of Spatial Chain Using Dual Lie Algebra and Derivative of Twist

Consider a spatial kinematic chain of the bodies $C_k, k = \overline{0,m}$ where the relative motion of the rigid body C_k with respect to C_{k-1} is given by the proper orthogonal tensor ${}^{k-1}\underline{\mathbf{R}}_k \in \underline{\mathbf{SO}}_3^{\mathbb{R}}$. The relative motion properties of the body C_m with respect to C_0 are described by the orthogonal dual tensor:

$$\underline{\mathbf{R}} = {}^0\underline{\mathbf{R}}_1 {}^1\underline{\mathbf{R}}_2 \ldots {}^{m-1}\underline{\mathbf{R}}_m \tag{29}$$

Instantaneous dual angular velocity (dual twist) of the rigid body C_m, with respect to the body C_0, will be given by the equation

$$_0\underline{\boldsymbol{\omega}}_m = \text{vect}\dot{\underline{\mathbf{R}}}\underline{\mathbf{R}}^T \tag{30}$$

It follows from (29) and (30) that:

$$_0\underline{\boldsymbol{\omega}}_m = \underline{\boldsymbol{\Omega}}_1 + {}^0\underline{\mathbf{R}}_1\underline{\boldsymbol{\Omega}}_2 + \cdots + {}^0\underline{\mathbf{R}}_1{}^1\underline{\mathbf{R}}_2 \ldots {}^{m-2}\underline{\mathbf{R}}_{m-1}\underline{\boldsymbol{\Omega}}_m \tag{31}$$

where

$$\underline{\boldsymbol{\Omega}}_k = \text{vect}\left({}^{k-1}\dot{\underline{\mathbf{R}}}_k{}^{k-1}\underline{\mathbf{R}}_k^T\right) \tag{32}$$

Using the notation

$$\underline{\boldsymbol{\omega}}_k = {}^0\underline{\mathbf{R}}_1{}^1\underline{\mathbf{R}}_2 \ldots {}^{k-2}\underline{\mathbf{R}}_{k-1}\underline{\boldsymbol{\Omega}}_k \tag{33}$$

the Eq. (31) becomes:

$$_0\underline{\boldsymbol{\omega}}_m = \underline{\boldsymbol{\omega}}_1 + \underline{\boldsymbol{\omega}}_2 + \cdots + \underline{\boldsymbol{\omega}}_m \tag{34}$$

where $\underline{\boldsymbol{\omega}}_k$ is the dual twist of the relative motion of the body C_k with respect to the body C_{k-1} observed from the body C_0.

We denote $\underline{\boldsymbol{\omega}}_p^{[n]} = {}^0\underline{\mathbf{R}}_1{}^1\underline{\mathbf{R}}_2 \ldots {}^{p-2}\underline{\mathbf{R}}_{p-1}\underline{\boldsymbol{\Omega}}_p^{(n)}$ the nth order derivative of the relative dual twist $\underline{\boldsymbol{\Omega}}_p$, resolved in the body frame of C_0.

In order to determine the nth order accelerations field of a rigid body C_m, we have to determine the $_0\underline{\boldsymbol{\omega}}_m^{(n)}$, $n \in \mathbb{N}^*$.

Theorem 4. *The following equation holds*:

$$\begin{aligned} _0\underline{\boldsymbol{\omega}}_m^{(n)} &= \underline{\boldsymbol{\omega}}_1^{(n)} + \mathbf{p}_n(\underline{\boldsymbol{\omega}}_1)\underline{\boldsymbol{\omega}}_2 + \mathbf{p}_n(\underline{\boldsymbol{\omega}}_1 + \underline{\boldsymbol{\omega}}_2)\underline{\boldsymbol{\omega}}_3 + \cdots \\ &+ \mathbf{p}_n(\underline{\boldsymbol{\omega}}_1 + \underline{\boldsymbol{\omega}}_2 + \cdots + \underline{\boldsymbol{\omega}}_{m-1})\underline{\boldsymbol{\omega}}_m; \forall n \in \mathbb{N} \end{aligned} \tag{35}$$

where $\mathbf{p}_n(\boldsymbol{\omega})$ *are polynomials of the differential operator* $\mathbf{D} = \frac{d}{dt}$, *with coefficients in the non-commutative ring of Euclidian dual tensors*

$$\mathbf{p}_n(\underline{\boldsymbol{\omega}}) = \sum_{k=0}^{n} C_n^k \underline{\boldsymbol{\Phi}}_{n-k}\mathbf{D}^{[k]}, \tag{36}$$

where C_n^k *is the binomial coefficient,* $\mathbf{D}^{[k]}\boldsymbol{\omega}_p = \boldsymbol{\omega}_p^{[k]}$ and $\underline{\boldsymbol{\Phi}}_p$ *are dual tensors*

$$\underline{\boldsymbol{\Phi}}_p = \underline{\mathbf{R}}^{(p)}\underline{\mathbf{R}}^T, \underline{\mathbf{R}} \in \underline{\mathbf{SO}}_3^{\mathbb{R}}, p = \overline{0,n} \tag{37}$$

which follow the recurrence relationship:

$$\begin{cases} \underline{\boldsymbol{\Phi}}_{p+1} = \dot{\underline{\boldsymbol{\Phi}}}_p + \underline{\boldsymbol{\Phi}}_p\tilde{\underline{\boldsymbol{\omega}}} \\ \underline{\boldsymbol{\Phi}}_0 = \underline{\mathbf{I}} \end{cases}, p \in \mathbb{N}. \tag{38}$$

Other equivalent forms of the Eqs. (35) are:

$$ {}_0\underline{\boldsymbol{\omega}}_m^{(n)} = \underline{\boldsymbol{\omega}}_1^{[n]} + \mathbf{p}_n({}_0\underline{\boldsymbol{\omega}}_1)\underline{\boldsymbol{\omega}}_2 + \mathbf{p}_n({}_0\underline{\boldsymbol{\omega}}_2)\underline{\boldsymbol{\omega}}_3 + \cdots + \mathbf{p}_n({}_0\underline{\boldsymbol{\omega}}_{m-1})\underline{\boldsymbol{\omega}}_m, \forall n \in \mathbb{N} \tag{39} $$

From Eq. (39) on obtain the closed form, free of coordinate formula:

$$ \begin{aligned} {}_0\underline{\boldsymbol{\omega}}_m^{(n)} &= \underline{\boldsymbol{\omega}}_1^{[n]} + \underline{\boldsymbol{\omega}}_2^{[n]} + \cdots + \underline{\boldsymbol{\omega}}_m^{[n]} \\ &+ \sum_{p=2}^{m} \sum_{k=1}^{n} C_n^k \underline{\boldsymbol{\Phi}}_k\left(\underline{\boldsymbol{\omega}}_1 + \underline{\boldsymbol{\omega}}_2 + \cdots + \underline{\boldsymbol{\omega}}_{p-1}\right)\underline{\boldsymbol{\omega}}_p^{[n-k]}, \forall n \in \mathbb{N} \end{aligned} \tag{40} $$

where

$$ \begin{aligned} &\underline{\boldsymbol{\Phi}}_k\left(\underline{\boldsymbol{\omega}}_1 + \underline{\boldsymbol{\omega}}_2 + \cdots + \underline{\boldsymbol{\omega}}_{p-1}\right) \\ &= \sum_{k_1+k_2+\cdots+k_{p-1}=k} C_n^{k_1,k_2,\ldots,k_{p-1}} \underline{\boldsymbol{\Phi}}_{k_1}(\underline{\boldsymbol{\omega}}_1)\underline{\boldsymbol{\Phi}}_{k_2}(\underline{\boldsymbol{\omega}}_2)\ldots\underline{\boldsymbol{\Phi}}_{k_{p-1}}\left(\underline{\boldsymbol{\omega}}_{p-1}\right) \end{aligned} \tag{41} $$

and $C_n^{k_1,k_2,\ldots,k_{p-1}} = \frac{n!}{k_1!k_2!,\ldots,k_{p-1}!}$ is the monomial coefficient.

5 Conclusions

Using the isomorphism between the Lie group of the rigid displacements $\mathbb{SE}_3$ and the Lie group of the orthogonal dual tensors $\underline{\mathbb{SO}}_3$, a new general method for the study of the higher-order accelerations field is described. It is proved that all information regarding the properties of the distribution of higher-order accelerations is contained in the nth order derivatives of the dual twist of the rigid body. These derivatives belong to the Lie algebra associated to the Lie group $\underline{\mathbb{SO}}_3$. For the case of the spatial kinematic chains, equations that allow the determination of the nth order accelerations field are given, using a Brockett-like formula. The obtained results are of interest for theoretical kinematics, jerk with jounce analysis in the case of parallel manipulators, control theory and sensitivity analysis in multibody kinematics.

References

1. Angeles, J.: The Application of Dual Algebra to Kinematic Analysis. Computational Methods in Mechanical Systems **161**, 3–32 (1998)
2. Ball, R.S.: The Theory of Screws. Cambridge University Press, New York (1900)
3. Bokelberg, E.H., Hunt, K.H., Ridley, P.R.: Spatial motion I points of inflection and the differential geometry of screws. Mech. Mach. Theory **27**(1), 1–15 (1992)
4. Condurache, D., Burlacu, A.: Orthogonal dual tensor method for solving the AX=XB sensor calibration problem. Mech. Mach. Theory **104**, 382–404 (2016)
5. Everett, J.D.: On the kinematics of a rigid body. Q. J. Pure Appl. Math. **13**, 33–66 (1875)
6. Hunt, K.H.: Kinematic Geometry of Mechanisms, Oxford University Press, Oxford (1978)
7. Karger, A.: Singularity analysis of serial robot-manipulators. ASME J. Mech. Des. **118**(4), 520–525 (1996)

8. Lerbet, J.: Analytic geometry and singularities of mechanisms, ZAMM. Z. Angew. Math. Mech. **78**(10b), 687–694 (1999)
9. Müller, A.: Higher derivatives of the kinematic mapping and some applications. Mech. Mach. Theory **76**, 70–85 (2014)
10. Phillips, J.: Freedom in Machinery. Introducing Screw Theory, vol. 1, Cambridge University Press, Cambridge (1984)
11. Rico, J.M., Gallardo, J., Duffy, J.: Screw theory and higher order kinematic analysis of open serial and closed chains. Mech. Mach. Theory **34**(4), 559–586 (1999)
12. Samuel, A.E., McAree, P.R., Hunt, K.H.: Unifying screw geometry and matrix transformations. Int. J. Robot. Res. **10**(5), 454–472 (1991)
13. Veldkamp, G.R.: Canonical systems and instantaneous invariants in spatial kinematics. J. Mech. **3**(3), 329–388 (1967)

An Algorithm for Trajectory Generation in Redundant Manipulators with Joint Transmission Accommodation

Kristopher Wehage[1(✉)] and Bahram Ravani[2]

[1] Jet Propulsion Laboratory, Pasadena, CA 91109, USA
kristopher.t.wehage@jpl.nasa.gov
[2] University of California – Davis, Davis, CA 91516, USA
bravani@ucdavis.edu

Abstract. Trajectory generation for manipulators involves generating incremental updates of joint variables to achieve a desired end–effector motion. The Jacobian matrix maps incremental joint motion to incremental end–effector motion in a linear fashion and is typically used in manipulator trajectory generation algorithms. In the case of a redundant manipulator, the Jacobian matrix is not square or invertible and therefore algorithms based on pseudoinverses and their variations are commonly used for trajectory generation. These methods either are computationally not efficient or do not utilize all the joints in motion generation and therefore do not completely exploit the redundancy of the manipulator. The method presented in this paper is a simple method that maximizes transmission of all joint variables onto a desired end–effector motion trajectory. The method is based on aligning the null–space of an augmented Jacobian matrix with the path of the desired end–effector motion, from which a linear combination of joints that projects fully onto the desired end–effector trajectory is obtained. In this manner, all joints of the redundant manipulator are used to generate the end–effector trajectory accommodating the ability of each joint in terms of its motion transmission.

Keywords: Redundant manipulators · Trajectory generation
Differential kinematics · Transmission accommodation

1 Introduction

Redundant manipulators with 7 or 8 Degrees Of Freedom (DOF) are finding increased applications in industrial settings (see, for example [1,2]) due to their increased dexterity and ability to avoid bifurcations and singular configurations. Kinematic algorithms used for trajectory generation linearly relate the incremental joint motions to the desired end-effector motion from a nearby location

K. Wehage—This work was done as a private venture and not in the author's capacity as an employee of the Jet Propulsion Laboratory, California Institute of Technology.

J. Lenarcic and V. Parenti-Castelli (Eds.): ARK 2018, SPAR 8, pp. 92–99, 2019.
https://doi.org/10.1007/978-3-319-93188-3_11

using the Jacobian matrix. In general the kinematic equations for a single arm serial manipulator can be written as $f(\underline{q}, \Phi_{fe}) = 0$ where $\underline{q}$ is the vector of joint variables and Φ_{fe} represents a parametrization of the end-effector configuration which is typically described as a screw or an element of the special Euclidean group SE(3). The matrix Φ_{fe} indicates an active transformation from an inertial world frame f to the end-effector frame e. In incremental trajectory generation for a manipulator, a linearized form of this equation is obtained using Taylor series expansion and neglecting the higher order terms:

$$f(\underline{q}, \delta\underline{q}, \Phi_{fe}) = f(\underline{\bar{q}}, \Phi_{ef}) + J\delta\underline{q} = 0 \tag{1}$$

In the above equation, the term $f(\underline{\bar{q}}, \Phi_{fe})$ represents the incremental movement of the end-effector from its existing location to a nearby location. If the end-effector nearby location is specified as Φ^*_{fe} (see for example, [3], pp. 164–168), then the position error E can be expressed as:

$$E = \Phi^*_{fe}\Phi^{-1}_{fe} - I \tag{2}$$

The non-zero terms of E when put in screw coordinate can be represented as $\underline{\mathbf{e}}$. The linearized kinematic equations then become:

$$\underline{\mathbf{e}} = J\delta\underline{q} \tag{3}$$

For redundant manipulators the Jacobian matrix J is rank deficient or not a square matrix. A common technique to resolve rank deficiency is to take the generalized inverse or MoorePenrose pseudo-inverse of the Jacobian as:

$$\delta\underline{q} = (J^T J)^{-1} J^T \underline{\mathbf{e}} = J^+ \underline{\mathbf{e}} \tag{4}$$

which effectively finds a non-linear least squares solution [4] with the terms that are minimed being quadratic in joint velocities and therefore proportional to the kinetic energy of the system. A variation of the generalized inverse method is the use of weighted or damped least squares method:

$$\delta\underline{q} = (J^T W J + W)^{-1} J^T W \underline{\mathbf{e}} \tag{5}$$

where W is a diagonal matrix of weighting factors [5–7]. The weights can also be selected based on the inertia of the links [8]. The generalized inverse methods, in general, not only require many additional floating point operations (FLOPS) but may also sometimes fail [3]. Despite these shortcomings such methods and their variations are still widely used in robotic literature for kinematic control or incremental trajectory generation of redundant manipulators (see, for example, [9–11]).

In this paper, we present a method that at each configuration of the manipulator numerically determines how well each joint is aligned with a direction orthogonal to the motion of the end-effector from its current position to the desired nearby configuration. The approach aligns the tangent vector in the constraint manifold of the manipulator with the particular path of interest and results in a method that utilizes all the joints maximizing their motion transmission abilities.

2 Incremental Trajectory Generation

In incremental or kinematic trajectory generation for robot manipulators, the motion path of the end-effector is first parameterized in terms of successive incremental orientations and positions. There are a number of approaches for this process in the literature including using methods from Computer Aided Geometric Design such as quaternion based spherical interpolation (SLERP) [12], smooth invariant interpolation of rotations [13], or cubic Hermite splines [14] to name a few techniques. The differential kinematic equations are then used to map each successive motion path of the end-effector to the corresponding joint variables, thereby incrementally updating the joint coordinates to move the robot through its desired trajectory.

The differential kinematic equations map joint velocities to the end-effector velocity using the Jacobian matrix. As stated earlier the Jacobian matrix for a redundant manipulator is not full rank. It can be permuted and block-partitioned into coefficients associated with its primary and secondary constraint equations and dependent and independent joint coordinates [15]. The method presented in this paper does not require an analyst to select dependent and independent joint variables or primary and secondary constraint equations but rather automatically obtains the optimal split for each of its incremental configurations throughout the trajectory generation process.

Applying LU factorization with complete row and column pivoting permutes and partitions the Jacobian matrix J into the following form:

$$P_R J P_C^T = \begin{bmatrix} J^{pd} & J^{pi} \\ J^{sd} & J^{si} \end{bmatrix} \tag{6}$$

and factors the largest nonsingular submatrix J^{pd}, into L and U. Here P_R and P_C^T are row and column permutation matrices, the superscripts p and s indicate terms associated with primary and secondary constraint equations, and the subscripts i and d indicate terms associated with independent and dependent variables. Applying LU decomposition with complete pivoting to the differential kinematic equations yields:

$$\begin{bmatrix} J^{pd} & J^{pi} \\ J^{sd} & J^{si} \end{bmatrix} \begin{Bmatrix} \delta \underline{q}^d \\ \delta \underline{q}^i \end{Bmatrix} = \begin{Bmatrix} -\underline{\mathbf{e}}^p \\ -\underline{\mathbf{e}}^s \end{Bmatrix} \tag{7}$$

The independent joint coordinates $\underline{q}^i$ may be set to any arbitrary value and a change in dependent joint coordinates $\underline{q}^d$ can be found from the row-space as:

$$\delta \underline{q}^d = -U^{-1} L^{-1} (\underline{\mathbf{e}}^p + J^{pi} \delta \underline{q}^i) \tag{8}$$

where L and U are the LU factors of J^{pd}. The incremental changes in the joint variables, $\delta \underline{q}^i$, exert the smallest over-all effect on a change in incremental configuration of the manipulator and therefore can be used in accommodation of best transmission advantage in resolving redundancy as discussed in the next section. Uicker et al. [3] suggest a simple way of solving Eq. (8) to resolve redundancy.

The process involves setting $\delta \underline{q}^i$ equal to zero and solving the reduced system of equations:

$$\delta \underline{q}^d = -U^{-1} L^{-1} \underline{\mathbf{e}}^p \tag{9}$$

Unfortunately this approach does not fully exploit the redundancy of the manipulator because setting $\delta \underline{q}^i = \underline{0}$ makes the manipulator nonredundant in terms of its independent variable. The approach presented in this paper re-establishes the distribution of independent and dependent joint variables to achieve the best transmission accommodation at each incremental configuration.

3 New Redundancy Resolution Method

The new redundancy resolution method begins with computing the right null-space of an augmented Jacobian matrix J with one column appended to represent the desired instantaneous screw motion of the end-effector. A convenient representation of the row-space of J is found from:

$$L^{-1} U^{-1} \left[J^{pd} \; J^{pi} \right] = \left[I \; B^{di} \right] \tag{10}$$

which allows trivially computing the right null-space of J, N, as:

$$N = \begin{bmatrix} -B^{di} \\ I \end{bmatrix} \tag{11}$$

The LU decomposition with complete pivoting algorithm overwrites the row–space of the matrix J with its LU factors and the residual matrix U_R as:

$$\left[LU \; U^{-1} J^{pi} \right] = \left[LU \; U_R \right] \tag{12}$$

The quantity $-B^{di}$ is computed efficiently using backward substitution as:

$$- B^{di} \leftarrow -L^{-1} U_R \tag{13}$$

QR decomposition [16] is then applied to the matrix N as:

$$N = ZR = \begin{bmatrix} Z^{di} \\ R^{-1} \end{bmatrix} R = \begin{bmatrix} Z^{di} \\ Z^{ii} \end{bmatrix} R \tag{14}$$

In Eq. 14, R is a transformation matrix and Z is an orthonormal basis. The matrix R effectively takes a linear combination of directions along the manipulator's constraint manifold and orthogonalizes and normalizes the columns of Z. The performance of the LU decomposition in the above computations can be enhanced using a heuristic application of complete pivoting as described in [15].

The resulting Z matrix is analogous to a unit orthogonal direction-cosine matrix on the manipulator's constraint manifold where each column corresponds to a direction along the manifold and each row corresponds to unique joint variable. Each of the entries in Z corresponds to a projection of a joint onto one orthogonal direction tangent to the constraint manifold of the manipulator configuration. The matrix Z is partitioned into Z^{di} and Z^{ii} (see Eq. (14)) with each row corresponding, respectively, to a unique dependent or independent joint variable. The null-space basis Z of the Jacobian matrix, therefore, represents two orthogonal tangent vectors on the constraint manifold of the manipulator, which is schematically illustrated in Fig. 1. At this stage neither of the two tangent vectors is aligned with the desired path of the end-effector which is shown by the black dashed line in Fig. 1(b). The necessary information to align the null-space basis of the Jacobian of the manipulator to the desired path is found in the last row of Z. When the tangentspace is aligned with the desired endeffector path, the entry in the first column of the row associated with the path will project fully onto the path. The entry in the corresponding second column will be zero. A 2×2 orthonormal rotation matrix, C, can therefore be applied to rotate Z to align the tangent vectors as:

$$Z' = ZC \tag{15}$$

Note that both Z and Z' matrices lie in the right null-space of the Jacobian matrix J and represent a set of orthonormal basis vectors aligned with the mechanism's constraint manifold.

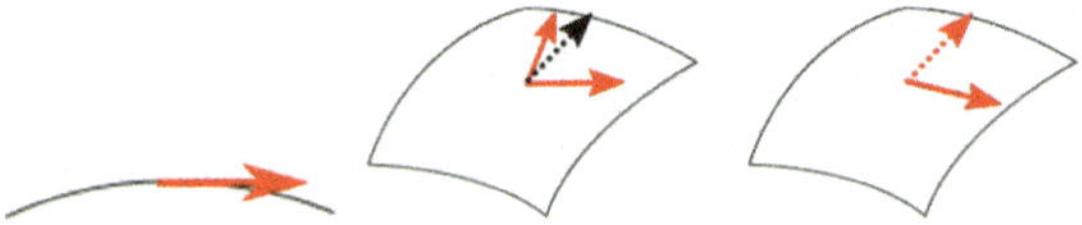

(a) (b) (c)

(a) a one-dimensional manifold; (b) a two dimensional manifold with an arbitrary set of tangent vectors; (c) a two dimensional manifold with one tangent vector aligned with path direction

Fig. 1. Basis vectors along one and two-dimensional manifolds.

The matrix Z' now has the first column aligned with the target end-effector path. The first column represents a linear combination of the joint velocities that fully project onto the instantaneous screw motion of the path. The second column of Z' represents a tangent vector that is orthogonal to the target path. The linear combination of joint velocities indicated in the second column of Z' imparts no motion along the target path. Once the first column of Z' has been aligned to the target path, the entries in the first column indicate how to proportion joint velocities to impart maximum transmission onto the path. The method is illustrated by an example in the next section.

4 Example

This section provides a brief illustration of the method for redundancy resolution by applying it to a 7R serial manipulator depicted in Fig. 2 consisting of seven

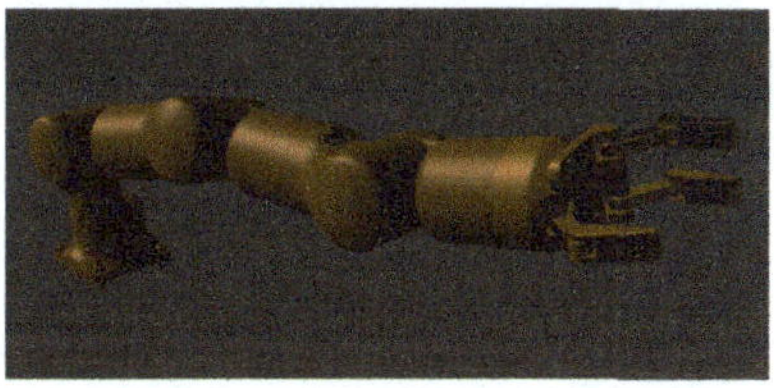

Fig. 2. A 7R redundant manipulator

revolute joints labeled A through G. In order to determine the Jacobian matrix for this manipulator the shape matrices (see [3]) are formed. The numerical values of these matrices are presented in Eq. 16.

$$\Phi^{1h+} = \begin{bmatrix} 1 & 0 & 0 & 0 \\ 0 & 1 & 0 & 0.045 \\ 0 & 0 & 1 & 0 \\ 0 & 0 & 0 & 1 \end{bmatrix}, \Phi^{2h+} = \begin{bmatrix} 1 & 0 & 0 & 0 \\ 0 & 1 & 0 & 0.09 \\ 0 & 0 & 1 & 0.02 \\ 0 & 0 & 0 & 1 \end{bmatrix}, \Phi^{3h+} = \begin{bmatrix} 1 & 0 & 0 & 0 \\ 0 & 1 & 0 & 0 \\ 0 & 0 & 1 & -0.085 \\ 0 & 0 & 0 & 1 \end{bmatrix}, \Phi^{4h+} = \begin{bmatrix} 1 & 0 & 0 & 0 \\ 0 & 1 & 0 & 0 \\ 0 & 0 & 1 & -0.125 \\ 0 & 0 & 0 & 1 \end{bmatrix},$$
$$\Phi^{5h+} = \begin{bmatrix} 1 & 0 & 0 & 0 \\ 0 & -1 & 0 & 0 \\ 0 & 0 & -1 & -0.085 \\ 0 & 0 & 0 & 1 \end{bmatrix}, \Phi^{6h+} = \begin{bmatrix} 1 & 0 & 0 & 0 \\ 0 & 0 & -1 & -0.125 \\ 0 & 1 & 1 & 0 \\ 0 & 0 & 0 & 1 \end{bmatrix}, \Phi^{7h+} = \begin{bmatrix} 1 & 0 & 0 & 0 \\ 0 & 1 & 0 & 0.085 \\ 0 & 0 & 1 & 0 \\ 0 & 0 & 0 & 1 \end{bmatrix} \tag{16}$$

The augmented Jacobian matrix J for the 7R manipulator is given as:

$$J = \left[\underline{\mathbf{h}}_A \; \underline{\mathbf{h}}_B \; \underline{\mathbf{h}}_C \; \underline{\mathbf{h}}_D \; \underline{\mathbf{h}}_E \; \underline{\mathbf{h}}_F \; \underline{\mathbf{h}}_G \middle| \underline{\mathbf{h}}_{path} \right] \tag{17}$$

where the terms $\underline{\mathbf{h}}_{A-G}$ represent the influence coefficient matrices associated with revolute joints A–G, and the quantity $\underline{\mathbf{h}}_{path}$ represents the influence coefficient matrix associated with the instantaneous screw motion of the end-effector.

Inserting numerical values for a specific configuration yields:

$$J = \left[\begin{array}{ccccccc|c} 0.0000 & -0.0134 & -0.1356 & -0.0256 & 0.1756 & -0.0123 & -0.0936 & 0.2853 \\ 0.0000 & 0.0200 & -0.0000 & 0.1821 & 0.0073 & -0.3956 & -0.0074 & 0.6905 \\ 0.0000 & -0.1343 & 0.0136 & 0.1766 & -0.0106 & -0.1770 & 0.0040 & -0.0611 \\ 0.0000 & 0.9950 & -0.0978 & -0.9791 & 0.0604 & 0.9981 & -0.0584 & 0.2548 \\ 1.0000 & 0.0000 & 0.1986 & -0.1947 & -0.0038 & -0.0040 & 0.2024 & -0.6369 \\ 0.0000 & -0.0998 & -0.9751 & 0.0585 & 0.9981 & -0.0604 & -0.9775 & -0.1671 \end{array} \right] \tag{18}$$

An orthonormalized null–space basis for J is obtained as

$$Z = \begin{array}{r} A \\ B \\ C \\ D \\ E \\ F \\ G \\ path \end{array} \begin{bmatrix} -0.2296 & 0.1972 \\ 0.0255 & -0.0278 \\ -0.3567 & -0.7338 \\ -0.5108 & 0.2346 \\ 0.1345 & -0.5346 \\ -0.5013 & 0.2115 \\ 0.5165 & 0.1783 \\ -0.1491 & 0.0677 \end{bmatrix} \tag{19}$$

As discussed previously, the null-space basis in Eq. (19) is not unique and needs to be rotated to align the tangent vectors.

$$Z' = ZC = Z \begin{bmatrix} -0.9107 & -0.4132 \\ 0.4132 & -0.9107 \end{bmatrix} = \begin{matrix} A \\ B \\ C \\ D \\ E \\ F \\ G \\ path \end{matrix} \begin{bmatrix} 0.2906 & -0.0847 \\ -0.0348 & 0.0148 \\ 0.0217 & 0.8156 \\ 0.5621 & -0.0026 \\ -0.3434 & 0.4313 \\ 0.5439 & 0.0146 \\ -0.3967 & -0.3758 \\ 0.1638 & 0. \end{bmatrix} \tag{20}$$

The first column of Z' represents a linear combination of joint velocities AG that project fully onto the instantaneous screw motion of the path. The combination of quantities in the second column of Z' reconfigure the robot while maintaining the artificial constraint at the end–effector. Therefore, the first column of Z' represents the linear combination of joint variables that maximizes projection onto the desired trajectory with no wasted effort orthogonal to the desired trajectory.

$$\delta \underline{q}_{A-G} = \left\{ Z'_{[A,0]} \; Z'_{[B,0]} \; Z'_{[C,0]} \; Z'_{[D,0]} \; Z'_{[E,0]} \; Z'_{[F,0]} \; Z'_{[G,0]} \right\}^T \frac{\delta q_{path}}{Z'_{[path,0]}} \tag{21}$$

With this information, a first-order update of the joint variables $\delta \underline{q}$ can be applied using Eq. (21). It should be noted that this approach requires approximately 40% fewer floating–point operations than the pseudo-inverse and least squares methods [15, 16].

5 Conclusions and Future Work

The method for redundancy resolution presented in this paper represents a greedy algorithm for obtaining an incremental update of joint variables at any time interval. It is greedy in the sense that it maximizes transmission of joint variables onto the desired trajectory at a given time interval with no consideration of the state of the manipulator at the next time interval(s). It should be noted that it is possible that the optimal projection onto the desired trajectory at one time interval may put the robot in a configuration where one or more of the joints may have a low ability to impart transmission onto the desired trajectory at a later time interval. The work presented here may be further extended by developing optimal robot path tracking algorithms (i.e. following a specified path while minimizing an objective function), which will be investigated in the authors' future work.

Acknowledgements. This work was supported by the International Research Training Group (IRTG2057) with the University of Kaiserslautern, Germany.

References

1. Guizzo, E., Ackerman, E.: How rethink robotics built its new Baxter robot worker. IEEE Spectrum (2012)
2. Kreutz-Delgado, K., Long, M., Seraji, H.: Kinematic analysis of 7-DOF manipulators. Int. J. Robot. Res. **11**(5), 469–481 (1992)
3. Uicker, J.J., Ravani, B., Sheth, P.N.: Matrix Methods in the Design Analysis of Mechanisms and Multibody Systems. Cambridge University Press, Cambridge (2013)
4. Whitney, D.E.: Resolved motion rate control of manipulators and human prostheses. IEEE Trans. Man-Mach. Syst. **10**(2), 47–53 (1969)
5. Nakamura, Y., Hanafusa, H.: Inverse kinematic solutions with singularity robustness for robot manipulator control. J. Dyn. Syst. Meas. Contr. **108**(3), 163–171 (1986)
6. Wampler, C.W., Leifer, L.J.: Applications of damped least-squares methods to resolved-rate and resolved-acceleration control of manipulators. J. Dyn. Syst. Meas. Contr. **110**(1), 31–38 (1988)
7. Seraji, H., Colbaugh, R.: Improved configuration control for redundant robots. J. Rob. Syst. **7**(6), 897–928 (1990)
8. Wang, J., Li, Y., Zhao, X.: Inverse kinematics and control of a 7-DOF redundant manipulator based on the closed-loop algorithm. Int. J. Adv. Rob. Syst. **7**(4), 1–9 (2010)
9. Branz, F., Antonello, A., Carron, A., Carli, R., Francesconi, A.: Kinematics and control of redundant robotic arm based on dielectric elastomer actuators. In: SPIE Smart Structures and Materials+ Nondestructive Evaluation and Health Monitoring, p. 943023. International Society for Optics and Photonics (2015)
10. Müller, A.: A robust inverse dynamics formulation for redundantly actuated PKM. In: 13th World Congress in Mechanism and Machine Science, Guanajuato, Mexico, pp. 19–25 (2011)
11. Hollerbach, J.M., Suh, K.C.: Redundancy resolution of manipulators through torque optimization. IEEE J. Rob. Autom. **3**(4), 308–316 (1987)
12. Shoemake, K.: Animating rotation with quaternion curves. In: ACM SIGGRAPH Computer Graphics, vol. 19, pp. 245–254. ACM (1985)
13. Park, F.C., Ravani, B.: Smooth invariant interpolation of rotations. ACM Trans. Graph. (TOG) **16**(3), 277–295 (1997)
14. Zha, X.F., Du, H.: Generation and simulation of robot trajectories in a virtual cad-based off-line programming environment. Int. J. Adv. Manuf. Technol. **17**(8), 610–624 (2001)
15. Wehage, K.T., Wehage, R.A., Ravani, B.: Generalized coordinate partitioning for complex mechanisms based on kinematic substructuring. Mech. Mach. Theory **92**, 464–483 (2015)
16. Golub, G.H., Van Loan, C.F.: Matrix Computations. Johns Hopkins Studies in the Mathematical Sciences. Johns Hopkins University Press (1996)

Evaluation of Dynamic Relaxation to Solve Kinematics of Concentric Tube Robots

Quentin Peyron[1,2(✉)], Kanty Rabenorosoa[2], Nicolas Andreff[2], and Pierre Renaud[1]

[1] ICUBE-AVR, UDS-CNRS-INSA, Strasbourg, France
[2] FEMTO-ST Institute, Univ. Bourgogne Franche-Comté/CNRS, Besançon, France
quentin.peyron@univ-fcomte.fr,
{kanty.rabenorosoa,nicolas.andreff}@femto-st.fr,
pierre.renaud@insa-strasbourg.fr

Abstract. Concentric tube robots are based on the deformation of elastic pre-curved tubes mounted in a telescopic manner. Their kinematic model consists in a boundary value problem which must be solved during analysis and design. When arbitrary properties and number of the tubes are considered, this model must be solved numerically. We consider in this paper the use of dynamic relaxation to perform this resolution. Its performances in terms of accuracy and computation time are assessed in a case study involving a two-tube CTR. Robustness of the method tuning to variations of CTR behaviour as encountered during a deployment is finally assessed.

Keywords: Concentric tube robot · Continuum robot kinematics
Dynamic relaxation · Boundary value problem

1 Introduction

Concentric tube robots (CTR) consist in telescopic assemblies of pre-curved and elastic tubes, which have been used extensively in minimally invasive surgery [3]. The interactions between the tubes and with their environment create internal wrenches which deform the tubes and place the robot in a certain equilibrium configuration. Acting on the rotation and the translation of the tubes at their base allows a modification in the equilibrium configuration which is used to control the robot tip pose and the backbone shape. The kinematic model of the CTR consists therefore in equilibrium equations issued from continuum mechanics, which must be solved for analysis and design.

As they are composed of elastic beams performing large displacements, CTRs are modelled using either energy considerations [11] or Cosserat rod theory [4]. The equilibrium equations to consider then form a multi-point boundary value problem (BVP), of which the complexity depends on the number and the properties of the tubes and on the considered environment in interaction with the

J. Lenarcic and V. Parenti-Castelli (Eds.): ARK 2018, SPAR 8, pp. 100–107, 2019.
https://doi.org/10.1007/978-3-319-93188-3_12

CTR. For a two-tube robot with constant and planar pre-curvature deploying in free space, an analytical solution of the model can be obtained as demonstrated in [4]. When more complex geometries and larger number of tubes are considered, the equilibrium configuration must be computed numerically. Several numerical methods have been used for this purpose that need to handle the presence of boundary conditions at different locations along the CTR backbone. Initial and final boundary conditions are managed independently in a two-step process called the shooting method in [9]. The Galerkin and the LocattoIIIA methods are proposed in [5,6] which discretize the CTR along its backbone. The BVP is then transformed into a set of non-linear equations solved with root finding methods. The nodes are placed automatically on the backbone in order to optimize the method accuracy and computation time. Discretization is also considered in [1] where the tubes are represented with a 3D finite element model. Dynamics of the robot are then integrated using a Bathe time integration scheme in order to obtain the CTR equilibrium configuration.

We propose here to consider a discretization of the CTR, but as an initial step to manually control node locations. This allows us to potentially express any boundary condition at a given location, as in the case of contacts with the environment for instance. We show in the following that, after discretization, the problem of solving the CTR kinematics is similar to the one involved in the shape finding of discrete elastic structures such as tensegrity mechanisms [2,8]. The Dynamic Relaxation (DR) method has been successfully used to solve such form-finding problem, and has been already extended to finite element model of rods and plates [10]. We consider therefore in this paper the use of DR to compute the equilibrium configuration of the discrete CTR. This paper is organized as followed. The kinematic model of CTR to be solved is first of all briefly described in Sect. 2. CTR kinematics in the absence of external load is then considered for sake of simplicity. DR is implemented in Sect. 3. A first evaluation of the DR performances in terms of accuracy and computation time is finally performed in Sect. 4, before concluding.

2 Kinematic Model of CTR

Evaluation of the DR requires first the derivation of the CTR kinematic model. We consider the CTR to be composed of n tubes numbered from the innermost to the outermost. The deployed length and the base angle of tube i are denoted respectively L_i and α_i. The tubes are actuated at the arc-length $s = -\beta_i$, where β_i is the transmission length [5] of tube i. The robot is considered as composed of n sections, the number of tubes being constant along each one. The sections are indexed from the distal end to the proximal one, and the length of the i-th section is denoted ΔL_i.

The local configuration of the robot is defined using a Bishop frame $\mathcal{R}_B$ which z axis is the tangent of the robot backbone. A frame $\mathcal{R}_i$ is attached to each tube i, that is obtained by applying a rotation $(\theta_i(s), \mathbf{z_B})$ to $\mathcal{R}_B$, with $\theta_i(s)$ the twist angle of tube i at arc length s. In the remainder of this paper, we will use when needed subscripts (B and i) to indicate the frame in which vectors are projected.

CTR kinematics are classically described using the curvature of the robot backbone $\boldsymbol{u}(s)$, the pre-curvature vector of each tube i denoted $\hat{\boldsymbol{u}}_i(s)$ for tube i and the bending and torsional stiffnesses $k_{ib}(s)$ and $k_{it}(s)$, which form a stiffness matrix $\boldsymbol{K}_i(s)$ such that $\boldsymbol{K}_i(s) = \text{diag}(k_{ib}(s), k_{ib}(s), k_{it}(s))$ [11]. The dependences in s are not mentioned in the following of the paper for sake of compactness. As explained in [11], the equilibrium equations of the robot can be obtained first by expressing the balance of moments at each cross section of the robot:

$$\sum_{i=1}^{m} \boldsymbol{K}_j \boldsymbol{u}_B = \sum_{i=1}^{m} {}^B\boldsymbol{R}_i \boldsymbol{K}_i (\hat{\boldsymbol{u}}_i - \hat{u}_{iz}\boldsymbol{e}_z) \tag{1}$$

and second by applying the Euler-Lagrange formula to the total potential energy of the robot, giving:

$$k_{it}\theta_i'' = g_{\theta_i} = k_{it}\hat{u}_{iz}' + k_{it}'(\theta_i' - \hat{u}_{iz}) + \boldsymbol{u}_B^T \frac{\partial^B \boldsymbol{R}_i}{\partial \theta_i} \boldsymbol{K}_i \hat{\boldsymbol{u}}_i \tag{2}$$
$$i = 1...n$$

where ${}^B\boldsymbol{R}_i$ is the rotation matrix describing the rotation of $\mathcal{R}_i$ with respect to $\mathcal{R}_B$, $[...]'$ is a derivative according to the arc length s and $\boldsymbol{e}_z = [0\,0\,1]^T$. These differential equations are constrained by boundary conditions modelling the proximal actuation of each tube and the free tube extremities. Assuming that the torsional curvature is constant along the transmission lengths as introduced in [5], these conditions write:

$$\begin{cases} \theta_i(0) = \alpha_i - \beta_i\theta_i'(0) \\ \theta_i'(L_j) = \begin{cases} 0 & , \quad i \neq j \\ \hat{u}_{iz} & , \quad i = j \end{cases} \end{cases} \quad i = 1...n \tag{3}$$

The kinematic model of the CTR is therefore formulated, in this situation, as a so-called two-point boundary value problem (BVP) composed of the moment balance (1), the energetic equilibrium (2) and the boundary conditions (3).

3 Dynamic Relaxation Method for CTR

Solving in the general case the CTR kinematic model must be performed numerically. To do this, we discretize first of all the robot geometry in a number N of points along its backbone. The points are placed as denoted for illustration on Fig. 1, a point k being located at the arc length s_k. The number of points for section j is designated by N_j so that $N = \sum_{j=1}^{n} N_j$. The twist angles defining the equilibrium configuration of the robot are then evaluated at each point, as illustrated on the figure for point 5 and point $N_3 + 3$. The distance in arc length between point k and point $k+1$ is denoted h_k, so that $h_k = s_{k+1} - s_k$.

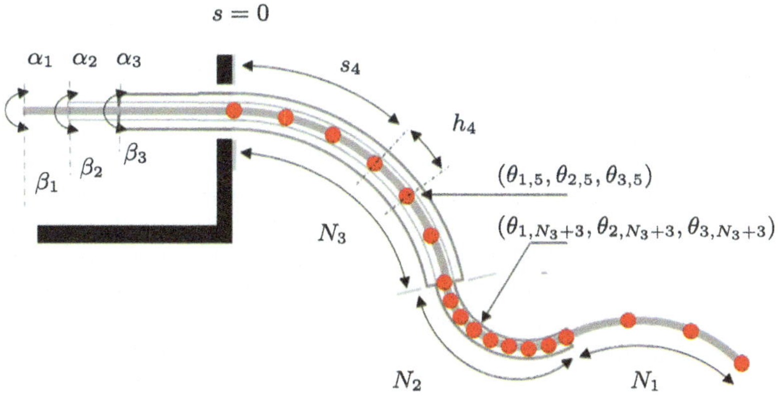

Fig. 1. Representation of the CTR after discretization. Here $n = 3$ and $[N_1 \quad N_2 \quad N_3] = [4 \quad 8 \quad 6]$.

The equilibrium Eq. (2) can then be written using discrete formulation of derivatives. The second-order derivative of $\theta_i(s)$ is replaced by a central second-order finite difference, so that (2) becomes:

$$k_{it,k}\frac{h_k\theta_{i,k-1} - (h_k + h_{k-1})\theta_{i,k} + h_k\theta_{i,k-1}}{h_k h_{k-1}^2} - g_{\theta_i,k} = 0 \qquad (4)$$
$$\text{with } i = 1...n \quad , \quad k = 2...N-1$$

where subscript k denotes an evaluation at point k. The initial and final boundary conditions (3) are then included to the central finite differences evaluated at robot extremities:

$$\begin{aligned} &k_{it,1}\frac{\beta_i\theta_{i,2} - (\beta_i + h_1)\theta_{i,1} + h_1\alpha_i}{h_1\beta_i^2} - g_{\theta_i,1} = 0 \\ &k_{it,N}\frac{\theta_{i,Nf} - 2\theta_{i,N} + \theta_{i,N-1}}{h_N^2} - g_{\theta_i,N} = 0 \\ &\theta_{i,Nf} = \hat{u}_{iz,N}h_N + \theta_{i,N} \\ &i = 1...n \end{aligned} \qquad (5)$$

Gathering the twist angle in a state vector $\boldsymbol{X}$ such that:

$$\boldsymbol{X} = \left[\theta_{1,1} \ldots \theta_{1,N}\ \theta_{2,1} \ldots \theta_{n,N}\right]^T, \qquad (6)$$

the boundary-value problem is thus transformed into a set of non-linear equations of the form:

$$\boldsymbol{G}(\boldsymbol{X}) = \boldsymbol{0} \qquad (7)$$

The resulting discrete model is similar to the ones involved in discrete elastic structures, and can be solved using DR. With DR, state variables are artificially considered as functions of time. The principle is to initially place the robot in an arbitrary shape, which means an initial value of $\boldsymbol{X}$ is guessed. The CTR is then relaxed virtually until it reaches an equilibrium. The dynamics of the state variables according to the virtual time are imposed to fit a second order damped differential equation:

$$\boldsymbol{M}\ddot{\boldsymbol{X}} + \boldsymbol{D}\dot{\boldsymbol{X}} + \boldsymbol{G}(\boldsymbol{X}) = \boldsymbol{0} \tag{8}$$

where $\boldsymbol{M}$ and $\boldsymbol{D}$ are called the fictitious mass matrix and damping matrix respectively, and are usually diagonal matrices [10]. The equilibrium configuration resulting from the DR is then the steady state of (8). The computational efficiency of DR, in terms of numerical stability and computation time relies in the selection of diagonal elements of $\boldsymbol{M}$ and $\boldsymbol{D}$ [10].

The mass matrix $\boldsymbol{M}$ is classically determined in order to maximize integration time-step while ensuring numerical stability, which includes integration stability and convergence of the dynamical system. We consider the integration with the `ode45` solver of Matlab (The MathWorks Inc., Natick, USA) which processes a 4-th order Runge-Kutta method. This implicit integration scheme automatically optimizes the integration time step. Consequently, following the same arguments as [12], the modulus of the mass matrix coefficients do not need to be tuned and can be chosen as unitary. Their sign must be chosen equal to the diagonal components of the gradient of $\boldsymbol{G}$ [2]. According to [10], the damping coefficients are determined to minimize the virtual time at which the steady state is reached. Since this virtual time affects the DR performances, the question of their selection is considered in the following.

4 Method Evaluation

A realistic case study is chosen for the method evaluation using the CTR geometry given in [7] and described by the parameters gathered in Table 1.

4.1 Initial Validation of the DR

Obviously, DR is of interest only if it can allow us to solve accurately the CTR kinematic model. Given the numerical values of $\boldsymbol{G}$ components, the diagonal components of $\boldsymbol{M}$ and $\boldsymbol{D}$ are in an initial step chosen equal to -1 and -10 respectively. The dynamical system is then integrated using `ode45` on a virtual time interval large enough to reach the steady state. The time interval is determined by trials and errors process in the following. The initial guess is chosen as $\theta_{i,k} = \alpha_i, \forall k$ for tube i.

Since the accuracy of the discrete derivatives involved in (4) depends on the number of discretization points N_1 and N_2, so does the accuracy of the computed equilibrium. With $[N_1, N_2] = [4, 100]$ and the steady state defined by a threshold on $\|\boldsymbol{G}(\boldsymbol{X})\|$ equal to 1.10^{-6}, that we will keep in the following of the paper, the equilibrium configuration is reached within a virtual time interval of 1.10^6s. Comparing the DR results with the analytical solution of two-tube CTR fully overlapped provided in [4], the maximum relative error is of approximately 0.5%, validating thus the ability of the DR to compute accurate CTR equilibria.

	Tube 1	Tube 2
α_i (rad)	$\pi/4$	0
L_i (mm)	117	100
β_i (mm)	0	0
$\hat{u}_{ix}(\text{mm}^{-1})$	1/60	1/60
k_{ib} (N.mm^2)	1	1
k_{it} (N.mm^2)	1/1.3	1/1.3

Table 1. Geometrical and mechanical properties of the two tubes.

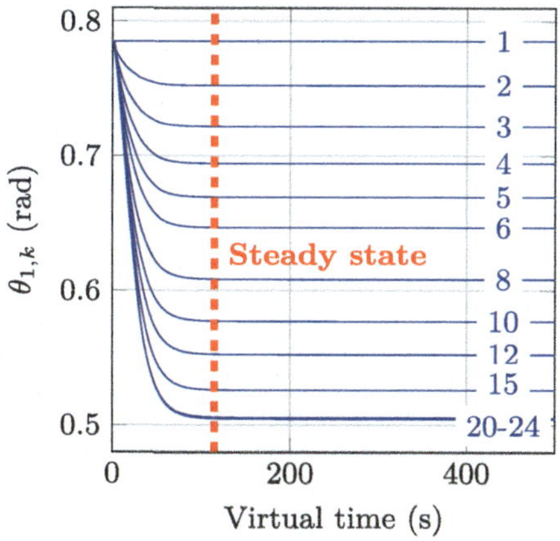

Fig. 2. Evaluation of torsion angles of tube 1 using DR. Index k on the curve.

4.2 Adjustement of DR Parameters

Selection of the damping coefficients in DR is important to control the steady step computation time [10]. The main approach to determine them consists in linearizing (8) and to identify each of its equations to a standard second order differential equation with damping factor ξ and corner pulsation ω_0. The optimal damping corresponds then to $\xi = 1$, and writes for equations numbered k:

$$D_k = 2\omega_{0,k} M_k \tag{9}$$

Different methods exist in the literature to calculate ω_0. In [12], (8) is integrated once without damping. This results in an oscillating response which fundamental pulsation is measured and used as corner pulsation in (9). This method has been proved to provide efficient damping matrices in several application cases of structural mechanics [12] and is consequently chosen here.

The optimal damping matrix is determined for the two-tube CTR described before with $[N_1, N_2] = [4, 20]$, which ensures acceptable accuracy of 2% and reasonable computation times. The dynamical system is integrated without damping on a virtual time interval of 2000 s so that several oscillations are generated to identify ω_0. Using the damping matrix obtained with (9), Fig. 2 is generated by applying DR. The torsion angles converge at the virtual time 114 s, to compare with 9500 s considering the arbitrary values used in Sect. 4.1. Such virtual time is obtained after approximately 5 s of simulation on an Intel Core I5-6300HQ CPU running at 2.3 GHz.

4.3 Robustness of the DR Parameter Selection

It would be interesting to be able to compute several equilibrium configurations of CTR with the same mass and damping matrices in order to simplify the

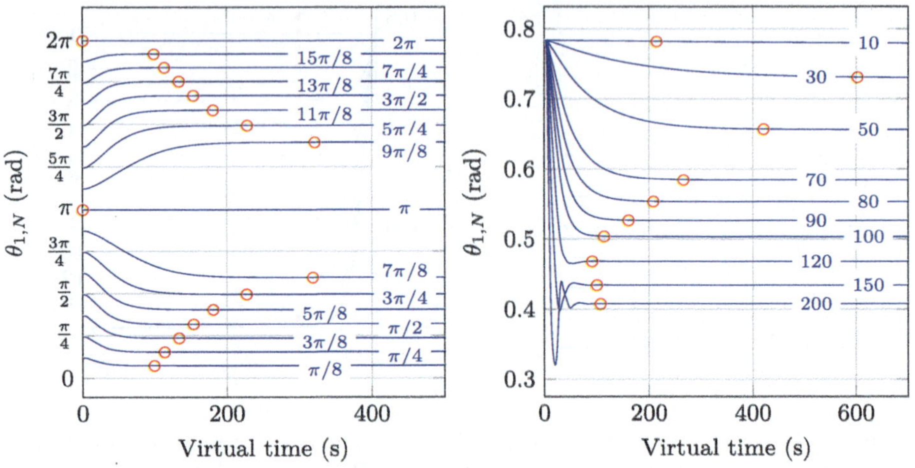

(a) Variations of α_1 (rad). (b) Variations of L_2 (mm).

Fig. 3. Evolution of the virtual time responses obtained with DR for several values of actuation parameters. Steady state times localized with a red circle on each curve. Parameter values written on the curves.

use of DR during the kinematic analysis of a CTR. This would be particularly interesting for the simulation of CTR deployments, where several equilibria must be computed for consecutive values of the actuation inputs.

The mass and damping coefficients determined in the previous section are here used to solve the CTR kinematic model for several values of the angle α_1 and of the deployed length L_2 with DR. The Fig. 3a shows that the method, starting from the initial guess proposed previously, is able to solve the CTR kinematics for all values of α_1, even for the unstable configuration where the tubes are placed in opposition [7]. The virtual time at which the equilibrium is reached equals 321 s for $\alpha_1 = \pi \pm \pi/8$ and 114 s for $\alpha_1 = \pm\pi/8$, indicating an increase in computation time in the neighbourhood of the unstable configuration. The Fig. 3b shows the ability to solve the CTR kinematics for all values of L_2 at the cost of increased virtual response time, which are reached however with acceptable computation time of maximum 13s for $L_2 = 30$ mm.

5 Conclusion

In this paper we have proposed to use dynamic relaxation to solve kinematic model of CTR. From our case study, it appears first that the numerical method is able to compute accurately equilibrium configurations of CTR. Second, computation time can be minimized dramatically by choosing an appropriate damping matrix to reach acceptable values even when CTR geometry is modified during a deployment.

Such initial encouraging results open several perspectives. On one hand, more efficient integration methods in term of computation time could be used to reach the steady-state, such as the explicit integration schemes. On the other hand the efficiency of the damping may be improved by applying tuning methods all along the integration, as proposed in [13]. This could largely improve DR efficiency. Further use to solve CTR kinematics in complex situations with interactions will then be evaluated.

Acknowledgement. This work was supported by the French National Agency for Research within the Biomedical Innovation program (NEMRO ANR-14-CE17-0013), the Investissements d'Avenir (Robotex ANR-10-EQPX-44, Labex CAMI ANR-11-LABX-0004 and Labex ACTION ANR-11-LABX-0001-01) and Aviesan France Life Imaging infrastructure.

References

1. Baek, C., Yoon, K., Kim, D.-N.: Finite element modeling of concentric-tube continuum robots. Struct. Eng. Mech. **57**(5), 809–821 (2016)
2. Barnes, M.R.: Form finding and analysis of tension structures by dynamic relaxation. Int. J. Space Struct. **14**(2), 89–104 (1999)
3. Burgner-Kahrs, J., Rucker, D.C., Choset, H.: Continuum robots for medical applications: a survey. IEEE Trans. Robot. **31**(6), 1261–1280 (2015)
4. Dupont, P.E., Lock, J., Itkowitz, B., Butler, E.: Design and control of concentric-tube robots. IEEE Trans. Robot. **26**(2), 209–225 (2010)
5. Gilbert, H.B., Neimat, J., Webster, R.J.: Concentric tube robots as steerable needles: achieving follow-the-leader deployment. IEEE Trans. Robot. **31**(2), 246–258 (2015)
6. Girerd, C., Rabenorosoa, K., Renaud, P.: Combining tube design and simple kinematic strategy for follow-the-leader deployment of concentric tube robots. In: Advances in Robot Kinematics 2016. Springer Proceedings in Advanced Robotics, pp. 23–31. Springer, Cham (2018)
7. Ha, J., Park, F.C., Dupont, P.E.: Elastic stability of concentric tube robots subject to external loads. IEEE Trans. Biomed. Eng. **63**(6), 1116–1128 (2016)
8. Lewis, W.J., Jones, M.S., Rushton, K.R.: Dynamic relaxation analysis of the non-linear static response of pretensioned cable roofs. Comput. Struct. **18**(6), 989–997 (1984)
9. Lock, J., Laing, G., Mahvash, M., Dupont, P.E.: Quasistatic modeling of concentric tube robots with external loads. In: 2010 IEEE/RSJ International Conference on Intelligent Robots and Systems (IROS), pp. 2325–2332, October 2010
10. Rezaiee-Pajand, M., Alamatian, J.: The dynamic relaxation method using new formulation for fictitious mass and damping. Struct. Eng. Mech. **34**(1), 109–133 (2010)
11. Rucker, D.C., Webster, R.J., Chirikjian, G.S., Cowan, N.J.: Equilibrium conformations of concentric-tube continuum robots. Int. J. Robot. Res. **29**(10), 1263–1280 (2010)
12. Rushton, K.R.: Dynamic-relaxation solutions of elastic-plate problems. J. Strain Anal. **3**(1), 23–32 (1968)
13. Zhang, L.C., Kadkhodayan, M., Mai, Y.W.: Development of the maDR method. Comput. Struct. **52**(1), 1–8 (1994)

Iterative Method for the Inverse Kinematics of a 3-Limb Parallel Mechanism with 3-DOF Using a 6-Limb Mechanism with 6-DOF

Xinghai Liang(✉) and Yukio Takeda

Department of Mechanical Engineering, School of Engineering, Tokyo Institute of Technology, Meguro-Ku, Tokyo, Japan
{liang.x.ab, takeda.y.aa}@m.titech.ac.jp

Abstract. In general, constraint equations are needed for the inverse kinematics (IK) of the parallel mechanism (PM) with less than 6 degrees of freedom (DOF) and dependent parameters describing the pose of the moving platform. However, the applicability of such an approach is restricted to the mechanism configuration from which the constraint equations are derived. In order to solve this problem, this paper presents an iterative method for solving the IK of a 3-limb parallel mechanism with 3 DOF. Therefore, a kinematically equivalent mechanism, which is a virtual model composed of a 6-limb parallel mechanism with 6 DOF under the active joint motion constraint, is introduced. In this method, an iterative procedure for the IK of a 3-PRS parallel mechanism using the IK of the kinematically equivalent mechanism is proposed. Numerical examples demonstrate the convergence process during iteration, the ability to obtain all solutions, and the applicability to various structural parameters of the proposed method.

Keywords: Lower-mobility parallel mechanism · Inverse kinematics
Iterative method · Kinematically equivalent mechanism

1 Introduction

As widely known, the inverse kinematics (IK) of a fully parallel mechanism (PM) is relatively straightforward compared to that of a serial mechanism. For instance, the joints variables are not difficult to determine when the all the pose parameters of the moving platform are known. However, determination of the pose of the moving platform is not intuitive for all types of parallel mechanisms such as some lower-mobility parallel mechanisms (LMPM) with less than 6 DOF.

The kinematic feature of the LMPM considered in this paper is the dependency among the six parameters (three position coordinates and three Euler angles) describing the output pose of the moving platform. Because of this feature, only a limited number (the same as the DOF of the mechanism) of pose parameters of the moving platform can be explicitly specified in the inverse kinematics and the values of the rest pose parameters are determined by the kinematic constraint conditions. For instance, the 3-RPS PM (R, P, S refer to revolute joint, prismatic joint, and spherical joint separately), which was proposed by Hunt in 1983 [1], was also called zero-torsion PM [2]. In order

J. Lenarcic and V. Parenti-Castelli (Eds.): ARK 2018, SPAR 8, pp. 108–115, 2019.
https://doi.org/10.1007/978-3-319-93188-3_13

to solve the IK problem of such 3-RPS PM using conventional methods, firstly the constraint equations need to be derived according to its geometric condition. Secondly, the dependent pose parameters are obtained using three desired output pose parameters and the relationship among the pose parameters of the moving platform. Then, the IK can be solved for the known pose. Following the procedure as mentioned above, Lee [3] chose two of the Euler angles and the Z-coordinate of the moving platform as independent pose parameters. Bürüncük [4] derived the relationship among the pose parameters using the Rodrigues formula. Sokolov [5] confirmed that for the 3-RPS PM there are four solutions for the IK for a given position. Bonev [2] converted the relation between position and orientation into two simple equations using modified Euler Angles. However, the formulas used for the IK in these studies were derived from geometric conditions and thus are limited to mechanisms with special configurations such as those with symmetrically arranged limbs and the same joint arrangements in each limb. This means if the configuration of the mechanism changes, the method may not be suitable or need to be reformulated. Therefore, the applicability of these methods is restricted, and a general approach is needed.

The IK for fully PM with 6 DOF composed of six serial limbs, each having two spherical (S) joints or a spherical joint and a universal (U) joint, such as 6-S(U)PS or 6-PS(U)S, where P stands for an active prismatic joint, is well known to be quite easy to solve. As discussed in [6], a 3-RPS mechanism and a 6-SPS mechanism with three double spherical joints in the moving platform are considered to be kinematically equivalent if the three pairs of the prismatic joints in 6-SPS mechanism are driven in pairs. Following a similar concept of kinematic equivalence, the transmission index for a 3-PRS [7] mechanism and a 3-SPR mechanism [8] were proposed based on the pressure angle approach defined for 6-PSS or 6-SPS mechanisms [9], respectively.

This paper presents a novel method for solving IK problem of 3-PRS PM using an iterative procedure for the kinematically equivalent mechanism, a 6-PSS PM. Section 2 shows the kinematic equivalent modeling and procedure for solving IK of 3-PRS PM. Section 3 shows numerical results of solving IK with different parameters and discusses the applicability of the proposed method. Section 4 summarizes the results.

2 Iterative Method for the Inverse Kinematics of the 3-PRS PM

2.1 Mechanism Configuration and IK Based on Constraint Equations

The 3-PRS PM as shown in Fig. 1 is taken as an example to demonstrate the proposed method. This mechanism is composed of a base and a moving platform that are connected to each other by three limbs of identical structure. Each limb consists of an actuated prismatic joint, a passive revolute joint, and a passive spherical joint. Two coordinate systems are considered. One is fixed on the base O-*XYZ* and the other one is fixed on the moving platform P-*uvw*, where P refers to the reference point of the moving platform.

The constraint equations of 3-PRS PM are usually derived from (1) the trigonometric relationships of the positions of the joints on the base and moving platform,

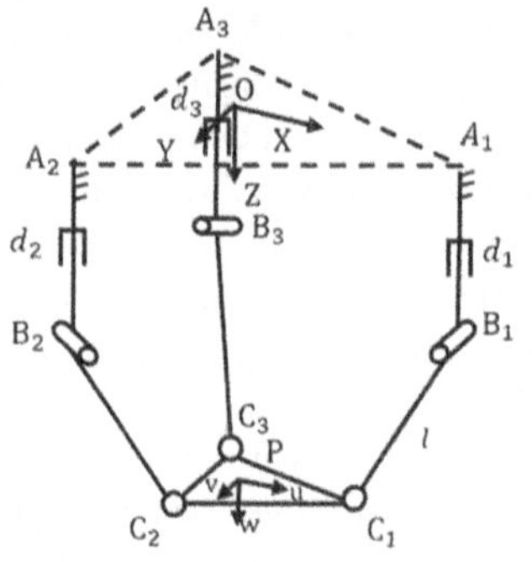

Fig. 1. 3-PRS parallel mechanism (PM)

(2) the constraints imposed by the revolute joints, and (3) the orthogonal conditions of the orientation matrix. In literature, many studies are concerned with PMs with symmetrically arranged limbs, in which the base and moving platform are set to be equilateral triangles and the three limbs are identical to each other. As a result, closed-form formulas of the IK are successfully obtained. In some research, the actuator layout angle (the angle between the actuated prismatic joint and the base) [10] and different limb arrangements [11] were considered in the derivation of the constraint equations. As pointed out in [12], the IK of a 3-SPR PM, which is the kinematic inversion of 3-RPS PM, is much more complicated than that of the 3-RPS PM. To our best knowledge, however, there have not been developed any effective methods for solving the IK of the 3-PRS and 3-PSR PMs with general structural parameters.

2.2 Kinematically Equivalent PM Composed of Six Limbs

Figure 2 shows two kinematically equivalent mechanisms of the 3-PRS PM. In the 6-PSS PM shown in Fig. 2(a), a pair of S-S chains imposes the same constraint as an R-S chain. Let us consider a case, where kinematic dimensions of 6-PSS PM are selected so that the locations of C_i in P-*uvw* coordinate system and the distance l between C_i and line $B_{1,i}B_{2,i}$ (i = 1,2,3) are the same as those of 3-PRS PM and B_iC_i (= l) in 3-PRS PM, respectively. The distance between $A_{1,i}$ and $A_{2,i}$ is denoted as t, which corresponds to the axial length of the revolute joint in the real machine of the LMPM and affects its motion/force transmissibility and constrainability [7, 8]. When both displacements of the active joints $q_{1,i}$ and $q_{2,i}$ in the 6-PSS PM are the same as the equivalent displacement d_i in the 3-PRS PM (Fig. 1), the motions of these two mechanisms will be the same. In this context, we call these mechanisms kinematically equivalent to each other under the active joint motion constraint. Assume the 3-PUS PM shown in Fig. 2 (b) to exhibit additional actuation of the first revolute joints of the universal joints in each limb (with $\theta_{B'i}$ denoting the related displacements), which is 3-PRRS. The motion of the moving platform when the active prismatic joints are driven while keeping $\theta_{B'i} = 0$ in the 3-PRRS PM is completely the same as that of the 3-PRS PM. In other words, these mechanisms are considered kinematically equivalent under the active joint motion constraint in the 3-PRRS PM. In the present paper, we focus on the former case

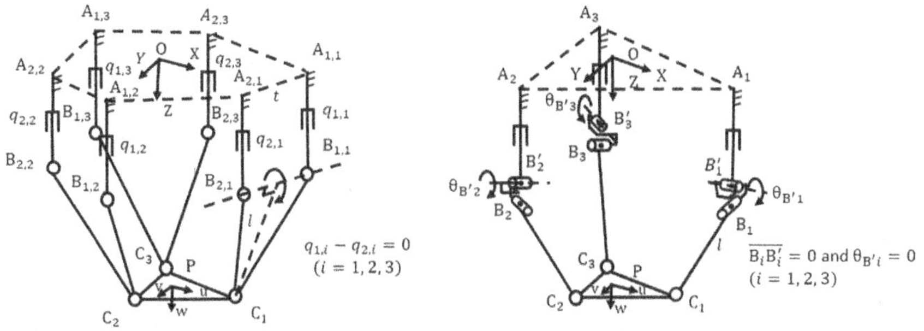

(a) 6-PSS PM (b) 3-PUS PM

Fig. 2. Parallel mechanisms composed of 6-DOF limbs with active joint displacement constraints as kinematically equivalent mechanisms to 3-PRS parallel mechanism

to develop an iterative method for the IK of the 3-PRS PM based on the IK for the 6-PSS PM. The input displacements $\left[q_{1,1}, q_{2,1}, q_{1,2}, q_{2,2}, q_{1,3}, q_{2,3}\right]^T$ result from the solution of the IK of the 6-PSS PM and can be easily obtained as the intersection of a sphere centered at C_i with a radius $B_{j,i}C_i$ (j = 1,2) and a line $A_{j,i}B_{j,i}$. The IK problem for the 3-PRS PM can thus be defined as finding the input displacements $\left[q_{1,1}, q_{2,1}, q_{1,2}, q_{2,2}, q_{1,3}, q_{2,3}\right]^T$ satisfying $q_{1,i} = q_{2,i}$ and the three dependent pose parameters for the given three independent pose parameters using the kinematically equivalent 6-PSS PM. In this paper, the position coordinates $\boldsymbol{P} = \left[p_x, p_y, p_z\right]^T$ are considered the independent pose parameters while the three parameters representing the orientation of the moving platform are considered the dependent pose parameters.

2.3 Iterative Method for the Inverse Kinematics

The infinitesimal displacement relationship between the input and output of the 6-PSS PM is written as

$$\begin{bmatrix} \Delta \boldsymbol{P} \\ \Delta \boldsymbol{\Theta} \end{bmatrix} = \boldsymbol{J}_{6-\mathrm{PSS}}^{-1} \Delta \boldsymbol{q}, \Delta \boldsymbol{q} = \left[\Delta q_{1,1}\ \Delta q_{2,1}\ \Delta q_{1,2}\ \Delta q_{2,2}\ \Delta q_{1,3}\ \Delta q_{2,3}\right]^T, \quad (1)$$

while $\Delta \boldsymbol{\Theta} = \left[\varDelta\Theta_x\ \varDelta\Theta_y\ \varDelta\Theta_z\right]^T$ and $\boldsymbol{J}_{6-\mathrm{PSS}}$ represent the infinitesimal orientation vectors of the moving platform and the Jacobian matrix, respectively. For kinematic equivalence to 3-PRS PM, the following conditions should be approximately satisfied:

$$\boldsymbol{J}_{6-\mathrm{PSS}}^{-1} \begin{bmatrix} \varDelta q_{1,1} \\ \varDelta q_{2,1} \\ \varDelta q_{1,2} \\ \varDelta q_{2,2} \\ \varDelta q_{1,3} \\ \varDelta q_{2,3} \end{bmatrix} = \begin{bmatrix} 0 \\ 0 \\ 0 \\ \varDelta\Theta_x \\ \varDelta\Theta_y \\ \varDelta\Theta_z \end{bmatrix}, q_{1,i} - q_{2,i} = 0 (i = 1, 2, 3). \quad (2)$$

These can be reformulated as

$$\begin{bmatrix} \boldsymbol{J}_{6-\mathrm{PSS}}^{-1} & \begin{matrix} \boldsymbol{O}_{3\times 3} \\ \begin{matrix} -1 & 0 & 0 \\ 0 & -1 & 0 \\ 0 & 0 & -1 \end{matrix} \end{matrix} \\ \begin{matrix} 1 & -1 & 0 & 0 & 0 & 0 \\ 0 & 0 & 1 & -1 & 0 & 0 \\ 0 & 0 & 0 & 0 & 1 & -1 \end{matrix} & \boldsymbol{O}_{3\times 3} \end{bmatrix}_{9\times 9} \begin{bmatrix} \Delta q_{1,1} \\ \Delta q_{2,1} \\ \Delta q_{1,2} \\ \Delta q_{2,2} \\ \Delta q_{1,3} \\ \Delta q_{2,3} \\ \Delta\Theta_x \\ \Delta\Theta_y \\ \Delta\Theta_z \end{bmatrix} = \begin{bmatrix} 0 \\ 0 \\ 0 \\ 0 \\ 0 \\ 0 \\ q_{2,1} - q_{1,1} \\ q_{2,2} - q_{1,2} \\ q_{2,3} - q_{1,3} \end{bmatrix}. \quad (3)$$

Using the solution $\boldsymbol{\Delta\Theta} = [\Delta\Theta_x\, \Delta\Theta_y\, \Delta\Theta_z]^T$ of Eq. (3), the orientation of the moving platform can be updated. The similar process is iterated until the solution converged. The flowchart of this iterative method is shown in Fig. 3.

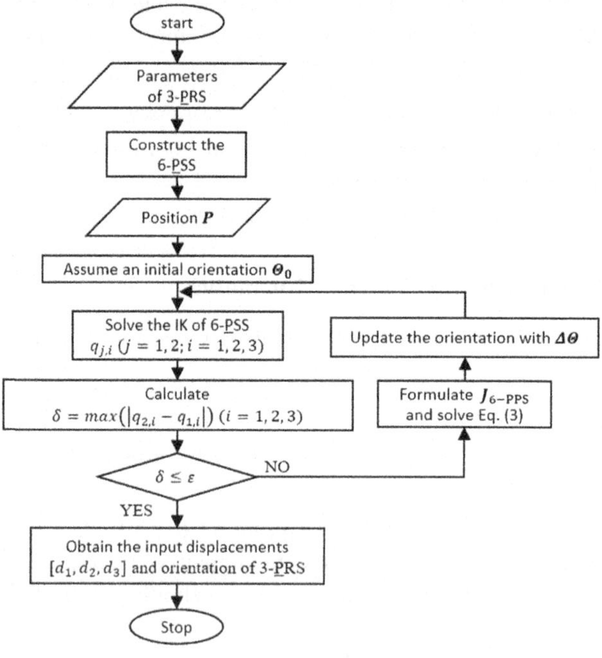

Fig. 3. The iteration process of the proposed method

Because the structural parameters of the 6-PSS PM are determined by the given 3-PRS PM and the IK of the 6-PSS PM is straightforward to solve, the proposed method is not necessary to formulate the constraint equations and is able to solve the IK problems of 3-PRS PMs with any structural parameters. Furthermore, based on the

Eq. (1), it is possible to use any three of the six parameters of the pose as the inputs to obtain the desired outputs of the pose.

3 Numerical Examples

In the following, numerical results of the IK of 3-PRS PMs with different structural parameters are shown. Without specified definition, the following conditions and values are used: the base and moving platform are assumed to be equilateral triangles (the radii of the circumcircles are 0.25 and 0.125, respectively), the prismatic joints are parallel to the *Z*-axis in O, the link length $l = 0.35$, and the distance between $A_{1,i}$ and $A_{2,i}$ $t = 0.4$.

Figure 4 shows the iterating process of this method for solving the IK for a specified position $\boldsymbol{P} = [0, 0.015, 0.45]^T$ and the initial orientation of the moving platform (45°, 5°, 0) set as Tilt-and-Torsion (T&T) angles which refers to modified Euler angles mentioned in [2]. In Fig. 4, the two horizontal reference lines in each figure indicate the two solutions of the inverse kinematics of 3-PRS PM conventionally calculated based on the constraint equations. As shown in the figure, the input displacements smoothly converge to the solution for orientation 1 following the proposed method.

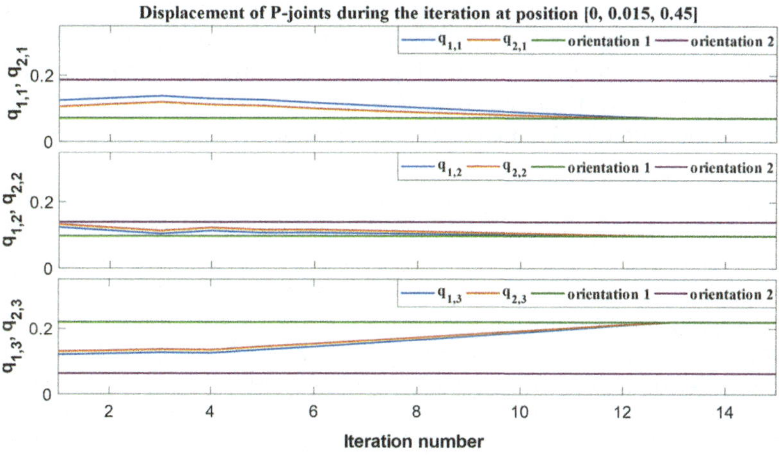

Fig. 4. Convergence process of the input displacements of the 6-PSS PM for solving the IK of 3-PRS PM at position [0, 0.015, 0.45]

All possible orientations of the moving platform for a given position can be obtained by the proposed method with different initial orientations. Figure 5 shows the iteration process of the input displacements $q_{1,1}$ and $q_{2,1}$ at the same position as shown in Fig. 4 with two different initial orientations. The orientations were defined by as T&T angles (φ, 5°, 0) and (φ = 135°, 315°), respectively. As shown in the figure, it is confirmed that the two possible solutions were successfully obtained by using different initial orientations.

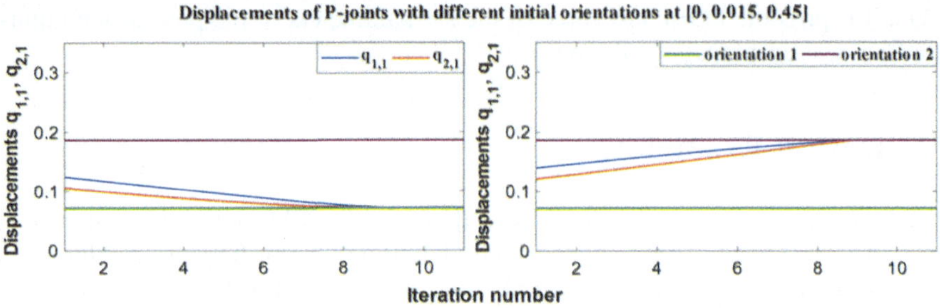

(a) φ=135°   (b) φ=315°

Fig. 5. Convergence process of the input displacements with two different initial orientations

The input displacements of the 3-PRS PM along a trajectory are shown in Fig. 6(a). The trajectory was given as a horizontal circle centered at $[0, 0, 0.45]^T$ with a radius of 0.015. In this case, except for the first position on the trajectory, the IK was solved at all positions using the orientation of the previous position as the initial orientation for iteration. As a result, an efficient convergence of iteration is achieved. As shown in the figure, it is also confirmed that the proposed method can be applied to the IK for trajectory tracking or path planning. Also, the input displacements of the 3-PRS mechanism with different structural parameters are shown in Fig. 6(b), (c) and (d). In Fig. 6(b) the layout angles of the prismatic joints are set to 60°, in (c) the prismatic joints are located asymmetrically on the base, and in (d) the order of the R and S joints in each limb is reversed. It can be concluded that the solutions of the IK of the 3-PRS PM with various structural parameters are successfully obtained by the proposed method.

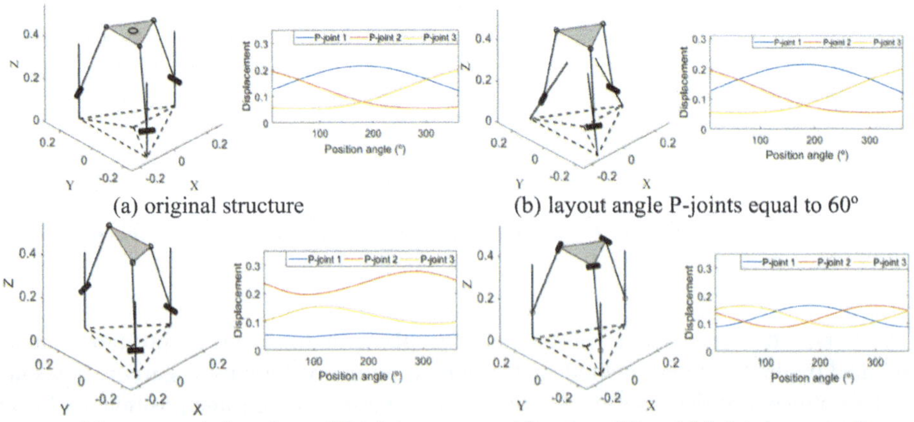

(a) original structure   (b) layout angle P-joints equal to 60°

(c) asymmetric locations of P-joints   (d) order of R and S joints is reversed

Fig. 6. Trajectory tracking based on IK of 3-PRS PM with varied structural parameters

4 Discussions and Conclusions

In this paper, a novel iterative method for solving the inverse kinematic problem of the lower-mobility parallel mechanism is proposed. The fundamental ideas of this method, modeling the kinematically equivalent mechanism by a 6-limb PM with 6 DOF and the iteration procedure to solve the IK problem, are demonstrated. The IK problems of the 3-PRS PM with various structural parameters were successfully solved using the proposed method. The results revealed that it is an effective approach to solve the IK of the 3-PRS PM. Moreover, without considering the geometric (trigonometric) conditions among the joints but only using the positions of them, this method can also be used in other types lower-mobility parallel mechanism with this unified model based on the concept of kinematic equivalence. Moreover, it has the potential to be used as a more general and simple method in parametric design, control and calibration in practice.

References

1. Hunt, K.H.: Structural Kinematics of in- parallel-actuated robot-arms. J. Mech. Transm. Autom. Des. **105**(4), 705–712 (1983)
2. Bonev I.A., Zlatanov D., Gosselin C.M.: Advantages of the modified euler angles in the design and control of PKMs. In: Parallel Kinematic Machines International Conference, pp. 171–188 (2002)
3. Lee, K.M., Shah, D.K.: Kinematic analysis of a three degrees of freedom in-parallel actuated manipulator. IEEE J. Robot. Autom. **4**(3), 354–360 (1988)
4. Bürüncük, K., Tokad, Y.: On the kinematic of a 3-DOF stewart platform. J. Field Robot. **16** (2), 105–118 (1999)
5. Sokolov, A., Xirouchakis, P.: Kinematics of a 3-DOF parallel manipulator with an R-P-S joint structure. Robotica **23**(2), 207–217 (2005)
6. Nanua, P., Waldron, K.J., Murthy, V.: Direct kinematic solution of a stewart platform. IEEE Trans. Robot. Autom. **6**(4), 438–444 (1990)
7. Takeda, Y., Liang, X.: Transmission index of lower-mobility parallel mechanism: case study on 3-PRS mechanism. In: ASME 2017 IDECT/CIE Conference. ASME, USA (2017)
8. Russo, M., Ceccarelli, M., Takeda, Y.: Force transmission and constraint analysis of a 3-SPR parallel manipulator. In: Proceedings of the Institution of Mechanical Engineers, Part C: Journal of Mechanical Engineering Science, 954406217750190 (2017)
9. Takeda, Y., Funabashi, H.: Motion Transmissibility of In-Parallel Actuated Manipulators. JSME Int. J. Ser. C, Dyn. Control Robot., Des. Manuf., 38(4), 749–755 (1995)
10. Li, Y., Xu, Q.: Kinematic analysis of a 3-PRS parallel manipulator. Robot. Comput.-Integrated Manuf. **23**(4), 395–408 (2007)
11. Li, Q., Chen, Z., Chen, Q., Wu, C., Hu, X.: Parasitic motion comparison of 3-PRS parallel mechanism with different limb arrangements. Robot. Comput.-Integr. Manuf. **27**(2), 389–396 (2011)
12. Lukanin, V.: Inverse kinematics, forward kinematics and working space determination of 3DOF parallel manipulator with S-P-R joint structure. Period. Polytech. Eng. Mech. Eng. **49** (1), 39–61 (2005)

Optimal Object Placement Using a Virtual Axis

Martin Weiß[(✉)]

Faculty of Computer Science and Mathematics,
Ostbayerische Technische Hochschule Regensburg, Regensburg, Germany
martin.weiss@oth-regensburg.de

Abstract. A basic task in the design of a robotic production cell is the relative placement of robot and workpiece. The fundamental requirement is that the robot can reach all process positions; only then one can think further optimization. Therefore an algorithm that automatically places an object into the workspace is very desirable. However many iterative optimization algorithms cannot guarantee that all intermediate steps are reachable, resulting in complicated procedures. We present a novel approach which extends a robot by a virtual prismatic joint - which measures the distance to the workspace - such that any TCP frames are reachable. This allows higher order nonlinear programming algorithms to be used for placement of an object alone as well as the optimal placement under some differentiable criterion.

Keywords: Optimization · Virtual joint · Inverse kinematics
Nonsmooth optimization · Workspace · Cartesian tasks

1 Problem Statement

We consider the following task: A robot should unload a storage box with a chessboard like structure containing $B_x \times B_y$ identical workpieces at positions P_{kl}, $k = 1, \ldots, B_x$, $l = 1, \ldots, B_y$, counted in the coordinate directions of the frame $C \in \mathbb{F}$ (where $\mathbb{F}$ denotes the set of all frames $\mathbb{R}^3 \times \mathrm{SO}(3)$) associated with the box at distances D_x and D_y. Think of test-tubes in medicine or small parts in general production. The cell setup is considered fixed, so only the placement of the box in the cell can be chosen, e.g. because a new order type.

Usually a pick-and-place operation is programmed at one corner only, the other position commands are computed from this corner position and the indices and distances. However it is difficult for the user to assess whether all positions are reachable because of nonlinearity and axis limits. Testing the corners is a heuristic that works in many cases but there is no guarantee, so one has to run time-consuming tests. When the process needs to work on an object from different sides or with different orientations the situation is even more complicated. So the user would like to have an algorithm that determines a

J. Lenarcic and V. Parenti-Castelli (Eds.): ARK 2018, SPAR 8, pp. 116–123, 2019.
https://doi.org/10.1007/978-3-319-93188-3_14

feasible object frame C near some initial guess C_0, maybe additionally optimizing one of the many known manipulability measures, see [8,12].

It is easy to check in a program whether a given frame C leads to reachable positions or not but it is difficult for a nonlinear optimizer to determine a direction that leads to a "more feasible" situation: Feasibility is a binary decision; the backward transformation will usually issue an error only, and abort.

Our idea is to introduce a virtual joint as a *slack variable* in terms of nonlinear programming (see [9]) into the optimization problem that measures the distance of a position from feasibility. This variable therefore has an intuitive geometric interpretation. We can also interpret the virtual axis as a homotopy variable similar to [4] which gives an "easy" solution for large values. Our slack variable is not generated by the standard procedure replacing an inequality constraint $g(x) \leq 0$ by the equality $g(x) + z = 0$, with the sign constraint $z \geq 0$.

Our approach has some similarity to the introduction of virtual axes for singularity avoidance in [10] or [7]. However we do not introduce a rotational joint to reduce velocities near singularities but rather use a prismatic joint to enlargen the mathematical workspace in the optimization process. In combination with a smoothing operation we can use standard optimization algorithms which require differentiability of order 1 like all algorithms based on gradient descent, or order 2 like Sequential Quadratic Programming (SQP), cf. [9]. Our approach is not related at all to voxelization of the workspace like in [11] and other algorithms aiming at collision free planning.

2 Virtual Axis Approach

For ease of exposition we choose a 6R robot resembling the well known Puma 560 but with more zeros in the parameters. We could extend all formulae to similar 6R real industrial robots. We use the DH convention

$$\mathrm{R}_z(q_i) \cdot \mathrm{T}_z(d_i) \cdot \mathrm{T}_x(a_i) \cdot \mathrm{R}_x(\alpha_i) =: \mathrm{R}_z(q_i) \cdot B_i =: A_i(q_i)$$

to get the wrist centre point (WCP) and tool centre point (TCP)

$$\begin{aligned} \mathsf{WCP}(q) &= A_1(q_1) \cdot A_2(q_2) \cdot A_3(q_3) \cdot A_4(q_4) \cdot A_5(q_5) \cdot A_6(q_6) \\ \mathsf{TCP}(q) &= \mathsf{WCP}(q) \cdot \mathsf{TOOL} \end{aligned}$$

expressed relative to the world coordinate system chosen as the axis 1 coordinate system. Note that $a_6 = d_6 = 0$ in our case so the DH chain ends in the WCP as the essential point for the backward transform. Figure 1 shows the robot data and the reference position. We use $l_{23} = 315$, $l_{35} = 365$ and a tool with $\mathrm{T}_z(t_z)$, $t_z = 100$, pointing in direction z_6 from the flange. Lengths are measured in [mm], angles in [rad]. Note that joint 3 is pointing upward for $q = 0$, so the stretched position corresponds to $q_3 = -\frac{\pi}{2}$. We assume joint limits $-\pi \leq q_{\min,i} \leq q_i \leq q_{\max,i} \leq \pi$, $i = 1, \ldots 6$. As usual, this type of robots has up to 8 discrete solutions of the backward transform for non-singular positions. We identify these 8 configurations with an integer $s \in \{0, \ldots, 7\}$.

i	θ_i	d_i	a_i	α_i	type
1	q_1	0	0	$\frac{\pi}{2}$	R
2	q_2	0	l_{23}	0	R
3	q_3	0	0	$-\frac{\pi}{2}$	R
4	q_4	l_{35}	0	$\frac{\pi}{2}$	R
5	q_5	0	0	$-\frac{\pi}{2}$	R
6	q_6	0	0	0	R

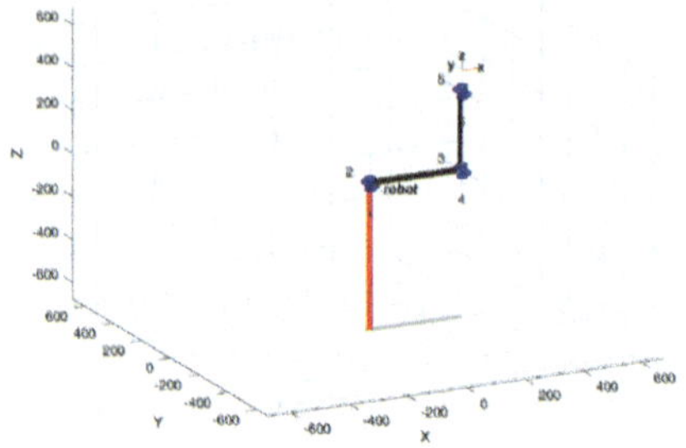

Fig. 1. DH parameters and reference position $q = 0$ for original robots

Infeasibility of the backward transform for a given frame F and configuration s may arise from two reasons with different severity: First, the WCP may be to far from the robot such that the triangle construction for q_3 fails. There is no remedy in this case. Second, even if axis values q exist such that $\mathsf{TCP}(q) = F$, these might violate the joint limits: $q_i \notin [a_i, b_i]$ for some i. This is no obstacle during the optimization process, only for a solution. So the second problem can be fixed by dropping the joint limits and allowing $q_i \in (-\pi, \pi]$, $i = 1, \ldots, 6$. We describe this difference by a *physical* and a *mathematical* workspace $\mathcal{W}_P$ and $\mathcal{W}_M$ which are the WCP frames under the two joint restrictions under consideration:

$$\mathcal{W}_P = \mathsf{WCP}\left(\prod_{i=1}^{6}[q_{\min,i}, q_{\max,i}]\right), \qquad \mathcal{W}_M = \mathsf{WCP}\left((-\pi,\pi]^6\right)$$

We have $\mathcal{W}_P \subseteq \mathcal{W}_M$ and $\mathcal{W}_P \neq \mathcal{W}_M$ in general but $\mathcal{W}_M$ is still a bounded set, which is the first problem.

In order to use optimization algorithms which may leave the feasible set $\mathcal{W}_M$, our goal is to define a *virtual robot* which has a solution for the backward transform for any frame $F \in \mathbb{F}$ and any configuration s. So we associate to our *original robot* a *virtual robot* with an additional *virtual prismatic joint* between joints 3 and 4, which has no joint limits. Any WCP in $\mathbb{R}^3$ is reachable then. The variable of the virtual joint will be denoted v, the other joints keep their names giving a combined joint variable $\tilde{q} = (q_1, q_2, q_3, v, q_4, q_5, q_6) \in \mathbb{R}^7$. DH parameters of the virtual robot are shown in Fig. 2.

Sufficient conditions for our approach are stated as two assumptions:

Assumption 1 Reachability of $\mathbb{R}^3$: The mapping of the original joints and the virtual joint to the WCP position is surjective onto $\mathbb{R}^3$.

Assumption 2 Reachability of SO(3): Joints 4, 5, 6 form a central wrist parametrizing all of $SO(3)$, i.e. the mapping $(-\pi,\pi]^3 \to SO(3)$, $(q_4, q_5, q_6) \mapsto \mathrm{R}_z(q_4) \cdot B_4 \cdot \mathrm{R}_z(q_5) \cdot B_5 \cdot \mathrm{R}_z(q_6)$ is surjective.

In our case Assumption 1 is satisfied because for any fixed q_3 value the virtual joint generates an infinite line with the WCP, the second joint rotates the line

i	θ_i	d_i	a_i	α_i	type
1	$\tilde{q}_1$	0	0	$\frac{\pi}{2}$	R
2	$\tilde{q}_2$	0	l_{23}	0	R
3	$\tilde{q}_3$	0	0	$-\frac{\pi}{2}$	R
4	0	v	0	0	P
5	$\tilde{q}_4$	l_{35}	0	$\frac{\pi}{2}$	R
6	$\tilde{q}_5$	0	0	$-\frac{\pi}{2}$	R
7	$\tilde{q}_6$	0	0	0	R

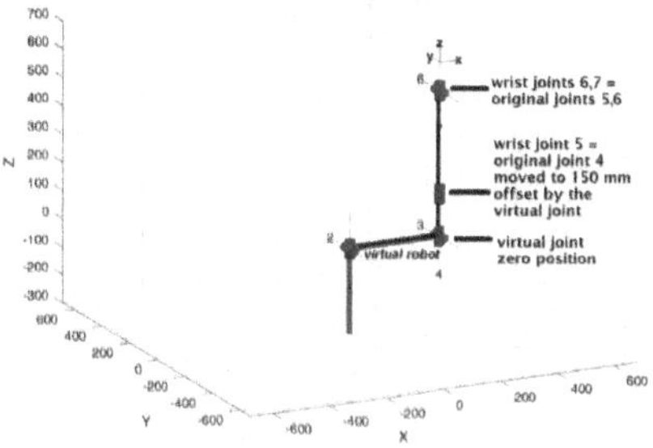

Fig. 2. DH parameters and reference position $\tilde{q} = (0, 0, 0, 150, 0, 0, 0)$ for virtual robot

through a plane, the first joint rotates the plane through all $\mathbb{R}^3$. Central wrists satisfying Assumption 2 – for unbounded joint variables – are the most common choice in industry. Assumption 2 also guarantees an 8-solution kinematics. Using the notation WCP^v to distinguish the forward transform of the virtual robot we denote

$$\mathcal{W}_V = \mathsf{WCP}^v \left((-\pi, \pi]^3 \times \mathbb{R} \times (-\pi, \pi]^3] \right).$$

Under our assumptions we get $\mathcal{W}_V = \mathbb{F}$. We call such a robot a *dextrous robot* because the dextrous workspace in $\mathbb{R}^3$ (points reachable with all orientations, see [2]) and reachable workspace (points reachable with at least one orientation) coincide, and are all of $\mathbb{R}^3$.

3 Backward Transform with Redundancy Resolution

However we have introduced redundancy in our kinematics so we have to define a backward transform giving unique results. The virtual robot backward transform sets the virtual joint to the smallest absolute value such that a solution exists. In our case this is the distance between the WCP position and $\mathcal{W}_M$ which is a hollow sphere for our robot so calculations are simple. Algorithm 1 uses the backward transform of the original robot: Note that both $l_{23} > l_{35}$ and $l_{23} < l_{35}$ lead to an empty interior, but $l_{23} < l_{35}$ results in negative v values. Also note that for our robot the stretched position for axis 3 is $q_3 = \pm\frac{\pi}{2}$, not 0 as in most industrial robots. This stretched position is used for all WCP positions outside $\mathcal{W}_M$.

For robots other than our simple one the computation of q_1, as well as the definition of the stretched position and the computation of v have to be adapted.

4 Smoothness Properties

The dexterity of the virtual robot makes all TCP frames feasible. Also the joint values $\tilde{q}$ depend continuously on the TCP frame, if we keep the configuration fixed and avoid singularities or the original robot. Inside $\mathcal{W}_M$ continuity is clear from the usual backward transformation formulae. When the WCP reaches the boundary of $\mathcal{W}_M$ the triangles for the elbow-up and elbow-down configuration degenerate to a line and coincide, hence $q_3 \to -\frac{\pi}{2}$ which is the same value as when approaching from outside. Also $q_2 = \operatorname{atan2}(z, r)$ depends continuously on the wrist centre point. However the dependence is not differentiable: When the wrist centre point enters $\mathcal{W}_M$ from outside the solution for q_3 applies the cosine theorem to get $c = \cos(\varphi)$ near $\phi = \pi$. Computing $\varphi(c) = \operatorname{atan2}(\pm\sqrt{1-c^2}, c)$ we get an infinite slope at $c = 1$, see Fig. 3. This non-differentiability can affect all other axes.

Algorithm 1. virtual robot backward transform

Input: WCP frame F parametrized by $P = [x, y, z] \in \mathrm{I\!R}^3$, $Q \in \mathrm{SO}(3)$
Input: configuration coded as $s \in \{0, \ldots, 7\}$
Output: virtual robot axes $\tilde{q} \in (-\pi, \pi]^3 \times \mathrm{I\!R} \times (-\pi, \pi]^3]$ with $\mathsf{WCP}(\tilde{q}) = F$ where $\tilde{q}$ corresponds to configuration s
  if solution $q \in (-\pi, \pi]^6$ of original backward transform exists **then**
    return essentially this solution as $\tilde{q} = (q_1, q_2, q_3, 0, q_4, q_5, q_6)$
  else
    compute q_1 using the original backward transform: $q_1 = \operatorname{atan2}(y, x)$
    $d := ||P||_2$, distance of WCP from robot base
    if P is outside of the hollow sphere, $d > l_{23} + l_{35}$ **then**
      $v := d - l_{23} - l_{35}$
      put joints 2 and 3 in stretched position $q_2 := \operatorname{atan2}(z, r)$, $q_3 := -\frac{\pi}{2}$
    else
      P is in the empty interior of the hollow sphere, $d < |l_{23} - l_{35}|$
      $v := d - l_{23} + l_{35}$
      put joints 2 and 3 in stretched position $q_2 := \operatorname{atan2}(z, r)$, $q_3 := +\frac{\pi}{2}$
    end if
  end if

In order to get C^2 behaviour of all angles we modify $\varphi(c)$ on some interval $[1-\epsilon, 1]$ with a polynomial of degree 5 with C^2 transition at $c = 1-\epsilon$ and first and second derivative 0 at $c = 1$, see the red graph in Fig. 3, right. This is a basic idea

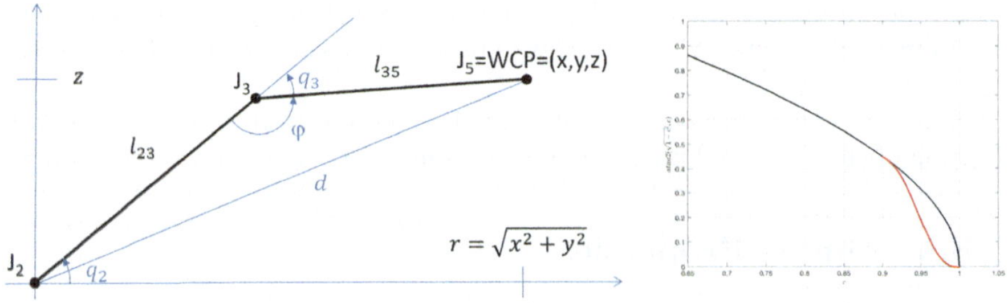

Fig. 3. Backward transform and smoothing

in nonsmooth optimization [1]. Of course this distorts the backward transform and one has to check whether smoothing affects the optimal solution but the hope is that the optimum is sufficiently inside the workspace, and smoothing is only a temporary help for the optimization algorithm.

Figure 4 shows this behaviour for a motion of the virtual robot TCP from L_0 along the x axis to L_1 in height $z = 215$ with constant orientation Q with

$$L_0 = \begin{bmatrix} 500 \\ 0 \\ 215 \end{bmatrix}, \qquad L_1 = \begin{bmatrix} 1300 \\ 0 \\ 215 \end{bmatrix}, \qquad Q = \begin{bmatrix} -1 & 0 & 0 \\ 0 & 1 & 0 \\ 0 & 0 & -1 \end{bmatrix}$$

in the elbow-up configuration such that the tool is pointing downwards. The WCP crosses the boundary $\partial\mathcal{W}_M$ at

$$\partial x = \sqrt{(l_{23} + l_{35})^2 - (215 + t_z)^2} = \sqrt{680^2 - 315^2} \approx 602.6.$$

Axes q_1, q_4, q_6 do not move. All joints are continuous indeed with infinite slope of q_2, q_3, q_5 at ∂x. The left picture shows the axis values without smoothing, the right a zoom into the q_3 behaviour without and with smoothing. The virtual axis remains at $v = 0$ in $\mathcal{W}_M$, then v grows linearly. So the virtual joint $v(x)$ is also non-differentiable at ∂x but has essentially the same behaviour like $x \mapsto \max\{0, x\}$ at $x = 0$. Forming the square $x \mapsto (\max\{0, x\})^2$ creates a C^1 function, we will use this trick in our objective function.

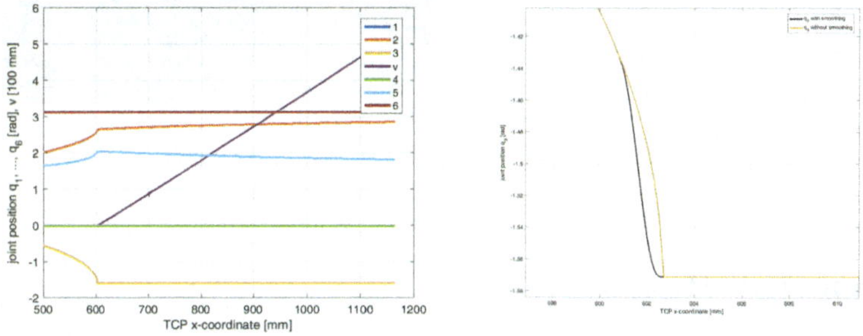

Fig. 4. Joint values for motion through workspace boundary

5 Formulation of the Optimization Problem

We assume that poses must be reached with the same configuration; this is quite usual for Cartesian task. We parametrize the corner frame as $C = \mathrm{Trans}(x, y, z) \cdot \mathrm{R}_z(\alpha) \cdot \mathrm{R}_y(\beta) \cdot \mathrm{R}_x(\gamma)$. These parameters or a subset thereof constitute our optimization variables. Denoting $\tilde{q}^{(k,l)} = \tilde{q}^{(k,l)(C)} = \tilde{q}^{(k,l)}(x, y, z, \alpha, \beta, \gamma)$ the joint values obtained by the virtual robot backward transform for grid position (k, l) in the box, $q_i^{(k,l)}$ and $v^{(k,l)}$ the original and virtual joints, we may optimize

$$\min_{x,y,z,\alpha,\beta,\gamma} \sum_{k=1}^{B_x}\sum_{l=1}^{B_y}(v^{(k,l)})^2$$
$$\text{under } q_{\min,i} \le q_i^{(k,l)} \le q_{\max,i}$$
$$i = 1,\ldots,6, k = 1,\ldots,B_x, l = 1,\ldots,B_x$$

We can also add constraints on the frame parameters $(x, y, z, \alpha, \beta, \gamma)$. The square in the objective function makes the objective function C^1. To make advantage of this however we must use the C^2 smoothing of the backward transform at the workspace boundary so that the constraints are C^2 as well. With $|v^{(k,l)}|^3$ we could even obtain a C^3 objective function.

Adding some manipulability criterion from [12] or [8] to the objective function can optimize feasibility and manipulability in combination. However one has to use appropriate weighting because in extreme cases the optimizer might tolerate some infeasible points in exchange for high manipulability at other points.

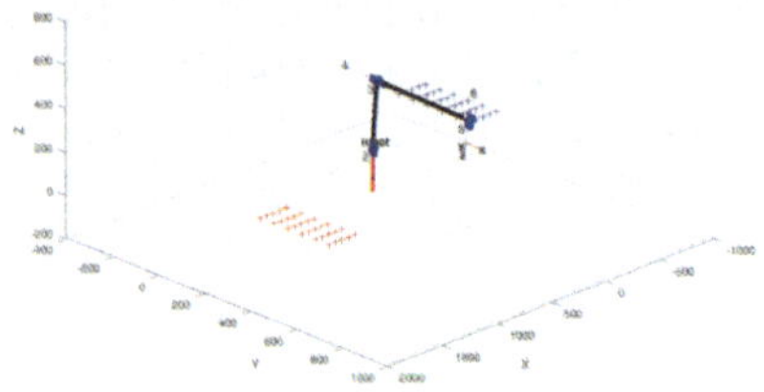

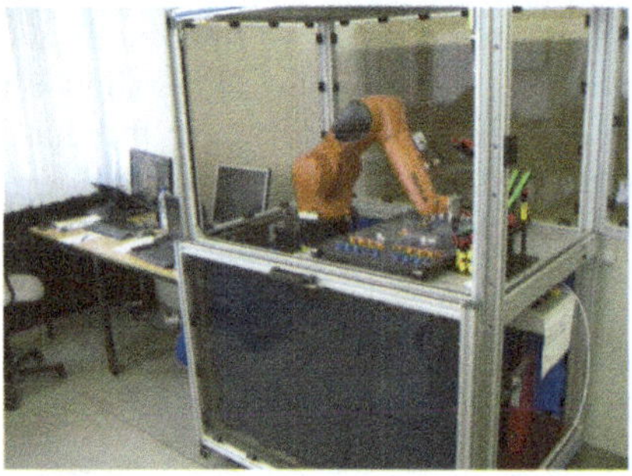

Fig. 5. Optimization results and demonstrator setup

6 Numerical Results

We have tested our optimization procedure with the solvers implemented in the MATLAB `fmincon` command. We obtained optimal solutions both with the default interior point algorithm and the SQP algorithms. However, in many cases the SQP algorithm required only 10 iterations, about half the iterations of the interior point algorithm. Computation time was below 10 s on a standard laptop with $B_x \cdot B_y = 5 \cdot 6 = 30$ grid point in the box. Figure 5 shows some typical run where a box is drawn from far outside the workspace (red) to the interior (blue). The results were almost independent from differentiability properties. The interpretation is: Even when crossing the critical workspace boundary the majority of grid points is away from the vertical slope, dominating the numerical derivatives. When sufficiently inside the workspace the algorithm hardly ever enters the infeasible region again.

We have also tested the algorithm with success for a more complicated geometry: The robot makes the moves for the popular board game "Settlers of Catan" with real tokens on a hexagonal structure on a playfield simulated on a screen, see [5]. However some manual intervention is necessary as the robot cannot reach all positions with the same configuration.

7 Conclusions

The approach presented in this contribution opens a way to non-interior-point optimization algorithms for the most common class of industrial 6R robots. Obviously the approach can also be used for the placement of a robot in a fixed work cell; we only need to consider the robot's base coordinate system as the variable. Furthermore it seems promising to use the idea for the optimization of redundant tasks like in [6] or redundant robots and to compare the results.

New algorithms for non-differentiable problems [3] should be investigated in comparison to higher order approaches, including estimates for the convergence of the solution of the smoothed problem to the original problem's solution.

References

1. Bertsekas, D.P.: Nondifferentiable optimization via approximation. Math. Program. Study **3**, 1–25 (1975)
2. Craig, J.: Introduction to Robotics: Mechanics and Control. Addison Wesley, Reading (1986)
3. Curtis, F.E., Overton, M.L.: A sequential quadratic programming algorithm for nonconvex, nonsmooth constrained optimization. SIAM J. Optim. **22**, 474–500 (2012)
4. Garcia, C., Zangwill, W.: Pathways to Solutions, Fixed Points and Equilibria. Prentice-Hall, Englewood Cliffs (1981)
5. KUKA Robot Group: Robots Play Board Games - Students Win Big (2016). https://www.youtube.com/watch?v=oCQPWv_ky2c
6. Léger, J., Angeles, J.: Off-line programming of six-axis robots for optimum five-dimensional tasks. Mech. Mach. Theor. **100**, 155–169 (2016)
7. Leontjevs, V., Flores, F.G., Lopes, J., Kecskemethy, A.: Singularity avoidance by virtual redundant axis and its application to large base motion compensation of serial robots In: Proceedings of the RAAD 2012 21st International Workshop on Robotics in Alpe-Adria-Danube Region, Naples (2012)
8. Merlet, J.P.: Jacobian, manipulability, condition number, and accuracy of parallel robots. J. Mech. Des. **128**(1), 199–206 (2006)
9. Nocedal, J., Wright, S.J.: Numerical Optimization. Springer, New York (2006)
10. Reiter, A.: Ein Beitrag zur Singularitätsvermeidung bei Industrierobotern durch Einführung virtueller Achsen. Master Thesis, Johannes Kepler University Linz (2015)
11. Vahrenkamp, N., Asfour, T., Dillmann, R.: Robot placement based on reachability inversion. In: IEEE International Conference on Robotics and Automation (2013)
12. Yoshikawa, T.: Manipulability of robotic mechanisms. Int. J. Robot. Res. **4**(2), 3–9 (1985)

The Forward Kinematics of Doubly-Planar Gough-Stewart Platforms and the Position Analysis of Strips of Tetrahedra

Josep M. Porta and Federico Thomas(✉)

Institut de Robòtica i Informàtica Industrial (CSIC-UPC), 08028 Barcelona, Spain
{porta,fthomas}@iri.upc.edu

Abstract. A strip of tetrahedra is a tetrahedron-tetrahedron truss where any tetrahedron has two neighbors except those in the extremes which have only one. The problem of finding all the possible lengths for an edge in the strip compatible with a given distance imposed between the strip end-points has been revealed of relevance due to the large number of possible applications. In this paper, this is applied to solve the forward kinematics of 6-6 Gough-Stewart platforms with planar base and moving platform, a problem which is known to have up to 40 solutions (20 if we do not consider mirror configurations with respect to the base as different solutions).

Keywords: Position analysis · Closed-form solutions
Distance geometry · Gough-Stewart platform

1 Introduction

During the last years, different methods have been developed, based on the use of Distance Geometry, to solve position analysis problems which includes the inverse kinematics and the forward kinematics of serial and parallel robots, respectively. The strategy followed in all these methods is simple: the problem is reduced to find some distances so that, once known, the problem can be trivially solved by a sequence of trilaterations. Position analysis problems are thus essentially reduced to find a set of closure polynomials whose variables are the unknown distances. Since no arbitrary reference frames have to be introduced to obtain these polynomials, these methods can be classified as intrinsic.

Contrarily to what happens when using kinematic loop-closure equations to solve a position analysis problem [1], the number of generated equations using distance-based methods does not depend on the number of kinematic loops. Nice examples of this fact are the planar Watt-Baranov trusses [2] and a remarkable class of spatial variable geometry trusses [3]. Actually, the number of unknown distances that have to be introduced, which is independent of the number of

J. Lenarcic and V. Parenti-Castelli (Eds.): ARK 2018, SPAR 8, pp. 124–132, 2019.
https://doi.org/10.1007/978-3-319-93188-3_15

kinematic loops, is the most relevant factor in these methods. Obviously, this number has to coincide with the number of generated equations in well-defined problems.

This paper can be seen as a continuation of the work presented in [4] where a simple procedure to obtain the distances between the end-points of a strip of tetrahedra is detailed. The resulting formula can be manipulated to obtain any edge length of the strip as the roots of a polynomial whose coefficients depend on the distance between the strip end-points and all other edge lengths. This technique was applied to solve the forward kinematics of a 3-4 and a 4-4 Gough-Stewart platform because in these two cases the problem can be solved by introducing a single distance or, in other words, the problem can be reduced to the position analysis of a single strip of tetrahedra.

The algorithm presented in [4] assumes that the tetrahedra in the strip have no orientation constraints. Although the technique can be extended to take into account such constraints, here we still adhere to this assumption as our main goal is explaining how the method extends to the case in which we have to consider several strips of tetrahedra at the same time.

To solve the forward kinematics of the general 6-6 Gough-Stewart platform, using a sequence of trilaterations, we have to introduce three unknown distances. Indeed, let us consider the parallel platform in Fig. 1. If the unknown distances $d_{2,7}$, $d_{3,7}$, $d_{2,9}$ are taken as variables, we can apply a trilateration process to locate P_{10}, P_{11}, and P_{12}. As a consequence, since the distances $d_{4,10}$, $d_{5,11}$, $d_{6,12}$ are known, three closure conditions can straightforwardly be obtained. A single univariate condition can then be derived by elimination. Now, observe that, in the general case, the orientation of the moving platform is constrained with respect to that of the base. Nevertheless, if the moving platform is planar, its

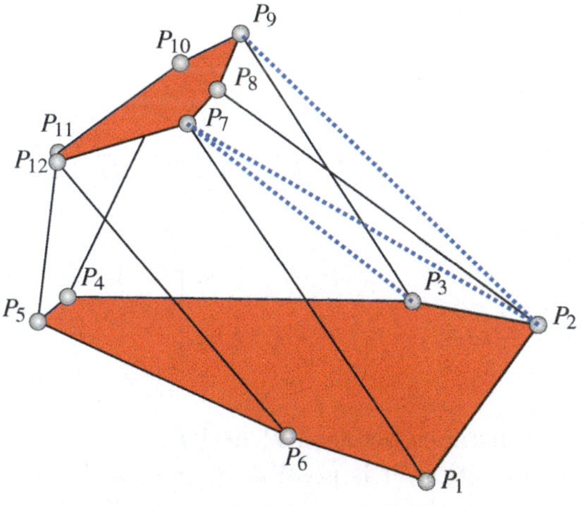

Fig. 1. A 6-6 Gough-Stewart platform. If the distances of the dotted blue segments are known, its forward kinematics can be straightforwardly solved by a sequence of trilaterations.

mirror reflection can be described as a rotation in three dimensions. In this case, no orientation constraints have to be considered and the problem can be entirely described in terms of distances. In this paper, we study the particular case in which both base and moving platform are planar.

Algebraic geometric methods have shown that the forward kinematics of the general 6-6 Gough-Stewart platform has up to 40 solutions [5–9], all of which can be real [10]. Husty [7] obtained a 40th-degree univariate polynomial by finding the greatest common divisor of the intermediate polynomials of degree 320. Innocenti [8] derived it from the two 56th-degree univariate polynomials. Dhingra *et al.* [9] used Gröbner-Sylvester hybrid method to obtain a 40th-degree polynomial directly from the 64×64 Sylvester's matrix formed by the calculated Gröbner basis. Any of these methods can obviously be applied to solve the particular case in which the base and the platform are planar, but this case has interest on its own, both from practical and theoretical points of view. This is why Lee and Shim presented a method adapted to solve its forward kinematics [11], and Borras *et al.* studied its singularities [12].

The paper is organized as follows. Section 2 gives some basic facts on the distance geometry of strips of tetrahedra. These concepts are needed to understand the procedure, detailed in Sect. 3, to generate the three distance-based closure conditions associated with a 6-6 Gough-Stewart platform. For comparison purposes, Sect. 4 solves the same problem as the one reported in [11] using these three closure conditions. Finally, Sect. 5 offers some conclusions and prospects for further research.

2 Background

Let us consider two tetrahedra that share the face defined by points P_i, P_j, and P_k. The squared distance between their apexes, say P_l and P_m, can be expressed as follows (see [13,14] for details):

$$s_{l,m} = \frac{2}{D(i,j,k;i,j,k)} \Bigg(D(i,j,k,l;i,j,k,m)\big|_{s_{l,m}=0} \pm \sqrt{D(i,j,k,l;i,j,k,l)\, D(i,j,k,m;i,j,k,m)} \Bigg), \quad (1)$$

where

$$D(i_1,\ldots,i_n;j_1,\ldots,j_n) = 2\,\left(-\tfrac{1}{2}\right)^n \begin{vmatrix} 0 & 1 & \ldots & 1 \\ 1 & s_{i_1,j_1} & \ldots & s_{i_1,j_n} \\ 1 & \vdots & \ddots & \vdots \\ 1 & s_{i_n,j_1} & \ldots & s_{i_n,j_n} \end{vmatrix}, \quad (2)$$

$s_{i,j}$ being the squared distance between P_i and P_j.

By iterating this operation, it is possible to obtain the distance between the end-points of a strip of tetrahedra as a function of all edge-lengths in the strip. The result is an expression with nested square roots. As explained in [3,4], these squared roots can be cleared to obtain a polynomial expression in terms of all involved distances.

3 Obtaining Closure Polynomials

We can derive the strip of tetrahedra that appears in Fig. 2 from the 6-6 Gough-Stewart platform in Fig. 1. The end-points of this strip can either be, on one side, P_4, P_5, or P_6, and, on the other, either P_{10}, P_{11}, or P_{12}. Then, since only $s_{2,7}$, $s_{3,7}$, and $s_{2,9}$ are unknown distances, using the procedure mentioned in the previous section and detailed in [3,4], we have that:

$$s_{i,j} = f_{i,j}(s_{2,7}, s_{3,7}, s_{2,9}), \tag{3}$$

where $i \in \{4, 5, 6\}$ and $j \in \{10, 11, 12\}$. Thus, in particular, we have

$$s_{4,10} = f_{4,10}(s_{2,7}, s_{3,7}, s_{2,9}), \tag{4}$$
$$s_{5,11} = f_{5,11}(s_{2,7}, s_{3,7}, s_{2,9}), \tag{5}$$
$$s_{6,12} = f_{6,12}(s_{2,7}, s_{3,7}, s_{2,9}). \tag{6}$$

Since the distances on the left-hand sides of the above three equations are known, we have three equations in three unknowns. Now, since the right-hand sides contain radicals, we have to use the procedure mentioned above to clear them [3,4].

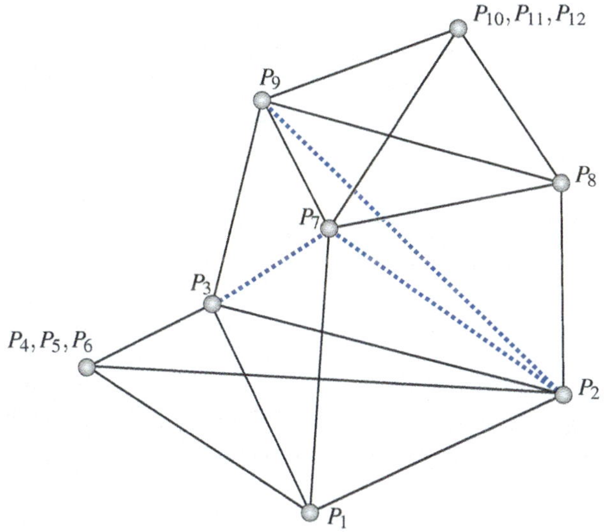

Fig. 2. This strip of five tetrahedra can be derived from the 6-6 Gough-Stewart platform in Fig. 1. The distances in dotted blue lines are unknown.

Assuming that we are dealing with a Gough-Stewart platform with coplanar leg attachments in general position within their planes, we obtain the following system of three polynomial closure conditions, of degree 12 each, in three variables

$$\left.\begin{aligned} p_1(s_{2,7}, s_{3,7}, s_{2,9}) &= 0 \\ p_2(s_{2,7}, s_{3,7}, s_{2,9}) &= 0 \\ p_3(s_{2,7}, s_{3,7}, s_{2,9}) &= 0 \end{aligned}\right\} \tag{7}$$

Now, we can eliminate, for example, $s_{3,7}$ from $p_1(\cdot)$ and $p_2(\cdot)$, and from $p_1(\cdot)$ and $p_3(\cdot)$. Then, we obtain two polynomials in $s_{2,7}$ and $s_{2,9}$, say $p_{1,2}(s_{2,7}, s_{2,9})$ and $p_{1,3}(s_{2,7}, s_{2,9})$. These two polynomials, of degree 128 each, factor into four terms of degree 20, 28, 36 and 44.

By computing the resultant of pairs of these factors, one from each polynomial, to eliminate $s_{2,7}$, we obtain polynomials in one variable: $s_{2,9}$. These univariate polynomials also factorize into terms of different degrees as detailed in Table 1:

Table 1. Degrees of all polynomial factors in $s_{2,9}$ resulting from computing the resultant of the four factors of $p_{1,2}(s_{3,7}, s_{2,9})$ and $p_{1,3}(s_{3,7}, s_{2,9})$. The multiplicity of each factor appears in parenthesis, if not 1.

		$p_{1,2}(s_{3,7},s_{2,9})$			
		20	**28**	**36**	**44**
$p_{1,3}(s_{3,7},s_{2,9})$	**20**	20 380	60 500	60 660	100 780
	28	60 500	2 (32) 60 628	116 892	2 (32) 116 1020
	36	60 660	116 892	2 (16) 60 1108	2 (16) 116 1340
	44	100 780	2 (32) 116 1020	2 (16) 116 1340	2 (16) 2 (32) 2(32) 100 1548

While some quadratic factors in Table 1 are repeated (actually, there are only three different quadratic factors), all other factors of higher degree are different.

Since we know that the maximum number of assembly modes of a general Gough-Stewart platform is 40, all factors of higher degree than this bound necessarily encompass extraneous roots. Moreover, the quadratic factors with multiplicity 16 and 32 are neither part of the solution because this would mean that the analyzed platform is quadratically solvable; in other words, it would be *trilaterable* [15]. In this context, trilaterable means that it would be possible to generate a single strip of tetrahedra involving all vertices such that all edge lengths in it are known, which is impossible in this problem. As a result, only one factor survives: the one of degree 20 which appears when factorizing the resultant of the factors of degree 20 of both $p_{1,2}(\cdot)$ and $p_{1,3}(\cdot)$.

Alternatively, to conclude that the 20th-degree factor is the only factor containing the desired solutions, we have also generated a different set of strips of tetrahedra containing three different variable edges, except P_2, P_9. Then, we have verified that the sought factor is the common factor resulting from using both sets of strips.

Once the roots of the 20th-degree factor in $s_{2,9}$ are computed, each of these roots can be substituted back in any of the two factors of degree 20 of $p_{1,2}(\cdot)$ or $p_{1,3}(\cdot)$ to obtain the corresponding solutions for $s_{3,7}$. Then, the results can be substituted in either $p_1(\cdot) = 0$, $p_2(\cdot) = 0$, or $p_3(\cdot) = 0$, to finally obtain the solutions for $s_{2,7}$.

4 Example

To properly compare our results with those reported in [11], we use here the same example. Nevertheless, it is important to notice that the length of the sixth leg in this latter reference is incorrect. Using the same notation as in Fig. 1, instead of $s_{6,12} = 117.805089939638$, it should be $s_{6,12} = 119.2390086191201$. Then, the considered Gough-Stewart platform defines the following distance matrix:

$$\mathbf{S} = \begin{bmatrix}
0.0 & 3844.0 & 3965.0 & 3208.0 & 2545.0 & 218.0 & 9889.0 & s_{1,8} & s_{1,9} & s_{1,10} & s_{1,11} & s_{1,12} \\
3844.0 & 0.0 & 121.0 & 1844.0 & 2421.0 & 3194.0 & s_{2,7} & 14977.47 & s_{2,9} & s_{2,10} & s_{2,11} & s_{2,12} \\
3965.0 & 121.0 & 0.0 & 1129.0 & 1684.0 & 3029.0 & s_{3,7} & s_{3,8} & 24340.67 & s_{3,10} & s_{3,11} & s_{3,12} \\
3208.0 & 1844.0 & 1129.0 & 0.0 & 101.0 & 1850.0 & s_{4,7} & s_{4,8} & s_{4,9} & 23700.59 & s_{4,11} & s_{4,12} \\
2545.0 & 2421.0 & 1684.0 & 101.0 & 0.0 & 1301.0 & s_{5,7} & s_{5,8} & s_{5,9} & s_{5,10} & 18569.53 & s_{5,12} \\
218.0 & 3194.0 & 3029.0 & 1850.0 & 1301.0 & 0.0 & s_{6,7} & s_{6,8} & s_{6,9} & s_{6,10} & s_{6,11} & 14217.94 \\
9889.0 & s_{2,7} & s_{3,7} & s_{4,7} & s_{5,7} & s_{6,7} & 0.0 & 196.0 & 2378.0 & 2845.0 & 2554.0 & 2020.0 \\
s_{1,8} & 14977.47 & s_{3,8} & s_{4,8} & s_{5,8} & s_{6,8} & 196.0 & 0.0 & 1258.0 & 1753.0 & 2106.0 & 1768.0 \\
s_{1,9} & s_{2,9} & 24340.67 & s_{4,9} & s_{5,9} & s_{6,9} & 2378.0 & 1258.0 & 0.0 & 197.0 & 1600.0 & 1802.0 \\
s_{1,10} & s_{2,10} & s_{3,10} & 23700.59 & s_{5,10} & s_{6,10} & 2845.0 & 1753.0 & 197.0 & 0.0 & 853.0 & 1125.0 \\
s_{1,11} & s_{2,11} & s_{3,11} & s_{4,11} & 18569.53 & s_{6,11} & 2554.0 & 2106.0 & 1600.0 & 853.0 & 0.0 & 58.0 \\
s_{1,12} & s_{2,12} & s_{3,12} & s_{4,12} & s_{5,12} & 14217.94 & 2020.0 & 1768.0 & 1802.0 & 1125.0 & 58.0 & 0.0
\end{bmatrix},$$

where $\mathbf{S}(i,j) = s_{i,j} = d^2_{i,j}$.

Following the steps detailed in the previous section, the 20th-degree polynomial in $s_{2,9}$ is:

$$\begin{aligned}
& s_{2,9}^{20} - 3.8124\times10^{5}\, s_{2,9}{}^{19} + 1.0636\times10^{10}\, s_{2,9}{}^{18} + 1.2896\times10^{16}\, s_{2,9}{}^{17} - 2.7847\times10^{21}\, s_{2,9}{}^{16} \\
& + 3.0704\times10^{26}\, s_{2,9}{}^{15} - 2.2140\times10^{31}\, s_{2,9}{}^{14} + 1.1390\times10^{36}\, s_{2,9}{}^{13} - 4.3622\times10^{40}\, s_{2,9}{}^{12} \\
& + 1.2726\times10^{45}\, s_{2,9}{}^{11} - 2.8611\times10^{49}\, s_{2,9}{}^{10} + 4.9798\times10^{53}\, s_{2,9}{}^{9} - 6.7075\times10^{57}\, s_{2,9}{}^{8} \\
& + 6.9694\times10^{61}\, s_{2,9}{}^{7} - 5.5643\times10^{65}\, s_{2,9}{}^{6} + 3.3970\times10^{69}\, s_{2,9}{}^{5} - 1.5529\times10^{73}\, s_{2,9}{}^{4} \\
& + 4.7760\times10^{76}\, s_{2,9}{}^{3} - 5.1150\times10^{79}\, s_{2,9}{}^{2} - 2.6566\times10^{83}\, s_{2,9} + 8.9169\times10^{86}.
\end{aligned}$$

The positive reals roots of this polynomial are: 7451.80, 9587.28, 17271.4, 24044.1, 24511.3, 24579.0, 26132.2, 27680.5, 28332.5, 28809.9, 44848.6, and 251456. Obviously, complex or negative real squared distances have no physical meaning in our problem and they can be discarded. Moreover, these ten solutions do not necessarily lead to real solutions for $s_{2,7}$ and $s_{3,7}$. In this case, only the roots 24511.3 and 24579.0 lead to real configurations of the strip of tetrahedra. The corresponding parallel platform configurations appear in Fig. 3. They coincide with the two real solutions reported in [11].

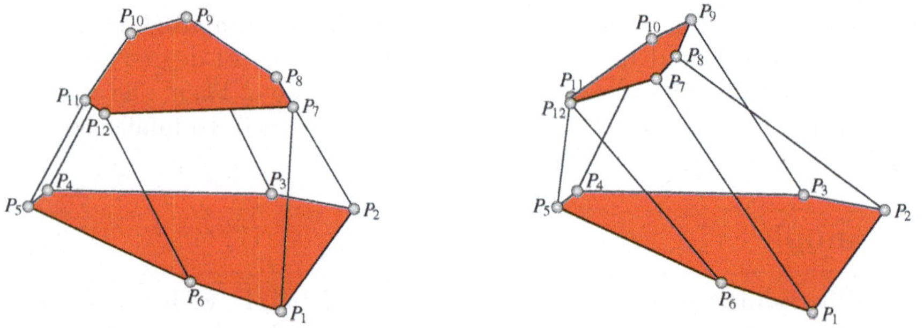

Fig. 3. Forward kinematics solutions of the analyzed 6-6 Gough-Stewart platform. The two mirror configurations with respect to the fixed base are also solutions but they are not represented.

In this example, the quadratic factors appearing in Table 1 are:

$$s_{2,9}^2 - 49578.5\, s_{2,9} + 598299000.0 = 0, \quad (8)$$

$$s_{2,9}^2 - 32415.2\, s_{2,9} + 188224000.0 = 0, \quad (9)$$

$$s_{2,9}^2 - 48923.3\, s_{2,9} + 586592000.0 = 0. \quad (10)$$

The last factor is identically equal to $D(2,3,9;2,3,9)$ which vanishes when P_2, P_3, and P_9 are aligned [16]. For symmetry reasons, some of the other terms of higher degree should correspond to other alignments. Nevertheless, since they are expressed in terms of $s_{2,9}$, instead of a variable edge length connecting two of the three points involved in an alignment, their expressions are very complicated.

5 Conclusions

It has been shown how to derive three closure polynomials for 6-6 Gough-Stewart platforms with planar base and moving platform in terms of three unknown distances, and how the solution to this system of equations permits solving the forward kinematics of the analyzed Gough-Stewart platform by a trivial sequence of trilaterations.

The elimination process to generate the 20th-degree final polynomial in a single distance from the three closure polynomials leads to extraneous factors.

The geometric meaning of some of them have been identified and that of some others, conjectured. The analysis of the groups of substitution of finite order associated with the symmetries of the analyzed strips of tetrahedra could play a fundamental role in the identification of the geometric meaning of all these factors. They actually seem to be correlated with the Galois groups of the obtained factors. In the same way that a knot polynomial is a knot invariant in the form of a polynomial whose coefficients encode some of the properties of a given knot, the study of all polynomial factors derived from a set of strips of tetrahedra in terms of invariants opens a fascinating new field in Kinematics.

Acknowledgements. This work has been partially supported by the Spanish Ministry of Economy and Competitiveness through projects DPI2014-57220-C2-2-P, DPI2017-88282-P, and MDM-2016-0656.

References

1. Sandor, G.: On the loop equation in kinematics. In: Transactions of the 7th Conference on Mechanisms, pp. 49–56. Purdue University, West Lafayette (1962)
2. Rojas, N., Thomas, F.: Closed-form solution to the position analysis of Watt-Baranov trusses using the bilateration method. ASME J. Mech. Robot. **3**(3), 031001 (2011). 10 pages
3. Porta, J.M., Thomas, F.: Closed-form position analysis of variable geometry trusses. Mech. Mach. Theor. **109**, 14–21 (2017)
4. Thomas, F., Porta, J.M.: Closure polynomials for strips of tetrahedra. In: Advances in Robot Kinematics, ARK 2016, Grasse, France (2016)
5. Ronga, F., Vust, T.: Gough-Stewart platforms without computer? In: Proceedings of the 1992 International Conference in Real Analytic and Algebraic Geometry, Walter de Gruyter, pp. 196–212 (1995)
6. Wampler, C.: Forward displacement analysis of general six-in-parallel SPS (Stewart) platform manipulators using soma coordinates. Mech. Mach. Theor. **31**(3), 331–337 (1996)
7. Husty, M.L.: An algorithm for solving the direct kinematics of general Stewart-Gough platforms. Mech. Mach. Theor. **31**(4), 365–380 (1996)
8. Innocenti, C.: Forward kinematics in polynomial form of the general Stewart platform. In: 25th Biennal Mechanisms Conference, Atlanta, DETC98/MECH-5894 (1998)
9. Dhingra, A., Almadi, A., Kohli, D.: A Gröbner-Sylvester hybrid method for closed-form displacement analysis of mechanisms. In: 25th Biennial Mechanisms Conference, Atlanta, DECTC98/MECH-5969 (1998)
10. Dietmaier, P.: The Stewart-Gough platform of general geometry can have 40 real postures. In: Lenarcic, J., Husty, M.L. (eds.) Advances in Robot Kinematics: Analysis and Control. Springer, Dordrecht (1998)
11. Lee, T.-Y., Shim, J.-K.: Algebraic elimination-based real-time forward kinematics of the 6-6 Gough-Stewart platform with planar base and platform. In: Proceedings of the IEEE International Conference on Robotics and Automation, pp. 1301–1306 (2001)
12. Borras, J., Thomas, F., Torras, C.: Singularity-invariant leg rearrangements in doubly-planar Gough-Stewart-Gough platforms. In: Robotics: Science and Systems 2010, Zaragoza, Spain, 27–30 June 2010

13. Porta, J.M., Thomas, F., Ros, L., Torras, C.: A branch-and-prune algorithm for solving systems of distance constraints. In: Proceedings of the 2003 IEEE International Conference on Robotics and Automation, Taipei, Taiwan (2003)
14. Rojas, N., Thomas, F.: The closure condition of the double banana and its application to robot position analysis. In: Proceedings of the International Conference on Robotics and Automation, pp. 4641–4646 (2013)
15. Porta, J.M., Ros, L., Thomas, F.: On the trilaterable six-degree-of-freedom parallel and serial manipulators. In: Proceedings of the 2005 IEEE International Conference on Robotics and Automation, Barcelona, Spain (2005)
16. Thomas, F., Ros, L.: Revisiting trilateration for robot localization. IEEE Trans. Rob. **21**(1), 93–101 (2005)

Six-Bar Linkage Design System with a Parallelized Polynomial Homotopy Solver

Jeffrey Glabe and J. Michael McCarthy(✉)

University of California, Irvine, Irvine, CA 92697, USA
{jglabe,jmmccart}@uci.edu

Abstract. The kinematic synthesis of six-bar linkages involves the solution of polynomial design equations that can have a total degree in the billions and thus requires polynomial homotopy continuation to find solutions. Here we present a six-linkage design system integrated with a parallel implementation of the polynomial homotopy solver Bertini that provides new design capabilities. In this implementation, we specify six task positions and two regions for the base pivots. The resulting design equations have total degree $\approx 5.9 \times 10^9$. The structure of the design system is presented and an example demonstrates its performance.

1 Introduction

The kinematic synthesis of six-bar linkages is limited by the size of the polynomial systems that must be solved to determine the link dimensions. Polynomial homotopy continuation systems, such as *Bertini* [1], PHCpack [2] and POLSYS-GLP [3], have demonstrated the ability to reliably solve large systems of linkage design equations, [3–5]. In order to bring this solution capability to designers, we have integrated the polynomial homotopy solver *Bertini* into a six-bar linkage design system. This paper derives the design equations and how their solution is integrated with a parallel implementation of *Bertini*.

2 The Design System

The design system uses Python [6] to (i) read the designer's task data from the solid modeling system and populate the design equations for solution by the polynomial homotopy solver, (ii) collect and evaluate the solutions, and (iii) display them for selection by the designer. Figure 1 shows the structure of such a design system.

3 The Design Equations

The goal of the design system is to compute of the dimensions of Watt 1 six-bar linkages that guide an end-effector through a user-defined set of six task positions

J. Lenarcic and V. Parenti-Castelli (Eds.): ARK 2018, SPAR 8, pp. 133–140, 2019.
https://doi.org/10.1007/978-3-319-93188-3_16

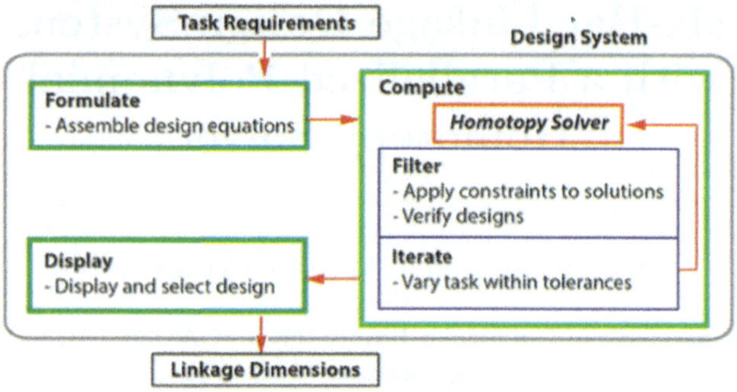

Fig. 1. The design system has three functions: (i) use the task data to populate of the design equations, (ii) collect and evaluate solutions, and (iii) display the realizable designs to the user for selection and output.

with fixed pivots in user-specified regions. This linkage consists of two four-bar loops, Fig. 2. The first loop, $ACDB$, has fixed pivots A and B and coupler CGD, and the second $GHFD$ has the end-effector attached to the link GH. In order to define the design equations, we follow [7].

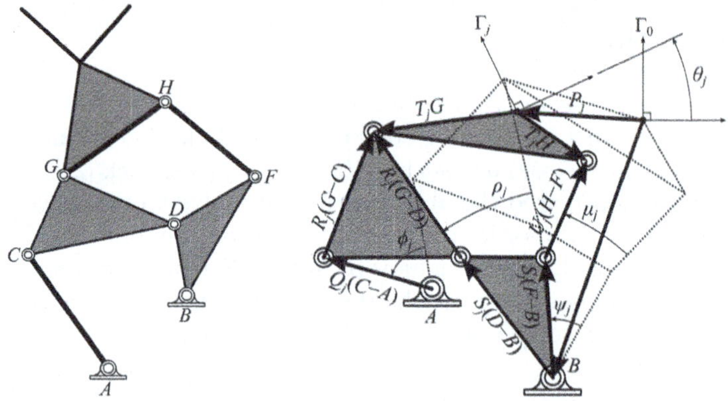

(a) Watt 1 six-bar linkage.

(b) Displaced position.

Fig. 2. A Watt 1 six-bar motion generator displaced relative to the initial task position.

The six task frames are defined by the world frame Γ_0 and five relative positions, $\Gamma_j, j = 1, \ldots, 5$, where P_j is the origin of the task frame and θ_j is its angle relative to the x-axis,

$$\Gamma_j = (x_j + iy_j, e^{i\theta_j}) = (P_j, T_j). \tag{1}$$

The design equations are obtained from the three vector loop equations that define the geometry of the six-bar linkage. Each task position defines three sets of loop equations $\mathcal{A}_j$, $\mathcal{B}_j$, and $\mathcal{C}_j$,

$$\mathcal{A}_j = \begin{cases} A + Q_j(C - A) + R_j(G - C) - T_jG &= P_j \\ \bar{A} + \bar{Q}_j(\bar{C} - \bar{A}) + \bar{R}_j(\bar{G} - \bar{C}) - \bar{T}_j\bar{G} &= \bar{P}_j \end{cases} \quad j = 1, \ldots, 5, \tag{2}$$

$$\mathcal{B}_j = \begin{cases} B + S_j(D - B) + R_j(G - D) - T_jG &= P_j \\ \bar{B} + \bar{S}_j(\bar{D} - \bar{B}) + \bar{R}_j(\bar{G} - \bar{D}) - \bar{T}_j\bar{G} &= \bar{P}_j \end{cases} \quad j = 1, \ldots, 5, \tag{3}$$

$$\mathcal{C}_j = \begin{cases} B + S_j(F - B) + U_j(H - F) - T_jH &= P_j \\ \bar{B} + \bar{S}_j(\bar{F} - \bar{B}) + \bar{U}_j(\bar{H} - \bar{F}) - \bar{T}_j\bar{H} &= \bar{P}_j \end{cases} \quad j = 1, \ldots, 5. \tag{4}$$

where Q_j, R_j, S_j, and U_j are the rotations of each joint relative to the initial configuration of the linkage, Fig. 2, and $\bar{Q}_j, \bar{R}_j, \bar{S}_j, \bar{U}_j$ are their respective complex conjugates,

$$\begin{aligned} (Q_j, \bar{Q}_j) &= (e^{i\phi_j}, e^{-i\phi_j}), \quad (R_j, \bar{R}_j) = (e^{i\rho_j}, e^{-i\rho_j}), \\ (S_j, \bar{S}_j) &= (e^{i\psi_j}, e^{-i\psi_j}), \quad (U_j, \bar{U}_j) = (e^{i\mu_j}, e^{-i\mu_j}). \end{aligned} \tag{5}$$

The relative rotation operators and their conjugates must further satisfy the normalization conditions

$$Q_j\bar{Q}_j = 1, \quad R_j\bar{R}_j = 1, \quad S_j\bar{S}_j = 1, \quad U_j\bar{U}_j = 1, \qquad j = 1, \ldots, 5. \tag{6}$$

This yields 50 equations in the 10 unknown joint coordinates C, D, F, G and H, and the 40 joint angles Q_j, R_j, S_j, and $U_j, j = 1, \ldots, 5$ and their conjugates.

These equations can be simplified by solving for the joint rotation angles, $(Q_j, \bar{Q}_j)$, in the loop equations $\mathcal{A}_j$, and substituting the results into the normalization conditions to obtain,

$$(P_j - A + R_j(C - G) + T_jG)(\bar{P}_j - \bar{A} + \bar{R}_j(\bar{C} - \bar{G}) + \bar{T}_j\bar{G}) = (C - A)(\bar{C} - \bar{A}), j = 1, \ldots 5. \tag{7}$$

A similar calculation computes $(R_j, \bar{R}_j)$ using its loop equations $\mathcal{C}_j$ and normalization conditions to define,

$$(P_j - B + S_j(B - D) + T_jG)(\bar{P}_j - \bar{B} + \bar{S}_j(\bar{B} - \bar{D}) + \bar{T}_j\bar{G}) = (G - D)(\bar{G} - \bar{D}), j = 1, \ldots 5. \tag{8}$$

Also substitute $(R_j, \bar{R}_j)$ into (7) to obtain,

$$\begin{aligned} ((G - D)(P_j - A + T_jG) + w_j(C - G))((\bar{G} - \bar{D})(\bar{P}_j - \bar{A} + \bar{T}_j\bar{G}) + \bar{w}_j(\bar{C} - \bar{G})) \\ = (C - A)(\bar{C} - \bar{A})(G - D)(\bar{G} - \bar{D}), j = 1, \ldots 5, \end{aligned} \tag{9}$$

where

$$w_j = P_j - B + S_j(B - D) + T_jG, \quad \bar{w}_j = \bar{P}_j - \bar{B} + \bar{S}_j(\bar{B} - \bar{D}) + \bar{T}_j\bar{G}. \tag{10}$$

Finally, solve the loop equations $\mathcal{B}_j$ for $(U_j, \bar{U}_j)$ and substitute the result into its normalization conditions. This yields the equations,

$$(P_j - B + S_j(B-F) + T_jH)(\bar{P}_j - \bar{B} + \bar{S}_j(\bar{B}-\bar{F}) + \bar{T}_j\bar{H}) = (H-F)(\bar{H}-\bar{F}), j = 1,\ldots 5. \quad (11)$$

The Eqs. 7, 9 and 11 and the normalization conditions for $(S_j, \bar{S}_j)$ form a system of 20 polynomials in the 10 unknown joint coordinates and the five joint angles S_j and their five conjugates. This polynomial system has a total degree of $\approx 5.90 \times 10^9$ and a multi-homogeneous degree of 1,998,720.

This polynomial system is converted to design equations for our design system by using *Bertini* to solve a generic system with randomly assigned task parameters. This calculation yielded a generic set of 5,743 roots that can be used in a parameter homotopy. This one-time calculation required seven hours on a Linux cluster assembled from four Lenovo workstations, each with dual Intel Xeon 2 GHz processors running 16 threads each for a total of 128 threads.

4 Parallel Solution of the Design Equations

The parameter homotopy obtained from the design equations was integrated into our design system for parallel computation on a Linux cluster. A Python script populates this system with the user-defined task, and manages the execution of *Bertini* to obtain solutions to the design equations. This consists of distributing the data necessary for path tracking from the starting point to a solution, and then collecting the solution data.

The Linux cluster communicates between the computer nodes using the University of California, Irvine, local network. Each computer generates SSH keys, which are sent to every other computer in the network to allow each node to communicate autonomously with every other node without a password. One node is selected as the head to coordinate the other nodes by distributing starting points and collecting solutions, while simultaneously performing its own path tracking.

Bertini generates intermediate output files as it solves a parameter homotopy, which results in a large number of file operations. This is managed by a shared network hard drive on a separate computer system running OS/X. Figure 3 shows the cluster's structure.

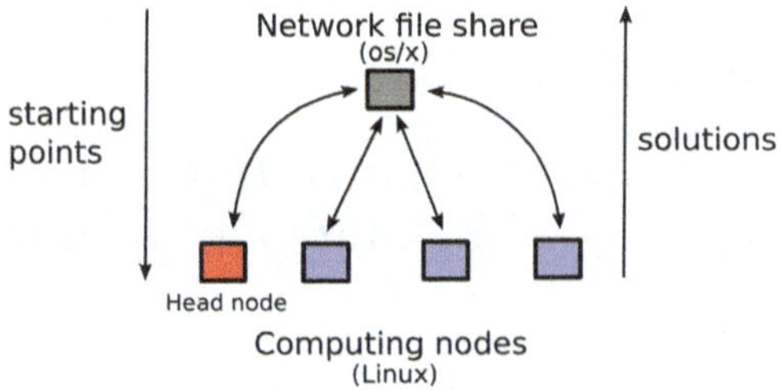

Fig. 3. Parallel computation is managed through network file sharing. Each not receives starting points and returns the resulting solutions.

Bertini was installed separately on each node of the cluster using *flex* and *bison* programs, as well as the *gmp*, *mpfr* and the *openmpi* libraries that are part of the Ubuntu Linux Package Management System [8]. *Bertini* is configured to use the make build automation tool to compile.

5 Filter and Iterate

Each solution obtained from the parameter homotopy provides the joint coordinates $\{A, B, C, D, F, G, H\}$ of the linkage in the first task position and the set of relative crank angles $\{S_1, S_2, ..., S_5\}$ that position the system in the other task positions. The size of each solutions is measured against packaging constraints $\{x_{min}, x_{max}\}$ and $\{y_{min}, y_{max}\}$ provided by the designer. That is, the set of joint coordinates are evaluated to ensure

$$x_{min} \leq x_i \leq x_{max} \quad y_{min} \leq y_i \leq y_{max}, \qquad (x_i, y_i) \in \{A, B, C, D, F, G, H\} \tag{12}$$

This is a rapid calculation that eliminates designs that are impractical.

Solutions that fit the packaging constraints are analyzed to verify smooth movement through the task positions. We determine the initial crank angle S_0 of the crank BD relative to the x-axis, and add it to the relative angles S_j to obtain

$$\{\Psi_1, \Psi_2, \Psi_3, ..., \Psi_6\} \quad = \quad \{S_0, S_1 + S_0, S_2 + S_0, ..., S_5 + S_0\} \tag{13}$$

Then, for each set of Ψ_i, the linkage is analyzed to determine the position of the end-effector in each assembly. Realizable linkages are those that reach each of the task positions in a single assembly. This required for successful operation of the linkage.

This process of generating and evaluating solutions is iterated for variations of the task parameters obtained by randomize selection from within tolerance zones around the nominal task parameters. The size of the tolerance zones and the number of iterations are specified by the designer.

6 Candy Coating Mechanism

In order to demonstrate the performance of this design system, we seek a linkage that selects an item, moves it through three stations, and returns it to the original location. Furthermore the location of the base of the device is specified. See Fig. 4. The data for this design is presented in Table 1. We chose mounting pivots for the coating mechanism to be $A = (228.15\,\text{mm}, 584.34\,\text{mm})$ and $B = (104.31\,\text{mm}, 463.84\,\text{mm})$ with a tolerance of ± 10 mm in x and y directions.

For this calculation, the design system ran on the cluster using only two Lenovo Workstations. Each solution of the parameter homotopy required approximately three minutes of computation time. 220 iterations produced 55,130 solutions of which only 12,581 candidates met the packaging constraints. The performance of each these candidates was evaluated and yielded 26 realizable designs. The time required for these design calculations was 41 h, the majority of which is devoted to the verification calculations.

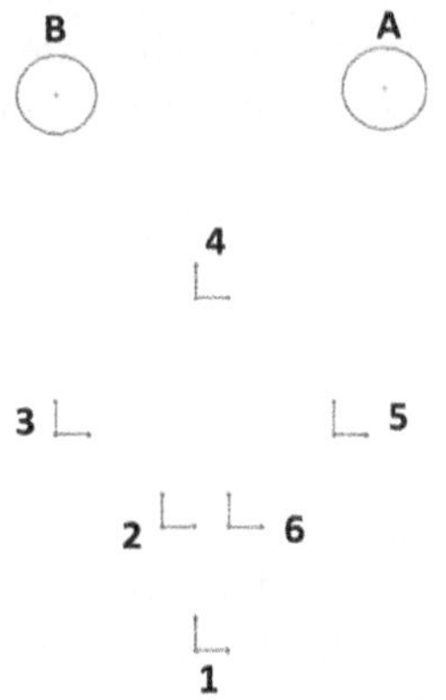

Fig. 4. The six task positions and fixed pivot locations A and B of the candy coating machine

Table 1. Prescribed task positions of the candy coater.

Task position	Angle (in degrees)	Angle tolerance (± degrees)	Location (x, y) in mm	Coordinate tolerance (± mm)
1	0.060	0.1	(20.00, −12.27)	2.0
2	0.034	0.1	(−39.99, 127.77)	2.0
3	359.99	0.1	(−127.21, 192.00)	2.0
4	359.99	0.1	(19.06, 318.34)	2.0
5	0.058	0.1	(129.14, 211.72)	2.0
6	359.9	0.1	(38.21, 120.79)	2.0

Nine of the verified designs were found to have rotating driving cranks, which is convenient, and were of similar in size and shape. Table 2 lists the coordinates of one of these nine designs and Fig. 5 is a solid model illustrating the device. Figure 6 illustrates the motion of this device.

Table 2. Joint coordinates of the candy coater.

Pivot	Location (x, y) in mm
A	(228.15, 584.34)
B	(104.31, 463.84)
C	(127.55, 682.36)
D	(25.32, 55.46)
F	(56.88, 311.73)
G	(342.81, 655.56)
H	(−2.61, 590.24)

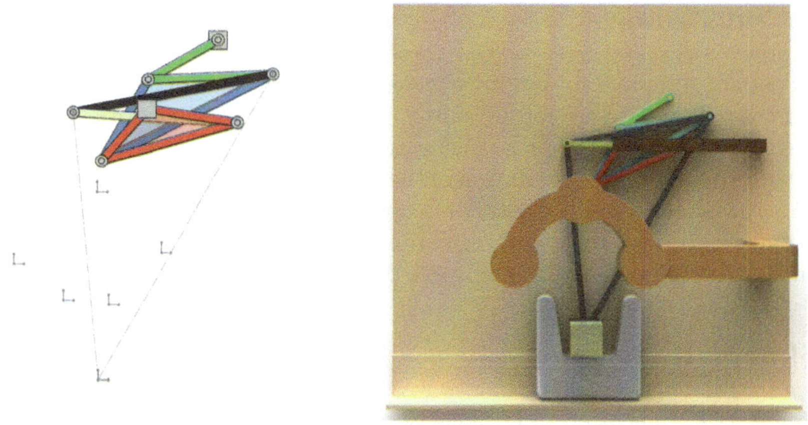

(a) Candy coater, first task position (b) Solid model of the candy coater

Fig. 5. Solid models of the candy coater six-bar linkage

(a) (b) (c)

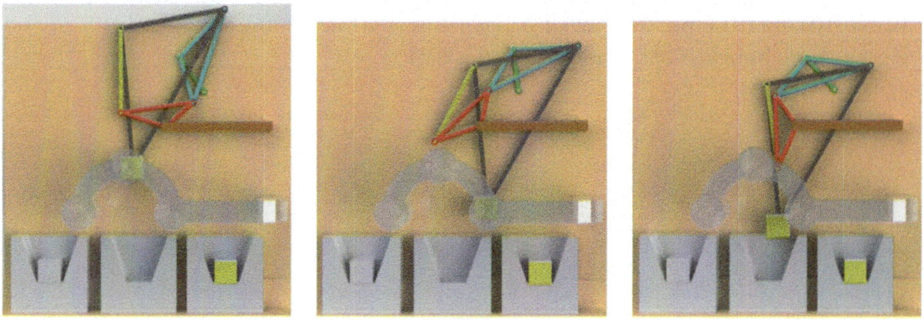

(d) (e) (f)

Fig. 6. Single frames from an animation demonstrating the linkage in each of the six task positions

7 Conclusion

This paper presents a six-bar linkage design system that includes the parallel execution of polynomial homotopy solver. The design equations for a Watt 1 six-bar linkage with two specified base pivots and six task positions have a total degree of $\approx 5.90 \times 10^9$. The generic solution of these equations yields a parameter homotopy with 5,743 roots. A Linux cluster of Lenovo workstations was implemented to solve this parameter homotopy, and verify the resulting designs. An example design task required 41 h to compute 26 realizable designs.

References

1. Bates, D.J., Hauenstein, J.D., Sommese, A.J., Wampler, C.W.: Numerically Solving Polynomial Systems with Bertini. SIAM Press, Philadelphia (2013)
2. Bliss, N., Sommars, J., Verschelde, J., Yu, X.: Solving polynomial systems in the cloud with polynomial homotopy continuation. In: Gerdt, V., Koepf, W., Seiler, W., Vorozhtsov, E. (eds.) Computer Algebra in Scientific Computing. Lecture Notes in Computer Science. Springer, Cham (2015)
3. Su, H., McCarthy, J.M., Sosonkina, M., Watson, L.: Algorithm 857: POLSYS-GLP-a parallel general linear product homotopy code for solving polynomial systems of equations. ACM Trans. Math. Softw. **32**(4), 561–579 (2006)
4. Su, H., McCarthy, J.M., Watson, L.: Generalized linear product homotopy algorithms and the computation of reachable surfaces. J. Comput. Inf. Sci. Eng. **4**, pp. 226–234 (2004)
5. Plecnik, M.M., McCarthy, J.M.: Computational design of Stephenson II six-bar function generators for 11 accuracy points. J. Mech. Robot. **8**, 9 (2016)
6. Rossum, G.: Python Reference Manual. CWI (Centre for Mathematics and Computer Science), Amsterdam (1995)
7. Plecnik, M., McCarthy, J. M., and Wampler, C. W.: Kinematic synthesis of a watt I six-bar linkage for body guidance. In: Lenarčič, J., Khatib, O. (eds.) Advances in Robot Kinematics, pp. 317–325. Springer International Publishing (2014)
8. Xerus, X.: Ubuntu 16.04. Canonica Ltd. (2016)

Algebraic Analysis of a 3-RUU Parallel Manipulator

Thomas Stigger[1], Abhilash Nayak[2], Stéphane Caro[3], Philippe Wenger[3], Martin Pfurner[1(✉)], and Manfred L. Husty[1]

[1] Unit Geometry and CAD, University of Innsbruck, 6020 Innsbruck, Austria
{thomas.stigger,martin.pfurner,manfred.husty}@uibk.ac.at

[2] École Centrale de Nantes, Laboratoire des Sciences du Numérique de Nantes (LS2N), UMR CNRS 6004, 1 rue de la Noë, 44321 Nantes, France
abhilash.nayak@ls2n.fr

[3] CNRS, Laboratoire des Sciences du Numérique de Nantes (LS2N), École Centrale de Nantes, UMR CNRS 6004, 1 rue de la Noë, 44321 Nantes, France
{philippe.wenger,stephane.caro}@ls2n.fr

Abstract. Constraint equations of a parallel manipulator can be used to analyze their kinematic behaviour. This paper deals with the determination of the algebraic constraint equations of a 3-RUU parallel manipulator with two approaches. The first one is based on the manipulator geometry and the second one uses the Linear Implicitization Algorithm. The obtained constraint equations through the former approach can be given a geometrical interpretation while the latter approach is less prone to missing physical constraints. Both the ideals of constraint polynomials should lead to the same variety. Furthermore, the simplest set of equations is chosen to solve the direct kinematics problem. For the manipulator under study, it turns out that its direct kinematics problem leads to a factorisable univariate polynomial and a translational operation mode appears.

Keywords: 3-RUU · Kinematic analysis · Direct kinematics
Algebraic geometry

1 Introduction

For theoretical and practical purposes, the kinematic analysis of a parallel manipulator (PM) is essential to understand its motion behavior. Kinematic constraints can be transformed via Study's kinematic mapping into algebraic constraint equations. Every configuration of the PM is thereby mapped to a point in a projective space, $\mathbb{P}^7$ [4,5]. Consequently, well developed concepts of algebraic geometry [2] can be used to interpret the algebraic constraint equations to obtain necessary information about the PM.

In that vein, many PMs were investigated using algebraic geometry concepts. Resultant methods were adopted to solve the direct kinematics of Stewart-Gough

J. Lenarcic and V. Parenti-Castelli (Eds.): ARK 2018, SPAR 8, pp. 141–149, 2019.
https://doi.org/10.1007/978-3-319-93188-3_17

platforms [3]. A complete kinematic analysis including the characterization of operation modes, solutions to direct kinematics and determination of singular poses was performed for the 3-RPS PM [10,11], the 3-RPS cube PM [8] and 3-PRS PMs with different arrangements of prismatic joints [6]. In the foregoing papers, the prismatic joints were considered to be actuated, which makes the analysis inherently algebraic. A more challenging kinematic analysis of an over-constrained 4-RUU PM with square base and moving platform was accomplished by decomposing it into two 2-RUU PMs [7]. The constraint equations of a 3-RUU PM are derived in this paper and its direct kinematics problem is solved. Nevertheless, a complete characterization of the manipulator operation modes has not been obtained yet.

The paper is organized as follows: Sect. 2 describes the manipulator architecture. Section 3 deals with the derivation of algebraic constraint equations with two approaches and their comparison. Section 4 presents the solutions to direct kinematics for arbitrary design parameters and hints the recognition of a translational operation mode.

2 Manipulator Architecture

The 3-RUU PM is shown in Fig. 1a. Each limb consists of a revolute joint and two universal joints mounted in series with the first revolute joint as the active joint. The moving platform and the fixed base form equilateral triangles with vertices C_i and A_i, respectively, $i = 1, 2, 3$. The unit vectors of the revolute joint axes within the i-th limb are denoted as $\mathbf{s}_{ij}$, $i = 1, 2, 3$; $j = 1, ..., 5$. $\mathbf{s}_{i5}$ and $\mathbf{s}_{i1}$ are tangent to the circumcircles (with centers P and O) of the moving platform and the base triangles, respectively. Vectors $\mathbf{s}_{i1}$ and $\mathbf{s}_{i2}$ are always parallel, so are vectors $\mathbf{s}_{i3}$ and $\mathbf{s}_{i4}$. The origin of the fixed coordinate frame, $\mathscr{F}_O$ is at O and the z_O-axis lies along the normal to the base plane whereas the origin of the moving coordinate frame $\mathscr{F}_P$ is at P and the z_P-axis lies along the normal to the moving platform plane. x_O and x_P axes are directed along OA_1 and PC_1, respectively. r_0 and r_1 are the circumradii of base and the moving platform, respectively. a_1 and a_3 are the link lengths. θ_{i1} is the angle of rotation of the first revolute joint about the axis represented by vector $\mathbf{s}_{i1}$ measured from the base plane whereas θ_{i2} is the angle of rotation of the second revolute joint about the axis represented by vector $\mathbf{s}_{i2}$ measured from the first link.

3 Constraint Equations

The constraint equations of the 3-RUU PM are derived using a geometrical approach and the Linear Implicitization Algorithm (LIA) [12]. First, canonical constraint equations for a limb of the PM are derived by attaching fixed and moving coordinate frames to the two extreme joints of a RUU limb as shown in Fig. 1b. Each U-joint is characterized by two revolute joints with orthogonal and intersecting axes and Denavit-Hartenberg (DH) convention is used to parameterize each limb. $\mathscr{F}_0$ and $\mathscr{F}_1$ are the fixed and the moving coordinate frames with their corresponding z-axes along the first and the last revolute joint axes,

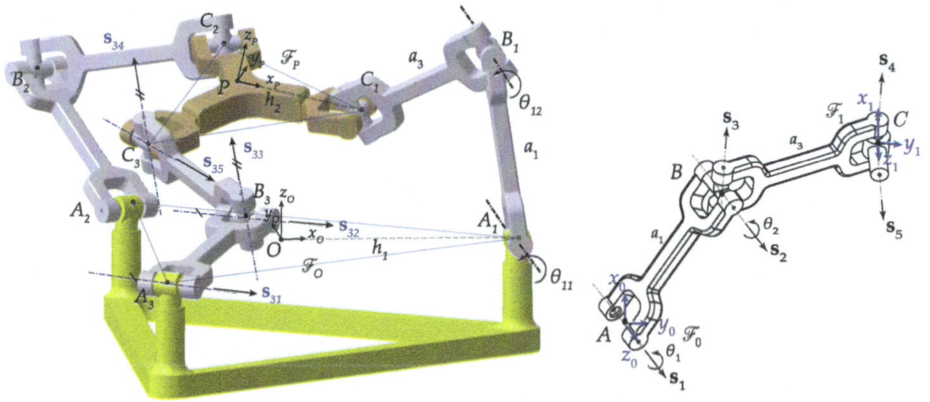

(a) The 3-RUU PM in a general configuration

(b) A RUU limb

Fig. 1. (a) The 3-RUU PM in a general configuration. (b) A RUU limb

respectively. Later on, general constraint equations are derived for the whole manipulator.

3.1 Derivation Using a Geometrical Approach

Canonical Constraints. In order to derive the geometric constraints for a RUU limb, the homogeneous coordinates[1] of points A, B, C ($\mathbf{a}, \mathbf{b}, \mathbf{c}$, respectively) and vectors $\mathbf{s}_j$, $j = 1, ..., 5$, shown in Fig. 1b are expressed as follows:

$$
\begin{aligned}
&{}^0\mathbf{a} = [1,0,0,0]^T \qquad {}^0\mathbf{b} = [1, a_1\cos(\theta_1), a_1\sin(\theta_1), 0]^T \qquad {}^1\mathbf{c} = [1,0,0,0]^T \\
&{}^0\mathbf{s}_1 = [0,0,0,1]^T \qquad {}^0\mathbf{s}_2 = [0,0,0,1]^T \qquad {}^0\mathbf{s}_3 = [0, \cos(\theta_1+\theta_2), \sin(\theta_1+\theta_2), 0]^T \\
&{}^0\mathbf{s}_4 = [0, \cos(\theta_1+\theta_2), \sin(\theta_1+\theta_2), 0]^T \qquad {}^1\mathbf{s}_5 = [0,0,0,1]^T
\end{aligned} \tag{1}
$$

where θ_1 and θ_2 are the angles of rotation of the first and the second revolute joints.

Study's kinematic mapping is used to express the vectors $\mathbf{c}$ and $\mathbf{s}_5$ in the fixed coordinate frame $\mathscr{F}_0$, using the transformation matrix ${}^0\mathbf{T}_1$ consisting of Study parameters x_i and y_i, $i = 0, 1, 2, 3$:

$$
{}^0\mathbf{c} = {}^0\mathbf{T}_1\,{}^1\mathbf{c} \qquad \text{and} \qquad {}^0\mathbf{s}_5 = {}^0\mathbf{T}_1\,{}^1\mathbf{s}_5,
$$

$$
\text{where } {}^0\mathbf{T}_1 = \frac{1}{\Delta}\begin{bmatrix} \Delta & 0 & 0 & 0 \\ d_1 & {x_0}^2+{x_1}^2-{x_2}^2-{x_3}^2 & -2\,x_0x_3+2\,x_1x_2 & 2\,x_0x_2+2\,x_1x_3 \\ d_2 & 2\,x_0x_3+2\,x_1x_2 & {x_0}^2-{x_1}^2+{x_2}^2-{x_3}^2 & -2\,x_0x_1+2\,x_2x_3 \\ d_3 & -2\,x_0x_2+2\,x_1x_3 & 2\,x_0x_1+2\,x_2x_3 & {x_0}^2-{x_1}^2-{x_2}^2+{x_3}^2 \end{bmatrix} \tag{2}
$$

[1] Left superscript k denotes the vector expressed in coordinate frame $\mathscr{F}_k$, $k \in \{0, 1\}$.

with $\Delta = {x_0}^2 + {x_1}^2 + {x_2}^2 + {x_3}^2 \neq 0$ and $d_1 = -2\,x_0y_1 + 2\,x_1y_0 - 2\,x_2y_3 + 2\,x_3y_2$, $d_2 = -2\,x_0y_2 + 2\,x_1y_3 + 2\,x_2y_0 - 2\,x_3y_1$, $d_3 = -2\,x_0y_3 - 2\,x_1y_2 + 2\,x_2y_1 + 2\,x_3y_0$. All vectors are now expressed in the base coordinate frame $\mathscr{F}_0$ and hence the geometric constraints can be derived. The following constraints are already satisfied:

1. The first and the second revolute joint axes are parallel: $\mathbf{s}_1 = \mathbf{s}_2$
2. Third and fourth revolute joint axes are parallel: $\mathbf{s}_3 = \mathbf{s}_4$
3. $\overrightarrow{AB}$ is perpendicular to the first and the second revolute joint axes: $(\mathbf{b} - \mathbf{a})^T\mathbf{s}_1 = 0$
4. The second revolute joint axis is perpendicular to the third revolute joint axis: $\mathbf{s}_2^T\mathbf{s}_3 = 0$
5. Length of the link AB is a_1: $||\mathbf{b} - \mathbf{a}||_2 = a_1$

The remaining geometric constraints are derived as algebraic equations[2]: The second revolute joint axis, the fifth revolute joint axis and link BC lie in the same plane. In other words, the scalar triple product of the corresponding vectors is null:

$$g_1 : (\mathbf{b} - \mathbf{c})^T(\mathbf{s}_2 \times \mathbf{s}_5) = 0 \tag{3}$$

Vector $\overrightarrow{BC}$ is perpendicular to the third and the fourth revolute joint axes:

$$g_2 : (\mathbf{b} - \mathbf{c})^T\mathbf{s}_4 = 0 \tag{4}$$

The fourth and the fifth revolute joint axes are perpendicular:

$$g_3 : \mathbf{s}_4^T\mathbf{s}_5 = 0 \tag{5}$$

Length of the link BC is a_3:

$$g_4 : ||\mathbf{b} - \mathbf{c}|| - a_3 = 0 \tag{6}$$

Furthermore, Study's quadric equation $\mathscr{S} : x_0y_0 + x_1y_1 + x_2y_2 + x_3y_3 = 0$ must be taken into account. The five geometric relations $g_1, g_2, g_3, g_4, \mathscr{S}$ describe the RUU limbs of the PM under study. As a matter of fact, when the first revolute joint is actuated, each limb has four DoF and it should be possible to describe it by only two constraint equations. Equations (4) and (5) contain the passive joint variable v_2 along with the active joint variable v_1. Eliminating v_2 from g_2 and g_3 results in an equation that amounts to g_1. Therefore, the two constraint equations in addition to the Study quadric describing a RUU limb are g_1 and g_4, namely Eqs. (3) and (6). The polynomials g_1, g_4 and $\mathscr{S}$ define an ideal, which is a subset of all polynomials in the Study parameters:

$$\mathscr{I}_1 = \langle g_1, g_4, \mathscr{S} \rangle \subseteq k[x_0, x_1, x_2, x_3, y_0, y_1, y_2, y_3]. \tag{7}$$

[2] Cosine and sine of angles are substituted by tangent half-angles to render the equations algebraic; $\cos(\theta_i) = \frac{1-v_i^2}{1+v_i^2}$ $\sin(\theta_i) = \frac{2v_i}{1+v_i^2}$ where $v_i = \tan(\theta_i/2)$, $i = 1, 2$.

Explicitly these polynomials take the form:

$$
\begin{aligned}
g_1 :=& \left((x_0x_1 - x_2x_3)\left(v_1{}^2 - 1\right) + (-2x_0x_2 - 2x_1x_3)\,v_1\right)(x_0^2 + x_1^2 + x_2^2 + x_3^2)a_1 \\
&- 2((x_0^2 + x_3^2)(x_1y_1 + x_2y_2) + 2(x_1^2 + x_2^2)(x_0y_0 + x_3y_3))(v_1^2 - 1) = 0, \qquad (8)\\
g_4 :=& -\left(x_0{}^2 + x_1{}^2 + x_2{}^2 + x_3{}^2\right)\left(v_1{}^2 + 1\right)a_1{}^2 + \left(4\left(y_1x_0 - y_0x_1 + y_3x_2 - y_2x_3\right)v_1{}^2\right. \\
&+ 8\left(-x_0y_2 + x_1y_3 + x_2y_0 - x_3y_1\right)v_1 + 4\left.\left(y_2x_3 - y_3x_2 - y_1x_0 + y_0x_1\right)\right)a_1 \\
&+ \left(\left(x_0{}^2 + x_1{}^2 + x_2{}^2 + x_3{}^2\right)a_3{}^2 - 4\left(y_2{}^2 + y_3{}^2 + y_0{}^2 + y_1{}^2\right)\right)\left(v_1{}^2 + 1\right) = 0. \qquad (9)
\end{aligned}
$$

General Constraints. g_1 and g_4 are the constraint equations of an RUU limb with specially adapted coordinate systems. To assemble the PM one has to transform these equations so that the limbs get into the positions of Fig. 1a. It is well known [9] that the necessary transformations are linear in the image space coordinates. Due to lack of space these transformations are only shown for the derivation of the constraint equations using the LIA in Sect. 3.2 (Eq. 14). One ends with six constraint equations $g_{i1}, g_{i4}, i = 1, 2, 3$ which form together with $\mathscr{S} = 0$ and the normalization condition $\mathscr{N} : x_0^2 + x_1^2 + x_2^2 + x_3^2 - 1 = 0$ an ideal

$$
\mathscr{I} = \langle g_{11}, g_{14}, g_{21}, g_{24}, g_{31}, g_{34}, \mathscr{S}, \mathscr{N} \rangle \subseteq k[x_0, x_1, x_2, x_3, y_0, y_1, y_2, y_3] \qquad (10)
$$

3.2 Derivation Using a Linear Implicitization Algorithm

Canonical Constraints. The canonical pose of a RUU limb is chosen such that the rotation axes coincide with the z-axes and the common normals of these axes are in the directions of the x-axes of the coordinate systems in order to derive the canonical constraint equations using LIA. It computes implicit equations of lowest possible degree out of parametric equations by comparing coefficients with an arbitrary system of implicit equations with the same degree. An extended explanation is given in [12]. To describe the RUU kinematic chain using the usual Denavit-Hartenberg (DH) parameters, the following 4×4 matrices are defined: $\mathbf{T} = \mathbf{M_i}.\mathbf{G_i}, i = 1, \ldots, 5$, where the $\mathbf{M_i}$-matrices describe a rotation about the z-axis with u_i as the rotation angle. The $\mathbf{G_i}$-matrices describe the transformation of one joint coordinate system to the next.

$$
\mathbf{M_i} = \begin{bmatrix} 1 & 0 & 0 & 0 \\ 0 & \cos(u_i) & -\sin(u_i) & 0 \\ 0 & \sin(u_i) & \cos(u_i) & 0 \\ 0 & 0 & 0 & 1 \end{bmatrix}, \qquad \mathbf{G_i} = \begin{bmatrix} 1 & 0 & 0 & 0 \\ a_i & 1 & 0 & 0 \\ 0 & 0 & \cos(\alpha_i) & -\sin(\alpha_i) \\ d_i & 0 & \sin(\alpha_i) & \cos(\alpha_i) \end{bmatrix}. \qquad (11)
$$

The parameters in $\mathbf{G_i}$ are DH parameters encoding the distance along x-axis a_i, the offset along z-axis d_i and the twist angle between the axes α_i. The DH parameters for the RUU limb are $\alpha_2 = \frac{\pi}{2}, \alpha_4 = -\frac{\pi}{2}, d_1 = a_2 = d_2 = d_3 = a_4 = d_4 = \alpha_1 = \alpha_3 = 0$. Computing the Study-Parameters based on the transformation matrix $\mathbf{T}$ yields the parametric representation of the limb [5].

Applying LIA yields the following quadratic canonical constraint equations $\mathscr{S}$, f_1 and f_2:

$$\mathscr{J} = \langle f_1, f_2, \mathscr{S} \rangle \subseteq k[x_0, x_1, x_2, x_3, y_0, y_1, y_2, y_3], \tag{12}$$

where

$$f_1 := ((x_0x_1 - x_2x_3)(v_1{}^2 - 1) - (2x_0x_2 + 2x_1x_3)v_1)a_1 + 2(v_1{}^2 + 1)(x_0y_0 + x_3y_3) = 0$$

$$\begin{aligned} f_2 := &-(x_0{}^2 + x_1{}^2 + x_2{}^2 + x_3{}^2)(v_1{}^2 + 1)a_1{}^2 + (4(y_1x_0 - y_0x_1 + y_3x_2 - y_2x_3)v_1{}^2 \\ &+ 8(-x_0y_2 + x_1y_3 + x_2y_0 - x_3y_1)v_1 + 4(y_2x_3 - y_3x_2 - y_1x_0 + y_0x_1))a_1 \\ &+ ((x_0{}^2 + x_1{}^2 + x_2{}^2 + x_3{}^2)a_3{}^2 - 4(y_2{}^2 + y_3{}^2 + y_0{}^2 + y_1{}^2))(v_1{}^2 + 1) = 0 \end{aligned} \tag{13}$$

General Constraints. To obtain the constraint equations of the whole mechanism from the canonical constraint equations, coordinate transformations are applied in the base and moving platform. To facilitate the comparison of the constraint equations derived by two different approaches, the coordinate transformations should be consistent with the global frames $\mathscr{F}_O$ and $\mathscr{F}_P$ as shown in Fig. 1a. The necessary transformations can be done directly in the image space $\mathbb{P}^7$ [9] by the mapping

$$\begin{bmatrix} x_0 \\ x_1 \\ x_2 \\ x_3 \\ y_0 \\ y_1 \\ y_2 \\ y_3 \end{bmatrix} \mapsto \begin{bmatrix} 2(v_0{}^2 + 1)x_0 \\ -2v_0{}^2x_1 + 4v_0x_2 + 2x_1 \\ 2(v_0{}^2 + 1)x_3 \\ 2v_0{}^2x_2 + 4v_0x_1 - 2x_2 \\ ((r_0 - r_1)x_1 + 2y_0)v_0{}^2 - 2x_2(r_0 - r_1)v_0 + (-r_0 + r_1)x_1 + 2y_0 \\ ((r_0 - r_1)x_0 - 2y_1)v_0{}^2 + 4v_0y_2 + (r_0 - r_1)x_0 + 2y_1 \\ ((-r_0 - r_1)x_2 + 2y_3)v_0{}^2 - 2(r_0 + r_1)x_1v_0 + (r_0 + r_1)x_2 + 2y_3 \\ ((r_0 + r_1)x_3 + 2y_2)v_0{}^2 + 4v_0y_1 + (r_0 + r_1)x_3 - 2y_2 \end{bmatrix}, \tag{14}$$

where $v_0 = \tan(\gamma_i), i = 1, 2, 3,\ \gamma_1 = 0, \gamma_2 = \frac{2\pi}{3}$ and $\gamma_3 = \frac{4\pi}{3}$. The general constraint equations are obtained by transforming the f_i of Eq. 12 with Eq. 14. The transformed equations are denoted $f_{i1} = f_{i2} = 0, i = 1, 2, 3$, and determine together with $\mathscr{S} = 0$ and $\mathscr{N} = 0$, the ideal $\mathscr{J}$:

$$\mathscr{J} = \langle f_{11}, f_{12}, f_{21}, f_{22}, f_{31}, f_{32}, \mathscr{S}, \mathscr{N} \rangle \subseteq k[x_0, x_1, x_2, x_3, y_0, y_1, y_2, y_3] \tag{15}$$

3.3 Ideal Comparison

A careful observation of the ideals spanned by the canonical constraint polynomials of both approaches reveals that $g_4 = f_2$ and $g_1 = f_1(x_0^2 + x_1^2 + x_2^2 + x_3^2) - 2(x_0^2 + x_2^2)(v_1^2 + 1)\mathscr{S}$. Since $x_0^2 + x_1^2 + x_2^2 + x_3^2$ cannot be null, these ideals are the same. Thus, it follows that the ideals $\mathscr{I}$ and $\mathscr{J}$ spanned by the constraint equations of the whole manipulator are also contained in each other: $\mathscr{I} \subseteq \mathscr{J} \subseteq \mathscr{I}$. Since $\mathscr{I}$ and $\mathscr{J}$ determine the same ideal, the variety of the constraint polynomials must be the same [2]. Therefore, the set of constraint equations derived in Sect. 3.2 is used for further computations as it contains only quadratic equations.

4 Direct Kinematics: Numerical Examples

Because of the complexity of the manipulator, it is not possible to compute the direct kinematics without using some numerical values. In the following subsections, the following arbitrary values are assigned to the design parameters of the manipulator, $a_1 = 3$, $a_3 = 5$, $r_0 = 11$, $r_1 = 7$.

Identical Actuated Joints. Assuming the actuated joint angles are equal, $\theta_{i1} = \frac{\pi}{2}, i = 1, 2, 3$ for simplicity, the system of constraint equations in Eq. (15) yields the following real solutions and the corresponding manipulator poses are shown in Fig. 2.

$$(a) \quad \left\{x_0 = \frac{\sqrt{23023}}{154}, y_3 = -\frac{3}{2}x_0, x_3 = -\frac{3\sqrt{77}}{154}, y_0 = \frac{3}{2}x_3, x_1 = x_2 = y_1 = y_2 = 0\right\},$$
$$(b) \quad \left\{x_0 = -\frac{\sqrt{23023}}{154}, y_3 = -\frac{3}{2}x_0, x_3 = \frac{3\sqrt{77}}{154}, y_0 = \frac{3}{2}x_3, x_1 = x_2 = y_1 = y_2 = 0\right\},$$

$$(c) \quad \{x_0 = 1, x_1 = x_2 = x_3 = y_0 = y_1 = y_2 = y_3 = 0\},$$
$$(d) \quad \{x_0 = 1, x_1 = x_2 = x_3 = y_0 = y_1 = y_2 = 0, y_3 = -3\}. \qquad (16)$$

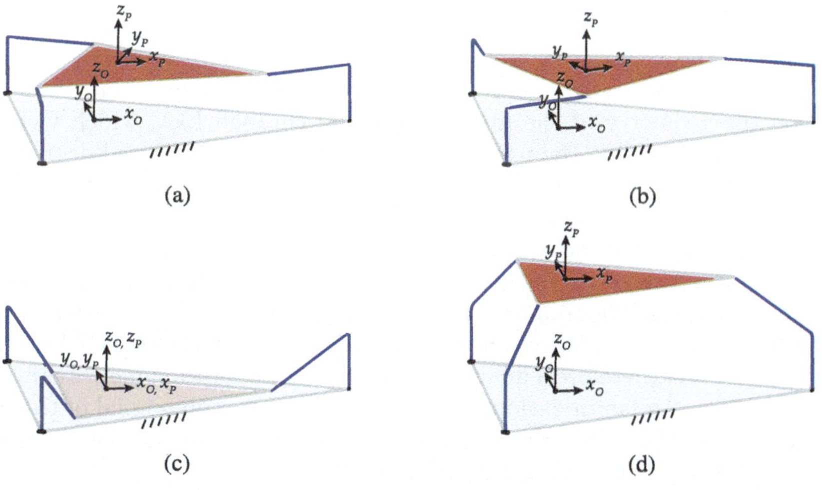

Fig. 2. A numerical example: solutions to direct kinematics corresponding to (16)

Different Actuated Joints. Substituting distinct arbitrary inputs, setting $x_0 = 1$ and computing a Groebner basis of the resulting polynomials with pure lexicographic ordering yields a univariate polynomial

$$x_3 \cdot P(x_3) = 0, \qquad \text{where} \quad degree(P(x_3)) = 80. \tag{17}$$

Translational Operation Mode. The univariate polynomial of the previous section shows that this manipulator exhibits two operation modes. The one corresponding to $x_3 = 0$ yields pure translational motions of the moving platform with the identity as the orientation, similar to the motion of the famous delta robot [1]. From $\mathscr{S}$ follows also $y_0 = 0$. The set of original constraint equations reduces to

$$\begin{aligned}&[(3y_3 - {y_1}^2 - {y_2}^2 - {y_3}^2 - 4y_1)\,{t_1}^2 - 6\,(y_1 + 2)\,t_1 - {y_1}^2 - {y_2}^2 - {y_3}^2 - 4y_1 - 3y_3,\\ &-(2{t_2}^2 + 3t_2 + 2)\,y_2\sqrt{3} + (-{y_1}^2 - {y_2}^2 - {y_3}^2 + 2y_1 + 3y_3)\,{t_2}^2 + (3y_1 - 12)\,t_2 - {y_1}^2\\ &- {y_2}^2 - {y_3}^2 + 2y_1 - 3y_3,\, (2{t_3}^2 + 3t_3 + 2)\,y_2\sqrt{3} + (-{y_1}^2 - {y_2}^2 - {y_3}^2 + 2y_1 + 3y_3)\,{t_3}^2\\ &+ (3y_1 - 12)\,t_3 - {y_1}^2 - {y_2}^2 - {y_3}^2 + 2y_1 - 3y_3].\end{aligned} \tag{18}$$

This system of equations yields a quadratic univariate in one of the y_i variables, which gives a parametrization of the motion as a function of the input variables $v_{i1} = \tan(\theta_{i1}/2)$, $i = 1, 2, 3$.

5 Conclusion

In this paper, the constraint equations of a 3-RUU PM were derived by two different approaches: geometrical approach, where all possible constraints were listed based on the geometry of the manipulator and through LIA, which yields the constraints by specifying the parametric equations and the desired degree. Both approaches have benefits and disadvantages such that it is possible to miss a constraint by merely observing the manipulator geometry while it is hard to interpret the physical meaning of the equations derived through LIA. However, it turns out that the ideals spanned by the constraint polynomials with both approaches are the same. As a result, the simplest set of equations was chosen for further analysis. Due to the complexity of the mechanism, a primary decomposition of these ideals is not possible and therefore a final answer to possible operation modes can not be given. However, the factorization of the final univariate polynomial of the direct kinematics algorithm gives strong evidence that this manipulator has a translational and a general three DoF operation mode.

Acknowledgements. This work was supported by the Austrian Science Fund (FWF I 1750-N26) and the French National Research Agency (ANR Kapamat #ANR-14-CE34-0008-01).

References

1. Clavel, R.: Delta, a fast robot with parallel geometry. In: Burckhardt, C.W. (ed.) Proceedings of the 18th International Symposium on Industrial Robots, pp. 91–100. Springer, New York (1988)
2. Cox, D.A., Little, J., O'Shea, D.: Ideals, Varieties, and Algorithms: An Introduction to Computational Algebraic Geometry and Commutative Algebra, 3rd edn. Springer, New York (2007)
3. Husty, M.: An algorithm for solving the direct kinematics of general Stewart-Gough platforms. Mech. Mach. Theory **31**(4), 365–379 (1996)
4. Husty, M., Schröcker, H.P.: Kinematics and algebraic geometry. In: 21st Century Kinematics, pp. 85–123. Springer, London (2012)
5. Husty, M.L., Pfurner, M., Schröcker, H.P., Brunnthaler, K.: Algebraic methods in mechanism analysis and synthesis. Robotica **25**, 661–675 (2007)
6. Nurahmi, L., Caro, S., Wenger, P.: Operation modes and singularities of 3-PRS parallel manipulators with different arrangements of P-joints. In: DETC 2015. ASME (2015)
7. Nurahmi, L., Caro, S., Wenger, P., Schadlbauer, J., Husty, M.: Reconfiguration analysis of a 4-RUU parallel manipulator. Mech. Mach. Theory **96**, 269–289 (2016)
8. Nurahmi, L., Schadlbauer, J., Husty, M., Wenger, P., Caro, S.: Kinematic analysis of the 3-RPS cube parallel manipulator. ASME J. Mech. Rob. **7**(1), 1–11 (2015)
9. Pfurner, M.: Analysis of spatial serial manipulators using kinematic mapping. Doctoral thesis, University Innsbruck (2006)
10. Schadlbauer, J., Nurahmi, L., Husty, M., Wenger, P., Caro, S.: Operation Modes in Lower-Mobility Parallel Manipulators. In: IAK: Proceedings of the International Conference, Lima, Peru, 9–11 September 2013, pp. 1–9. Springer, Cham (2015)
11. Schadlbauer, J., Walter, D.R., Husty, M.L.: The 3-RPS parallel manipulator from an algebraic viewpoint. Mech. Mach. Theory **75**, 161–176 (2014)
12. Walter, D.R., Husty, M.L.: On implicitization of kinematic constraint equations. Mach. Des. Res. **26**, 218–226 (2010)

References

Singularities

Kinematic Analysis of Planar Tensegrity 2-X Manipulators

Matthieu Furet[1,2], Max Lettl[1], and Philippe Wenger[1,2(✉)]

[1] Ecole centrale de Nantes, 44321 Nantes, France
max.lettl@gmx.de
[2] Laboratoire des Sciences du Numérique de Nantes (LS2N), CNRS, 44321 Nantes, France
{matthieu.furet,Philippe.Wenger}@ls2n.fr

Abstract. This paper analyzes the kinematics of planar tensegrity manipulators made of two Snelson's X-shape mechanisms in series. The variable instantaneous center of rotation of each mechanism renders the kinematic analysis of the resulting manipulator more challenging. A general formulation of the direct kinematics is set. A method is proposed to solve the inverse kinematic problem in a symbolic way and up to four inverse kinematic solutions are found. The singularities of the manipulator are shown to divide the joint space into two singularity-free components, showing for the first time a planar positioning manipulator that can be *cuspidal*. The workspace is determined and plotted for different values of the geometric parameters.

Keywords: Tensegrity · Kinematics · Singularity · Workspace Cuspidal

1 Introduction

A tensegrity structure is an assembly of compressive elements (bars) and tensile elements (cables, springs) held together in equilibrium [1,3]. Tensegrity is known in architecture and art for more than a century [2] and is suitable for modeling living organisms [4]. Tensegrity mechanisms have been more recently studied for their promising properties in robotics such as low inertia, natural compliance and deployability [5–7]. A tensegrity mechanism is obtained when one or several elements are actuated. This work falls withing the context of the AVINECK project involving biologists and roboticists with the main objective to model and design bird necks. Accordingly, a class of planar tensegrity manipulators made of a series assembly of several Snelson's X-shape mechanisms [8] i.e. crossed four-bar mechanisms with springs along their lateral sides, has been chosen as a suitable candidate for a preliminary planar model of a bird neck, see Fig. 1. First investigations on the kinematics of such manipulators turn out to be more challenging than expected with interesting properties, which has motivated the work presented in this paper. Snelson's X-shape mechanisms have been studied

J. Lenarcic and V. Parenti-Castelli (Eds.): ARK 2018, SPAR 8, pp. 153–160, 2019.
https://doi.org/10.1007/978-3-319-93188-3_18

by a number of researchers, either as a single mechanism [5,7,9] or assembled in series [10–12]. A planar two-degree-of-freedom manipulator is obtained with a series assembly of two such mechanisms. The manipulator can be driven with tendons threaded through the spring attachment points like in [12], or it can be actuated with a rotary motor at a base joint of each mechanism. There are several possible actuation schemes that can involve over-actuation if stiffness needs also to be controlled [10]. The detailed actuation scheme is not reported here and will be the scope of further work. This paper focuses on the full kinematic analysis of the manipulator, which has never been done before to the best of the authors' knowledge. The variable instantaneous center of rotation of each mechanism renders the symbolic calculation of the inverse geometric problem more challenging, resulting in polynomials of excessive degree if the problem is not tackled with care. A practical method is provided that makes it possible to come up with a quartic polynomial, which turns out to be of minimal degree. The singularities of the manipulator are shown to divide the joint space into two singularity-free components or *aspects*, showing for the first time a planar positioning manipulator that is *cuspidal* i.e. that can perform non-singular solution changing motions. The workspace is calculated with its boundaries for different values of the geometric parameters.

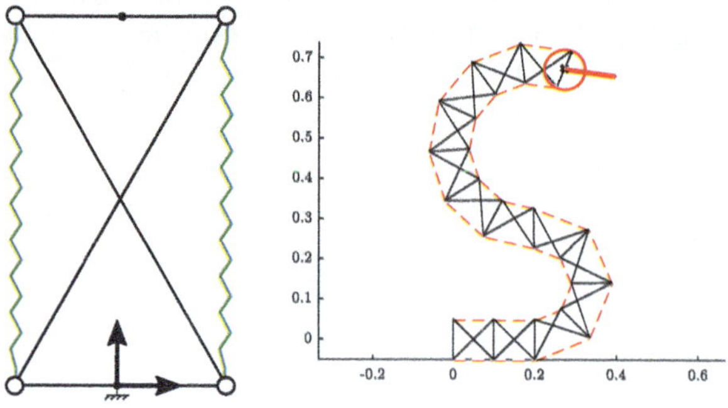

Fig. 1. Snelson's X-shape mechanism *(left)* and a series assembly of several such mechanisms, mimicking a bird neck *(right)*

2 Derivation of the Kinematic Equations

2.1 Manipulator Description and Parametrization

The manipulator studied consists of a series assembly of two identical crossed-bar mechanisms as shown in Fig. 2. Each mechanism i has a base bar and an upper bar of length b and two crossed bars of length L. Note that the mechanism assembly condition is $L > b$. The springs, shown in green in Figs. 1 and 2, are of length l_{ij}. We need to define a suitable joint variable that describes each mechanism configuration without any ambiguity. In [12], the mechanism configuration was chosen as the orientation α_i of the upper bar with respect to the base bar, but such a choice is allowed for mechanism motions restricted to $-\pi < \alpha_i < \pi$. For a matter of completeness, we are interested in allowing each mechanism to move within its full range. In this case,$-2\pi < \alpha_i < 2\pi$ and α_i is not appropriate to define the mechanism configuration. Let us introduce a line segment of length l_i that links the middle points of the top and base bars of each mechanism i (shown in red dotted line in Fig. 2). The angle between this line and the direction orthogonal to the base bar is referred to as θ_i. For the symmetric design considered here, it can be shown that $\alpha_i = 2\theta_i$ and when $-\pi < \theta_i < \pi$, the mechanism makes a full turn. Assuming that each mechanism remains always in the crossed-bar assembly mode, the manipulator configuration can thus be fully defined with (θ_1, θ_2). Since the mechanism sides define an isosceles trapezoid, the length l_i of the line segment that links the middle points of the top and base bars can be expressed as follows:

$$l_i(\theta_i) = \frac{l_{i1} + l_{i2}}{2} = \sqrt{L^2 - b^2 \cos^2(\theta_i)} \tag{1}$$

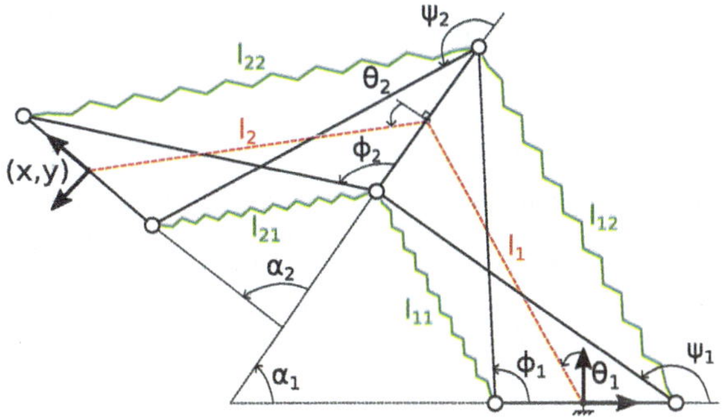

Fig. 2. Manipulator description

2.2 Direct Kinematics

We first want to establish the direct kinematic equations. The base frame is centered at the middle point of the base bar with the x-axis aligned along this bar. The reference point (x, y) is chosen as the middle point of the top-bar (Fig. 2). Note that any other location on the top bar would not affect the study, with the exception of one of the two extremities. In such a case, indeed, the input/output equations would be the same as in a simple 2R planar manipulator and the kinematic properties of the manipulator studied would not be captured. The direct kinematic equations of a manipulator with n mechanisms in series can be written as for a virtual serial nR manipulator with (varying) link lengths l_i and joint angles θ_i:

$$\begin{cases} x = \sum_{i=1}^{n} l_i(\theta_i) \cos(\frac{\pi - 2\theta_i}{2} + \sum_{j=1}^{n} 2\theta_j) \\ y = \sum_{i=1}^{n} l_i(\theta_i) \sin(\frac{\pi - 2\theta_i}{2} + \sum_{j=1}^{n} 2\theta_j) \end{cases} \tag{2}$$

For a manipulator made of two mechanisms in series with input variables θ_1 and θ_2, the direct kinematic equations can be put in the following form :

$$\begin{cases} x = -l_1(\theta_1)\sin(\theta_1) - l_2(\theta_2)\sin(2\theta_1 + \theta_2) \\ y = l_1(\theta_1)\cos(\theta_1) + l_2(\theta_2)\cos(2\theta_1 + \theta_2) \end{cases} \tag{3}$$

where l_1 and l_2 are defined in (1). Note that these equations assume that each mechanism is in its crossed-bar assembly-mode.

2.3 Inverse Kinematics

We would like now to derive the so-called characteristic polynomial, a univariate polynomial established to solve the inverse kinematics. The direct kinematic equations (3) have been derived in their most compact form, namely, the two output variables x and y are expressed as a function of the two input variables θ_1 and θ_2 like in a 2R serial manipulator. However, starting from these equations to derive a characteristic polynomial is not appropriate. Indeed, the square roots appearing in l_1 and l_2 should be first cleared out, with the consequence of raising artificially the degree of the resulting characteristic polynomial and providing spurious solutions. Accordingly, it is better to write x and y as functions of the crossed-bar angles ϕ_1 and ψ_2 as follows:

$$\begin{cases} x = -\frac{b}{2} + L\cos(\phi_1) + L\cos(\psi_2 + \alpha_1) + \frac{1}{2}b\cos(\alpha_1 + \alpha_2) \\ y = L\sin(\phi_1) + L\sin(\psi_2 + \alpha_1) + \frac{1}{2}b\sin(\alpha_1 + \alpha_2) \end{cases} \tag{4}$$

In addition, the two loop-closure equations below are derived:

$$\begin{cases} -2Lb\sin(\alpha_1)\sin(\phi_1) - 2Lb(\cos(\alpha_1) + 1)\cos(\phi_1) + 2b^2(\cos(\alpha_1) + 1) = 0 \\ 2Lb\sin(\alpha_2)\sin(\psi_2) + 2Lb(\cos(\alpha_2) + 1)\cos(\psi_2) + 2b^2(\cos(\alpha_2) + 1) = 0 \end{cases} \tag{5}$$

A set of four equations in four unknowns is thus available. A univariate polynomial can be obtained upon elimination of three of the four unknowns. There are several ways to do so with a computer algebra software. We have used Maple and its *Siropa* library [13]. This library was developed in the frame of a collaborative project on the kinematics of parallel manipulators, see e.g. [14,15] for more details on the implementation and use of these tools. It contains specific macro-functions that use efficient algebraic tools such as Groebner bases. Specifically, the *Projection* function was used to project the system of four equations in order to obtain one single equation in one single variable, chosen here as ϕ_1. The half-tangent substitution yields a factored polynomial, one of which defines the characteristic polynomial. The characteristic polynomial obtained turns out to be of degree 4 in $t = tan(\phi_1/2)$ and can be written as follows:

$$a_4t^4 + a_3t^3 + a_2t^2 + a_1t + a_0 = 0 \tag{6}$$

where:

$$a_4 = (b+1)^2(b^2y^2 + x^4 + 2x^2y^2 + y^4 + 4x^3 + 4xy^2 + 5x^2 + y^2 + 2x) \tag{7}$$

$$a_3 = 4y(b+1)(2b^2x + b^2 - 2x^2 - 2y^2 - 4x - 1) \tag{8}$$

$$a_2 = 2(b^4y^2 + b^2x^4 + 2b^2x^2y^2 + b^2y^4 + b^2x^2 - 10b^2y^2 + x^4 + 2x^2y^2 + y^4 - 3x^2 + 9y^2) \tag{9}$$

$$a_1 = 4y(b-1)(2b^2x - b^2 + 2x^2 + 2y^2 - 4x + 1) \tag{10}$$

$$a_0 = (b-1)^2(b^2y^2 + x^4 + 2x^2y^2 + y^4 - 4x^3 - 4xy^2 + 5x^2 + y^2 - 2x) \tag{11}$$

Note that L was set equal to 1 to simplify the calculations without loss of generality. For each solution ϕ_1, α_1 is then solved from the first equation in (5) (disregarding the solution $\alpha_1 = \pi$ corresponding to the parallelogram assembly):

$$\tan(\alpha_1/2) = \frac{-L\cos(\phi_1) + b}{L\sin(\phi_1)} \tag{12}$$

Finally, ψ_2 and α_2 are solved from system (4), which gives two solutions, one of which never satisfies the closed-loop equations (5). In total, thus, one comes up with a maximum number of four solutions. The inverse kinematics is solved for a manipulator defined by $L = 1$ and $b = 2/3$ at $x = 0.03$ and $y = 1.6$. Four solutions are found (see Fig. 3), showing that the characteristic polynomial (6) is of minimal degree.

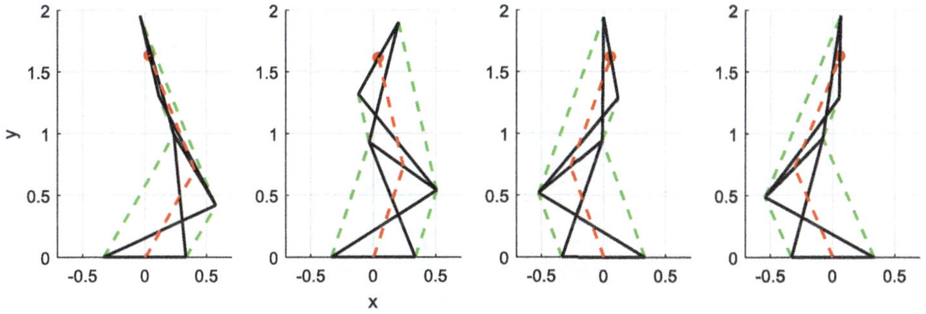

Fig. 3. The four inverse solutions at $x = 0.03$ and $y = 1.6$ ($L = 1$ and $b = 2/3$)

3 Singularity Analysis and Aspects

The Jacobian matrix of the manipulator is derived from system (3) and its determinant is calculated and can be put in the following form:

$$\begin{aligned} Det(J) = \; & (\sin(\theta_1)^2 b^2\cos(\theta_1) - (L^2 - b^2\cos(\theta_1)^2)(\cos(\theta_1) + 2\cos(\theta_2 + 2\theta_1))) \\ & (\cos(\theta_2 + 2\theta_1) b^2 \cos(\theta_2)\sin(\theta_2) - (L^2 - b^2\cos(\theta_2)^2)\sin(\theta_2 + 2\theta_1)) \\ & -(-\sin(\theta_2 + 2\theta_1) b^2\cos(\theta_2)\sin(\theta_2) - (L^2 - b^2\cos(\theta_2)^2)\cos(\theta_2 + 2\theta_1)) \\ & (\cos(\theta_1)^2 b^2\sin(\theta_1) - (L^2 - b^2\cos(\theta_1)^2)(\sin(\theta_1) + 2\sin(\theta_2 + 2\theta_1))) \end{aligned} \tag{13}$$

Solving $Det(J) = 0$ for θ_1 yields two solution sets. The first one is $\theta_1 = -\theta_2 \pm \pi$ and means that the two modules are coincident. It is equivalent to the fully-folded back configuration of a planar 2R serial manipulator. When the manipulator is in this singularity, the reference point is at the origin $x = 0$ and $y = 0$, whatever θ_1. The second solution is given by:

$$\theta_1 = atan2(-\sin(\theta_2)(4b^4\cos(\theta_2)^4 - 4L^2b^2\cos(\theta_2)^2 - L^2b^2 + L^4),\\ \cos(\theta_2)(4b^4\cos(\theta_2)^4 - 4b^4\cos(\theta_2)^2 - 4L^2b^2\cos(\theta_2)^2 + 3L^2b^2 + L^4)) \quad (14)$$

This solution is associated with the configuration where the two instantaneous centers of rotation (ICR) are aligned with the end-effector control point, like in the fully outstretched configuration of a planar 2R serial manipulator. In this singularity, the manipulator cannot produce an instantaneous motion along a line passing through the two mechanism ICR and the end-effector control point. The above singularities divide the joint space into singularity-free domains called *aspects* [16]. The singularity $\theta_1 = -\theta_2 \pm \pi$ produces two lines while the singularity defined by Eq. (14) gives a curve, as shown in Fig. 4. In the absence of joint limits, the opposite sides of the square are in fact coincident and the singularity curves divide the joint space into only two aspects. Since the manipulator admits up to four solutions, there are two solutions in each aspect, which means that the manipulators can move from one inverse kinematic solution to another without meeting a singularity, namely, it is *cuspidal* [18]. To the authors' knowledge, it is the first time that a planar positioning manipulator is shown to be cuspidal.

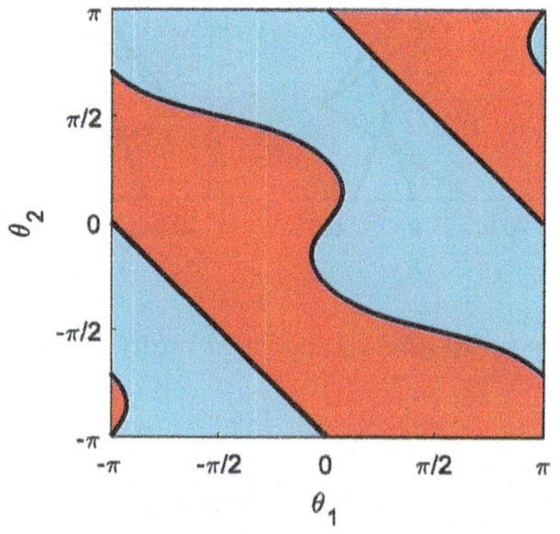

Fig. 4. Singularity curves in the joint space and aspects when $L = 1$ and $b = 2/3$

4 Workspace Analysis

When plotted into the workspace, the singularity curves define its boundaries. The equation of these curves can be obtained by deriving the discriminant of the characteristic polynomial (6). By doing so, a polynomial of degree 16 in x

and y is obtained. Its expression is quite large and shall not be reported here for lack of space. Figure 5 shows the plot of these curves for three examples. In the second and third figures, they divide the workspace into three regions. In the largest one, the manipulator admits two inverse kinematic solutions. In the two smaller regions, there are four solutions. Clearly, the 4-solution regions get larger when b is increased. The boundaries of these two 4-solution regions have three singular points: a node and two cusps. The existence of cusps confirms the fact that the manipulator is cuspidal. A non-singular solution changing motion can be defined by encircling one of the cusps [17,18]. Note that the two aspects map onto the whole workspace, namely, Fig. 5 also shows the image of any of the two aspects.

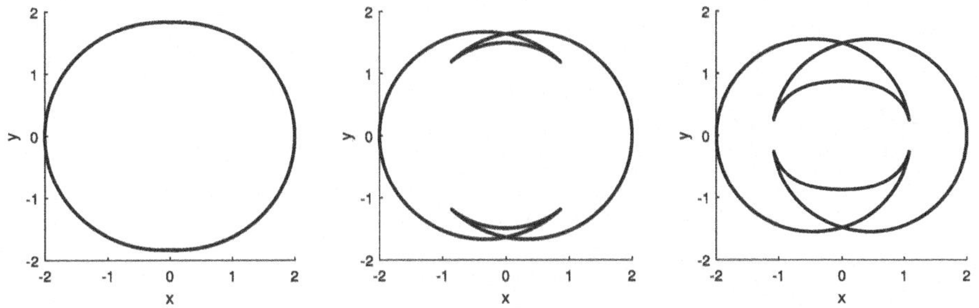

Fig. 5. Workspace boundaries when $L = 1$ and $b = 2/5$ *(left)*, $b = 2/3$ *(center)*, $b = 9/10$ *(right)*

5 Conclusion

The complete kinematic analysis of a planar manipulator made of two X-four bar mechanisms in series has been carried out. The manipulator was shown to have two or four solutions depending on its geometric parameter values. When it has four solutions, the manipulator turns out to be cuspidal, a result that was not expected. The influence of the geometric parameters on the shape and size of the workspace was analyzed. The influence of joint limits was not reported for lack of space and shall be studied in a near future as they also play an important role. Internal collisions were not considered since a suitable design by assembling the rods in different planes allows avoiding any physical interference. Future work will be conducted on the actuation strategy.

Acknowledgement. This work was conducted with the support of the French National Research Agency (AVINECK Project ANR-16-CE33-0025).

References

1. Fuller, R.B.: Tensile-integrity structures, United States Patent 3063521 (1962)
2. Skelton, R., de Oliveira, M.: Tensegrity Systems. Springer (2009)
3. Motro, R.: Tensegrity systems: the state of the art. Int. J. Space Struct. **7**(2), 75–83 (1992)
4. Levin, S.: The tensegrity-truss as a model for spinal mechanics: biotensegrity. J. Mech. Med. Biol. **2**(3), 375–388 (2002)
5. Arsenault, M., Gosselin, C.M.: Kinematic, static and dynamic analysis of a planar 2-dof tensegrity mechanism. Mech. Mach. Theory **41**(9), 1072–1089 (2006)
6. Crane, C., et al.: Kinematic analysis of a planar tensegrity mechanism with pres-stressed springs. In: Lenarcic, J., Wenger, P. (eds.) Advances in Robot Kinematics: Analysis and Design, pp. 419-427. Springer (2008)
7. Wenger, P., Chablat, D.: Kinetostatic analysis and solution classification of a planar tensegrity mechanism. In: Proceedings of the 7th International Workshop on Computational Kinematics, pp. 422-431. Springer (2017). ISBN 978-3-319-60867-9
8. Snelson, K.: Continuous Tension, Discontinuous Compression Structures, US Patent No. 3,169,611 (1965)
9. Boehler, Q., et al.: Definition and computation of tensegrity mechanism workspace. ASME J. Mech. Robot. **7**(4), 044502-1–044502-4 (2015)
10. Aldrich, J.B., Skelton, R.E.: Time-energy optimal control of hyper-actuated mechanical systems with geometric path constraints. In: 44th IEEE Conference on Decision and Control, pp. 8246-8253 (2005)
11. Chen, S., Arsenault, M.: Analytical computation of the actuator and cartesian workspace boundaries for a planar 2-degree-of-freedom translational tensegrity mechanism. J. Mech. Robot. **4**, 011010 (2012)
12. Bakker, D.L., et al.: Design of an environmentally interactive continuum manipulator. In: Proceedings of the 14th World Congress in Mechanism and Machine Science, IFToMM 2015, Taipei, Taiwan (2015)
13. Rouillier, F., et al.: Siropa Library V1, IDDN.FR.001.140015.000.S.P.2017.000.20600
14. Manubens, M., et al.: Cusp points in the parameter space of degenerate 3-RPR planar parallel manipulators. ASME J. Mech. Robot. **4**(4), 041003–8 (2012)
15. Moroz, G., et al.: On the determination of cusp points of 3-RPR parallel manipulators. Mech. Mach. Theory **45**(11), 1555–1567 (2011)
16. Borrel, P., Liegeois, A.: A study of manipulator inverse geometric solutions with application to trajectory planning and workspace determination
17. El Omri, J., Wenger, P.: How to recognize simply a non-singular posture changing manipulator. In: Proceedings of the 7th International Conference on Advanced Robotics, pp. 215-222 (1995)
18. Wenger, P.: Cuspidal and noncuspidal robot manipulators. Spec. Issue Robot. Geom. Robot. Sens. **25**(6), 677–690 (2007)

Rotational Mobility Analysis of the 3-RFR Class of Spherical Parallel Robots

David Corinaldi(✉), Luca Carbonari, Matteo-Claudio Palpacelli, and Massimo Callegari

DIISM, Polytechnic University of Marche, 60131 Ancona, Italy
{d.corinaldi,l.carbonari,m.palpacelli,m.callegari}@univpm.it
http://www.meccanicadellemacchine.it

Abstract. Spherical parallel manipulators (SPMs) are used to orient a tool in the space with three degrees of freedom exploiting the strengths of a multi-limb architecture. On the other hand, the performance of parallel kinematics machines (PKMs) is often affected by the occurrence of different kinds of singular configurations. The paper aims at characterizing a class of SPMs for which all singularities come to coincide and a single expression is able to describe all the singular configurations of the machines. The study is focused on a class of SPMs with 3-RFR topology (Revolute-Planar-Revolute pairs for each of the three limbs) addressing the mobility and singularity analysis by means of polynomial decomposition and screw theory. The neatness of the equations that are worked out, expressed in a robust formulation based on rotation invariants, allows a straightforward planning of singularity free tasks and simplifies the synthesis of dexterous machines.

Keywords: Mobility analysis · Polynomial decomposition
Singularity · Spherical parallel machines

1 Introduction

Spherical parallel manipulators (SPMs) are a specific class of parallel kinematics machines (PKMs) which are able to generate a three-dof spherical motion; they are sometimes also called three-dof orientational parallel manipulators or rotational parallel manipulators [1]. SPMs can be divided into two families: *overconstrained* or *hyperstatic* SPMs and *non-overconstrained* or *isostatic* SPMs. The main advantage of isostatic architectures is that they do not need the strict dimensional and geometric tolerances of overconstrained machines during manufacturing and assembly; thus they can work even when precise geometrical conditions are not accurately met, if at the expense of parasitic motions. Moreover, modular solutions characterized by three identical legs (symmetrical PKMs) are usually preferred for economic reasons. The present work is focused on the 3-RFR class of tripod SPMs which have an isostatic and symmetrical architecture.

J. Lenarcic and V. Parenti-Castelli (Eds.): ARK 2018, SPAR 8, pp. 161–169, 2019.
https://doi.org/10.1007/978-3-319-93188-3_19

This class of machines has already been studied by many researchers, with notable contributions by Karouia and Hervé [2] that first studied the mobility of this class of SPM by means of Lie algebra, Kong and Gosselin [3] that used screw theory for the synthesis of parallel wrists and Di Gregorio, who developed the kinematic relations of a number of SPMs [4–7].

After a first description of this class of manipulators, the paper is divided into two sections: the analysis of mobility and a comprehensive study on singular configurations, which are addressed by means of powerful geometrical tools such as polynomial decomposition and screw theory. The latter has been used to analyze the instantaneous mobility of the machines while the global mobility has been addressed by using algebraic geometry as done by Kong in [8]. In this work, the pose of the mobile platform (MP) is described by means of the Study parameters, which allow a polynomial representation of legs constraint equations. The subset of the Study parameters used to represent rotations are the same Euler-Rodrigues Parameters (ERPs) which are related to the invariants of the rotational matrix [9]. They allow a robust representation of the problem and an effective manipulation of constraints equations, therefore they have been used for pose and trajectory optimization by the same A.'s in [10,11].

The synthesis of PKMs is usually driven by the goal of maximizing stiffness and accuracy: Sect. 4 shows how the actuated joints can be chosen so as to obtain the coincidence of the singularity loci, therefore simplifying the avoidance of singular poses. Wrist configurations that cause singularities are identified and interpreted by means of *screw theory* [12]. This theory is based on the concept of the instant screw axis, represented in Plücker coordinates by a six-dimensional array: it is composed by the unit vector defining the direction of the axis and by its moment about the origin [13]. The present study has been developed according to the approach of Conconi and Carricato [14], by which the linear dependence of three screw moments is sufficient to identify all singularities, i.e., unactuated and actuated kinematic chain singularities and passive constraint singularities.

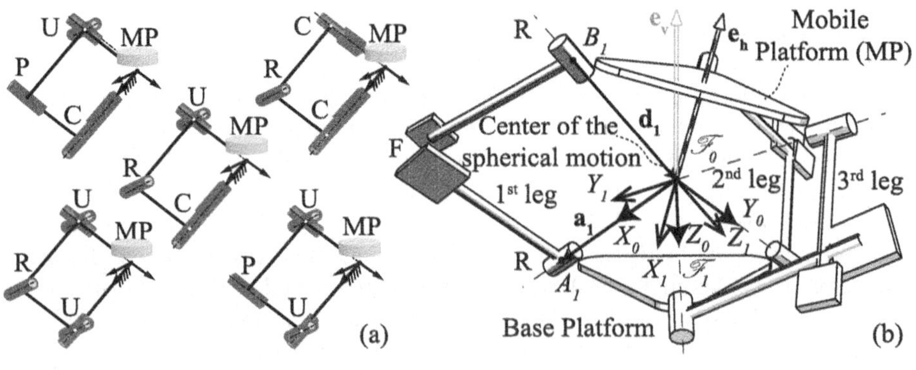

Fig. 1. The 3-RFR class of SPMs: (a) different leg topologies; (b) kinematic model.

2 The 3-RFR Class of Spherical Parallel Manipulators

Three-legged isostatic SPMs include the relevant class of 3-RFR parallel mechanisms, see Fig. 1(b): their limbs are attached to the fixed and mobile platforms by means of two revolute (R) pairs, whose axes intersect at the center of the spherical motion; in between three lower pairs are equivalent to a planar joint (F) whose plane contains the axes of the revolute joints and the center of the motion. Corinaldi et al. [15] showed that this architecture can yield a large orientational workspace characterized by good dexterity around the home configuration, i.e., when the reference frame $\mathscr{F}_1$ attached to the MP has the same orientation as the reference frame $\mathscr{F}_0$ fixed to the ground. The planar joint in the middle of each limb can be indifferently substituted by any of the PRR, RPR, PPR, PRP and RRR topologies, provided that some rules are observed: the directions of the prismatic joints must be parallel to the plane of motion, while the axes of the revolute pairs must be perpendicular to such plane, as shown in Fig. 1(a). Finally, such pairs may be merged in a cylindrical (C) joint when the axes of revolute and prismatic pairs are coaxial; in the same way two orthogonal revolute pairs are equivalent to a universal (U) joint. The following set of leg topologies arise: (RP)R(RR) = CRU, (RR)P(RR) = UPU, (RP)P(RR) = CPU, (RP)R(PR) = CRC and (RR)R(RR) = URU.

3 Global Mobility Verification

The pose of the MP is expressed by means of the Study parameters [16]. This notation was introduced by Study who used a superabundant set of eight parameters in the seven-dimensional projective space to map the Euclidean space through a one to one representation $\chi : \mathbf{T} \in SE(3) \mapsto \mathbf{z} \in \mathbb{P}^7$. The eight parameters y_i and r_i for $i = 0, 1, 2, 3$ are related through a quadric polynomial equation, namely the Study's quadric, and through a metric equation which ensures that the transformations actually represent rigid body motions:

$$\begin{array}{ll} \sigma_1 : r_0 y_0 + \mathbf{r}^T \mathbf{y}, & \sigma_1 = 0 \\ \sigma_2 : r_0^2 + \mathbf{r}^T \mathbf{r} - 1, & \sigma_2 = 0 \end{array} \tag{1}$$

having collected the parameters in the vectors $\mathbf{r} = [r_1\ r_2\ r_3]^T$ and $\mathbf{y} = [y_1\ y_2\ y_3]^T$ for terseness. By using the homogeneous notation, the transformation $\mathbf{T}$ between the mobile frame $\mathscr{F}_1$ and the fixed frame $\mathscr{F}_0$ of Fig. 2(a) can be written as usual in terms of the rotation matrix $\mathbf{Q}$, and the position vector $\mathbf{p}$ between their origins:

$$\begin{array}{ll} \mathbf{Q} = (r_0^2 - \mathbf{r}^T \mathbf{r})\mathbf{1} + 2\mathbf{r}\mathbf{r}^T + 2r_0\mathbf{R}, & \mathbf{R} = \mathrm{CPM}(\mathbf{r}) \\ \mathbf{p} = -2(r_0 \mathbf{y} + y_0 \mathbf{r} + \mathbf{Y}\mathbf{r}), & \mathbf{Y} = \mathrm{CPM}(\mathbf{y}) \end{array} \tag{2}$$

where CPM($\mathbf{r}$) stands for the cross-product matrix operator on the vector $\mathbf{r}$[1]. Besides the Study's quadric Eq. (1), additional algebraic relations are written

[1] That is, $\mathrm{CPM}(\mathbf{r}) = \partial(\mathbf{r} \times \mathbf{v})/\partial \mathbf{v}$, $\forall \mathbf{v}$, $\mathbf{r} \in \mathbb{R}^3$.

to describe the constraints imposed by joints' arrangement on the MP; in this way the mobility of the platform within its workspace is fully characterized, as done for the 3-URU PKM by Carbonari et al. in [17]. These relations, which are specific of each legs' architecture, are introduced in the following lines by making reference to the sketch in Fig. 2(a). The ith limb, for $i = 1, 2, 3$, is built so as to constrain the axes of the first and the last revolute joint on the plane π_i. Equivalently, the line ℓ_i, identified by the axis of such joint, has an intersection point with all the lines that lie on the plane π_i, for example the axis of the first revolute pair, here called m_i. The two revolute joints are connected through a planar joint that allows them a relative motion parallel to the plane π_i. This joint constraints the two axes to lie on the same plane for all wrist configurations. The two lines can be parametrized through their direction plus a passing point, as

$$\ell_i : \alpha_i \hat{\mathbf{d}}_i + \mathbf{p}, \quad m_i : \beta_i \hat{\mathbf{a}}_i \tag{3}$$

where $\alpha_i, \beta_i \in \mathbb{R}$ and the hat on the vectors indicates that they are unit vectors. The legs' constraints can be expressed by

$$\ell_i = m_i \;\rightarrow\; \alpha_i \hat{\mathbf{d}}_i + \mathbf{p} = \beta_i \hat{\mathbf{a}}_i \;\Rightarrow\; \Big[[\mathbf{Q}]_0 [\hat{\mathbf{d}}_i]_1 \;\; -[\hat{\mathbf{a}}_i]_0 \;\; [\mathbf{p}]_0 \Big] \Big[\alpha_i \; \beta_i \; 1 \Big]^T = \mathbf{0} \tag{4}$$

where in the second part of Eq. 4 the three column vectors composing the matrix are expressed in the ground frame $\mathscr{F}_0$. A non-trivial solution of this homogeneous equation can be found by letting the determinant of the respective matrix vanish. Substituting $\mathbf{Q}$ and $\mathbf{p}$ relationships Eq. (2) into Eq. 4, three polynomials in the Study parameters are obtained, one for each leg. After simplification [18], the three constraints can be expressed by the varieties of the following polynomials

$$\begin{aligned} g_1 &: r_0 y_2 + r_1 y_3 + r_2 y_0 + r_3 y_1, \;\; g_1 = 0 \\ g_2 &: r_0 y_3 + r_1 y_2 + r_2 y_1 + r_3 y_0, \;\; g_2 = 0 \\ g_3 &: r_0 y_1 + r_1 y_0 + r_2 y_3 + r_3 y_2, \;\; g_3 = 0 \end{aligned} \tag{5}$$

The vanishing set of polynomials g_1, g_2, and g_3, together with Study quadric Eq. (1) are not yet sufficient to fully describe the kinematics of the 3-RFR class of SPMs, because a set of actuation dependent equations must still be written. Nevertheless, such relations provide the searched information about the mobility of the MP [19]. In fact, when the four Study parameters y_0, y_1, y_2, and y_3 are null, the general transformation matrix $\mathbf{T}$ describes a pure rotation since the translation vector $\mathbf{p}$ vanishes. Under these conditions, the $\mathbf{Q}$ matrix fully describes the motion of the MP by means of the four scalar quantities in r_0 and $\mathbf{r}$, that correspond to the Euler-Rodrigues parametrization of rotations [13]. It can be easily verified that such condition satisfies both constraints Eq. (5) and the Study quadric relation $\sigma_1 = 0$, thus demonstrating that a pure rotational motion has been obtained by the 3-RFR SPM.

4 Singularity Analysis

Differently from serial robots, where all joints are actuated, parallel robots are composed by many passive joints, which makes their analysis more complex

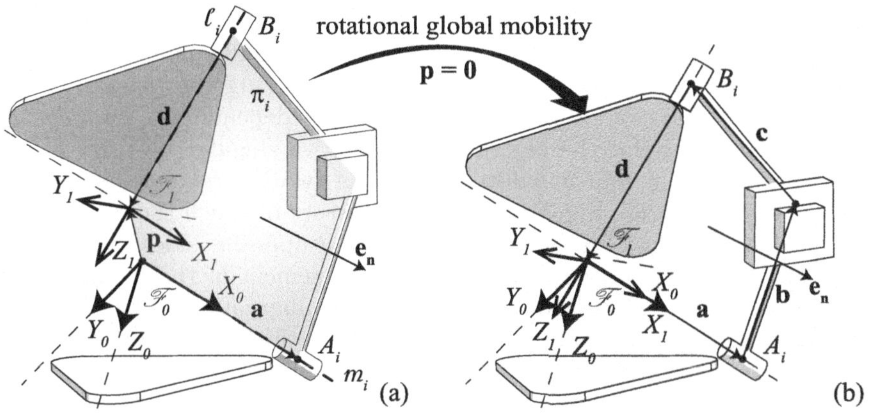

Fig. 2. (a) Kinematics of the ith leg for a generic assembly of the PKM class and (b) the building architecture that grants the rotational behavior.

and increases the number of singular configurations. Therefore the identification of these configurations and their parametric representation in terms of ERPs is useful for the singularity avoidance during path-planning and for a possible maximization of their dexterity during kinematic synthesis. The following section develops such kind of analysis through geometrical inspection with the aim of selecting the most suitable joints to be actuated in order to obtain a simple-to-use singularity relationship.

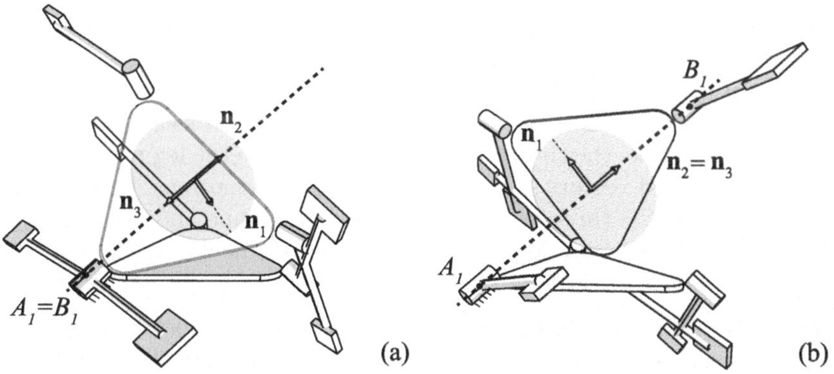

Fig. 3. Singular robot postures when a leg singularity occurs on the first limb.

Unactuated Kinematic Chain Singularities. Since the MP is subject to a pure rotation, the point B_i of Fig. 2(b) moves on the surface of a sphere; moreover, legs topology constraints the points A_i and B_i to have a relative planar motion. Keeping in mind these two facts, it is possible to identify the singular configurations of the legs with a geometric inspection of the mechanism. At this point no joint is actuated, thus kineto-static properties only depend on the kind, number, and mutual disposition of the joints composing the kinematic chain. When a subset of screws of a leg becomes linearly dependent a *leg singularity* occurs. In particular, for the case under study, two singular postures for each leg are identified that appear when the axes of two the revolute joints of the same leg become coaxial: Fig. 3(a) shows the posture in which the revolute pairs are coincident while in Fig. 3(b) they lie on the opposite sides of the sphere surface. Moreover, through an appropriate arrangement of the R and P joints which compose the planar pair, such singular configurations lead also to the linear dependence of the planar sub-chain pairs. In these configurations, the leg generates a new mobility constraint, i.e., it is unable to rotate around an axis perpendicular to the plane generated by the two revolute axes and the rotation axis of the planar pair. The relationships in terms of the ERPs can be obtained for $i = 1, 2, 3$ from

$$\hat{\mathbf{d}}_i^T \hat{\mathbf{a}}_i = 1 \quad \Rightarrow \quad \left([\mathbf{Q}]_0 \, [\mathbf{d}_i]_1\right)^T [\mathbf{a}_i]_0 = 1 \tag{6}$$

Passive-Constraint Singularities. A *passive-constraint singularity* (PCS) occurs when the dimension of the MP-motion space increases with respect to its global mobility value. Each leg develops on the MP a constraining force orthogonal to the plane of the leg and passing through the origin so that the MP motion is a three-dimensional rotation around the origin. When these forces become linearly dependent, e.g. they are parallel to the same plane, the dimension of the motion space increases from three to four and the mechanism is at a PCS. Unlike the previous case, there is at least a PCS posture for each pointing direction of MP, i.e. given a pointing vector, it is always possible to find a singular posture by rotating the MP around this axis. Two examples are shown in Fig. 4: case (a) shows a PCS posture for the vertical direction while case (b) sketches a different pointing direction. In both cases, the constraints imposed by each leg on the MP, i.e. the three forces perpendicular to the leg planes (shown in the figures as applied vectors in the center of spherical motion), become linearly dependent: in fact, the three vectors lie on the same plane and the twist system of MP gains a translational degree of mobility perpendicular to the force plane.

Actuated Kinematic Chain Singularities. The parallel kinematic chain turns into a machine once the actuated joints are specified. The total number of actuators is obviously equal to the dimension of the global mobility space, i.e. three; moreover the motors should be placed at the base platform, i.e. the ground, for practical considerations. A first layout of the motors is obtained through the actuation of the revolute joints at the base so that their action on the MP consists of a force perpendicular to the leg plane.

A better alternative solution is obtained by actuating the prismatic joints closest to the ground that compose the planar joint of each leg, once they have been brought back to the base platform. This choice brings many advantages; in

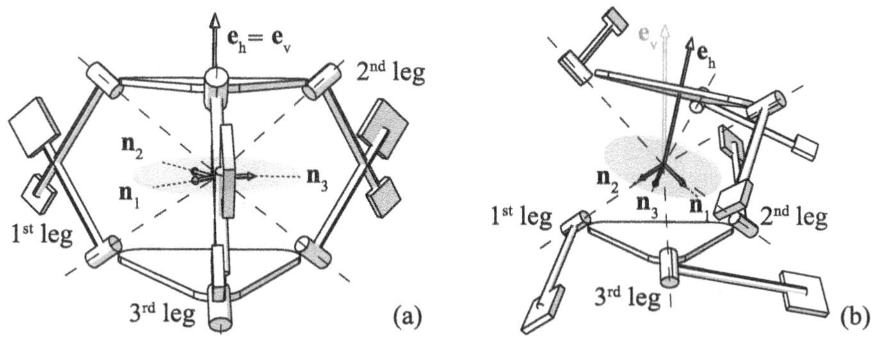

Fig. 4. Passive-constraint singularity postures for two different pointing directions.

fact, this time the wrench acting on the platform is given by a force that lies in the leg plane, whose direction and application point depend of the planar joint sub-chain. Anyhow, this force causes a moment on the MP that is perpendicular to the leg plane, i.e, it is in the same direction of the constraint force imposed by each leg on the MP. In this way, the condition of linear dependence of the wrench constraints becomes coincident with the linear dependence of the actuation wrenches. Furthermore, it should be noted that, if these moment vectors are computed as cross product between the axis of the revolute joint $\mathbf{a}_i$ and the vector $\mathbf{d}_i$, then their magnitudes vanish when the two revolute joints into the leg become coaxial, i.e. when a leg singularity occurs. Due to these reasons, the study of these three moments is sufficient to fully describe the singularities assumed by the SPMs with linear actuation fixed to the base. Within the class of 3-RFR spherical manipulators of Fig. 1, only three of them can have the prismatic joint fixed to the base, i.e., 3-CRU, 3-CPU and 3-CRC, whose architecture is illustrated in Fig. 5. As a matter of fact, the R and P joints of the first cylindrical joint can be switched without affecting the MP mobility because of the coaxiality of the direction of the P joint with the axis of the R joint. If the moments of the actuation forces are denoted by $\mathbf{n}_i$, for $i = 1, 2, 3$, then the singular configurations of the 3-RFR wrists with prismatic actuated joints are given by:

$$\mathbf{n}_1 \times \mathbf{n}_2 \cdot \mathbf{n}_3 = 0, \quad \mathbf{n}_i = -\mathbf{d}_i \times \hat{\mathbf{a}}_i \Rightarrow [\mathbf{n}_1]_0 = d[\mathbf{a}_i]_0 \times [\mathbf{Q}]_0[\mathbf{d}_i]_1 \tag{7}$$

where d is the constant magnitude of the vector $\mathbf{d}$. Writing the rotation matrix in terms of the ERPs using Eq. 2, it is obtained

$$\begin{aligned}\mathbf{n}_1 \times \mathbf{n}_2 \cdot \mathbf{n}_3 = d^3 &(r_1 + r_2 + r_3 - r_0)(-r_1 - r_2 + r_3 - r_0)\\ &\cdot (r_1 - r_2 - r_3 - r_0)(-r_1 + r_2 - r_3 - r_0)\end{aligned} \tag{8}$$

that vanishes when one of its factors become zero, indicating the robot poses for which a singular configuration is attained by the robot.

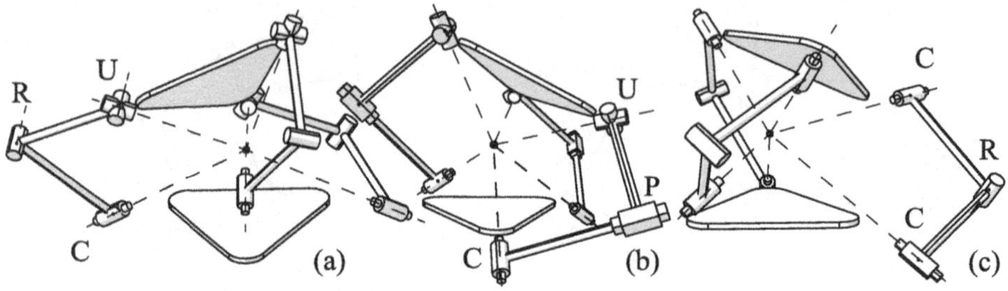

Fig. 5. Kinematic sketches of the (a) 3-CRU, (b) 3-CPU, (c) 3-CRC architectures.

5 Conclusions

The paper has characterized the rotational mobility of the 3-RFR class of SPMs and has shown that it is beneficial to actuate the prismatic joint of each leg, once it has been brought back to the frame. In this way, wrench constraints become linearly dependent only when the actuation wrenches do and thus the study of wrist singularities can be performed by simply studying their moments. Furthermore, such moments represent the row vectors of the direct Jacobian matrix of the 3-CPU that map the angular velocity vector of the MP into the actuated joint rates. This means that for this particular robot a single 3×3 Jacobian matrix is able to describe all the possible singularities. This feature can be exploited for posture optimization, e.g. when the 3-CPU robot is functionally redundant with respect to the task to be performed [15]. During path planning, when singularities must be avoided, the simple structure of Eq. (8) may represent a beneficial aspect for the considered class of manipulators. The interested reader can find a mapping of singularities in the Cartesian space of pointing directions in reference [20].

References

1. Li, Q.C., Huang, Z.: A family of symmetrical lower-mobility parallel mechanisms with spherical and parallel subchains. J. Rob. Syst. **20**(6), 297–305 (2003)
2. Karouia, M., Hervé, J.: A three-DOF tripod for generating spherical rotation. In: Advances in Robot Kinematics, pp. 395–402. Springer, Dordrecht (2000)
3. Kong, X., Gosselin, C.: Type synthesis of 3-DOF spherical parallel manipulators based on screw theory. J. Mech. Des. **126**(1), 101–108 (2004)
4. Di Gregorio, R.: Kinematics of a new spherical parallel manipulator with three equal legs: the 3-URC wrist. J. Rob. Syst. **18**(5), 213–219 (2001)
5. Di Gregorio, R.: A new parallel wrist using only revolute pairs: the 3-RUU wrist. Robotica **19**(3), 305–309 (2001)
6. Di Gregorio, R.: Kinematics of the 3-UPU wrist. Mech. Mach. Theory **38**(3), 253–263 (2003)
7. Di Gregorio, R.: The 3-RRS wrist: a new, simple and non-overconstrained spherical parallel manipulator. J. Mech. Des. **126**(5), 850–855 (2004)

8. Kong, X.: Reconfiguration analysis of a 3-DOF parallel mechanism using Euler parameter quaternions and algebraic geometry method. Mech. Mach. Theory **74**, 188–201 (2014)
9. Dai, J.S.: Euler-Rodrigues formula variations, quaternion conjugation and intrinsic connections. Mech. Mach. Theory **92**, 144–152 (2015)
10. Corinaldi, D., Carbonari, L., Callegari, M.: Optimal motion planning for fast pointing tasks with spherical parallel manipulators. IEEE Rob. Autom. Lett. **2**(3), 735–741 (2018)
11. Corinaldi, D., Callegari, M., Palpacelli, M.-C., Palmieri, G., Carbonari, L.: Dynamic optimization of pointing trajectories exploiting the redundancy of parallel wrists. In: Proceedings of the IDETC/CIE 41st Mechanisms and Robotics Conference, ASME, Cleveland, vol. 5A (2017)
12. Hunt, K.H.: Kinematic Geometry of Mechanisms, vol. 7. Oxford University Press, Oxford (1978)
13. Angeles, J.: Fundamentals of Robotic Mechanical Systems, vol. 2. Springer, New York (2002)
14. Conconi, M., Carricato, M.: A new assessment of singularities of parallel kinematic chains. IEEE Trans. Rob. **25**(4), 757–770 (2009)
15. Corinaldi, D., Angeles, J., Callegari, M.: Posture optimization of a functionally redundant parallel robot. In: Advances in Robot Kinematics 2016, vol. 4, pp. 101–108. Springer, Cham (2018)
16. Selig, J.M.: Geometric Fundamentals of Robotics, 2nd edn. Springer, New York (2004)
17. Carbonari, L., Corinaldi, D., Palpacelli, M.-C., Palmieri, G., Callegari, M.: A novel reconfigurable 3-URU parallel platform. In: Advances in Service and Industrial Robotics, vol. 49, pp. 63–73. Springer, Cham (2018)
18. Cox, D.A., Little, J., O'Shea, D.: Using Algebraic Geometry, vol. 185. Springer, New York (2006)
19. Walter, D.R., Husty, M.L., Pfurner, M.: A complete kinematic analysis of the SNU 3-UPU parallel robot. Contemp. Math. **496**, 331 (2009)
20. Corinaldi, D.: Task optimization of functionally redundant parallel kinematics machines. Ph.D. thesis, Polytechnic University of Marche (2017)

Randomized Planning of Dynamic Motions Avoiding Forward Singularities

Ricard Bordalba(✉), Lluís Ros(✉), and Josep M. Porta(✉)

Institut de Robòtica i Informàtica Industrial (CSIC-UPC),
08028 Barcelona, Catalonia
{rbordalba,ros,porta}@iri.upc.edu

Abstract. Forward singularities, also known as direct, or actuator singularities, cause many problems to the planning and control of robot motions. They yield position errors and rigidity losses of the robot, and generate unbounded actions in typical control laws. To circumvent these issues, this paper proposes a randomized kinodynamic planner for computing trajectories avoiding such singularities. Given initial and final states for the robot, the planner attempts to connect them by means of a dynamically-feasible, singularity-free trajectory that also respects the force limits of the actuators. The performance of the strategy is illustrated in simulation by means of a parallel robot performing a highly-dynamic task.

Keywords: Kinodynamic planning · Singularity
Constrained system · RRT

1 Introduction

The kinodynamic motion planning problem constitutes an important research theme in robotics. The problem entails computing trajectories connecting two given states, both with a prescribed position and velocity of a robot. The trajectories must be free of collisions and have to be consistent with the system dynamics, i.e., they can only involve acceleration profiles compatible with the limited forces deliverable by the actuators. The problem has been extensively studied for open-chain robots, and general practical solutions have been given [14], prominent ones being those based on randomized methods [15]. In comparison, the topic has received much less attention in the case of robots with closed kinematic chains (see [8] and references therein). These robots complicate the problem substantially because their state space is, in general, an implicitly-defined manifold not admitting a global parameterization. As a result, the extension of randomized methods like [15] to the closed-chain case is difficult since (1) it is not easy to come up with a mechanism to uniformly sample an implicitly-defined state space, (2) the numerical integration of the motion equations alone (disregarding the loop closure constraints) generates trajectories that drift away from such space, and (3) the mechanism exhibits forward singularities, whose traversal may compromise the control of the robot at execution time [2,5,10,16]. The

J. Lenarcic and V. Parenti-Castelli (Eds.): ARK 2018, SPAR 8, pp. 170–178, 2019.
https://doi.org/10.1007/978-3-319-93188-3_20

first two difficulties were recently addressed by Bordalba et al. [8], who showed that the iterative construction of manifold atlases provides effective sampling and numerical integration schemes over the state space. The purpose of this paper is to address the third difficulty: to make the planner in [8] robust to the presence of forward singularities.

It is well known that in a forward singularity the velocities of the actuators do not determine the full configuration velocity of the robot [4–6]. Such singularities, however, also have an impact on the system dynamics. The robot becomes underactuated, being unable to traverse these configurations under arbitrary accelerations. In their proximity, in fact, the inverse dynamic problem yields very large or unbounded motor actions, which can cause controllability issues or even a breakage of the mechanism. To circumvent these issues, several strategies have been proposed. For example, one can use actuation redundancy to avoid all, or almost all forward singularities [17], or design specific control laws able to traverse the singularities with bounded actions and consistent accelerations [10,12]. Although both approaches are interesting, they also come with a few caveats. The use of additional actuators yields more complex robot designs, and current motion controllers used to traverse singularities tend to be intricate and may not cope with all possible degeneracies of the forward kinematic map [16]. An alternative is to avoid the singularities at the planning stage, but current planners for this purpose cannot cope with dynamic constraints. As described in [3], such planners avoid the singularities by treating them as obstacles, or by defining a singularity-free configuration space. While the former approach requires the configuration space to be globally parametrizable, the latter remains applicable to general robot designs. This justifies the strategy we present in this paper, which extends the work in [3] to also satisfy dynamic constraints.

Our approach relies on a system of equations characterizing the singularity-free state space of a robot (Sect. 2). The solution set of this system is a smooth manifold diffeomorphic to the classic state space, but with all forward singularities removed. We also adapt the dynamic equations of the robot to define an action-varying vector field on this manifold, which we use to steer the robot between two states, while avoiding forward singularities (Sect. 3). We illustrate the planning method on a parallel robot that has to perform a highly dynamic task (Sect. 4) and draw a few conclusions from the study (Sect. 5).

2 Dynamics on the Singularity-Free State Space

Let us assume that the configuration of our robot is described by means of a tuple $\boldsymbol{q} = (\boldsymbol{a}, \boldsymbol{r})$ of n_q generalised coordinates, where $\boldsymbol{a}$ contains the n_a actuated degrees of freedom of the robot, and $\boldsymbol{r}$ encompasses the remaining coordinates in $\boldsymbol{q}$. Since the robot contains closed kinematic chains, the components of $\boldsymbol{q}$ will not be independent; they will be subject to a system of n_e equations

$$\boldsymbol{\Phi}(\boldsymbol{q}) = \boldsymbol{0}, \tag{1}$$

where $\boldsymbol{\Phi}(\boldsymbol{q})$ is a differentiable vector function defined by the loop closure constraints. Generically, then, the configuration space $\mathcal{C}$ of the robot will be a manifold of dimension $d_{\mathcal{C}} = n_q - n_e$ formed by all points $\boldsymbol{q}$ satisfying Eq. (1).

Although we have defined $\mathcal{C}$, the planning of dynamic motions is better solved in the state space $\mathcal{X}$ of the robot. In closed-chain systems, this space is traditionally defined as the set of pairs $(\boldsymbol{q}, \dot{\boldsymbol{q}})$ that satisfy Eq. (1) and its time derivative

$$\boldsymbol{\Phi}_q \cdot \dot{\boldsymbol{q}} = \mathbf{0}, \tag{2}$$

where $\boldsymbol{\Phi}_q = \frac{\partial \boldsymbol{\Phi}}{\partial \boldsymbol{q}}$. However, since we wish to generate motions avoiding forward singularities, we shall define $\mathcal{X}$ using an additional constraint. Recall here that a forward singularity is a point $\boldsymbol{q}$ for which the $n_e \times n_e$ Jacobian $\boldsymbol{\Phi}_r = \partial \boldsymbol{\Phi}/\partial \boldsymbol{r}$ is rank deficient [4]. To exclude these singularities from $\mathcal{X}$, thus, we need to enforce $\det(\boldsymbol{\Phi}_r) \neq 0$. This can be satisfied by introducing the constraint

$$b \cdot \det(\boldsymbol{\Phi}_r) = 1, \tag{3}$$

where b is a newly-defined auxiliary variable. Note that if we let $d = \det(\boldsymbol{\Phi}_r)$, then Eq. (3) can be written as $b \cdot d = 1$, which defines a hyperbola that never crosses the $d = 0$ axis of the (b, d) plane. In our formulation, thus, $\mathcal{X}$ will be the set of points $\boldsymbol{x} = (\boldsymbol{q}, \dot{\boldsymbol{q}}, b)$ satisfying the system formed by Eqs. (1–3). For short, this system will hereafter be written as

$$\boldsymbol{F}(\boldsymbol{x}) = \mathbf{0}. \tag{4}$$

Now note that the time evolution of the robot is determined by its motion equations, which can be obtained using the Euler-Lagrange formalism for example. In our case, the fact that $\boldsymbol{\Phi}_r$ is full rank for all $\boldsymbol{x} \in \mathcal{X}$ guarantees that $\boldsymbol{\Phi}_q$ is also full rank on $\mathcal{X}$. This allows the motion equations to be written in the explicit form $\ddot{q} = \boldsymbol{f}(\boldsymbol{q}, \dot{\boldsymbol{q}}, \boldsymbol{u})$ for all $\boldsymbol{x} \in \mathcal{X}$, where $\boldsymbol{u} \in \mathbb{R}^{n_a}$ is the vector of actuator forces of the robot. By defining $\boldsymbol{v} = \dot{\boldsymbol{q}}$, we can then write this equation in the first-order form used in numerical integration schemes:

$$\begin{bmatrix} \dot{\boldsymbol{q}} \\ \dot{\boldsymbol{v}} \end{bmatrix} = \begin{bmatrix} \boldsymbol{v} \\ \boldsymbol{f}(\boldsymbol{q}, \boldsymbol{v}, \boldsymbol{u}) \end{bmatrix}. \tag{5}$$

The evolution of b is also governed by a differential equation, which can be obtained by taking the time derivative of Eq. (3) to yield

$$\dot{b} = -\frac{b}{\det(\boldsymbol{\Phi}_r)} \cdot \frac{d}{dt}(\det(\boldsymbol{\Phi}_r)). \tag{6}$$

By noting that $\dot{\boldsymbol{x}} = (\dot{\boldsymbol{q}}, \dot{\boldsymbol{v}}, \dot{b})$, Eqs. (5) and (6) can be compactly written as

$$\dot{\boldsymbol{x}} = \boldsymbol{g}(\boldsymbol{q}, \boldsymbol{v}, b, \boldsymbol{u}). \tag{7}$$

Note that for each value of $\boldsymbol{u}$, Eq. (7) defines a vector field over $\mathcal{X}$, which can be used in conjunction with Eq. (4) to integrate the motion of the robot forward in time, using proper numerical methods [19]. Given two states of $\mathcal{X}$, $\boldsymbol{x}_s$ and $\boldsymbol{x}_g$, the goal of our planner is then to find an action trajectory $\boldsymbol{u}(t)$ such that the trajectory $\boldsymbol{x}(t)$ determined by Eqs. (4) and (7) for $\boldsymbol{x}(0) = \boldsymbol{x}_s$ satisfies $\boldsymbol{x}(t_f) = \boldsymbol{x}_g$ for some time $t_f > 0$. For any time $t \in [0, t_f]$, moreover, we shall require $\boldsymbol{x}(t)$ to stay inside a region $\mathcal{X}_{\text{feas}} \subset \mathcal{X}$ of non-collision states, and we shall confine $\boldsymbol{u}(t)$ to a bounded subset $\mathcal{U} \in \mathbb{R}^{n_a}$ given by the actuator force limits.

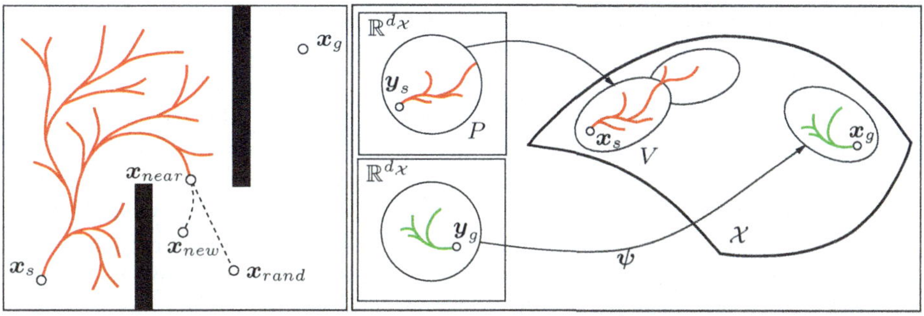

Fig. 1. Left: Extension process of an RRT. Right: Construction of an RRT on an implicitly-defined state space manifold.

3 Planning in the Singularity-Free State Space

To explain the planner in an easy way, we follow [7]. Suppose initially that the variables in $\boldsymbol{x}$ are independent; i.e., that Eq. (4) does not apply, so that $\mathcal{X}$ can be thought of as $\mathbb{R}^{2n_q+1}$. In this situation, we can look for a trajectory connecting $\boldsymbol{x}_s$ with $\boldsymbol{x}_g$ by constructing a rapidly-exploring random tree (RRT) over $\mathcal{X}$ [15]. The RRT is rooted at $\boldsymbol{x}_s$ and it is grown incrementally towards $\boldsymbol{x}_g$ while staying inside $\mathcal{X}_{\text{feas}}$. Every tree node stores a feasible state $\boldsymbol{x} \in \mathcal{X}_{\text{feas}}$, and every edge stores the action $\boldsymbol{u} \in \mathcal{U}$ needed to move between the connected states. This action is assumed to be constant during a fixed time Δt. The expansion of the RRT proceeds by applying three steps repeatedly (Fig. 1, left). First, a state $\boldsymbol{x}_{rand} \in \mathcal{X}$ is randomly selected; then, the RRT state $\boldsymbol{x}_{near}$ that is closest to $\boldsymbol{x}_{rand}$ is computed according to some metric; finally, a movement from $\boldsymbol{x}_{near}$ towards $\boldsymbol{x}_{rand}$ is performed by applying an action $\boldsymbol{u} \in \mathcal{U}$ during Δt seconds. The movement from $\boldsymbol{x}_{near}$ towards $\boldsymbol{x}_{rand}$ is simulated by integrating Eq. (7) numerically, which yields a new state $\boldsymbol{x}_{new}$ that may or may not be in $\mathcal{X}_{\text{feas}}$. In the former case, $\boldsymbol{x}_{new}$ is added to the RRT, and in the latter it is discarded. To test whether $\boldsymbol{x}_{new} \in \mathcal{X}_{\text{feas}}$, $\boldsymbol{x}_{new}$ is checked for collisions by using standard algorithms [13], and the joint positions are computed to check whether they stay within bounds. The action $\boldsymbol{u}$ applied is typically chosen as the one from $\mathcal{U}$ that brings the robot closer to $\boldsymbol{x}_{rand}$. One can either try all possible values in $\mathcal{U}$ (if it is a discrete set) or only those of n_s random points on $\mathcal{U}$ (if it is continuous). To force the RRT to extend towards $\boldsymbol{x}_g$, $\boldsymbol{x}_{rand}$ is set to $\boldsymbol{x}_g$ once in a while, stopping the whole process when a RRT leaf is close enough to $\boldsymbol{x}_g$. Usually, however, a solution trajectory can be found more rapidly if two RRTs respectively rooted at $\boldsymbol{x}_s$ and $\boldsymbol{x}_g$ are grown simultaneously towards each other. The expansion of the tree rooted at $\boldsymbol{x}_g$ is based on the integration of Eq. (7) backwards in time.

When the coordinates in $\boldsymbol{x}$ are constrained by Eq. (4), $\mathcal{X}$ becomes an non-linear manifold of dimension $d_\mathcal{X} = 2d_\mathcal{C}$. This fact complicates the generation of RRTs over $\mathcal{X}$ because there is no straightforward way to randomly select points $\boldsymbol{x}$ satisfying Eq. (4), and the numerical integration of Eq. (7) easily drifts away from $\mathcal{X}$ when standard methods for ordinary differential equations are used. However, these two issues can be circumvented by constructing an atlas of $\mathcal{X}$ in parallel to the RRT [8].

An atlas is a collection of charts mapping $\mathcal{X}$ entirely, where each chart is a local diffeomorphism $\boldsymbol{\psi}$ from an open set $P \subseteq \mathbb{R}^{d_{\mathcal{X}}}$ of parameters to an open set $V \subset \mathcal{X}$ (Fig. 1, right). The V sets can be thought of as partially-overlapping tiles covering $\mathcal{X}$, in a way that every $\boldsymbol{x} \in \mathcal{X}$ lies in at least one set V. Assuming that an atlas is available, the problem of sampling $\mathcal{X}$ boils down to generating random values $\boldsymbol{y}$ in the P sets, since these values can always be projected to $\mathcal{X}$ using $\boldsymbol{x} = \boldsymbol{\psi}(\boldsymbol{y})$. Also, the atlas allows the conversion of the vector field defined on $\mathcal{X}$ by Eq. (7) into one in the coordinate spaces P, which permits the integration of Eq. (7) using local coordinates [19]. As a result, the RRT motions satisfy Eq. (4) by construction, eliminating any drift from $\mathcal{X}$ to machine precision.

The construction of the atlas is incremental. The atlas is initialized with two charts covering $\boldsymbol{x}_s$ and $\boldsymbol{x}_g$, respectively (Fig. 1, right). Then, these charts are used to pull the expansion of the RRT, which in turn adds new charts to the atlas as needed, until $\boldsymbol{x}_s$ and $\boldsymbol{x}_g$ become connected. To be able to construct the charts, the method requires $\mathcal{X}$ to be smooth, i.e., with a well-defined tangent space at all of its points. However, using the formulation in Sect. 2, we can see that this property holds. Observe that all functions defining $\boldsymbol{F}(\boldsymbol{x})$ are differentiable, so that the Jacobian of $\boldsymbol{F}(\boldsymbol{x})$,

$$\boldsymbol{F}_{\boldsymbol{x}} = \boldsymbol{F}_{q,\dot{q},b} = \left[\begin{array}{c|c|c} \boldsymbol{\Phi}_q & \boldsymbol{0} & \boldsymbol{0} \\ \hline \frac{\partial \boldsymbol{\Phi}_q}{\partial q} \cdot \dot{\boldsymbol{q}} & \boldsymbol{\Phi}_q & \boldsymbol{0} \\ \hline b \cdot \frac{\partial \det(\boldsymbol{\Phi}_r)}{\partial q} & \boldsymbol{0} & \det(\boldsymbol{\Phi}_r) \end{array}\right], \tag{8}$$

is well defined at every $\boldsymbol{x} \in \mathcal{X}$. This Jacobian, moreover, is full rank for all $\boldsymbol{x} \in \mathcal{X}$ because each of the diagonal blocks contains a non-vanishing minor of maximal size (since $\det(\boldsymbol{\Phi}_r) \neq 0$ for all $\boldsymbol{x} \in \mathcal{X}$, $\boldsymbol{\Phi}_q$ is guaranteed to be full rank on $\mathcal{X}$). By the implicit function theorem, these facts imply that $\mathcal{X}$ is smooth.

4 A Weight Throwing Task

The previous planner has been implemented in C and it has been integrated into the CUIK Suite [18]. We next illustrate its performance on planning a weight throwing task for a 5-bar robot with the geometry shown in Fig. 2. The angles q_1 and q_5 are actuated, allowing to control the (x, y) position of point Q, which is taken as the end effector. Various versions of this mechanism have been used both in research and the industry, including the DEXTAR and RAPI-MOD models from ÉTS de Montréal [9] and Gridbot Technologies [1]. In our version,

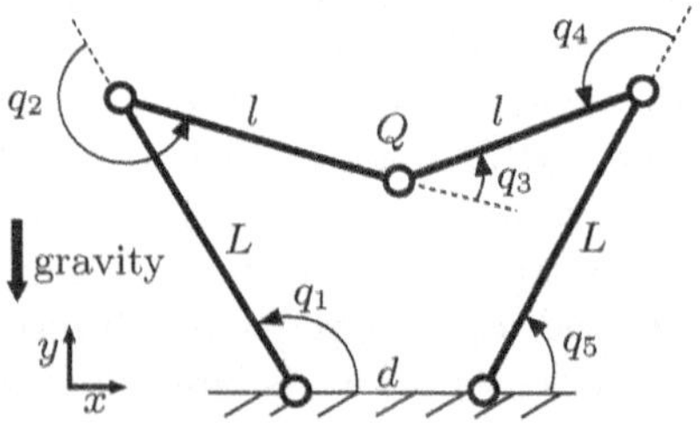

Fig. 2. A 5-bar robot.

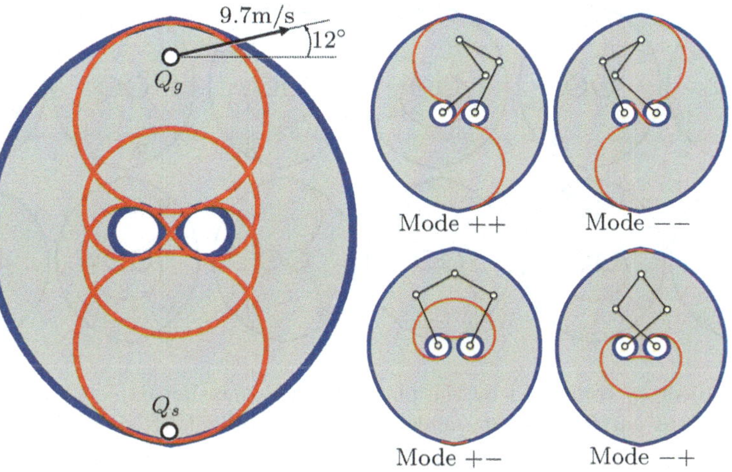

Fig. 3. Left: Forward singularities of the 5-bar robot (in red). Right: the same singularities, classified according to the different working modes.

the base distance is $d = 0.6472$ [m], and the proximal and distal link lengths are $L = 1.14$ [m] and $l = 0.9$ [m], respectively. With these dimensions, the workspace of Q is the grey area shown in Fig. 3, left. The task to be planned consists in picking an object of 1 [kg] located at point Q_s, to later throw it from point Q_g with the shown velocity. To increase the difficulty of the task, we limit the motor torques to ± 15 [Nm], forcing the planner to generate pendulum-like motions that increase the kinetic energy progressively, until the required velocity at Q_g is achieved. The mass of each link is 0.5 [kg] and the moments of inertia of the proximal and distal links (with respect to their center of gravity) are of 0.0541 [kg·m^2] and 0.0338 [kg·m^2], respectively.

We recall from [9] that the forward singularities of this robot occur when the distal links get aligned, i.e., when q_3 is either 0 or π. The positions of Q for which such an event occurs are depicted in red in Fig. 3, left. From this figure it appears that the workspace is severely limited by the presence of forward singularities, but note that each position of Q can be attained by up to four inverse kinematic solutions of the robot. Each solution corresponds to one working mode identified by the signs of q_4 and q_2 in the range $[-\pi, \pi]$. If we separate the singularities according to these signs, larger singularity-free regions arise (Fig. 3, right).

Some existing planners require the use of only one working mode [11], but the full motion range of the robot can only be exploited when working mode changes are allowed. This is especially true in highly dynamic motions like the ones we require, in which the rapid movements of the robot, and its inertia, will frequently favor the mode changes. Our planner is beneficial in this respect, because it works in the state space manifold of the robot, on which the changes will occur in a natural way.

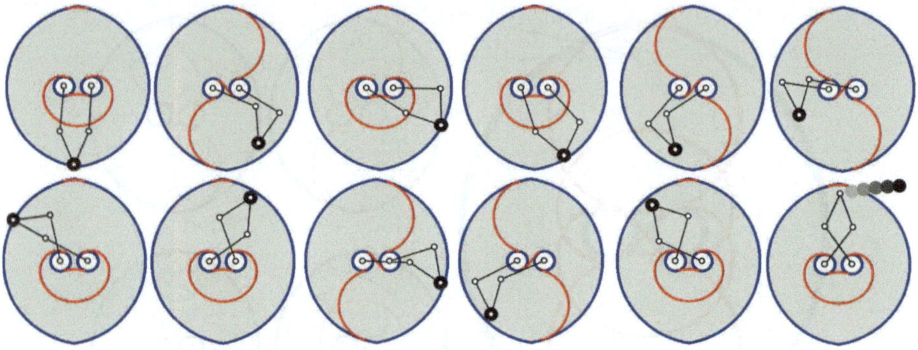

Fig. 4. Trajectory computed by the planner. Each snapshot depicts the forward singularities of the current working mode. See also youtube/ENKX0wrZAKM.

Figure 4 shows the trajectory obtained by the planner, which lasts for 7.5 s, and takes 150 CPU seconds to be computed on average, on a Macbook Pro with an Intel i7 processor. Note how the robot begins to oscillate to the right, to later go left, climb to the top-most position, and finally complete a full revolution about the base to gain momentum before throwing the load. See the video in youtu.be/ENKX0wrZAKM for an animation. As we appreciate, many working mode changes occur during the move and the robot never crosses the forward singularity locus (the sign of q_3 is always kept constant). The video also shows that, if we plan the same task without singularity avoidance constraints (Eqs. (3) and (6)), we obtain trajectories that easily cross the singularities. As we mentioned in the introduction, this fact would hinder the control of the robot during task execution.

5 Conclusions

This paper has proposed a randomized planner to compute dynamic trajectories avoiding forward singularities. The planner relies on a system of equations characterizing the singularity-free state space of the robot, and on an ordinary differential equation describing the robot dynamics in such space (Eqs. (4) and (7), respectively). A great advantage of the planned trajectories is that they can be stabilized by means of simple computed-torque control laws [2], because the inverse dynamic problem always yields bounded actions in their vicinity. A second advantage is that, when executing the trajectories, it suffices to sense the positions and velocities of the actuators in order to keep track of the full robot state. The remaining coordinates can always be recovered with the forward kinematic maps, which never degenerate in non-singular configurations. Our current efforts are focused on validating the planned trajectories in real robots.

Acknowledgements. Work partially funded by the Spanish Ministry of Economy and Competitiveness under projects DPI2014-57220-C2-2-P and DPI2017-88282-P.

References

1. RAPI-MOD: Ultrafast scara robot. http://gridbots.com/rapi_mov.html
2. Aghili, F.: A unified approach for inverse and direct dynamics of constrained multibody systems based on linear projection operator: applications to control and simulation. IEEE Trans. Robot. **21**(5), 834–849 (2005)
3. Bohigas, O., Henderson, M.E., Ros, L., Manubens, M., Porta, J.M.: Planning singularity-free paths on closed-chain manipulators. IEEE Transa. Robot. **29**(4), 888–898 (2013)
4. Bohigas, O., Manubens, M., Ros, L.: Singularities of non-redundant manipulators: a short account and a method for their computation in the planar case. Mech. Mach. Theory **68**, 1–17 (2013)
5. Bohigas, O., Manubens, M., Ros, L.: Singularities of robot mechanisms: numerical computation and avoidance path planning. Springer, Cham (2016). http://www.springer.com/us/book/9783319329208
6. Bohigas, O., Zlatanov, D., Ros, L., Manubens, M., Porta, J.M.: A general method for the numerical computation of manipulator singularity sets. IEEE Trans. Robot. **30**(2), 340–351 (2014)
7. Bordalba, R., Porta, J.M., Ros, L.: Randomized kinodynamic planning for cable-suspended parallel robots. In: Cable-Driven Parallel Robots, pp. 195–206. Springer, Cham (2018)
8. Bordalba, R., Ros, L., Porta, J.M.: Randomized kinodynamic planning for constrained systems. In: IEEE International Conference on Robotics and Automation (2018). To appear
9. Bourbonnais, F., Bigras, P., Bonev, I.A.: Minimum-time trajectory planning and control of a pick-and-place five-bar parallel robot. IEEE/ASME Trans. Mechatron. **20**(2), 740–749 (2015)
10. Briot, S., Arakelian, V.: Optimal force generation in parallel manipulators for passing through the singular positions. Int. J. Robot. Res. **27**(8), 967–983 (2008)
11. Cortés, J., Siméon, T., Laumond, J.P.: A random loop generator for planning the motions of closed kinematic chains using PRM methods. In: IEEE International Conference on Robotics and Automation, pp. 2141–2146 (2002)
12. Hill, R.B., Six, D., Chriette, A., Briot, S., Martinet, P.: Crossing type 2 singularities of parallel robots without pre-planned trajectory with a virtual-constraint-based controller. In: IEEE International Conference on Robotics and Automation (ICRA), pp. 6080–6085 (2017)
13. Jiménez, P., Thomas, F., Torras, C.: 3D collision detection: a survey. Comput. Graph. **25**(2), 269–285 (2001)
14. LaValle, S.M.: Planning Algorithms. Cambridge University Press, New York (2006). http://planning.cs.uiuc.edu/
15. LaValle, S.M., Kuffner, J.J.: Randomized kinodynamic planning. Int. J. Robot. Res. **20**(5), 378–400 (2001)
16. Özdemir, M.: Removal of singularities in the inverse dynamics of parallel robots. Mech. Mach. Theory **107**, 71–86 (2017)
17. Parsa, S.S., Boudreau, R., Carretero, J.A.: Reconfigurable mass parameters to cross direct kinematic singularities in parallel manipulators. Mech. Mach. Theory **85**, 53–63 (2015)

18. Porta, J.M., Ros, L., Bohigas, O., Manubens, M., Rosales, C., Jaillet, L.: The Cuik Suite: analyzing the motion of closed-chain multibody systems. IEEE Robot. Autom. Mag. **21**(3), 105–114 (2014)
19. Potra, F.A., Yen, J.: Implicit numerical integration for Euler-Lagrange equations via tangent space parametrization. J. Struct. Mech. **19**(1), 77–98 (1991)

Analysis of Kinematic Singularities for a Serial Redundant Manipulator with 7 DOF

Zijia Li, Mathias Brandstötter(✉), and Michael Hofbaur

Institute for Robotics and Mechatronics, JOANNEUM RESEARCH, Lakeside B08a, 9020 Klagenfurt, Austria
zijia.li@ricam.oeaw.ac.at,
{Mathias.Brandstoetter,Michael.Hofbaur}@joanneum.at

Abstract. In this work, the kinematic singularities of serial manipulators with 7 rotational joints are analyzed and their effects on the possible end effector movement are studied. We obtain the numerical kinematic singularities through algebraic varieties and demonstrate this on the kinematically redundant serial manipulator KUKA LBR iiwa. The algebraic equations for determining the variety are derived by taking the determinant of the 6-by-6 submatrix of the Jacobian matrix of the forward kinematics. By using the primary decomposition, the singularities can be decomposed and classified. Further analysis of the kinematic singularities including the inverse kinematics of the redundant manipulator provide us valuable insights. Firstly, there are kinematic singularities where the inverse kinematics has no effect on the self-motion and cannot be used to avoid obstacles. Secondly, there are kinematic singularities, which lead to a single closed-loop connection with the serial redundant manipulator, so that a kinematotropic mechanism is achieved.

Keywords: Kinematotropic mechanism
Computational algebraic geometry · Primary decomposition

1 Introduction

A kinematically redundant manipulator possesses more joints than those strictly required to execute its task. This increases the robot's level of dexterity, enabling it to avoid joint limits and obstacles in its workspace. Moreover, the surplus freedom enables optimization regarding joint torques, energy or, in general, suitable performance indices. The motivation for the study of kinematic singularities for redundant serial manipulators is to look deeply into the algebraic varieties.

It is well known that if the end effector of a serial manipulator is fixed the kinematic chain is equivalent to a closed loop mechanism. Single-loop kinematotropic mechanisms [4,10,13] can be obtained in this way. A kinematotropic mechanism [14] should have two different positive dimensional configuration subset (e.g., dim $= 1, 2$), hence a kinematically redundant manipulator will be a

J. Lenarcic and V. Parenti-Castelli (Eds.): ARK 2018, SPAR 8, pp. 179–186, 2019.
https://doi.org/10.1007/978-3-319-93188-3_21

good candidate to achieve the goal of variable-DOF. This is because a redundant manipulator already has a positive dimensional configuration set. The critical point is that we need to know the position and orientation (pose) that we should fix the end effector such that the produced single-loop mechanism can have variable-DOF. At the same time, by fixing such a pose, we can expect that the inverse kinematics solution (the configuration set for the mechanism) contain a kinematic singularity (at least one configuration) where the kinematotropic mechanism can change its DOF. In other words, a kinematic singularity is a necessary condition for constructing kinematotropic mechanisms.

In this paper, we first present the computation of kinematic singularities varieties for serial redundant manipulators. An algebraic primary decomposition gives a classification of the kinematic singularities. Using this primary decomposition, we compute the intersection of their corresponding varieties. By using this intersection of singularity varieties, single-loop kinematotropic mechanisms can be found. Since it is a necessary condition, it cannot work for all singularities. There are other singularities where the inverse kinematics has no influence on the self-motion and therefore cannot be used to avoid obstacles. Once the manipulator is at such a posture, it needs to pass the singularity to achieve an effect on the self-motion. This characteristic will be demonstrated by studying the structure of the KUKA LBR iiwa robots. Nevertheless, the method can be applied for other redundant serial manipulators as well. For manipulators like KUKA LBR iiwa, the Gröbner basis computation is fast. This leads to the same argument for the efficient inverse kinematics solution as in [7], according to which the inverse kinematics becomes simpler at kinematic singularities. This does not conflict the closed-form inverse kinematics solution for a general pose as discussed in [3].

2 Preliminaries

We give a recall on the forward kinematics mapping via dual quaternions following [6]. The structural design of manipulators is determined by its geometrical invariants of the Denavit-Hartenberg parameter set [2], i.e., twist angles, orthogonal distances and offsets. However, once the manipulator is assembled, all rotational axes are known. A nonzero dual quaternion h (its conjugate $\overline{h}$) represents a rotation (around an axis) if and only if $h\overline{h}$ and $h + \overline{h}$ are strictly real and its primal vectorial part is nonzero.

We use a formulation of the forward kinematics in the language of dual quaternions in this paper, based on the fact that SE(3) is isomorphic to the multiplicative group of dual quaternions with nonzero real norm modulo multiplication by nonzero real scalars. In the isomorphism described in [8, Sect. 2.4], the rotation with axis determined by $\mathbf{i}$ and angle q, corresponds to the dual quaternion $(\cos(\frac{q}{2}) - \sin(\frac{q}{2})\mathbf{i})$, which is projectively equivalent to $(1 - \tan(\frac{q}{2})\mathbf{i})$. The forward kinematics mapping is therefore

$$(1 - t_1 h_1)(1 - t_2 h_2) \cdots (1 - t_n h_n) \equiv h_e, \qquad (1)$$

where $h_1, \ldots, h_n$ are dual quaternions specifying the rotation axes in the initial position of the robot, $t_i = \tan(\frac{q_i}{2})$, h_e is a dual quaternion specifying a pose of the end-effector, and "$\equiv$" means projectively equivalent. The set G of all n-tuple $(t_1, \ldots, t_n) \in (\mathbb{P}^1)^n$ fulfilling (1) is called the *configuration set* (inverse kinematics) of the robot arm with respect to a fixed pose h_e. To solve (1), the equation must be multiplied with the conjugate of h_e from both sides. The expansion of both sides to 8 dimensional vectors gives us a system of equations. In the following we analyze these equations with computer algebra tools, e.g. Gröbner basis, to understand the kinematics of redundant serial manipulators.

3 Kinematics Computation

In this section, kinematic properties for redundant manipulators are determined using computer algebra. With the help of Gröbner basis we can find all the inverse kinematics solutions for a given pose of the end effector. It is essential, that if a pose is given, the manipulator becomes a closed loop. In this case, we also call the inverse kinematics solution a configuration set. To compute the Jacobian matrix, which is widely used in relevant literature, we use our algebraic kinematic approach. The kinematic singularities for a serial manipulator, algebraically speaking, are the configurations or postures where the Jacobian matrix is rank deficient. With the help of the determinant we will construct the algebraic variety for an industrial serial redundant manipulator. We are particularly interested in the configuration set, which has a kinematic singularity. Results are presented in the last subsection dealing with the KUKA LBR iiwa, also to demonstrate our analysis.

3.1 Configuration Set

The dimension of the configuration set is called the *mobility* of the manipulator with a given fixed pose h_e. In this work we are interested in robot structures and fixed poses h_e that lead to mobility of one or higher. Let $L = (h_1, \ldots, h_n)$ be dual quaternions specifying the joints to compute the configuration set by Algorithm 1 [11].

Algorithm 1. *Computation of the configuration set*

Input: An nR robot L with one initial configuration for a given end effector pose, namely, n lines h_1, h_2, ..., h_n.

Output: A list of equations in $t_1, t_2, \ldots, t_n$ such that its solution set equals the *configuration set* of L for the end effector pose.

1: Write $F := (1 - t_1 h_1)(1 - t_2 h_2) \cdots (1 - t_n h_n) \in \mathbb{DH}$.
2: Take the seven coefficients of polynomial F except the first coefficient and add them to list E.
3: Supplement E by one more polynomial, namely, $(t_1^2 + 1)(t_2^2 + 1) \cdots (t_n^2 + 1)u - 1$.
4: Compute the Gröbner basis G of the elimination ideal of E with respect to u.
5: **Return** G – the elimination ideal.

3.2 Kinematic Singularities

The set of kinematic singularities is an algebraic variety (see [5] and the references there) which is defined by the Jacobian determinant of the forward kinematics. It is also well-known (see [9]) that the Jacobian matrix is singular if and only if all rotation axes are contained in a linear line complex. In our kinematic setting, we can represent oriented lines as purely vectorial dual quaternions with norm 1. For $s = 1, \ldots, n$, each h_i is the dual quaternion corresponding to the i-th rotation axis at the initial configuration, where $h_i^2 = -1$. Moreover, t_i is the tangent of half of the i-th rotation angle. If one expresses the position of a rotation axis in the coordinate frame of a middle link (the j-th and $j+1$-th axes) of the serial chain, then the corresponding oriented lines can be easily computed by

$$\begin{aligned}
H_1 &= (1 + t_j h_j) \cdots (1 + t_2 h_2) h_1 (1 - t_2 h_2) \cdots (1 - t_j h_j), \\
&\vdots \\
H_{j-1} &= (1 + t_j h_j) h_{j-1} (1 - t_j h_j), \\
H_j &= h_j, \\
H_{j+1} &= h_{j+1}, \\
H_{j+2} &= (1 - t_{j+1} h_{j+1}) h_{j+2} (1 + t_{j+1} h_{j+1}), \\
&\vdots \\
H_n &= (1 + t_{j+1} h_{j+1}) \cdots (1 + t_{n-1} h_{n-1}) h_n (1 - t_{n-1} h_{n-1}) \cdots (1 + t_{j+1} h_{j+1}).
\end{aligned}$$

For instance, we have

$$\begin{aligned}
&H_1 = (1 + t_3 h_3)(1 + t_2 h_2) h_1 (1 - t_2 h_2)(1 - t_3 h_3), \quad H_2 = (1 + t_3 h_3) h_2 (1 - t_3 h_3), \\
&H_3 = h_3, \quad H_4 = h_4, \quad H_5 = (1 + t_4 h_4) h_5 (1 - t_4 h_4), \\
&H_6 = (1 + t_4 h_4)(1 + t_5 h_5) h_6 (1 - t_5 h_5)(1 - t_4 h_4), \\
&H_7 = (1 + t_4 h_4)(1 + t_5 h_5)(1 + t_6 h_6) h_7 (1 - t_6 h_6)(1 - t_5 h_5)(1 - t_4 h_4).
\end{aligned}$$

The equations for the geometric condition that these lines lie in a linear line complex can be found as follows: they are determinants of all 6×6 submatrices (there are $\binom{n}{6}$ submatrices) of the Jacobian matrix which columns are the coefficients of $H_1, \ldots, H_n$. Each column has six entries which come from the Plücker coordinates of H_i, namely, the vectorial part of dual quaternion H_i. By the assumption of $h_i^2 = -1$, the scalar part of dual quaternion h_i is zero. In addition, the scalar part of the dual quaternion H_i also equals zero by quaternion multiplication.

4 Example

The LBR iiwa from KUKA is a comprehensive example for an analysis. We define the home position as depicted in Fig. 1 and note its Denavit-Hartenberg parameters in Table 1.

We take the initial position (home position) as:

$$h_1 = h_3 = h_5 = h_7 = \mathbf{k},\ h_2 = \mathbf{i},\ h_4 = \mathbf{i} - d_3\epsilon\mathbf{j},\ h_6 = \mathbf{i} - (d_3 + d_5)\epsilon\mathbf{j},$$

Then using the definition of the Jacobian matrix and taking the determinants of its submatrices into account. We can compute the variety of kinematic singularities which is generated by $\binom{7}{6} = 7$ polynomials:

$$\begin{aligned} I =< 8d_5(t_5-1)(t_5+1)t_4^2d_3^2t_2(t_3^2+1)^2, -32d_5t_4^2t_5t_6d_3^2t_2(t_3^2+1)^2, \\ -8d_5t_4d_3t_2(t_3^2+1)^2(t_5^2+1)(d_3t_4^2t_5^2t_6 - d_3t_4t_5^2t_6^2 - d_5t_4^2t_5^2t_6 \\ +d_3t_4^2t_6 + d_3t_4t_5^2 + d_3t_4t_6^2 - d_3t_5^2t_6 - d_5t_4^2t_6 - d_5t_5^2t_6 - d_3t_4 - d_3t_6 - d_5t_6), \\ -8d_3d_5^2t_4^2t_6(t_5^2+1)^2(t_4^2+1)(t_3-1)(t_3+1), 32d_5^2t_4^2d_3t_2t_3t_6(t_5^2+1)^2(t_4^2+1), 0, \\ -8d_5t_4d_3t_6(t_5^2+1)^2(t_4^2+1)(t_3^2+1)(d_3t_2t_3^2t_4^2 - d_5t_2^2t_3^2t_4 - d_5t_2t_3^2t_4^2 \\ +d_3t_2t_3^2 + d_3t_2t_4^2 + d_5t_2^2t_4 + d_5t_2t_3^2 - d_5t_2t_4^2 + d_5t_3^2t_4 + d_3t_2 + d_5t_2 - d_5t_4) > . \end{aligned}$$

After deleting the quadratic factors $t_i^2 + 1$ and the geometric parameters d_3, d_5 which do not vanish (when we consider real configurations), a primary decomposition of this ideal gives us six components:

$$< t_4 >, < t_2, t_6 >, < t_2, t_3 - 1 >, < t_2, t_3 + 1 >, < t_6, t_5 - 1 >, < t_6, t_5 + 1 > .$$

Remark 1. *Whenever we take $(1 - t_ih_i)$ to parametrize the rotation of the i-th axis, we ignore rotations of 180° in our configuration set. For this reason, one can use $(s_i - t_ih_i)$ to compute the singularities and will get homogeneous polynomials. We ignore the homogeneous case because all joint limits of the KUKA LBR iiwa are less than 180°. The real (or physical reachable) singularities are computed by using $(1 - t_ih_i)$.*

Fig. 1. KUKA LBR iiwa 7R.

i-th Link	a_i	d_i (mm)	α_i	θ_i
1	0	d_1	90	θ_1
2	0	0	−90	θ_2
3	0	d_3	90	θ_3
4	0	0	−90	θ_4
5	0	d_5	90	θ_5
6	0	0	−90	θ_6
7	0	d_7	0	θ_7

Table 1. Denavit-Hartenberg parameter set of KUKA LBR iiwa 7R.

We will invest the configuration set behavior in the following computation. For simplicity, we take the KUKA LBR iiwa 14 for our numerical computation of the kinematics, where $d_3 = 21/50$ and $d_5 = 2/5$:

First Singularity ($t_4 = 0$): Let us consider a pose of the end effector that can be reached by the singular posture

$$\{t_1 = \tfrac{3}{4}, t_2 = -\tfrac{5}{2}, t_3 = -1, t_4 = 0, t_5 = -1, t_6 = -\tfrac{3}{5}, t_7 = -\tfrac{1}{4}\}.$$

By the forward kinematics the pose can be calculated to be

$$\tfrac{5}{2} - \tfrac{2337}{800}\epsilon + \mathbf{i}\left(-\tfrac{19}{5} + \tfrac{1599}{2000}\epsilon\right) - \mathbf{j}\left(\tfrac{247}{80} + \tfrac{123}{125}\epsilon\right) + \mathbf{k}\left(\tfrac{95}{16} + \tfrac{123}{110}\epsilon\right).$$

Using Algorithm 1, all other configurations for this pose can be found. With primary decomposition of the ideal, the configuration set has four components:

$$\begin{aligned}
&< t_4, -5 + 2t_2, -4 + t_7, 3t_1 + 4, 5t_6 - 3, t_3t_5 - 1 >,\\
&< t_4, -5 + 2t_2, 1 + 4t_7, 3t_1 + 4, t_3 + t_5, 5t_6 + 3 >,\\
&< t_4, -4 + t_7, 5 + 2t_2, 4t_1 - 3, t_3 + t_5, 5t_6 - 3 >,\\
&< t_4, 1 + 4t_7, 5 + 2t_2, 4t_1 - 3, 5t_6 + 3, t_3t_5 - 1 >
\end{aligned}$$

It coincides to the pattern of 8 different components discussed in [1]. We can see that the singularities occur in every configuration (or all inverse kinematics solutions). The four self-motions (four components) have no effect on the obstacles avoidance.

Second Singularities ($t_2 = t_6 = 0$): Let us consider a pose that can be reached by the singular posture

$$\{t_1 = \tfrac{3}{4}, t_2 = 0, t_3 = -\tfrac{1}{3}, t_4 = \tfrac{1}{2}, t_5 = -\tfrac{1}{4}, t_6 = 0, t_7 = -\tfrac{1}{4}\}.$$

By the forward kinematics, the pose is

$$\tfrac{256}{192} - \mathbf{i}\left(-\tfrac{185}{384} - \tfrac{273}{1280}\epsilon\right) + \mathbf{j}\left(\tfrac{65}{28} + \tfrac{259}{1280}\epsilon\right) + \mathbf{k}\tfrac{15}{64}.$$

Using Algorithm 1, we can find all other configurations for the given end effector pose. With primary decomposition of the ideal, the configuration set has also four components (two 2-dimensional components and two 1-dimensional components):

$$\begin{aligned}
&< 2t_4 - 1, t_2, t_6, t_1t_3 + 3t_1 + 3t_3 - 1, 8t_5t_7 - 15t_5 - 15t_7 - 8 >,\\
&< 2t_4 + 1, t_2, t_6, 3t_1t_3 - t_1 - t_3 - 3, 15t_5t_7 + 8t_5 + 8t_7 - 15 >,\\
&< 2t_4 - 1, t_1t_3 + 3t_1 + 3t_3 - 1, 165t_2t_3 - 165t_2t_1 + 80t_3 - 64t_1 + 48, \ldots >,\\
&< 2t_4 + 1, 3t_1t_3 - t_1 - t_3 - 3, 165t_2t_3 - 165t_2t_1 - 80t_3 - 64t_1 + 48, \ldots >.
\end{aligned}$$

This coincides to the pattern of 8 different components stated in [1]. Through analysis we obtain that all singularities happen at the two 2-dimensional configuration sets. In addition, the 2-dimensional configuration set and the 1-dimensional configuration set also intersect in real. We thus achieve a kinematotropic single loop mechanism by this serial manipulator.

Remark 2. *We can find kinematotropic for the KUKA iiwa by computing the kinematic singularities. The special structure of this robot provides us the kinematotropic, but the proposed algorithm is still useful for other kinematotropic linkages. For instance, kinematotropic happens at special kinematic singularities with respect to mechanisms which have two Bennett conditions (see [13]). In addition, the ABB Yumi arm is also a 7R serial manipulator and contain these special kinematotropic singularities.*

Third Singularities ($t_2 = 0$ or $t_6 = 0$): Let us have a pose which can be reached by the singular posture

$$\left\{t_1 = \tfrac{3}{4}, t_2 = 0, t_3 = -1, t_4 = \tfrac{1}{2}, t_5 = -\tfrac{1}{4}, t_6 = 1, t_7 = -\tfrac{1}{4}\right\}.$$

The case of $t_6 = 0$ is similar to this. By forward kinematics the pose is

$$\tfrac{256}{192} - \mathbf{i}\left(-\tfrac{185}{384} - \tfrac{273}{1280}\epsilon\right) + \mathbf{j}\left(\tfrac{65}{28} + \tfrac{259}{1280}\epsilon\right) + \mathbf{k}\tfrac{15}{64}.$$

Using Algorithm 1, we can find all other configurations for the pose. With primary decomposition of the ideal, the configuration set has six components (all are 1-dimensional components):

$$< 2t_4 - 1, t_2, -4 + t_7, t_5 - 4, t_6 + 1, t_1t_3 - 7t_1 - 7t_3 - 1 >,$$
$$< 2t_4 + 1, t_2, -4 + t_7, t_5 - 4, t_6 + 1, 7t_1t_3 + t_1 + t_3 - 7 >,$$
$$< 2t_4 - 1, t_2, 1 + 4t_7, 4t_5 + 1, t_6 - 1, t_1t_3 - 7t_1 - 7t_3 - 1 >,$$
$$< 2t_4 + 1, t_2, 1 + 4t_7, 4t_5 + 1, t_6 - 1, 7t_1t_3 + t_1 + t_3 - 7 >,$$
$$< 2t_4 - 1, t_1t_3 - 7t_1 - 7t_3 - 1, -825t_1t_2 + 825t_2t_3 - 384t_1 - 400t_3 - 112, \ldots >,$$
$$< 2t_4 + 1, 7t_1t_3 + t_1 + t_3 - 7, -825t_1t_2 + 825t_2t_3 - 384t_1 + 400t_3 - 112, \ldots >$$

The set coincides to the pattern of 8 different components like before. We can find that all singularities appear at the intersection of the components. In addition, the first four motion components have no effect on the obstacles avoidance when we fix the end effector. We have pass the kinematic singularities to get the other components (last two) and to complete the obstacle avoidance.

Remark 3. *We can pass the kinematic singularities in our situations because there are enough motors to control the redundant manipulators. The crucial points (singularities) are the points where $t_2 = 0$ and $t_3 = \pm 1$.*

5 Conclusions

With the help of computer algebra, we can clearly comprehend the proper motion of the 7 DOF redundant manipulator (KUKA LBR iiwa) at a pose in which kinematic singularities appear. Either no effect motion mode or variable dimensional motion modes occur when the manipulator has a kinematic singularity in its configuration set (or inverse kinematics). Further research to analyze the kinematic singularities of other redundant serial manipulators could be done in a similar way. We believe that this algebraically kinematic analysis is very useful for advanced applications. In particular, a control strategy is required to go through the kinematic singularities in order to accomplish the task of bypassing obstacles when the manipulator is in a no-effect movement mode. A further investigation on kinematic singularities with higher differentials (e.g., [12]) will be one of our future works.

Acknowledgements. This research was funded by the Austrian Ministry for Transport, Innovation and Technology (BMVIT) within the framework of the sponsorship agreement formed for 2015–2018 under the project RedRobCo.

References

1. Bongardt, B.: Inverse kinematics of anthropomorphic arms yielding eight coinciding circles. In: Computational Kinematics, pp. 525–534. Springer (2018)
2. Denavit, J., Hartenberg, R.S.: A kinematic notation for lower-pair mechanisms based on matrices. Trans. ASME J. Appl. Mech. **22**, 215–221 (1955)
3. Faria, C., Ferreira, F., Erlhagen, W., Monteiro, S., Bicho, E.: Position-based kinematics for 7-dof serial manipulators with global configuration control, joint limit and singularity avoidance. Mech. Mach. Theory **121**, 317–334 (2018)
4. Galletti, C., Fanghella, P.: Single-loop kinematotropic mechanisms. Mech. Mach. Theory **36**(6), 743–761 (2001)
5. Hayes, M.J.D., Husty, H., Zsombor-Murray, P.J.: Singular configurations of wrist-partitioned 6R serial robots: a geometric perspective for users. Trans. Can. Soc. Mech. Eng. **26**, 41–56 (2002)
6. Hegedüs, G., Schicho, J., Schröcker, H.-P.: Factorization of rational curves in the study quadric. Mech. Mach. Theory **69**, 142–152 (2013)
7. Manfred, L.H., Pfurner, M., Schröcker, H.-P.: A new and efficient algorithm for the inverse kinematics of a general serial 6R manipulator. Mech. Mach. Theory **42**(1), 66–81 (2007)
8. Manfred, L.H., Schröcker, H.-P.: Algebraic geometry and kinematics. In: Ioannis, Z.E., Sottile, F., Theobald, T. (eds.) Nonlinear Computational Geometry, The IMA Volumes in Mathematics and its Applications, vol. 151, pp. 85–107. Springer (2010)
9. Karger, A.: Singularity analysis of serial robot manipulators. ASME J. Mech. Des. **118**, 505–525 (1996)
10. Kong, X., Pfurner, M.: Type synthesis and reconfiguration analysis of a class of variable-dof single-loop mechanisms. Mech. Mach. Theory **85**, 116–128 (2015)
11. Li, Z., Schicho, J.: A technique for deriving equational conditions on the Denavit-Hartenberg parameters of 6R linkages that are necessary for movability. Mech. Mach. Theory **94**, 1–8 (2015)
12. Müller, A., Piipponen, S.: On regular kinematotropies. In: Proceedings of the 14th World Congress in Mechanism and Machine Science, Taipei, Taiwan (2015)
13. Pfurner, M., Kong, X.: Algebraic analysis of a new variable-dof 7R mechanism. In: New Trends in Mechanism and Machine Science, pp. 71–79. Springer (2017)
14. Wohlhart, K.; Kinematotropic linkages. In: Recent Advances in Robot Kinematics, pp. 359–368 (1996)

A Geometric Method of Singularity Avoidance for Kinematically Redundant Planar Parallel Robots

Nicholas Baron[1(✉)], Andrew Philippides[1(✉)], and Nicolas Rojas[2(✉)]

[1] University of Sussex, Brighton BN1 9RH, UK
{n.baron,andrewop}@sussex.ac.uk
[2] Imperial College London, London SW7 1NA, UK
n.rojas@imperial.ac.uk

Abstract. Methods for avoiding singularities of closed-loop robot mechanisms have been traditionally based on the value of the determinant or the condition number of the Jacobian. A major drawback of these standard techniques is that the closeness of a robot configuration to a singularity lacks geometric, physical interpretation, thus implying that it is uncertain how changes in the robot pose actually move further away the mechanism from such a problematic configuration. This paper presents a geometric approach of singularity avoidance for kinematically redundant planar parallel robots that eliminates the disadvantages of Jacobian-based techniques. The proposed method, which is based on the properties of instantaneous centres of rotation, defines a mathematical distance to a singularity and provides a reliable way of moving the robot further from a singular configuration without changing the pose of the end-effector. The approach is demonstrated on an example robot mechanism and the reciprocal of the condition number of the Jacobian is used to show its advantages.

1 Introduction

A parallel robot is a closed-loop kinematic chain mechanism where the end-effector is connected to the base via two or more independent kinematic chains. A key feature of parallel robots is that once the actuated joints are locked, the mechanism becomes, in general, rigid. However, it is well known that there are particular configurations, known as singularities, in which the total number of degrees of freedom changes instantaneously and consequently the robot loses its rigidity. Herein, the term *singularity* will refer to the so-called forward kinematics singularity, which describes the instance where different branches of the direct position problem meet and the determinant of the Jacobian equals zero [1,2]. Singular configurations are problematic because the task wrench cannot be resisted by active joint torques, or equivalently the end-effector is able to move despite the input velocities of the active joints being zero [3]. In fact, these problems are not just confined to the exact point at which the robot is in a

J. Lenarcic and V. Parenti-Castelli (Eds.): ARK 2018, SPAR 8, pp. 187–194, 2019.
https://doi.org/10.1007/978-3-319-93188-3_22

singularity; the robot can exhibit such behavior when close enough to a singular configuration. This is a major issue in terms of the control of the robot and also for robot safety.

The traditional method of singularity analysis is to calculate the Jacobian of the robot, that is, the matrix that describes the relationship between the input velocities of the actuated joints and the output velocity of the end-effector, and then to compute an index, such as the reciprocal of the condition number or the determinant [4]. The point at which the robot is in a singularity is given by the instance when the value of this index is zero. The problem with using Jacobian indexes as means for performing the singularity analysis is that they have no straightforward geometric interpretation; meaning that for any non-zero value, it is not clear how close the robot is to a singular configuration. Resultantly, by just computing the value of such indexes, there is no physical understanding of the robot's proximity to a singularity and how to move away from it. Many methods of singularity avoidance have been proposed that are based on the calculation of Jacobian indexes [5,6].

In this paper, a simple, geometric method of singularity avoidance for kinematically redundant planar parallel robots is proposed. Kinematically redundant parallel robots are architectures with more actuators than required by the task workspace, but which all contribute to make the robot rigid when locked without serially connected subsets [7,8]. The suggested method, which is based on the properties of instantaneous centres of rotation, has a direct geometric interpretation, providing a physical insight of the robot's proximity to a singularity. These characteristics make it possible to compute a new configuration of the robot for which the distance to a singularity is greater, and yet the pose of the end-effector remains the same.

The paper is structured as follows. In Sect. 2, the concept of instantaneous centres of rotation and its use to determine singularities of planar parallel manipulators is discussed. In Sect. 3, the geometric method of singularity avoidance is proposed which involves, firstly, computing the robot's proximity to a singularity using the properties of instantaneous centres of rotation and, secondly, manipulating its configuration for moving further away from it. The method is demonstrated numerically on an example robot mechanism in Sect. 4, in which the reciprocal of the condition number of the Jacobian is also calculated for comparison. Finally, we conclude in Sect. 5.

2 Instantaneous Centres of Rotation and Singularities

The Instantaneous Centre of Rotation (ICR) between two rigid bodies that are moving relative to one another in the same plane is defined as the instantaneous location of a coincident pair of points of the two bodies of which the absolute velocities are equal [9]. This concept can be applied to any kinematic chain with mobility one in order to find the ICR between any two of the links in the system. Following this, the singularity analysis can be carried out on a planar parallel robot, that in general has a mobility zero when the actuators are locked, by considering the robot as a composition of its mobility one sub-mechanisms; these are the kinematic chains of mobility one within the robot that are obtained by removing one or more of the links that connect the base to the platform. The mobility one sub-mechanisms of a parallel robot can be identified by using, for instance, the extended Chebychev-Kutzbach-Grübler formula [10]

$$M = F - \sum_{i=1}^{\lambda} t_i, \tag{1}$$

where $\lambda = J - L + 1$ is the number of independent loops in the kinematic chain, t_i is the motion type of the ith independent loop ($t_i = 3$ in the planar case), J is the total number of joints, L is the number of links, and F is the total number of degrees of freedom of the joints.

A singularity in a planar parallel robot is produced in the instance where the ICR between the platform and the base of each mobility one sub-mechanism coincide, since in these configurations the platform is able to, instantaneously, rotate about such a point [8]. It therefore follows that a robot's proximity to a singular configuration can be determined by calculating the position of the ICRs between the platform and the base for each of these sub-mechanisms and computing the content of the simplex formed by them—*i.e.*, length in the case of two points. It can be shown that for planar parallel robots of up to three independent loops, the geometric conditions that cause two of these ICRs to coincide are the same that cause any pairing of these ICRs to coincide, hence it is only necessary to calculate the ICR between the platform and the base for two of the mobility one sub-mechanisms. Additionally, if the vector between these ICRs is found, the robot can be reconfigured such that the ICRs are moved further apart from each other, thus moving the robot further away from a singularity.

3 Geometric Method of Singularity Avoidance

The aim of the method is to calculate the vector between the ICRs between the base and the platform of two of the mobility one sub-mechanisms of the kinematically redundant planar parallel robot, and then subsequently reconfigure the robot such that one of these ICRs moves further away from the other. The advantage of kinematically redundant mechanisms is that it is possible to reconfigure at least one of the legs without manipulating the pose of the end-effector. This means that it is possible to change the positions of one of the ICRs between the platform and the base of one of the mobility one sub-mechanisms without manipulating the pose of the end-effector. Therefore, if the vector between this ICR and another ICR (one that is independent of the configuration of the redundant leg) is known, then the former ICR can be repositioned such that it moves further away from the latter ICR, and thus the robot moves further away from a singularity.

This is best understood by examining an example robot mechanism; herein, the kinematically redundant planar parallel robot proposed in [8] is used, in this case with revolute-prismatic-revolute (RPR) legs. A diagram of this robot is shown in Fig. 1; it consists of a platform, a base, four RPR legs and an additional ternary link that is referred to as the *moving base*; the underline in this convention means that the corresponding joint is actuated. The moving base is joined to the base via a passive revolute joint and to the platform via two of the legs, the other two legs connect the platform to the base.

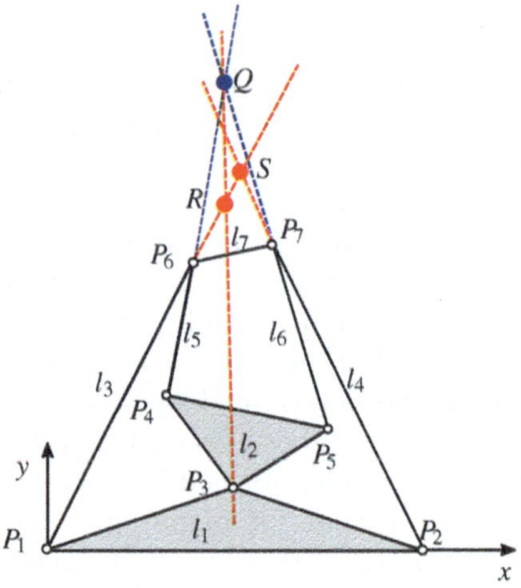

Fig. 1. A kinematic diagram of the kinematically redundant planar parallel robot proposed in [8] with RPR legs. R and S are the ICR(1,7) for the sub-mechanisms where links 4 and 5 are removed, respectively.

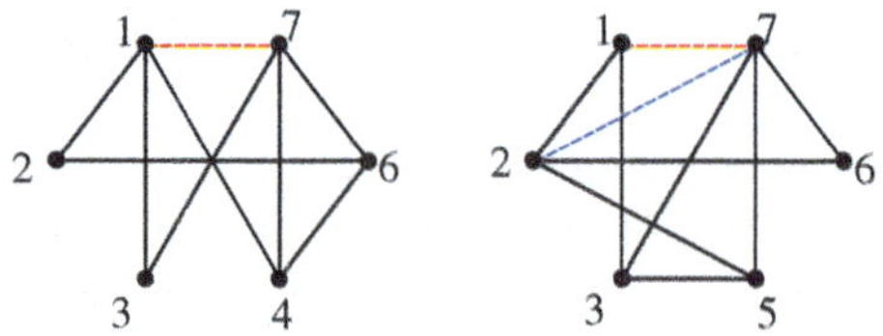

Fig. 2. Circle diagrams for the $M = 1$ sub-mechanisms with link 5 removed (left) and link 4 removed (right).

Firstly, the ICRs between the platform and the base for the mobility one sub-mechanisms of the robot need to be calculated. Herein, the ICR between two links i and j will be denoted by ICR(i,j); for example, according to the notation of Fig. 1, the ICR between the base, link 1, and the platform, link 7, for each of the mobility one sub-mechanisms of this robot is denoted by ICR(1,7). In order to determine the position of ICR(1,7) for each sub-mechanism, a book-keeping system through the use of so-called *Circle Diagrams* is used [11]. The links of the mechanism are denoted by number and a filled line between two links is drawn if the ICR between the two links is known—*i.e.*, the links share a joint. A dotted line is drawn between two links to indicate a desired, unknown ICR; it is possible to find the position of this ICR if the dotted line is a common side of two triangles of which the other sides are filled lines.

In the analyzed robot, five sub-mechanisms which have mobility one can be identified. However, it is only necessary to calculate the ICRs between the platform and the base of two of these sub-mechanisms as the resulting conditions are equivalent; these mechanisms are that where Link 5 is removed, which will be referred to as SM-A, and that where Link 4 is removed, which will be referred to as SM-B. The circle diagrams for SM-A and SM-B are displayed in Fig. 2.

For SM-A, it can be seen by inspecting the circle diagram that the ICR between the platform and the base, ICR(1,7), point $\mathbf{S}$ in Fig. 1, is positioned at the point of intersection between the lines which pass along links 3 and 4. For SM-B, the determination of ICR(1,7) is as follows. In the circle diagram, the dotted line drawn between links 1 and 7 is not the common side of two otherwise known triangles therefore an additional unknown ICR must be found first, this is ICR(2,7) and its position is located at the point of intersection between the lines that pass along links 5 and 6, point $\mathbf{Q}$ in Fig. 1. ICR(1,7) can then be found by drawing two lines; one which passes through Q and P_3, and another which passes along link 3. The point at which these lines intersect is the position of ICR(1,7) for SM-B, this is point $\mathbf{R}$ in Fig. 1.

3.1 Moving Away from a Singular Configuration

In what follows, the horizontal and vertical components of a point $\mathbf{P}_i$ in a given reference frame are denoted by x_i and y_i, and the gradient of the line which passes through points $\mathbf{P}_i$ and $\mathbf{P}_j$ is denoted by $m_{i,j}$. Then, from a robot configuration, the coordinates of $\mathbf{R}$, $\mathbf{Q}$, and $\mathbf{S}$ can be computed as

$$\mathbf{S} = \begin{bmatrix} \frac{1}{m_{1,6}-m_{2,7}}(y_2 - m_{2,7}x_2 - y_1 + m_{1,6}x_1) \\ \frac{m_{1,6}}{m_{1,6}-m_{2,7}}(y_2 - m_{2,7}x_2 - y_1 + m_{1,6}x_1) + y_1 - m_{1,6}x_1 \end{bmatrix}, \tag{2}$$

$$\mathbf{Q} = \begin{bmatrix} \frac{1}{m_{4,6}-m_{5,7}}(y_5 - m_{5,7}x_5 - y_4 + m_{4,6}x_4) \\ \frac{m_{4,6}}{m_{4,6}-m_{5,7}}(y_5 - m_{5,7}x_5 - y_4 + m_{4,6}x_4) + y_4 - m_{4,6}x_4 \end{bmatrix}, \tag{3}$$

$$\mathbf{R} = \begin{bmatrix} \frac{1}{m_{1,6}-m_{3,Q}}(y_3 - m_{3,Q}x_3 - y_1 + m_{1,6}x_1) \\ \frac{m_{1,6}}{m_{1,6}-m_{3,Q}}(y_3 - m_{3,Q}x_3 - y_1 + m_{1,6}x_1) + y_1 - m_{1,6}x_1 \end{bmatrix}. \tag{4}$$

It can be seen by inspecting the mechanism that the pose of the platform must necessarily change if the length of either link 3 or link 4 is modified. However, changing the lengths of link 5 and link 6 does not necessarily alter the pose of link 7. This means that point $\mathbf{R}$ is the ICR whose position can be altered in order to avoid singular configurations. Let $\mathbf{d}$ be the vector from $\mathbf{S}$ to $\mathbf{R}$. Hence, the robot can be moved further from a singular configuration by moving the position of $\mathbf{R}$ along the direction of this vector; this can be achieved by multiplying $\mathbf{d}$ by a scalar $\lambda > 1$, that is,

$$\mathbf{R}^* = \mathbf{S} + \lambda \mathbf{d}, \tag{5}$$

where $\mathbf{R}^*$ is the new position of $\mathbf{R}$.

Having computed $\mathbf{R}^*$, the new position of $\mathbf{Q}$, say $\mathbf{Q}^*$, needs to be determined. The only requirement of $\mathbf{Q}^*$ is that it has to lie on the line which passes through $\mathbf{R}^*$ and $\mathbf{P}_3$. Its position is determined by calculating the equation of this line and then selecting a feasible gradient of link 5, $m_{4,6}$. The aim of the method therefore is to obtain a value for $m_{4,6}$ that produces the desired value of $\mathbf{R}^*$. The first step is to calculate an expression of $\mathbf{P}_4$ in terms of $m_{4,6}$. $\mathbf{P}_4$ is positioned at one of the points of intersection between the line of gradient $m_{4,6}$ that passes through the centre of $\mathbf{P}_6$ and the circle whose centre is that of $\mathbf{P}_3$ and whose radius is the distance between $\mathbf{P}_3$ and $\mathbf{P}_4$, $d_{3,4}$. Thus,

$$\mathbf{P}_4 = \begin{bmatrix} \frac{-b \pm \sqrt{b^2-4ac}}{2a} \\ m_{4,6}\frac{-b \pm \sqrt{b^2-4ac}}{2a} + y_6 - m_{4,6}x_6 \end{bmatrix}, \tag{6}$$

where $a = 1 + m_{4,6}^2$, $b = 2m_{4,6}(y_6 - m_{4,6}x_6 - y_3) - 2x_3$, and $c = x_3^2 + (y_6 - m_{4,6}x_6 - y_3)^2 - d_{3,4}^2$. The $\pm$ signs in Eq. (6) denote the two possible positions of $\mathbf{P}_4$, either are feasible however the length of link 5 must lie within the limits of the actuated prismatic joint.

The positions of the centres of the remaining unknown joints are found by solving the inverse kinematics problem—a method to do this is discussed in [8]. Then, with an expression for the position of $\mathbf{P}_5$ obtained, the gradient of link 6, $m_{5,7}$, is calculated in terms of $m_{4,6}$. Now, a value of $m_{4,6}$ is calculated such that the lines which pass through links 5 and 6 intersect at a point which lies on the line which passes through $\mathbf{P}_3$ and $\mathbf{R}^*$. Then,

$$\mathbf{Q}^* = \begin{bmatrix} \frac{c_{5,7}-c_{4,6}}{m_{4,6}-m_{5,7}} \\ m_{4,6}\frac{c_{5,7}-c_{4,6}}{m_{4,6}-m_{5,7}} + c_{4,6} \end{bmatrix} \tag{7}$$

where $c_{4,6} = y_6 - m_{4,6}x_6$ and $c_{5,7} = y_7 - m_{5,7}x_7$. Finally, the components of $\mathbf{Q}^*$ in terms of $m_{4,6}$ can be expressed as

$$y_{Q^*} - m_{3,R^*}x_{Q^*} - c_{3,R^*} = 0, \tag{8}$$

where, $c_{3,R^*} = y_3 - m_{3,R^*}x_{3,R^*}$ is the y-intercept of the line which passes through $\mathbf{P}_3$ and $\mathbf{R}^*$.

Equation (8) is a radical equation in $m_{4,6}$, the roots of which are the possible values of $m_{4,6}$. Since only one solution is required, root-finding algorithms, such as the Newton-Raphson method, can be used. The positions of $\mathbf{P}_4$ and $\mathbf{Q}^*$ are then found by substituting the obtained value back into Eqs. (6) and (7), respectively.

4 Numerical Example

In this section the method is carried out on an example mechanism with the same architecture that is examined in the previous section. Firstly, $\mathbf{d}$ is calculated for the mechanism with the following joint coordinates: $\mathbf{P}_1 = (0, 2)^T$, $\mathbf{P}_2 = (16, 2)^T$, $\mathbf{P}_3 = (9, 1)^T$, $\mathbf{P}_4 = (3.05, 5.36)^T$, $\mathbf{P}_5 = (14.47, 1.70)^T$, $\mathbf{P}_6 = (6, 6)^T$ and $\mathbf{P}_7 = (10, 6)^T$. Using these values, and by computing the necessary gradients, the positions of $\mathbf{S}$, $\mathbf{Q}$ and $\mathbf{R}$ are found using Eqs. (2)–(4) and, resultantly, $\mathbf{d}$ is obtained by subtracting $\mathbf{S}$ from $\mathbf{R}$. Hence, $\mathbf{S} = (8, 7.33)^T$, $\mathbf{Q} = (9.26, 6.71)^T$, $\mathbf{R} = (9.33, 8.22)^T$ and $\mathbf{d} = (1.33, 0.89)^T$. This configuration is displayed on the left hand side of Fig. 3. $\mathbf{R}^*$, the new position of $\mathbf{R}$, is then calculated with Eq. (5), for instance $\mathbf{R} = (12, 10)^T$ when $\lambda = 3$, subsequently, using Eqs. (6)–(8), a radical equation in terms of $m_{4,6}$ is obtained. A feasible solution of such an equation is $m_{4,6} = -0.124$.

Given the above value for $m_{4,6}$, the new position of $\mathbf{Q}$ is then $\mathbf{Q}^* = (10.48, 5.44)^T$ and consequently $\mathbf{d} = (4, 2.67)^T$. The positions of the centres of joints $\mathbf{P}_1$, $\mathbf{P}_2$, $\mathbf{P}_3$, $\mathbf{P}_6$ and $\mathbf{P}_7$ are unchanged, and the remaining joints are recalculated. In this example, the kinematic redundancy of the robot is utilised as the pose of the end-effector is not modified. This configuration is displayed on the right hand side of Fig. 3.

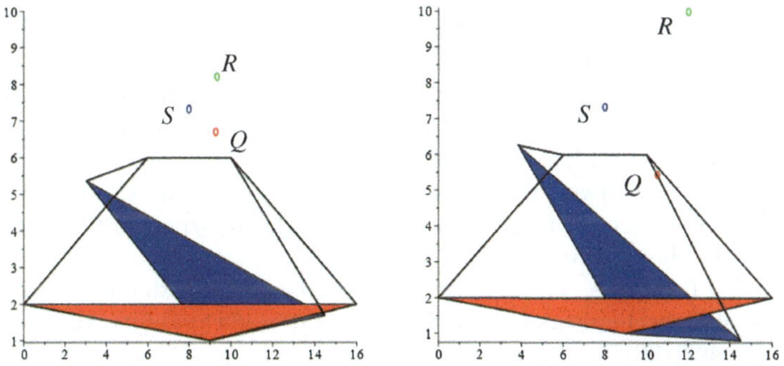

Fig. 3. Original and improved configurations of the robot (left and right, respectively).

The advantage of the proposed method of singularity avoidance is that it provides a straightforward interpretation of the robot's proximity to a singularity. The magnitude of $\mathbf{d}$ can be compared with the dimensions of the robot, or its resolution, such that there is a clear context to the size of this value. On the other hand, if we perform the above analysis using a traditional Jacobian-based index, such as the reciprocal of the condition number, we obtain a value of 1.094×10^{-2} for the original configuration and 2.118×10^{-2} for the second, improved configuration. It can be seen that these values provide no real insight into how close physically the robot is to a singularity.

5 Conclusion

A main problem with traditional methods of singularity analysis based on indexes of the Jacobian matrix, such as the reciprocal of the condition number, is the absence of a straightforward geometric interpretation, implying that, in essence, they are meaningless for all non-zero values. In this paper, a geometric method of singularity avoidance for kinematically redundant planar parallel robots, based on the properties of instantaneous centres of rotation, is introduced; the technique provides a physical understanding of the robot's proximity to a singularity and allows the definition of escaping trajectories that effectively deviate the robot from dangerous configurations. These characteristics, which are simple to implement, open the door to further research in geometric planners of singularity-free paths for closed-loop manipulators.

References

1. Gosselin, C., Angeles, J.: Singularity analysis of closed-loop kinematic chains. IEEE Trans. Robot. Autom. **6**(3), 281–290 (1990)
2. Rojas, N., Borras, J., Thomas, F.: The octahedral manipulator revisited. In: IEEE International Conference on Robotics and Automation, pp. 2293–2298 (2012)
3. O'Brien, J.F., Wen, J.T.: Kinematic control of parallel robots in the presence of unstable singularities. IEEE Trans. Rob. Autom. **1**, 354–359 (2001)
4. Merlet, J.-P.: Jacobian, manipulability, condition number, and accuracy of parallel robots. J. Mech. Des. **128**(1), 199–206 (2006)
5. Cha, S.-H., Lasky, T.A., Velinsky, S.A.: Singularity avoidance for the 3-RRR mechanism using kinematic redundancy. In: IEEE International Conference on Robotics and Automation, pp. 1195–1200 (2007)
6. Bhattacharya, S., Hatwal, H., Ghosh, A.: Comparison of an exact and an approximate method of singularity avoidance in platform type parallel manipulators. Mech. Mach. Theory **33**(7), 965–974 (1998)
7. Gosselin, C., Laliberté, T., Veillette, A.: Singularity-free kinematically redundant planar parallel mechanisms with unlimited rotational capability. IEEE Trans. Rob. **31**(2), 457–467 (2015)
8. Baron, N., Philippides, A., Rojas, N.: A kinematically redundant planar parallel robot with full rotatability (in preparation)
9. Daniali, H.M.: Instantaneous center of rotation and singularities of planar parallel manipulators. Int. J. Mech. Eng. Educ. **33**(3), 251–259 (2005)
10. Rojas, N., Dollar, A.M.: Gross motion analysis of fingertip-based within-hand manipulation. IEEE Trans. Rob. **32**(4), 1009–1016 (2016)
11. Hartenberg, R.S., Denavit, J.: Kinematic Synthesis of Linkages. McGraw-Hill, New York (1964). Chap. 4

The 3-PPPS Parallel Robot with U-Shape Base, a 6-DOF Parallel Robot with Simple Kinematics

Damien Chablat[1(✉)], Luc Baron[2], Ranjan Jha[3], and Luc Rolland[4]

[1] CNRS, LS2N, UMR 6004, 1 rue de la Noë, 44321 Nantes, France
Damien.Chablat@cnrs.fr
[2] Mechanical Engineering Department, École Polytéchnique de Montréal, Montreal, Québec H3C 3A7, Canada
Luc.Baron@polymtl.ca
[3] Biomedical Instrumentation Unit, CSIR-Central Scientific Instruments Organisation, Chandigarh, India
Ranjan.Jha@polymtl.ca
[4] School of Engineering and Computing, University of the West of Scotland, Paisley, Scotland, UK
Luc.Rolland@uws.ac.uk

Abstract. One of the main problems associated with the use of 6 DOF parallel robots remains the solving of their kinematic models. This is rarely possible to analytically solve their models thereby justifying the application of numerical methods. These methods are difficult to implement in an industrial controller and can cause solution bifurcations close to singularities resulting in following an unplanned trajectory. Recently, a 3-PPPS robot with U-shaped base was introduced where an analytical kinematic model can be derived. Previously, quaternion parameters were used to represent the orientation of the mobile platform. To allow for simpler model handling, this article introduces the use of Euler angles which have a physical meaning for the users. Compact writing of the direct and inverse kinematic model is thus obtained. Using algebraic and cylindrical decomposition for the workspace, this provides a simpler representation of the largest domain without singularity around the "home" configuration.

Keywords: 6 DOF parallel robot · Singularity · Kinematic model

1 Introduction

The first all-electric robot, namely the Stanford Arm, was created in 1969 by Victor David Scheinman in the Stanford Artificial Intelligence Laboratory (SAIL). By using 6-axis, (five revolute joints and one prismatic joint), this design permits a closed-form kinematics solution which could easily be calculated despite the limitations of the first computers. [1]. These robots are classified as

J. Lenarcic and V. Parenti-Castelli (Eds.): ARK 2018, SPAR 8, pp. 195–202, 2019.
https://doi.org/10.1007/978-3-319-93188-3_23

serial robots since all the bodies and joints form a serial chain. Thirty years later, the majority of industrial robots still have a simple architecture derived from this robot [2]. Also, for many years, serial robots have been designed with two distinct units: the robotic arm to position the end-effector and then the wrist to orient the end-effector, meaning (i) three articulations for translation and (ii) three pivot joints with concurrent axes for rotation. We can then summarize their kinematics characteristics as (i) a trivial direct kinematic model, (ii) an inverse kinematic model with only eight easily identifiable solutions (elbow up, elbow down. shoulder left, shoulder right) and (iii) a regular workspace (spherical, cylindrical or cubic). The main problem with serial robots is their lack of rigidity which is solved by more mass resulting in less accelerations. If the parallel robots can solve this problem, they bring new challenges like (i) the complexity of their kinematic models, (ii) the presence of singularities inside their workspace and especially (iii) a low ratio between the workspace and the size of the robot [3]. Seeking a compromise led to the creation of (i) hybrid robots, and (ii) parallel robots with several actuators per kinematics chain (not fully parallel). The Tripteron is the parallel robot with three degrees of freedom having the simplest kinematic model which is as a simple Cartesian serial robot of the PPP type [4]. For spatial parallel robots, finding architectures with simple kinematic models remains an open problem. The purpose of this article is to use the 3-PPPS robot with a U-shaped base where the Euler angles are implemented for the orientation of the mobile plate-form [9].

The outline of this paper is as follows. First, we introduce the 3-PPPS parallel robot design parameters with simplifications which permit to study its joint space and workspace in a three dimensional space. In the next section, we will define the inverse and the direct kinematic models. In the later section, we study its singular configuration to provide the definition of an aspect included the "home" pose of the robot.

2 Parameters and Kinematics of the 3-PPPS Parallel Robot with U-Shape Base

The results of the first work done at Monash University on the 3-PPPS introduced the MEPaM robot [5]. In the first design, the first actuators of the kinematics chains are located in orthogonal directions. For each leg, the three prismatic joints are placed orthogonally. With this design, it has been shown that the robot admits up to six solutions to the direct kinematic problem and it is able to follow non-singular assembly mode-changing trajectories. Another invariant is that the parallel singularity postures depend only on the orientation of the end-effector. Thus it is possible to study the workspace in a 3 dimensional space only. For the MEPaM manipulator, using a variable change, the joint space can also be studied in a 3 dimensional space. These properties allow us to calculate and display the domains without singularities of this robot. Another architecture was introduced where the two first actuated joints are located on the faces of a prism [5]. From the kinematic point of view, this design is simpler than the MEPaM,

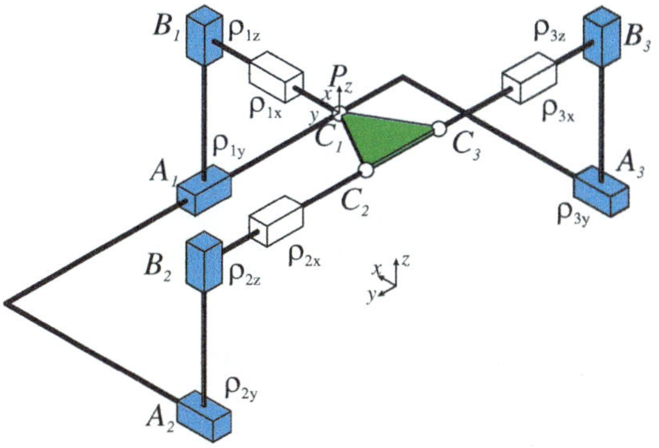

Fig. 1. The 3-PPPS parallel robot and its parameters in its "home" pose with the actuated prismatic joints in blue, the passive joints in white and the mobile platform in green

but the joint space needs to be analyzed in a five dimensional space. Recently a novel design was proposed where the actuated prismatic joints were located in a U-shaped base, as shown in Fig. 1 [9]. To define the end-effector orientation, the A, B and C angles define the rotation around the X, Y and Z axes respectively [8]. We will use the following rotation matrix $\mathbf{R} = \mathbf{R}(x)\mathbf{R}(y)\mathbf{R}(z)$

$$\mathbf{R} = \begin{bmatrix} C_\theta C_\sigma & -C_\theta S_\sigma & S_\theta \\ S_\phi S_\theta C_\sigma + C_\phi S_\sigma & -S_\phi S_\theta S_\sigma + C_\phi C_\sigma & -S_\phi C_\theta \\ -C_\phi S_\theta C_\sigma + S_\phi S_\sigma & C_\phi S_\theta S_\sigma + S_\phi C_\sigma & C_\phi C_\theta \end{bmatrix} \quad (1)$$

where $C_X = \cos(X)$ and $S_X = \sin(X)$ for $X = \phi$, θ or σ associated with the A, B and C angles respectively.

2.1 Geometric Parameters

Similarly to [9], the three kinematics chains are identical and comprised of two actuated prismatic joints plus one passive prismatic joint and a spherical joint (Fig. 1). The axes of the first three joints form an orthogonal reference frame. It may be noted that the placement order of the passive and active joints is not important. Thus, all the passive joints can be moved on the base of each leg to simplify the design. We have defined an origin A_i for each leg as

$$\mathbf{A}_1 = [2, 0, 0]^T, \quad \mathbf{A}_2 = [0, 2, 0]^T, \quad \mathbf{A}_3 = [0, -2, 0]^T \quad (2)$$

The coordinates of the point C_1 are ρ_{1x}, ρ_{1y} and ρ_{1z}, where the last two are actuated. The coordinates of C_2 and C_3 are obtained by a rotation around the z axis by $\pi/2$ and $-\pi/2$, respectively.

$$\mathbf{C}_1 = [\rho_{1x}, \rho_{1y}, \rho_{1z}]^T, \quad \mathbf{C}_2 = [-\rho_{2y}, \rho_{2x}, \rho_{2z}]^T, \quad \mathbf{C}_3 = [\rho_{3y}, -\rho_{3x}, \rho_{3z}]^T \quad (3)$$

The origin on the mobile equilateral platform is similar to the parameters used in [9], meaning on the mobile platform.

$$\mathbf{V}_1 = [0, 0, 0]^T, \quad \mathbf{V}_2 = [-\sqrt{3}/2, 1/2, 0]^T, \quad \mathbf{V}_3 = [-\sqrt{3}/2, -1/2, 0]^T \tag{4}$$

Only the mobile platform orientation is positioned at the "home" posture to obtain ϕ, θ, σ equal to zero. The moving platform coordinates, in the fixed reference frame, can be written as

$$\mathbf{W}_i = \mathbf{R}\mathbf{V}_i + \mathbf{P} \quad \text{where} \quad \mathbf{P} = [x\ y\ z]^T \tag{5}$$

2.2 Constraint Equations

From the nine equations coming from Eq. 5 and the position of points C_i, we can easily remove the passive joints, and thus obtain new constraint equations

$$\rho_{1y} - y = 0 \tag{6}$$
$$\rho_{1z} - z = 0 \tag{7}$$
$$C_\theta C_\sigma \sqrt{3} - C_\theta S_\sigma - 2(\rho_{2y} + x) = 0 \tag{8}$$
$$\sqrt{3}(S_\phi S_\sigma - C_\phi S_\theta C_\sigma) + C_\phi S_\theta S_\sigma + S_\phi C_\sigma + 2(\rho_{2z} - z) = 0 \tag{9}$$
$$C_\theta C_\sigma \sqrt{3} + C_\theta S_\sigma + 2(\rho_{3y} - x) = 0 \tag{10}$$
$$\sqrt{3}(S_\phi S_\sigma - C_\phi S_\theta C_\sigma) - C_\phi S_\theta S_\sigma - S_\phi C_\sigma + 2(\rho_{3z} - z) = 0 \tag{11}$$

We can notice that the symmetrical property of the mobile platform remains after the elimination of ρ_{1x}, ρ_{2x} and ρ_{3x}.

2.3 Change of Variables

We introduce a change of variables to reduce the joint space dimension from six to three where three coordinates are equal to zero [9].

$$\mu_{1x} = \rho_{1x} - \rho_{2y} \quad \mu_{1y} = \rho_{1y} - \rho_{1y} = 0 \quad \mu_{1z} = \rho_{1z} - \rho_{1z} = 0 \tag{12}$$
$$\mu_{2x} = \rho_{2x} - \rho_{1y} \quad \mu_{2y} = \rho_{2y} - \rho_{2y} = 0 \quad \mu_{2z} = \rho_{2z} - \rho_{1z} \tag{13}$$
$$\mu_{3x} = \rho_{3x} + \rho_{2y} \quad \mu_{3y} = \rho_{3y} - \rho_{2y} \quad \mu_{3z} = \rho_{3z} - \rho_{1z} \tag{14}$$

The remaining six parameters are three passive prismatic joints and three active prismatic joints. With this change of variables, the coordinates of the mobile platform are also translated but its orientation does not change.

$$x' = x - \rho_{2y} \quad y' = y - \rho_{1y} \quad z' = z - \rho_{1z} \tag{15}$$

This simplification is similar to the case where the three first prismatic joints are orthogonal as in [6]. If we apply the change of variable (12–14) in the constraint equations (6–11), we obtain the following

$$\left(\sqrt{3}C_\sigma - S_\sigma\right) C_\theta - 2x' = 0 \tag{16}$$

$$\left(-\sqrt{3}C_\phi S_\theta + S_\phi\right) C_\sigma + \left(\sqrt{3}S_\phi + C_\phi S_\theta\right) S_\sigma + 2\mu_{2z} = 0 \tag{17}$$

$$\left(\sqrt{3}C_\sigma + S_\sigma\right) C_\theta + 2\mu_{3y} - 2x' = 0 \tag{18}$$

$$\left(-\sqrt{3}C_\phi S_\theta - S_\phi\right) C_\sigma + \left(\sqrt{3}S_\phi - C_\phi S_\theta\right) S_\sigma + 2\mu_{3z} = 0 \tag{19}$$

3 Inverse and Direct Kinematics of the 3-PPPS Parallel Robot

Finding the solutions of the direct geometric model of a parallel robot is often a difficult problem, especially for robots with six degrees of freedom. If in [9] for the mechanism studied, a compact form was found, the use of Euler angles could lead to more complex expressions. With the Siropa library, it is possible to obtain the kinematic models by using the Gröbner based elimination method by changing the constraint equation in an algebraic form. A classical approach can be used to transform the trigonometric equations to algebraic ones as in [10].

3.1 Inverse Kinematic Model

Thanks to the choice of the origin on the mobile platform which is fixed at C_1, it is straightforward to find the joint positions of the first kinematics chain:

$$\rho_{1y} = y \quad \text{and} \quad \rho_{1z} = z \tag{20}$$

For the two other kinematics chain, the result keeps the symmetric property and becomes

$$\rho_{2y} = \frac{C_\theta\left(C_\sigma\sqrt{3} - S_\sigma\right)}{2} - x \tag{21}$$

$$\rho_{2z} = \frac{\sqrt{3}\left(C_\phi S_\theta C_\sigma - S_\phi S_\sigma\right) - C_\phi S_\theta S_\sigma - S_\phi C_\sigma}{2} + z \tag{22}$$

$$\rho_{3y} = -\frac{C_\theta\left(C_\sigma\sqrt{3} + S_\sigma\right)}{2} + x \tag{23}$$

$$\rho_{3z} = \frac{\sqrt{3}\left(C_\phi S_\theta C_\sigma - S_\phi S_\sigma\right) + C_\phi S_\theta S_\sigma + S_\phi C_\sigma}{2} + z \tag{24}$$

The computational cost to solve the inverse kinematics handles six trigonometric functions (sine and cosine), the square of three, 14 multiplications/divisions and 11 additions/subtractions if we used the code generation of Maple with the optimization function to create intermediary variables.

3.2 Direct Kinematic Model

The computation of the position does not depend on the choice of parameters for the orientation. For the position, we have the similar results than in [9].

$$x = \frac{{\mu_{2z}}^2 \mu_{3y} - \mu_{2z}\mu_{3y}\mu_{3z} - \mu_{3y}/2}{(\mu_{2z} - \mu_{3z} + 1)(\mu_{2z} - \mu_{3z} - 1)}$$
$$\pm \frac{\sqrt{({\mu_{2z}}^2 - \mu_{2z}\mu_{3z} + {\mu_{3z}}^2 - 3/4)({\mu_{2z}}^2 - 2\mu_{2z}\mu_{3z} + {\mu_{3y}}^2 + {\mu_{3z}}^2 - 1)}}{(\mu_{2z} - \mu_{3z} + 1)(\mu_{2z} - \mu_{3z} - 1)} \quad (25)$$

For the orientation, for any values of x, we obtain

$$\theta_1 = \pm \arccos\left(2\sqrt{\frac{\mu_{3y}^2 - \mu_{3y}x + x^2}{3}}\right) \quad \text{and} \quad \theta_2 = \pm(\pi - \theta_1) \quad (26)$$

$$\sigma_1 = \pm \arccos\left(\frac{\mu_{3y} - 2x}{2\sqrt{\mu_{3y}^2 - \mu_{3y}x + x^2}}\right) \quad \text{and} \quad \sigma_2 = \pm(\pi - \sigma_1) \quad (27)$$

$$\phi_1 = \pm \arccos\left(\sqrt{\frac{3/4 - \rho_{2z}^2 + \rho_{2z}\rho_{3z} - {\rho_{3z}}^2}{\rho_{3y}^2 - \rho_{3y}x + x^2}}\right) \qquad \phi_2 = \pm(\pi - \phi_1) \quad (28)$$

The computational cost to evaluate one solution comprises three arccos functions, three square roots, 17 multiplications, 8 divisions, 8 additions, and 12 subtractions. In [9], it was shown that there exists four direct kinematic solutions for any joint values. We have a similar result because two sets of θ can define the same mobile platform orientation. So, we keep that a value for θ, such that $\pi/2 \leq \theta \leq \pi$. We have 16 solutions considering all θ, σ and ϕ permutations but only four are solution of the constraint equations.

4 Workspace and Singular Configurations of the 3-PPPS Parallel Robot

From the constraint equations, we can easily determine the inverse Jacobian of the robot and its parallel singularities

$$C_\phi C_\theta^2 (S_\phi S_\theta S_\sigma - C_\phi C_\sigma) = 0 \quad (29)$$

As Eq. 29 is factored, we can isolate three surfaces as is shown in Fig. 2. The first term is represented by the blue surface, the second term by the green surface and the last one by the red surface.

We can also notice that the coordinate change does not change the singularity locus in the joint space, [9]. However, the computational cost is higher because in

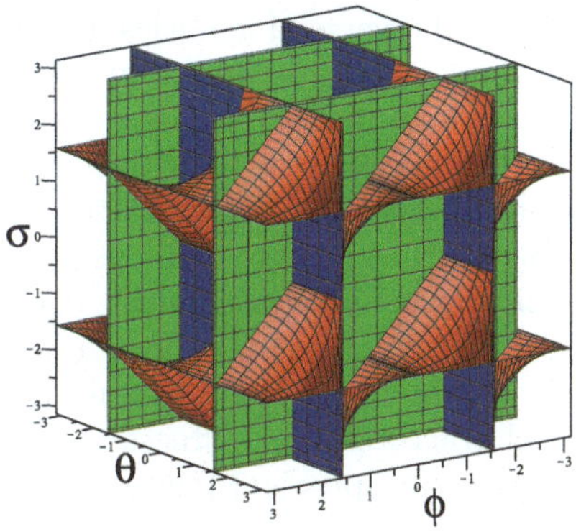

Fig. 2. The three singularity surface of the 3-PPPS parallel robot

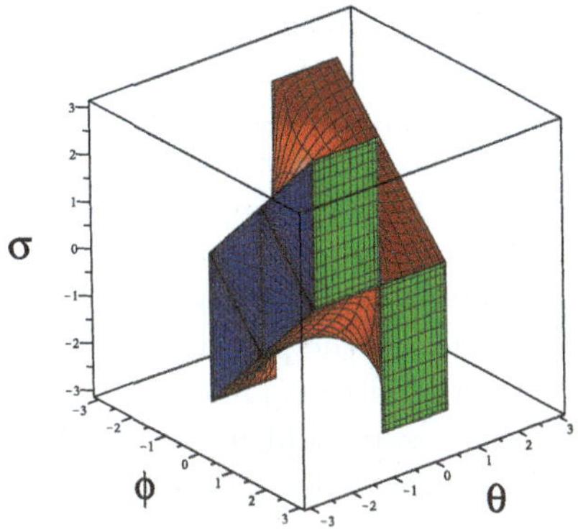

Fig. 3. The aspect of the 3-PPPS parallel robot around the "home" pose

the Gröbner based elimination, we have to remove six more parameters that are required with quaternions. Using the Cylindrical Algebraic Decomposition, the aspect is characterized including the robot "home" position, [7]. This is done by only two cells. The orientation workspace has the following distinct limitations which can be directly related to the platform rotations, Fig. 3:

- Along the θ and ϕ angles, the complete range of motion is from $-\pi/2$ until $\pi/2$;
- Along the ϕ angle, the complete motion ranges from $-\pi$ until π intersecting a plane at $\phi = \pi/2$, if $\alpha > 0$ then $-\pi/2 \leq \theta \leq 0$ the this plane segment looks like a rectangle;
- On the other side, accordingly, intersecting a plane at $\phi = -\pi/2$, if $\alpha < 0$ then $0 \leq \theta \leq pi/2$ meaning a symmetrical rectangle appears. These two symmetrical rectangles will evolve towards the other side as ϕ is changed from $\pi/2$ to $-\pi/2$. In the middle, intersecting a plane at $\phi = 0$, the workspace is a square with sides at $-\pi/2$ and $\pi/2$.

Hence, the θ range divides itself differently if α is positive or negative with traveling symmetrical rectangles. This results from the selected symmetries and axis positions. These traveling or sliding rectangles actually produce two interesting saddle points on the workspace limit where $\alpha = 0$ and $\theta = 0$. Hence, the graph solution is geometrically confirming the displacement limitations that can be intuitively understood from manipulator observation. These workspace distinct limits come from the passive prismatic joints.

5 Conclusions

In this article, we introduced kinematics modeling where the orientation representation of the 3-PPPS mobile platform implemented the Euler angles as it is often represented on many machine tools followed by a significant variable change. We can see that the equations of the direct and inverse kinematics model remain very compact. An estimate of the calculation cost is presented. Since the equations are quadratic, it is possible to obtain an analytical expression which is not possible for most parallel robots with 6 degrees of freedom. A representation of the aspect encompassing the "home" position is given. Conversely to the 5-axis Orthoglide [11] or 4-axis Kanuk [12], the 3-PPPS parallel robot is using 6 identical linear or prismatic actuators as a Gough-Stewart platform. Inside each kinematic chain, the passive and active joint order can be changed to obtain, for example, the passive joint fixed into the base without altering the singularity and aspect results. Future works will be done to characterize the stiffness model as well as to know the influence of the joint limits in the spherical joints.

References

1. Paul, R.P.: Robot Manipulators: Mathematics, Programming, and Control: The Computer Control of Robot Manipulators. MIT Press, Cambridge (1981)
2. Siciliano, B., Khatib, O.: Springer Handbook of Robotics. Springer New York, Inc., Secaucus (2007)
3. Merlet, J.P.: Parallel Robots. Springer, Heidelberg (2006)
4. Kong, X., Gosselin, C.M.: Type synthesis of linear translational parallel manipulators. In: Lenarčič, J., Thomas, F. (eds.) Advances in Robot Kinematic, pp. 453–462. Kluwer Academic Publishers, Dordrecht (2002)
5. Chen, C., Gayral, T., Caro, S., Chablat, D., Moroz, G.: A Six-Dof epicyclic-parallel manipulator. J. Mech. Roboti. **4**(4), 041011-1-8 (2012). American Society of Mechanical Engineers
6. Caro, S., Wenger, P., Chablat, D.: Non-singular assembly mode changing trajectories of a 6-DOF parallel robot. In: ASME Design Engineering Technical Conferences and Computers and Information in Engineering Conference IDETC/CIE, Chicago, USA, 12-15 August 2012
7. Manubens, M., Moroz, G., Chablat, D., Wenger, P., Rouillier, F.: Cusp points in the parameter space of degenerate 3-RPR planar parallel manipulators. ASME J. Mech. Robot. **10**(1) (2012)
8. Lamikiz, A., López de Lacalle, L.N., Ocerin, O., Díez, D., Maidagan, E.: The Denavit and Hartenberg approach applied to evaluate the consequences in the tool tip position of geometrical errors in five-axis milling centres. Int. J. Adv. Manuf. Technol. **37**, 122139 (2008)
9. Chablat, D., Baron, L., Jha, R.: Kinematics and workspace analysis of a 3PPPS parallel robot with U-shaped base. In: International Design Engineering Technical Conferences and Computers and Information in Engineering Conference (2017)
10. Moroz, G., Rouiller, F., Chablat, D., Wenger, P.: On the determination of cusp points of 3-RPR parallel manipulators. Mech. Mach. Theory **45**(11), 1555–1567 (2010)
11. Caro, S., Chablat, D., Lemoine, P., Wenger, P.: Kinematic analysis and trajectory planning of the Orthoglide 5-axis. In: Proceedings of the ASME 2015, International Design Engineering Technical Conferences and Computers and Information in Engineering Conference, Boston, USA, August 2015 (2015)
12. Rolland, L.: The manta and the kanuk: novel 4-DOF parallel mechanisms for industrial handling. In: 1999 Proceedings of the ASME Dynamic Systems and Control Division Conference, IMECE 1999, Nashville, vol. 67, pp. 831–844, 14–19 novembre 1999

On the Singularities of a Parallel Robotic System Used for Elbow and Wrist Rehabilitation

Iosif Birlescu[1], Doina Pisla[1](✉), Bogdan Gherman[1], Adrian Pisla[1], Calin Vaida[1], Giuseppe Carbone[2], and Nicolae Plitea[1]

[1] CESTER, Technical University of Cluj-Napoca, Cluj-Napoca, Romania
{iosif.birlescu, doina.pisla, bogdan.gherman, Calin.Vaida, Nicolae.Plitea}@mep.utcluj.ro, adrian.pisla@muri.utcluj.ro

[2] University of Cassino and South Latium, Cassino, Italy
carbone@unicas.it

Abstract. The paper presents the singularity analysis of a modular parallel robotic system designed for elbow and wrist post-stroke rehabilitation using the mathematical modeling based on the Study parameters of SE(3). The kinematics of the robotic system is achieved; the singularity conditions are identified and correlated with the human upper limb motion capabilities to develop a robust control system for a fail-safe rehabilitation robot.

Keywords: Singularity · Parallel robot · Study parameters · Rehabilitation

1 Introduction

Previous studies [1, 2] reveal a clear need for new rehabilitation devices in the near future in Europe, mainly due to the upward trend in the proportion of the elderly (who are the main target of stroke related disabilities) combined with an insufficient rehabilitation and medical personnel. The robotic systems for rehabilitation offer several advantages: the possibility of the therapist to work with multiple patients with the option of personalized training and monitoring and avoidance of stamina loss often incurred in the case of therapist leading exercises. Safety is an important issue in the functionality of these systems and the avoidance of singular configurations is one of the first conditions in developing a failsafe rehabilitation robotic system. Various methods are used in the singularities study such as vector methods and algebraic methods. The algebraic method with Study parameters of SE(3) based on dual quaternions was previously used in solving singularities for Stewart-Gough platform [3], the 3-RPS robot [4], and medical robots [5, 6]. This algebraic method leads to the determination of the singularity conditions in the active joint space, making them easy to implement in the robot control algorithm, and was used successfully for a complete singularity analysis in previous work [6]. The paper is structured as follows: Sect. 2 presents the PaRReEx robotic structure; Sect. 3 derives the forward kinematics using the Study parameters; Sect. 4 derives the singularities of the robotic system; Sect. 5 presents the conclusions of the work.

J. Lenarcic and V. Parenti-Castelli (Eds.): ARK 2018, SPAR 8, pp. 203–211, 2019.
https://doi.org/10.1007/978-3-319-93188-3_24

2 PaRReEx Modular Robotic System Description

The PaRReEx robotic system [7] has been designed with two parallel modules: one for the elbow rehabilitation (the elbow flexion and pronation/supination – as illustrated in Fig. 1(a) and the other for the wrist rehabilitation (flexion/extension and abduction/adduction – Fig. 1(b) with ranges of motion presented in Table 1 [8]. Each rehabilitation motion, as defined by the medical expert, uses two anchor points: a passive one which isolates the segment of interest, and an active one which is used to generate the motion. For the elbow, the passive anchor is fixed to the patient's arm and the active one is fixed to the patient's hand (palm). The elbow module is a 2 DoF parallel robot with two kinematic chains: the kinematic chain 1 (CK_{e1}), type $\underline{R}U$, starts at the origin of the fixed coordinate frame XYZ, is actuated by q_1 (with the rotation axis collinear with the X axis), and has two kinematic links *a* (of length l_0) and *b* (of length l_1) linking the active joint q_1 with the passive Cardan joint R_c and R_c with the origin of the mobile coordinate frame X′ Y′Z′, respectively. The second kinematic chain (CK_{e2}), type $\underline{R}R$, is actuated by the q_2 active joint and has a mechanical link *c* (of length l_1) that links the active joint q_2 with the passive revolute joint R_2 (the mechanical joint *b* is allowed to rotate freely through R_2) so that CK_{e2} intersects CK_{e1} in the origin of the mobile coordinate frame. With respect to the coordinate frames, the passive anchor point is displaced by l_0 on the Y axis, and the active anchor point is displaced by the same value l_0 on the Y' axis. The elbow rehabilitation module performs the pronation/supination of the forearm actuated by q_1 (motion transmitted to the active anchor through a gear G with ratio *n*), and the flexion/extension of the forearm actuated by q_2 (by lifting the link *c*). The wrist rehabilitation module (Fig. 3) is also a 2 DoF mechanism. A fixed coordinate frame $X^*Y^*Z^*$ is chosen at the intersection of the rotation axes of the active joint q_3 and the passive joint R_1' (with the passive anchor point fixed on the patients forearm displaced on the X^* axis by $-l_0'$), and a mobile frame $X^{*'}Y^{*'}Z^{*'}$ is chosen at distance *r* from the origin of the fixed frame (with its origin on the active anchor point located on the patient hand). The origin of the fixed frame serves as the rotation center of the wrist. Two kinematic chains are defined: one kinematic chain (CK_{w1}), $\underline{R}R$ type, actuated by q_3, with a passive revolute joint R_1' and a mechanical link *d* which links R_1 with the origin of the mobile frame $X^{*'}Y^{*'}Z^{*'}$ (at a distance *r*); the second kinematic chain (CK_{w2}) actuated by q_4 rotates the circular link *e* around the O^*Y^* axis of the fixed frame. The circular link *e* of CK_{w2} intersects with the link *d* (of CK_{w1}) through two rotation joints: R'_2 (which slides on the circular link *e*); and R'_3 (around the $O^{*'}X^{*'}$ of the moving frame). The wrist module performs flexion/extension of the hand actuated by q_3, and abduction/adduction actuated by q_4.

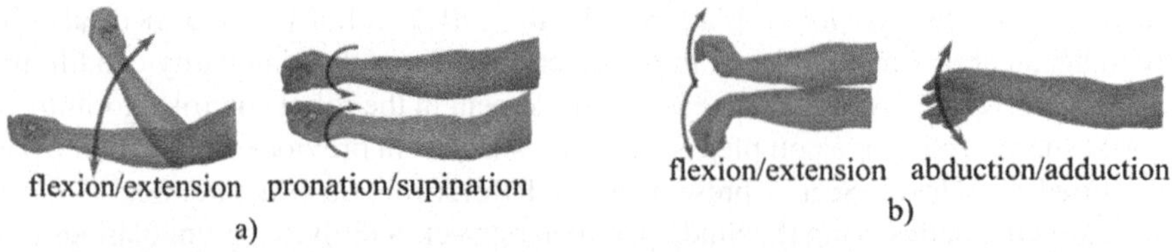

Fig. 1. Upper limb rehabilitation exercises. (a) elbow motions; (b) wrist motions.

Table 1. Motion ranges for the upper limb rehabilitation exercises

	Flex. (°)	Extens. (°)	Abd. (°)	Add. (°)	Pron. (°)	Supin. (°)
Elbow	~134	0	–	–	~74	~84
Wrist	~40	~55	~71	~79	–	–

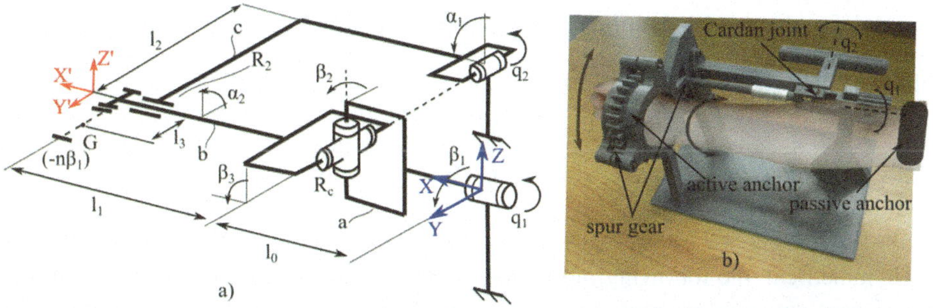

Fig. 2. The elbow rehabilitation module; (a) kinematic scheme; (b) 3D printed model

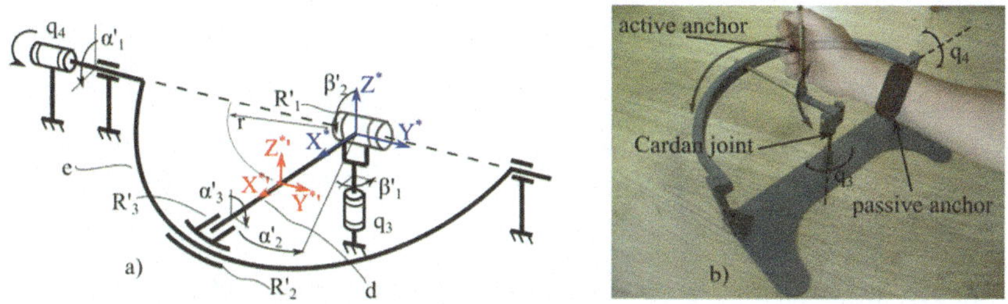

Fig. 3. The wrist rehabilitation module; (a) kinematic scheme; (b) 3D printed model.

3 Forward Kinematics

The kinematics of the PaRReEx robotic system was previously investigated in [2], using the vector method leading to analytic solutions. Since the algebraic method is based on quaternion representation of Euclidean displacement, parametric singularities (e.g. gimbal locks) are avoided. The algebraic method is used in this work to perform a rigorous singularity analysis to develop a robust control for a fail-safe robotic system (the method has been used successfully for a complete singularity analysis for another medical robot [6]).

The constraints of the kinematic chains are written in 4×4 homogenous matrix form by multiplying the transformation matrices (using the Denavit-Hartenberg convention) mapping the kinematic chains from the origin of the fixed frame to the origin of the moving frame. The elbow rehabilitation module kinematic chains (CK_{e1} and CK_{e2}) have the following constraints (Fig. 2):

$$K_{e1} = R_x(\beta_1) \cdot T_x(l_0) \cdot R_z(\beta_2) \cdot R_y(\beta_3) \cdot T_x(l_1) \tag{1}$$

$$K_{e2} = T_x(l_0) \cdot R_y(\alpha_1) \cdot T_x(l_1) \cdot R_x(\alpha_2) \tag{2}$$

And for the wrist rehabilitation module kinematic chains, CK_{w1} and CK_{w2} (see Fig. 3):

$$K_{w1} = R_z(\beta'_1) \cdot R_y(\beta'_2) \cdot T_x(r) \tag{3}$$

$$K_{w2} = R_y(\alpha'_1) \cdot R_z(\alpha'_2) \cdot T_x(r) \cdot R_x(\alpha'_3) \tag{4}$$

Since the 3 × 3 rotation matrices from the constraints (K_{e1}, K_{e2}, K_{e3} and K_{e4}) are not symmetrical, the following ratios were used to extract the Study parameters from a 4 × 4 homogeneous transformation matrix (although there are 4 ways to write the x_i parameters (as detailed in [3]), the first ratio may be used as long as the rotation matrix is not symmetrical):

$$x_0 : x_1 : x_2 : x_3 = 1 + a_{11} + a_{22} + a_{33} : a_{32} - a_{23} : a_{13} - a_{31} : a_{21} - a_{12} \tag{5}$$

$$\begin{aligned} y_0 : y_1 : y_2 : y_3 = {} & \frac{(t_x x_1 + t_y x_2 + t_z x_3)}{2} : \frac{(-t_x x_0 - t_y x_3 + t_z x_2)}{2} \\ & : \frac{(t_x x_3 - t_y x_0 - t_z x_1)}{2} : \frac{(-t_x x_2 + t_y x_1 - t_z x_0)}{2} \end{aligned} \tag{6}$$

where a_{ij} are the components of a 3 × 3 orthogonal matrix, and $\mathbf{t}[t_x, \ t_y, \ t_z]^T$ is a translation. The set of eight Study parameters $x_0 : x_1 : x_2 : x_3 : y_0 : y_1 : y_2 : y_3$ are composed of rotation parameters (x_i) and displacement parameters (y_i), that must satisfy the normalizing condition $\left(x_0^2 + x_1^2 + x_2^2 + x_3^2 = 1\right)$, and the Study quadric ($x_0y_0 + x_1y_1 + x_2y_2 + x_3y_3 = 0$) [3]. Using Eqs. (5) and (6) the Study parameters are extracted from each constraint equations (Eqs. (1)–(4)). Furthermore, the trigonometric functions are substituted with the tangent of the half angle formulae $\{sin(\beta_i) = 2t_i^2/(1 + t_i^2); cos(\beta_i) = 1 - t_i^2/(1 + t_i^2); \quad sin(\alpha_i) = 2s_i^2/(1 + s_i^2); cos(\alpha_i) = 1 - s_i^2/(1 + s_i^2), sin(\beta'_i) = 2t_i'^2/(1 + t_i'^2); cos(\beta'_i) = 1 - t_i'^2/(1 + t_i'^2); \ sin(\alpha'_i) = 2s_i'^2/(1 + s_i'^2); cos(\alpha'_i) = 1 - s_i'^2/(1 + s_i'^2)\}$ reducing the Study parameters to rational form. After factoring the denominator (common to every Study parameter), and dividing through the greatest common divisor, the Study parameters for the constraints of the kinematic are reduced to polynomial form. For the elbow rehabilitation module the Study parameters are:

$$\begin{aligned} & x_0 : x_1 : x_2 : x_3 : y_0 : y_1 : y_2 : y_3 = 2t_1t_2t_3 + 2 : -2t_2t_3 + 2t_1 : -2t_1t_3 + 2t_2 : 2t_1t_2 + 2t_3 : \\ & (l_0 + l_1)(-t_2t_3 + t_1) : -(l_0 + l_1)(t_1t_2t_3 + 1) : (l_0 - l_1)(t_1t_2 + t_3) : (l_0 - l_1)(t_1t_3 - t_2) \end{aligned} \tag{7}$$

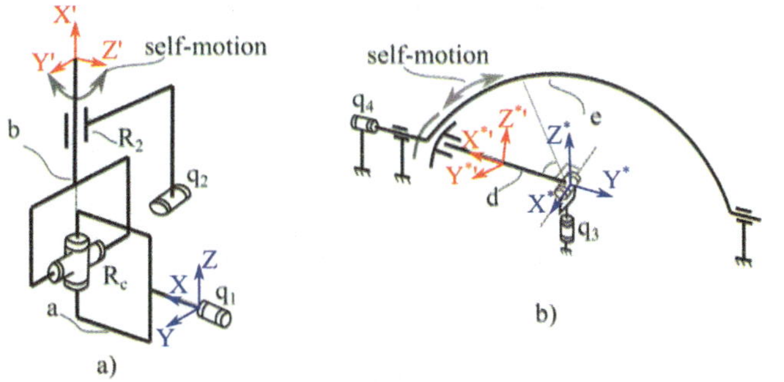

Fig. 4. Singularity poses; (a) elbow module singularity; (b) wrist module singularity.

$$\begin{aligned} & x_0 : x_1 : x_2 : x_3 : y_0 : y_1 : y_2 : y_3 = 2 : 2s_2 : 2s_1 : -2s_1s_2 : \\ & (l_0 + l_1)s_2 : -l_0 - l_1 : -(l_0 - l_1)s_1s_2 : -(l_0 - l_1)s_1 \end{aligned} \tag{8}$$

And for the wrist rehabilitation module:

$$x'_0 : x'_1 : x'_2 : x'_3 : y'_0 : y'_1 : y'_2 : y'_3 = 2 : -2t'_1t'_2 : 2t'_2 : 2t'_1 : -t'_2rt'_2 : -r : -rt'_1 : rt'_2 \tag{9}$$

$$\begin{aligned} & x'_0 : x'_1 : x'_2 : x'_3 : y'_0 : y'_1 : y'_2 : y'_3 = 2s'_1s'_2s'_3 - 2 : -2s'_1s'_2 - 2s'_3 : -2s'_2s'_3 - 2s'_1 : \\ & 2s'_1s'_3 - 2s'_2 : -(s'_1s'_2 + s'_3)r : -(s'_1s'_2s'_3 - 1)r : -(s'_1s'_3 - s'_2)r : -(s'_2s'_3 + s'_1)r \end{aligned} \tag{10}$$

The parameters t_1, s_1, t'_1, s'_1 are active joint parameters (being the tangents of the half angles of the active joints q_1, q_2, q_3 and q_4), while t_2, t_3, s_2, t'_2, s'_2, s'_3 are free rotation parameters, and l_0, l_1, r, are mechanical links lengths. To solve the forward kinematics of the two rehabilitation modules, each pair of constraints (Ke1 and Ke2 respectively Kw1 and Kw2) must have the same solution. The unknown variables from the constraints (the passive joint parameters) are eliminated by computing four lexdeg (elimination order) Groebner bases from the ideals of varieties generated by each of the eight equations describing the Study parameters in Eqs. (7)–(10). Let J_{e1} and J_{e2} denote the ideals for the elbow module constraints, and J_{w1} and J_{w2} denote the ideal for the wrist module constraints. The computed Groebner bases for the ideals J_{e1} and J_{e2} (for the elbow module) are G_{e1} (eliminating t_2, t_3) and G_{e2} (eliminating s'_2, s'_3) respectively, with the Study quadric contained in both bases (due to the length of the last polynomial $g_5 \in G_{e1}$ of degree 2, is presented only in general form):

$$\begin{aligned} G_{e1} = \langle & ((l_0 - l_1)x_3 - 2y_2, (l_0 - l_1)x_2 + 2y_3, (l_0 + l_1)x_1 \\ & - 2y_0, (l_0 + l_1)x_0 + 2y_1, g_5^2(t_1, y_0, y_1, y_2, y_3)\rangle \end{aligned} \tag{11}$$

$$\begin{aligned} G_{e2} = \langle & s_1(l_0 - l_1)y_1 - (l_0 + l_1)y_3, s_1(l_0 - l_1) + (l_0 + l_1)y_2, (l_0 - l_1)x_3 \\ & - 2y_2, (l_0 - l_1)x_2 + 2y_3, s_1(l_0 - l_1)x_1 + 2y_2, s_1(l_0 - l_1)x_2 + 2y_3\rangle \end{aligned} \tag{12}$$

The Groebner bases for the ideals J_{w1} and J_{w2} are G_{w1} (eliminating s_2', s_3') and $G_{e2}(s_2', s_3')$ respectively (also with the Study quadric contained in both bases):

$$G_{w1} = \langle t_1'y_1' - y_2', t_1'y_3' + y_0', rx_3' + 2y_2', rx_2' - 2y_3', rx_1' + 2t_1'y_3', rt_1'x_0' + 2y_2' \rangle \tag{13}$$

$$\begin{aligned} G_{w2} = \langle & rx_3' + 2y_2', rx_2' - 2y_3', rx_1' - 2y_0', rx_0' + 2y_1', \\ & s_1y_0^2 + (s_1^2 - 1)y_0y_2 + s_1y_1^2 + (-s_1^2 + 1)y_1y_3 - s_1y_2^2 - s_1y_3^2 \rangle \end{aligned} \tag{14}$$

Finally, the forward kinematics is solved (for the two modules) by computing a pure lexicographic Groebner base from the union of the bases describing the constraints for the module kinematic chains (together with the normalizing condition). For the elbow module, G_e^* is computed from the union of G_{e1} and G_{e2}, with the monomial ordering $y_0 > y_1 > y_2 > y_3 > x_0 > x_1 > x_2 > x_3$ and the normalizing condition $x_0^2 + x_1^2 + x_2^2 + x_3^2 = 1$. The base G_e^* contains a univariate polynomial $P_e^4(x_3, t_1, s_1)$ in x_3 of degree 4, with 2 solutions for the forward kinematics (the other 2 resulting from solving G * e are due to the introduction of the normalizing condition in the computation). However, only one solution is practical, the second one being a rotated version of the first one by a value of π around the X′ axis. For the wrist module, a Groebner base G_w^* was computed in the same manner, from the union of G_{w1} and G_{w2} (with monomial ordering $y_0' > y_1' > y_2' > y_3' > x_0' > x_1' > x_2' > x_3'$ and the normalizing condition $x_0'^2 + x_1'^2 + x_2'^2 + x_3'^2 = 1$). The base G_w^* also has a univariate polynomial $(P_w^4(x_3', t_1', s_1'))$ of degree 4 (describing 2 solutions for the forward kinematics with only one practical one). To find solutions for the forward kinematics the bases G_e^* and G_w^* must be evaluated with numerical values for the active joint parameters and the robot geometrical parameters. This in turn yields the numerical solutions for the Study parameters (which may be written back into a 4 × 4 matrix form to describe the total Euclidean displacement).

4 Singularities

The singularities for the PaRReEx robotic system are determined by the configurations where two or more solutions of the forward kinematics coincide (approach also described in [9]). The vanishing conditions of the discriminants of the two univariate polynomials $P_e^4 \in G_e^*, P_w^4 G_w^*$ are determined by computing the resultants $\mathrm{Res}_e(P_e^4, dP_e^4/dx_3)$ and $\mathrm{Res}_w(P_w^4, dP_w^4/dx_3')$ eliminating the last Study parameter. For the elbow rehabilitation module, a resultant $\mathrm{Res}_e = P_1 \cdot P_2 \cdot P_3 \cdot P_4$ has been computed with the factors P_i being also products of factors:

$$P_1 = s_1^4 t_1^2 (s_1^2 + 1) \tag{15}$$

$$P_2 = s_1^8 (t_1 - 1)^4 (t_1 + 1)^4 (s_1^2 + 1)^2 (s_1 - 1)^4 (s_1 + 1)^4 \tag{16}$$

$$P_{3,4} = (s_1^2 t_1^2 \mp 2s_1 t_1^2 + s_1^2 + t_1^2 \pm 2s_1 + 1)^4 \quad (17)$$

Equations (15)–(17) are not sufficient to determine whether or not the robotic system is in singular pose when the conditions are met (the factors vanish). The factors of the polynomial must be substituted in the original Groebner bases defining the constraints of the kinematic chains (G_{e1}, G_{e2}) and then the forward kinematics must be computed (the bases G_e^*) to check whether it yields double solution or not. The factors from Eq. (15) $t_1 = 0$, $s_1 = 0$ were substituted into the initial bases (G_{e1}, G_{e2}) but did not provide double solution for the forward kinematics (these are not singularities); $s_1^2 + 1 = 0$ has no real solution and it is not of interest. The factors $t_1 \pm 1 = 0$ (from Eq. (16)) were also substituted into the original bases without double solution on the forward kinematics; the substitution with $s_1 \pm 1 = 0$ yielded double solution only when the bases were also evaluated with $t_1 = 0$ (describing two singular poses). The same singularities are revealed by Eq. (17); solving P_3 and P_4 for t_1 yields $t_1 = \pm i(s_1 - 1)/(s_1 + 1)$, and $t_1 = \pm i(s_1 + 1)/(s_1 - 1)$ respectively. It is easy to see that Eq. (17) has real solutions only when $s_1 = \pm 1$ while $t_1 = 0$. This singular configuration describes a self-motion around the moving coordinate frame X′ axis when it points vertically ($q_2 = \pm\pi/2$) while $q_1 = 0$. Geometrically, this singularity is due to the alignment of one axis of rotation of the Cardan join R_c with the rotation axis of the revolute joint R_2 as illustrated in Fig. 4a.

For the wrist rehabilitation module the singularities are computed in the same manner. The resultant $Res_w = P'_1 \cdot P'_2 \cdot P'_3 \cdot P'_4 \cdot P'_5$ has the factors P'_l also a products of factors:

$$P'_1 = s_1'^2 t_1'^4 (t'_1 - 1)^2 (t'_1 + 1)^2 \quad (18)$$

$$P'_2 = (t_1'^2 + 1)^4 \quad (19)$$

$$P'_3 = t_1'^8 (t_1'^2 + 1)^8 (s'_1 - 1)^4 (s_1 + 1)^4 \quad (20)$$

$$P'_{4,5} = ((s_1'^2 + 1) t_1'^2 \pm 4 s'_1 t'_1 + s_1'^2 + 1)^4 \quad (21)$$

The factors from Eq. (18) ($s'_1 = 0, t'_1 = 0$) were substituted into (G_{w1}, G_{w2}) and did not yield double solution for the forward kinematics; the substitution with $t'_1 \pm 1 = 0$ yielded double solution only when the original bases were also evaluated with $s'_1 = \pm 1$ (resulting in a singularity). The same result is obtained from Eq. (21); solving for t'_1 yields $t'_1 = \pm(is_1'^2 - i \mp 2s_1)/(s_1'^2 + 1)$ as solution for P'_4 and $t'_1 = \pm(is_1'^2 - i \pm 2s_1)/(s_1'^2 + 1)$ as solution for P'_5, which have real values only if $s'_1 = \pm 1$ (leaving ± 1 the only possible values for t'_1). This singularity occurs when both active joints q_3, q_4 are actuated with $\pm\pi/2$ resulting in a self-motion of the moving frame around the X^* axis of the fixed frame. Geometrically this singularity occurs when the rotation axis of the revolute joint R'_1 is normal to the plane containing the curved link e (for this mechanism this is possible only if the link e is contained in Y^*Z^* plane). Equation (19) has no real solutions and is not of interest. The factor $t_1'^2 + 1 = 0$ (Eq. (20)) has no real solutions; substituting $s_1' \pm 1 = 0$ in the original bases does not yield double solution for the forward kinematics.

The elbow rehabilitation module is mounted on the upper limb with the elbow aligned with the Cardan joint R_c, the arm aligned with the link *a*, and the forearm aligned with the link *b*. According to Table 1 (from Sect. 2) the motion amplitude of flexion/extension of the forearm around the elbow is approx. [0°, 135°], while the motion of pronation/supination of the forearm is approx. [–74°, 83°]. Due to the constraints of the human upper limb kinematics, the only relevant singularity of the elbow module occurs when $q_1 = 0$ and $q_2 = -\pi/2$ (the pose is illustrated in Fig. 4(a) allowing self-motion of the forearm (in the form of pronation supination). This configuration should be avoided by imposing stroke limits (in the robot control) for the active joint q_2. The human wrist motion amplitude is approx. [–71°, 79°] for flexion/extension and [40°, 55°] for abduction/adduction. The wrist module should be mounted with its fixed frame origin near the wrist (with the active anchor point at the origin of the moving frame). Due to the limited motion amplitude of the wrist, it should never reach the singularities of the mechanism (which are at $q_3 = \pm\pi/2$, $q_4 = \pm\pi/2$).

5 Conclusion

The singularities in the active joint space of the PaRReEx modular robotic system were computed using Study parameters of SE(3). For the elbow module the singularities are carefully correlated with the rehabilitation exercises and will be implemented in the control algorithm to assure a safe use for the rehabilitation exercises. For the wrist module, in normal exercise condition the human hand should not reach a singular configuration of the robot due to the lower motion amplitude (of the hand), thus only the motion amplitude limits have to be implemented in the robot control.

Acknowledgement. The paper presents results from the research activities of the project ID 37_215, MySMIS code 103415 "Innovative approaches regarding the rehabilitation and assistive robotics for healthy ageing" co-financed by the European Regional Development Fund through the Competitiveness Operational Programme 2014–2020, Priority Axis 1, Action 1.1.4, through the financing contract 20/01.09.2016, between the Technical University of Cluj-Napoca and ANCSI as Intermediary Organism in the name and for the Ministry of European Funds.

References

1. Robotics 2020: Multi-Annual Roadmap for Robotics in Europe, Release B 02/12/2016
2. Gherman, B., et al.: Kinematic design of a parallel robot for elbow and wrist rehabilitation. In: The 12th IFToMM International Symposium on SYROM, Romania, 02–03 November 2017
3. Husty, M., et al.: Algebraic methods in mechanism analysis and synthesis. Robotica **25**(6), 661–675 (2007)
4. Schadlbauer, J., et al.: The 3-RPS parallel manipulator from an algebraic viewpoint. Mech. Mach. Theor. **75**, 161–176 (2014)
5. Vaida, C., et al.: Kinematic analysis of an innovative medical parallel robot using study parameters. Mechanisms and Machine Science, vol. 39, pp. 85–99. Springer, Cham (2016)

6. Schadlbauer, J., et al.: A complete analysis of singularities of a parallel medical robot. Advances in Robot Kinematics 2016, vol. 4, pp. 81–89. Springer, Cham (2018)
7. Gherman, B., et al.: A family of robotic systems for the rehabilitation of the upper limb. Patent Pending A00375/14.16.2017
8. Major, K.A., et al.: Ranges of motion as basis for robot-assisted post-stroke rehabilitation. Int. J. Bioflux Soc. Hum. Vet. Med. **8**(4), 192–196 (2016)
9. Srivatsan, R.A., Bandyopadhyay, S.: Analysis of constraint equations and their singularities. Advances in Robot Kinematics, pp 429–436. Springer, Cham (2014)

Kinematic Constraint Maps and C-space Singularities for Planar Mechanisms with Prismatic Joints

Seyed Vahid Amirinezhad(✉) and Peter Donelan

Victoria University of Wellington, Wellington 6012, New Zealand
{vahid.amirinezhad,peter.donelan}@sms.vuw.ac.nz

Abstract. Singularities of the configuration space for planar parallel mechanisms involving prismatic joint are investigated using a kinematic constraint map approach. Eliminating pose parameters from the singularity conditions together with the constraint equations leads to Grashof-type conditions. We illustrate the method by the planar RRRP and 3-RPR parallel mechanisms. In the latter case, conditions for higher order singularities are also determined.

Keywords: Kinematic constraint · Prismatic joint
C-space singularity · Grashof condition

1 Introduction

A mechanism singularity physically corresponds to a configuration where an instantaneous change occurs in its *degrees of freedom* (dofs). Mathematically, this is determined by a drop in rank of the Jacobian matrix of some kinematic mapping. Among the various types of kinematic singularities, those of the configuration or C-space are the most basic since at such a singularity there is no regular parametrisation that can be used to define inverse or forward kinematic singularities. One goal of this paper is to extend an approach introduced in [1] which is, in principle, unified across all types of manipulators and that takes into account the impact on singularities of design parameters in a given family of mechanisms.

The approach to singularities of parallel mechanisms due to Gosselin and Angeles [7] was based on combining input and output variables in an implicit (constraint) function. Zlatanov *et al.* [13] observed that it may not be possible to capture the full kinematics with just these limited variables—in particular, passive joints may play a role, giving rise to a greater variety of singularity classes and providing an explanation for unexpected motion observed in some mechanisms such as the 3UPU wrist [2]. However, this approach was purely local and did not provide global singularity information for a given mechanism. Park and Kim [9,10] developed an approach based on differential geometry that

J. Lenarcic and V. Parenti-Castelli (Eds.): ARK 2018, SPAR 8, pp. 212–220, 2019.
https://doi.org/10.1007/978-3-319-93188-3_25

enabled the identification of configuration space singularities, as well as actuator and end-effector singularities.

In [1], the authors introduced an approach based solely on joint constraints to define the C-space. This approach was illustrated by planar parallel mechanisms whose joints are revolute kinematic pairs. The goal was to provide a robust framework for kinematic singularities that takes into account the effect of design parameters. In particular, we derived Grashof-type conditions that define those choices of design parameters for which the C-space has a singularity.

In this paper, we extend the approach to include planar mechanisms with prismatic joints. We review the notion of *kinematic constraint map* (KCM) as the tool for defining the C-space. We then analyse two examples, the RRRP close 4-bar and the 3-RPR planar parallel mechanism and determine the associated Grashof-type conditions for C-space singularity in terms of design parameters. We also determine the additional conditions that result in a corank 2 singularity for the KCM of the 3-RPR.

2 Kinematic Constraints

With respect to a fixed reference frame R giving ambient space coordinates and a frame M in a rigid body moving in the plane, three *pose parameters* $(\theta, x, y)^T \in \mathbb{R}^3$ are necessary and sufficient to express the position and orientation of the M with respect to R. They represent an element of the Euclidean group $(A, \mathbf{a}) \in SE(2)$ (called the *absolute displacement* [11]), where A is a 2×2 matrix representing rotation by θ and $\mathbf{a} = (x, y)^T$ is the translation vector. Note that θ is unique only up to addition of multiples of 2π. The absolute displacement transforms coordinates $\mathbf{q}$ in M into reference frame coordinates $\mathbf{Q}$ by:

$$\mathbf{Q} = A\mathbf{q} + \mathbf{a}. \tag{1}$$

In a mechanism, suppose that two links $L_\pm$ are connected to each other by a joint, thereby constraining their relative motion. The two types of 1-dof joint in the plane are revolute (R) and prismatic (P). An R-joint permits rotation about a common fixed point while a P-joint permits translation along a common line. We express the corresponding constraints by equations on the pose parameters or, equivalently, the absolute displacements $(A_\pm, \mathbf{a}_\pm)$ of the joined links $L_\pm$.

The constraints imposed by an R-joint are given in [1]. This amounts to requiring the pose parameters for the two links to return the same coordinates when mapped to the reference frame. We follow [8] to obtain the constraint equations of a P-joint. The geometric object representing the joint is a line $\ell_\pm$ in each link given parametrically in the moving frames $M_\pm$ by, respectively:

$$\ell_\pm := \mathbf{p}_\pm + \tau_\pm \mathbf{r}_\pm, \tag{2}$$

where $\mathbf{p}_\pm$ is a point on the line and $\mathbf{r}_\pm$ is its direction.

The constraints imposed by the P-joint are that in the reference frame, the direction vectors $\mathbf{r}_\pm$ of the lines must be parallel and also parallel to the vector connecting the points $\mathbf{p}_{\pm}$. In the reference frame, Eq. 1 the lines transform to:

$$\Lambda_{\pm} := (A_{\pm}, \mathbf{a}_{\pm})(\mathbf{p}_{\pm} + \tau_{\pm}\mathbf{r}_{\pm}) = \mathbf{P}_{\pm} + \tau_{\pm}(A_{\pm}\mathbf{r}_{\pm}), \tag{3}$$

where $\mathbf{P}_{\pm} = A_{\pm}\mathbf{p}_{\pm} + \mathbf{a}_{\pm}$. Hence, we require:

$$(A_{+}\mathbf{r}_{+} \times A_{-}\mathbf{r}_{-})_z = 0, \qquad ([\mathbf{P}_{+} - \mathbf{P}_{-}] \times A_{-}\mathbf{r}_{-})_z = 0 \tag{4}$$

Here, $\times$ denotes 3-dimensional cross product and we treat 2-vectors as 3-vectors by appending a third coordinate $z = 0$, and the subscript z is a reminder that each equation is essentially scalar, the other components of the cross product being zero.

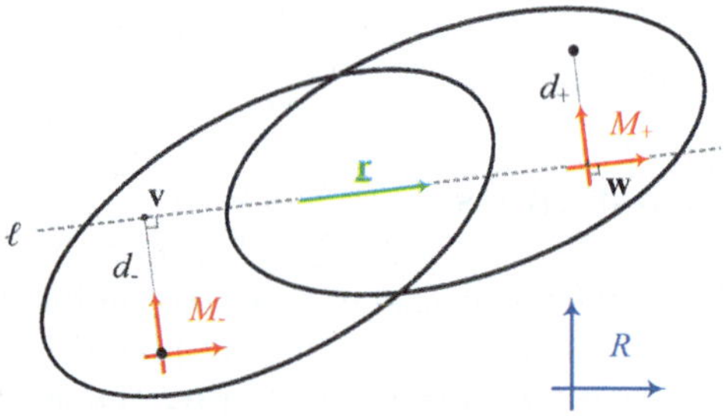

Fig. 1. A prismatic joint

It will simplify the constraint equations if we assign moving frames to the links as shown in Fig. 1 (cf. Denavit–Hartenberg parameters). Given that the frame M_- is already assigned, for M_+, choose the origin of coordinates on the line and set this to be $\mathbf{p}_+$; the x_+ axis lies along the line so that $\mathbf{r}_+ = (1, 0)^T$. Where necessary, choose the origin in M_+ so that the origin in a subsequent frame lies on the y_+ axis. The associated design parameter is the perpendicular distance d_- of the line from the origin in M_-.

Note that for a parallel mechanism, where a link is connected to more than one other, the choice of moving frame can only be made with respect to one of the joints and there will be additional design parameters relating to the second or subsequent joints.

3 Configuration Space and Singularities

A given mechanism consists of k rigid links connected by t kinematic pairs. Its *kinematic constraint map (KCM)* is the function F, whose components F_j are the constraint equations expressed in the form $F_j = 0$:

$$F : \mathcal{M} \times \mathcal{D} \longrightarrow \mathbb{R}^c \tag{5}$$

where $\mathcal{M} = [SE(2)]^{k-1}$ is its *pose space* (here we have assumed the reference frame is attached to a base link which is therefore immobile), $\mathcal{D} \subseteq \mathbb{R}^d$ its *design space* and $\mathbb{R}^c$ the *constraint space*. Here, d and c are the total number of design parameters and constraint equations, respectively, and if every joint has 1-dof, then $c = 2t$. Locally, $\mathcal{M}$ can be parametrised by $p = 3(k-1)$ pose parameters. We will assume that $c < p$, though in principle this is not required and the mechanism may be over-constrained.

For a fixed choice of design parameters $\delta \in \mathcal{D}$, the KCM is:

$$F_\delta : \mathcal{M} \longrightarrow \mathbb{R}^c \tag{6}$$

Denote by $\mathcal{C}_\delta := F_\delta^{-1}(\mathbf{0}_c) \subset \mathcal{M}$ the *configuration space* (C-space) of the specific mechanism determined by δ.

There is no need to include 'loop' constraints since these are algebraically a consequence of the joint constraint equations. From the form of the constraint equations, $\mathcal{C}_\delta$ is an analytic variety. If $\mathbf{0}_c$ is a regular value of F_δ, so that at every point of $\mathcal{C}_\delta$ the derivative of F_δ has maximum rank, then $\mathcal{C}_\delta$ is a smooth manifold (no singularities).

Our aim is to determine conditions on the pose and design parameters under which the C-space is, in fact, singular. Specifically, we first determine where the $c \times p$ Jacobian matrix $JF_\delta = [\partial f_j / \partial u_i]$ of first-order partial derivatives of the KCM in Eq. 6 with respect to pose parametrisations has rank $< c$. Here $i = 1, \dots, p$ indexes the pose parameters u_i, and $j = 1, \dots, c$ indexes the joint constraint equations. At a regular point $\mathbf{u}_r \in \mathcal{M}$, the Jacobian has full rank—its cokernel (the quotient of the range by the image) is trivial or dimension zero. At a singular point $\mathbf{u}_s \in \mathcal{M}$, this dimension will be a positive quantity κ, the *corank* or rank deficiency. κ provides a useful measure of how bad a singularity is and we can partition the singularity set:

$$\Sigma F_\delta = \{\mathbf{u}_s \in \mathcal{M} \mid \operatorname{corank} JF_\delta > 0\} \tag{7}$$

into subsets $\Sigma^\kappa F_\delta$ corresponding to corank $\kappa = 1, \dots, \mu = p - c$, the theoretical global mobility.

We are only interested in those singular points $\mathbf{u}_s$ that lie in the C-space. The second step is to solve the singularity conditions simultaneously with the constraint equations and to eliminate pose parameters to give a condition, solely involving design parameters, under which the C-space is singular. We call these *Grashof-type conditions*, following the terminology of Gibson *et al.* [3–5].

4 Planar RRRP Linkage

The first example is the closed loop linkage consisting of $k = 4$ rigid links connected by three R- and one P-joint, so that $t = 4$ also. Assume one of the links connected by the P-joint to be fixed and label this link 0. The reference frame R and the moving frames $M_i, i = 1, 2, 3$ are shown in Fig. 2. The free motion of each remaining link is represented by pose parameters $(\theta_i, x_i, y_i)^T; i = 1, 2, 3$. Design

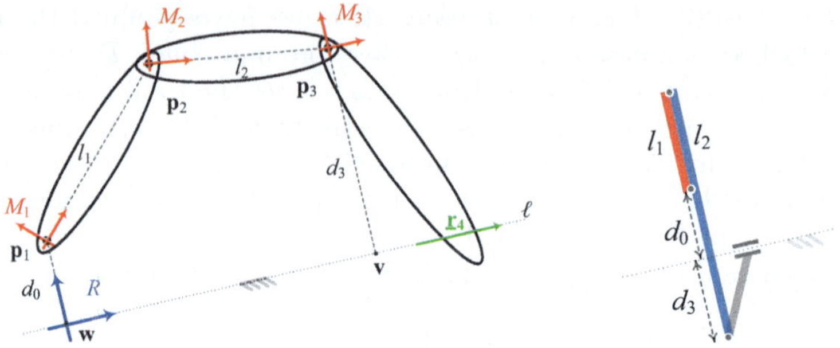

Fig. 2. RRRP linkage and C-space singularity configuration

parameters are lengths l_1, l_2 of the first and second links and the perpendicular distances d_0, d_3 in the other two.

Symbolically, the three R-joints give rise to (vector) constraint equations of the form:

$$A_i \mathbf{p}_i^i + \mathbf{a}_i = A_{i+1} \mathbf{p}_i^{i+1} + \mathbf{a}_{i+1} \tag{8}$$

where $i = 0, 1, 2$ and the joints are numbered in sequence with the links. The $\mathbf{p}_i^j$ denotes the coordinates of joint i with respect to frame M_j. The prismatic joint gives rise to two equations of the form Eq. 4.

In terms of the given choice of coordinates, the eight constraint equations give rise to the KCM $F : SE(2)^3 \times \mathcal{D} \mapsto \mathbb{R}^8$ where $\mathcal{D} \subseteq \mathbb{R}^4$ of the form:

$$\begin{aligned}(\theta_i, x_j, y_j)_{i=1,3} \longmapsto [&-x_1, d_0 - y_1, l_1 c_1 + x_1 - x_2, l_1 s_1 + y_1 - y_2, \\ &l_2 c_2 + x_2 - x_3, l_2 s_2 + y_2 - y_3, -s_3, d_3 + s_3 x_3 - c_3 y_3]^T\end{aligned} \tag{9}$$

where $c_i = \cos\theta_i$ and $s_i = \sin\theta_i, i = 1, 2, 3$. The positive corank condition on the resulting 8×9 Jacobian matrix can be reduced, by using row operations and deleting rows and columns consisting leading 1s, to considering the following 4×5 matrix:

$$JF_\delta^{\text{red}} = \begin{bmatrix} -1 & 0 & -l_1 s_1 & -l_2 s_2 & 0 \\ 0 & -1 & l_1 c_1 & l_2 c_2 & 0 \\ 0 & 0 & 0 & 0 & -c_3 \\ s_3 & -c_3 & 0 & 0 & c_3 x_3 + s_3 y_3 \end{bmatrix} \tag{10}$$

The singular set ΣF_δ is defined by the five 4×4 minors vanishing simultaneously and the equations yield:

$$\theta_1 = \theta_2 + n_1 \pi, \ \ \theta_2 = \theta_3 + \frac{\pi}{2} + n_2 \pi, \qquad n_1, n_2 \in \mathbb{Z} \tag{11}$$

Geometrically, the three R-joints must be collinear and perpendicular to the direction of the P-joint i.e. the first two links must be perpendicular to the translational axis of the prismatic joint.

Substituting Eq. 11 into Eq. 9 and eliminating $x_i, y_i, i = 1, 2, 3$, we obtain the Grashof-type condition:

$$d_0 \pm l_1 \pm l_2 \pm d_3 = 0 \tag{12}$$

The corresponding C-space singularity is illustrated in Fig. 2. This is the simplest case of the conditions found by Xue *et al.* [12] determining rotatability of components of closed chains with prismatic joints.

5 3-RPR Planar Parallel Manipulator

A fully general 3-RPR planar PM is shown in Fig. 3, along with a choice of moving and reference frames and design parameters. There are $k = 7$ (mobile) links hence 21 pose parameters, $t = 9$ joints and 12 design parameters, namely $d_i, i = 1, \ldots, 6$, base parameters b_1, b_2, β, and the platform parameters a_1, a_2, α. We make the non-triviality assumptions that $a_i, b_i > 0,\ i = 1, 2$. The local coordinates of all joints in the relevant moving frames are:

$$\begin{array}{llll} \mathbf{p}_1^R = (0,0)^T, & \mathbf{p}_2^R = (b_1,0)^T, \ \mathbf{p}_3^R = (b_{2x}, b_{2y})^T, & \mathbf{p}_i^- = (0, d_{i-3})^T; & i = 7,8,9 \\ \mathbf{p}_7^7 = (0,0)^T, & \mathbf{p}_8^7 = (a_1,0)^T, \ \mathbf{p}_9^7 = (a_{2x}, a_{2y})^T, & \mathbf{p}_i^+ = (0,0)^T; & i = 1,2,3 \\ \multicolumn{3}{l}{\mathbf{v}_i^- = (0, d_{i-3})^T, \ \mathbf{w}_i^+ = (0,0)^T, \ \underline{\mathbf{r}}_i^- = \underline{\mathbf{r}}_i^+ = (1,0)^T;} & i = 4,5,6 \end{array} \tag{13}$$

Each joint imposes a pair of constraint equations giving rise to a KCM of the form of $F : SE(2)^7 \times \mathcal{D} \mapsto \mathbb{R}^{18}$ where $\mathcal{D} \subset \mathbb{R}^{12}$:

$$\begin{aligned} (\theta_i, x_i, y_i)_{i=1,\ldots,7} \mapsto [&-x_1, -y_1, b_1 - x_2, -y_2, -x_3 + b_2 c_\beta, -y_3 + b_2 s_\beta, x_4 - x_7 - d_4 s_4, \\ &y_4 - y_7 + d_4 c_4, x_5 - x_7 - a_1 c_7 - d_5 s_5, y_5 - y_7 + d_5 c_5 - a_1 s_7, \\ &x_6 - x_7 - a_2 c_{\alpha,7} - d_6 s_6, y_6 - y_7 + d_6 c_6 - a_2 s_{\alpha,7}, s_{1,4}, s_{2,5}, s_{3,6}, \\ &d_1 + y_{1,4} c_1 - x_{1,4} s_1, d_2 + y_{2,5} c_2 - x_{2,5} s_2, d_3 + y_{3,6} c_3 - x_{3,6} s_3]^T \end{aligned} \tag{14}$$

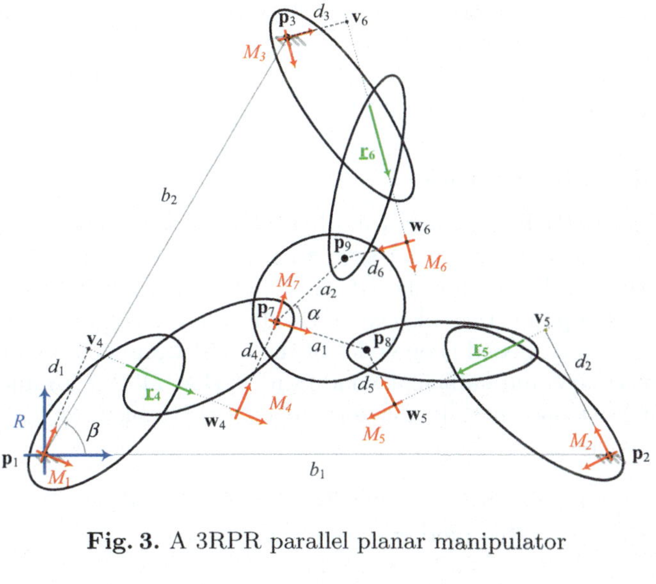

Fig. 3. A 3RPR parallel planar manipulator

where $s_{i,i+3} = \sin(\theta_i - \theta_{i+3})$, $x_{i,i+3} = x_i - x_{i+3}$, $y_{i,i+3} = y_i - y_{i+3}; i = 1, 2, 3$, $c_{\alpha,7} = \cos(\alpha + \theta_7)$, and $s_{\alpha,7} = \sin(\alpha + \theta_7)$.

The associated Jacobian is a 18×21 matrix. Rather than solving first for rank deficiency, we can eliminate pose parameters from via the KCM; clearly from $s_{i,i+3} = 0$, $\theta_i = \theta_i + 3 \mod \pi$ for $i = 1, 2, 3$. From this, for the purposes of rank computation the Jacobian can be reduced to the 3×6 matrix, JF_δ^{red}:

$$\begin{bmatrix} x_{1,4}c_1 + y_{1,4}s_1 & 0 & 0 & -s_1 & c_1 & 0 \\ 0 & x_{2,5}c_2 + y_{2,5}s_2 & 0 & -s_2 & c_2 & a_1 c_{2,7} \\ 0 & 0 & x_{3,6}c_3 + y_{3,6}s_3 & -s_3 & c_3 & a_2 c_{3,7,\alpha} \end{bmatrix} \tag{15}$$

where $c_{2,7} = \cos(\theta_2 - \theta_7)$ and $c_{3,7,\alpha} = \cos(\theta_3 - \theta_7 - \alpha)$.

5.1 Corank 1 Singularities

The conditions for a singularity are the vanishing of 20 3×3 minors of Eq. 15. Assuming F is well behaved, dimensional considerations (see, for example [6]) indicate that the corank 1 singular set ought to have (at least) codimension 4, so that four conditions on pose parameters are needed. There are three cases that can readily be seen to result in singularity. Clearly one of the non-trivial terms in the first three columns must vanish. When two of these terms vanish, finding two further conditions is straightforward. For example, if the first two terms vanish then the condition for singularity is realised by:

$$x_{1,4}c_1 + y_{1,4}s_1 = x_{2,5}c_2 + y_{2,5}s_2 = 0, \quad s_{1,2} = 0, \quad c_{2,7} = 0. \tag{16}$$

Further elimination of pose parameters from Eq. 16 results in the following Grashof-type condition on design parameters:

$$d_1 \pm d_4 \pm b_1 \pm d_2 \pm d_5 \pm a_1 = 0. \tag{17}$$

Geometrically, this relates to the lower closed 2-RPR loop comprised of the base, two legs and platform. It requires the two platform and two base anchor points to be collinear while the direction of the prismatic joints must be perpendicular to these. Similar configurations for the other two loops also correspond to singular configurations. This is illustrated in Fig. 4.

5.2 Corank 2 Singularities

As for the planar 3RRR [1], there are other components of the singular set with higher codimension but we do not give an exhaustive treatment here. Rather, we examine the conditions for the KCM to have a corank 2 singularity in the C-space. At such a point, the cokernel of the Jacobian has dimension 2. This can only happen for very special geometries of the 3-RPR and is a high codimension condition. Necessary and sufficient conditions is that all 2×2 minors of Eq. 15 vanish but at least one entry is non-zero. There are 45 such minors leading to 7 conditions as follows:

$$\begin{aligned} &x_{1,4}c_1 + y_{1,4}s_1 = x_{2,5}c_2 + y_{2,5}s_2 = x_{3,6}c_3 + y_{3,6}s_3 = 0; \\ &\theta_1 = \theta_2 = \theta_3 \mod \pi; \quad c_{2,7} = c_{3,7,\alpha} = 0. \end{aligned} \tag{18}$$

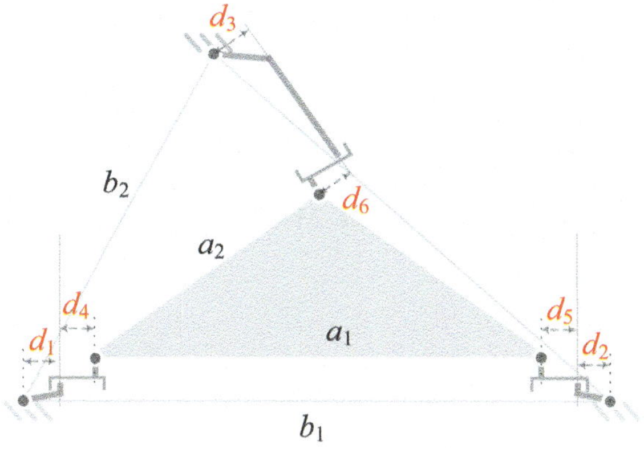

Fig. 4. A corank 1 C-space singularity configuration for 3-RPR

This is simply the union of the three closed-loop conditions giving corank 1 singularities. Therefore the three Grashof-type conditions Eq. 17 must hold simultaneously and these entail $\alpha = \beta = 0 \mod \pi$. These 5 equations correspond to four independent conditions.

References

1. Amirinezhad, S.V., Donelan, P.: Kinematic constraint maps, C-space singularities and generalised Grashof conditions. In: 2017 IDETC/CIE Conference. ASME (2017)
2. Di Gregorio, R.: Kinematics of the 3-UPU wrist. Mech. Mach. Theor. **38**(3), 253–263 (2003)
3. Gibson, C.G., Marsh, D.: On the linkage varieties of Watt 6-bar mechanisms-I. Basic geometry. Mech. Mach. Theor. **24**(2), 105–113 (1989)
4. Gibson, C.G., Newstead, P.E.: On the geometry of the planar 4-bar mechanism. Acta Applicandae Mathematicae **7**(2), 113–135 (1986)
5. Gibson, C.G., Selig, J.M.: Movable hinged spherical quadrilaterals-I. Mech. Mach. Theor. **23**(1), 13–18 (1988)
6. Golubitsky, M., Guillemin, V.: Stable Mappings and Their Singularities. Graduate Texts in Mathematics. Springer, Heidelberg (2012)
7. Gosselin, C., Angeles, J.: Singularity analysis of closed-loop kinematic chains. IEEE J. Robot. Autom. **6**(3), 281–290 (1990)
8. Haug, E.J.: Computer Aided Kinematics and Dynamics of Mechanical Systems, vol. 1. Allyn and Bacon, Boston (1989)
9. Park, F.C., Kim, J.W.: Manipulability of closed kinematic chains. J. Mech. Des. **120**, 542–548 (1998)
10. Park, F.C., Kim, J.W.: Singularity analysis of closed kinematic chains. J. Mech. Des. **121**(1), 32–38 (1999)

11. Uicker, J.J., Ravani, B., Sheth, P.N.: Matrix Methods in the Design Analysis of Mechanisms and Multibody Systems. Cambridge University Press, Cambridge (2013)
12. Xue, C., Ting, K.L., Wang, J.: Mobility criteria of planar single-loop n-bar chains with prismatic joints. J. Mech. Robot. **3**(1), 011011 (2011)
13. Zlatanov, D., Fenton, R.G., Benhabib, B.: Identification and classification of the singular configurations of mechanisms. Mech. Mach. Theor. **33**(6), 743–760 (1998)

Transversality and Its Applications to Kinematics

Seyed Vahid Amirinezhad[1(✉)], Peter Donelan[1], and Andreas Müller[2]

[1] School of Mathematics and Statistics, Victoria University of Wellington, Wellington 6012, New Zealand
{vahid.amirinezhad,peter.donelan}@sms.vuw.ac.nz
[2] Institute of Robotics, Johannes Kepler University Linz, Linz 4040, Austria
a.mueller@jku.at

Abstract. Transversality is a mathematical concept, widely used in singularity theory, which generalises the idea of regularity of a function. One of its uses is in providing a method for certifying that a property of a given set of functions or mappings is *generic*, which is to say that it holds for almost all members of that set. Its application in kinematics is illustrated via a prototype for showing that in a given class of parallel mechanisms, for almost all choices of design parameter the configuration space is non-singular and hence a manifold.

Keywords: Transversality · Kinematic mapping · Design parameter
Genericity

1 Introduction

For a given robot mechanism architecture, actual models are determined by fixing a number of design parameters. For serial mechanisms, the Denavit–Hartenberg parameters are widely used while the product-of-exponentials formulation [2] provides an alternative parametrisation in terms of screw coordinates. Among parallel mechanisms, the simplest examples are planar 4-bars, for which the design parameters are the link lengths, while more complex examples such as a planar 3-RRR have higher-dimensional spaces of design parameters [1]. It is reasonable to ask then how the design parameters affect the kinematics of the mechanism. This may involve determining whether certain properties are *generic*, which is to ask whether the property holds for almost all choices of design parameter. At a more refined level, one may look for bifurcation conditions. These correspond to non-generic choices of parameter across which there may be a significant change of behaviour. The classic case is the condition of Grashof [6] which distinguishes between planar 4-bars in terms of rotatability of links and, implicitly between those whose configuration space (C-space) has one or two connected components [4,11]. It is worth noting, however, that the requirement of symmetry or generation of special motion may be precisely what

J. Lenarcic and V. Parenti-Castelli (Eds.): ARK 2018, SPAR 8, pp. 221–229, 2019.
https://doi.org/10.1007/978-3-319-93188-3_26

renders a choice of design parameters non-generic so that, from the engineering point of view, genericity may *not* be a desirable characteristic.

Transversality is a mathematical concept, generalising that of regularity, by which we can seek to answer such questions in general. In particular, establishing transversality for functions with respect to certain submanifolds will enable us deduce not only genericity results but also information about the dimensions of a hierarchy of singularity sets. For example, if transversality could be verified for a kinematic mapping with respect to submanifolds that encode the rank (or its complement, the corank) of derivatives—in effect, Jacobian matrices of partial derivatives—then the set of configurations at which there is a singularity of a given corank will be a submanifold of specified dimension. Although not used extensively, these ideas have appeared already in the kinematics literature in the work of Pai [9], Tchoń [10] and others. More recently, applications of transversality to the analysis of singularities of mechanisms and related notions of genericity have been explored [3,8].

In this paper, we first give a brief introduction to transversality and some of the important theorems it gives rise to. To illustrate its application to kinematics we review the classical planar 4R mechanism. More general consideration of Grashof-type conditions, based on transversality considerations is then illustrated in both negative and positive senses by the extended Watt 7R mechanism, which has two degrees of freedom.

2 Transversality of Mappings

The appropriate mathematical language for transversality is that of differential topology (see, for example, [7]). The objects are (differentiable) manifolds—spaces that are everywhere locally parametrised by a fixed number of parameters, the *dimension* of the manifold—and differentiable functions or mappings between them. This encompasses the familiar spaces in kinematics such as Lie groups (the Euclidean group) and circles, spheres and their products including tori.

We start with the definition of regularity of a given function at a point of its domain:

Definition 1. *Given a differentiable function $f : M \to N$ between manifolds and a point $y \in N$. If for every $x \in M$ for which $f(x) = y$ the derivative df_x of f at x has maximum rank then y is called a regular value and each such x is called a regular point.*

Note that, for computational purposes, the derivative can be represented by a Jacobian matrix $Jf(x)$, which can be computed with respect to any choice of parametrisation and whose rank is invariant. Properly, we are considering the underlying linear map that best approximates f in a neighbourhood of x; the Jacobian is the matrix representation of this linear map with respect to choices of basis for the tangent spaces at x and y.

A fundamental theorem of topology states that if y is a regular value, so that the image of T_xM under the Jacobian is the whole of $T_{f(x)}N$, then

$P = f^{-1}(y) \subset M$ is a submanifold (or possibly empty). Moreover, the codimension of P is the dimension of N, or equivalently $\dim(P) = \dim(M) - \dim(N)$. This generalises the familiar result of linear algebra concerning the dimension of the solution space of an equation $A\mathbf{x} = \mathbf{b}$.

A key motivation for transversality is to generalise this situation to $f^{-1}(Q)$ where $Q \subset N$ is a submanifold. The idea is that for regularity, the image of the Jacobian $Jf(x)$ must be the whole of $T_{f(x)}N$, the tangent space at the image $f(x) = y$. Since T_yQ already fills up a subspace of T_yN, we now only need the image of the Jacobian $Jf(x)$, where $f(x) \in Q$, to span a *complement* to T_yQ.

Definition 2. *Define f to be transverse to Q (written $f \pitchfork Q$) if for all $x \in M$ such that $f(x) \in Q$,*

$$df_x(T_xM) + T_{f(x)}Q = T_{f(x)}N \tag{1}$$

In this case, an extension of the previous theorem [7] states that $P = f^{-1}(Q)$ is a submanifold of M and $\operatorname{codim} P = \operatorname{codim} Q$.

A point $y \in N$ which is the image of at least one point $x \in M$ is a *critical value* of f if it is not a regular value. So for at least some point $x \in M$, with $f(x) = y$, rank of $Jf(x) < \min(\dim M, \dim N)$. Critical points are scarce in the following sense [7]:

Theorem 1 (Sard's Theorem). *The set of critical values of $f : M \to N$, which is a subset of N, has measure zero.*

Here, a subset of a manifold has *measure zero* if it can be covered via parametrisations by at most a countably infinite union of hypercubes in $\mathbb{R}^n$ whose total volume can be made arbitrarily small. The idea of the proof is that in the neighbourhood of any critical point, f reduces volumes by at least an order of magnitude—think of the image of an interval $[-r, r]$ under the function $f(x) = x^2$. A property of points in a set that is true except on a set of measure zero is said to hold *almost everywhere*. So the theorem states that for almost all $y \in N$, y is a regular value. A corollary of Sard's Theorem is the following [7]:

Theorem 2 (Parametric Transversality Theorem). *Given a family of functions $F : M \times D \to N$, where D is a manifold of parameters, then for each fixed $d \in D$ there is a function $F_d : M \to N$ where, for $x \in M$, $F_d(x) = F(x, d)$. If $Q \subset N$ is a submanifold and $F \pitchfork Q$ then for almost all $d \in D$, $F_d \pitchfork Q$.*

This has direct interpretations in kinematics. We could take M as the joint space for a family of mechanisms, D as the space of design parameters and N the workspace, for example the Euclidean group. In this case, Q may be a submanifold that characterises some desired geometry for the end-effector or platform such as that it has a fixed orientation.

Alternatively, in the terminology of kinematic constraint maps (KCMs) [1], let M be the space of pose parameters for a class of mechanisms with $\dim M = m$, D as the space of design parameters with $\dim D = \delta$ and $N = \mathbb{R}^c$ the constraint space, where c is the total number of constraint equations. Here, the role of Q

is more limited as there is typically only one relevant choice of values (usually zero) for each equation so that $Q = \{\mathbf{0}\}$.

This corresponds to a special case of Theorem 2, when $Q = \{y_0\} \subset N$, that is a single point or 0-dimensional submanifold. In this case, $F \pitchfork Q$ if and only if y_0 is a regular value of F, since the second term in the defining condition 1 reduces to $\{0\}$. That is, the Jacobian has maximum rank for all $(x, d) \in M \times D$ for which $F(x, d) = y_0$. In this case, we can conclude that for almost all $d \in D$:

$$C_d := F_d^{-1}(y_0) \subset M \tag{2}$$

is a C^∞ submanifold with $\dim C_d = m - c$ (or else an empty set). In the context of KCMs, C_d is the configuration space for the choice of design parameters, $d \in D$. We illustrate this result in the following sections.

3 Planar 4R Linkage

To illustrate the application of Theorem 2 we consider the property 'that the configuration space for a given class of mechanism be a manifold' is generic. The simplest interesting case is the planar 4R mechanism, illustrated in Fig. 1. We follow the approach of [1].

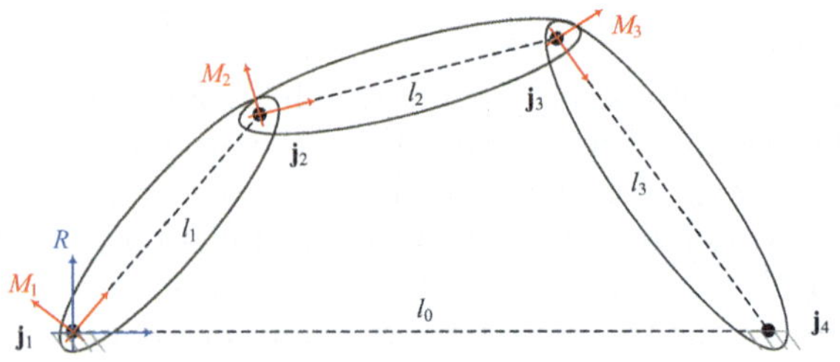

Fig. 1. 4-bar planar mechanism

There are three mobile links whose coordinates are given by moving frames $\mathcal{M}_1, \mathcal{M}_2, \mathcal{M}_3$ and a fixed link whose coordinates are expressed in the reference frame $\mathcal{R}$. The displacement of each moving frame with respect to the reference frame is represented by pose parameters $(\theta_i, x_i, y_i)^T, i = 1, \ldots, 3$. There are four design parameters $l_i, i = 0, \ldots, 3$ whilst the four revolute joints each impose two constraint equations. As a result, the KCM has the form $F : SE(2)^3 \times \mathbb{R}^4 \to \mathbb{R}^8$:

$$\begin{aligned} F(\theta_i, x_i, y_i, l_j)_{i=1,\ldots 3}^{j=0,\ldots 3} = [&x_1, y_1, l_1 c_1 + x_1 - x_2, l_1 s_1 + y_1 - y_2, \\ &l_2 c_2 + x_2 - x_3, l_2 s_2 + y_2 - y_3, l_3 c_3 + x_3 - l_0, l_3 s_3 + y_3]^T \end{aligned} \tag{3}$$

where $c_i = \cos\theta_i$ and $s_i = \sin\theta_i, i = 1, 2, 3$. The extended Jacobian of F with respect to both pose and design parameters is an 8×13 matrix. By row reduction and deletion of rows and columns with a leading 1, this has the same rank as the reduced 6×11 matrix:

$$JF^{\text{red}} = \begin{bmatrix} 1\,0\,0\,0\,0 & l_1 s_1 & 0 & 0 & -c_1 & 0 & 0 \\ 0\,1\,0\,0\,0 & -l_1 c_1 & 0 & 0 & -s_1 & 0 & 0 \\ 0\,0\,1\,0\,0 & l_1 s_1 & l_2 s_2 & 0 & -c_1 & -c_2 & 0 \\ 0\,0\,0\,1\,0 & -l_1 c_1 & -l_2 c_2 & 0 & -s_1 & -s_2 & 0 \\ 0\,0\,0\,0\,1 & l_1 s_1 & l_2 s_2 & l_3 s_3 & -c_1 & -c_2 & -c_3 \\ 0\,0\,0\,0\,0 & -l_1 c_1 & -l_2 c_2 & -l_3 c_3 & -s_1 & -s_2 & -s_3 \end{bmatrix} \tag{4}$$

This clearly has full rank since the last row cannot be identically zero since there is no solution to the simultaneous equations $\sin\theta = \cos\theta = 0$.

We have thus established that $F \pitchfork \{\mathbf{0}\}$ and thus for almost all $d \in \mathcal{D}$ the configuration space C_d has no singularity. Of course, in this case we already know that C_d is non-singular unless a certain equation relating to the Grashof condition holds for the design parameters, i.e. $l_0 \pm l_1 \pm l_2 \pm l_3 = 0$. This tells us directly that C_d is non-singular except on the closed subset of the positive octant defined by the union of these linear equations.

As an aside, we note that this Grashof-type condition can be found by solving the constraint equations $F = 0$ along with the equations that determine positive corank for its Jacobian JF, namely that all the degree 1 minors vanish simultaneously. Since all the equations are algebraic in x_i, y_i, c_i, s_i, $i = 1, 2, 3$ and subject to the additional conditions $c_i^2 + s_i^2 = 1$, we may use Gröbner bases to eliminate the pose parameters and obtain:

$$\begin{aligned} & l_0^8 - 4l_0^6 l_1^2 + 6l_0^4 l_1^4 - 4l_0^2 l_1^6 - 4l_0^6 l_2^2 + 4l_0^4 l_1^2 l_2^2 + 4l_0^2 l_1^4 l_2^2 + 6l_0^4 l_2^4 + 4l_0^2 l_1^2 l_2^4 \\ & \quad - 4l_0^2 l_2^6 - 4l_0^6 l_3^2 + 4l_0^4 l_1^2 l_3^2 + 4l_0^2 l_1^4 l_3^2 + 4l_0^4 l_2^2 l_3^2 - 40 l_0^2 l_1^2 l_2^2 l_3^2 + 4l_0^2 l_2^4 l_3^2 + 6l_0^4 l_3^4 \\ & \quad + 4l_0^2 l_1^2 l_3^4 + 4l_0^2 l_2^2 l_3^4 - 4l_0^2 l_3^6 + l_1^8 - 4l_1^6 l_2^2 + 6l_1^4 l_2^4 - 4l_1^2 l_2^6 + l_2^8 - 4l_1^6 l_3^2 + 4l_1^4 l_2^2 l_3^2 \\ & \quad + 4l_1^2 l_2^4 l_3^2 - 4l_2^6 l_3^2 + 6l_1^4 l_3^4 + 4l_1^2 l_2^2 l_3^4 + 6l_2^4 l_3^4 - 4l_1^2 l_3^6 - 4l_2^2 l_3^6 + l_3^8 = 0 \end{aligned} \tag{5}$$

which factorises as:

$$\begin{aligned} & (l_0 - l_1 - l_2 - l_3)(l_0 + l_1 - l_2 - l_3)(l_0 - l_1 + l_2 - l_3)(l_0 + l_1 + l_2 - l_3) \\ & \quad (l_0 - l_1 - l_2 + l_3)(l_0 + l_1 - l_2 + l_3)(l_0 - l_1 + l_2 + l_3)(l_0 + l_1 + l_2 + l_3) = 0 \end{aligned} \tag{6}$$

or, simply, the product of the linear Grashof-type conditions.

This union of 8 hyperplanes has codimension 1 as an algebraic variety in the design space. Furthermore, the factorisation yields non-generic mechanisms that includes some overconstrained cases. Consider the case $l_0 - l_1 - l_2 - l_3 = 0$, for instance; it is has a single shaky (infinitesimally mobile) configuration but is immobile.

4 Grashof-Type Conditions

It is reasonable to pose the question whether we can find Grashof-type conditions for singularity in the configuration space. We can undertake some dimensional analysis to indicate what is likely.

The set of points where the corank of the Jacobian matrix of a mapping $F : M \to N$ between manifolds is positive is called its *critical set*, denoted as follows:

$$\Sigma F := \{x \mid \operatorname{rank} JF(x) < \min(\dim M, \dim N)\} \tag{7}$$

This may be further refined by considering the sets of matrices of a fixed corank. Set $\dim M = m$, $\dim N = n$ and let us suppose that $m \geq n$; we are interested in the map $JF : M \to L(\mathbb{R}^m, \mathbb{R}^n)$ where the codomain is the set of linear functions between $\mathbb{R}^m$ and $\mathbb{R}^n$, i.e. $n \times m$ matrices. Define: $\Sigma^r = \{A \in L(\mathbb{R}^m, \mathbb{R}^n) \mid \operatorname{corank} A = r\}$. It can be shown [5] that these sets are submanifolds satisfying:

$$\operatorname{codim}(\Sigma^r) = r(m - n + r) \tag{8}$$

In the case that $JF \pitchfork \Sigma^r$, we would have $\Sigma^r F = JF^{-1}(\Sigma^r)$ a submanifold of M of the same codimension; in the simplest case, $r = 1$, this is $m - n + 1$.

Let us apply this to the setting of kinematic constraint maps. Here $F : M \times D \to \mathbb{R}^c$. Under the assumption of transversality on F, the *total* configuration space $C = F^{-1}(0) \subset M \times D$ has codimension c. For each $d \in D$ we have $C_d = C \cap (M \times \{d\})$, with the same codimension (provided it is a submanifold). Each C_d may have singularities and the simplest singularity set $\Sigma^1 F_d$ has codimension $m - c + 1$. Since the sum of the codimensions of C_d and $\Sigma^1 F_d$ in M is $(m - c + 1) + c > m$ we would not typically expect a given C_d to have any singularities.

On the other hand, in the total configuration space, the singularity set of interest is $\Sigma_1^1 F = J_1 F^{-1}(\Sigma^1)$, where $J_1 F : M \times D \to L(\mathbb{R}^m, \mathbb{R}^c)$ is the Jacobian map which treats D as the parameter space rather than as variables to be differentiated. Transversality then gives $\operatorname{codim} \Sigma_1^1 F = m - c + 1$. Under the additional assumption that C and $\Sigma_1^1 F$ meet transversely, their intersection is a submanifold satisfying:

$$\operatorname{codim}(C \cap \Sigma_1^1 F) = \operatorname{codim} C + \operatorname{codim} \Sigma_1^1 F = m + 1. \tag{9}$$

In other words, in $M \times D$, the singularity set is defined by $m + 1$ simultaneous (non-linear) equations. If it is possible to eliminate the m pose parameters we should arrive at just one equation in the d design parameters. If such a condition can be found, we call it the *Grashof-type condition* for the class of mechanisms. Note that there should be a single condition, independent of the number of links, joints or design parameters. This, of course, concurs with what we described in Sect. 3.

5 Extended Watt Mechanism

As shown in Fig. 2, this mechanism consists of six mobile links which can be considered as a combination of a planar 4-bar and a planar 5-bar mechanism that share common bodies 2 and 3. The frames $\mathcal{M}_i, i = 1, \ldots 6$ are assigned to the moving bodies whose poses are described by the parameters (θ_i, x_i, y_i), $i = 1, \ldots, 6$ measured in the reference frame $\mathcal{R}$ attached to fixed link. There are 11 design parameters $a_i, i = 0, \ldots, 6$, b_2, b_3, and α, β. It is assumed that all $a_i, b_i > 0$. Eight revolute joints each impose two constraint equations, and so the KCM has the form $F : SE(2)^6 \times \mathbb{R}^{11} \to \mathbb{R}^{16}$ given by:

$$
\begin{aligned}
[x_1, y_1, & -a_0 + x_2, y_2, -x_1 + x_3 - a_1c_1, -y_1 + y_3 - a_1s_1 - x_2 + x_4 - a_2c_2, \\
& - y_2 + y_4 - a_2s_2, -x_3 + x_5 - b_3c_{\beta,3}, -y_3 + y_5 - b_3s_{\beta,3}, -x_4 + x_6 - a_4c_4, \\
& - y_4 + y_6 - a_4s_4, -x_5 + x_6 - a_5c_5 + a_6c_6, -y_5 + y_6 - a_5s_5 + a_6s_6, \\
& -x_2 + x_3 - b_2c_{\alpha,2} + a_3c_3, -y_2 + y_3 - b_2s_{\alpha,2} + a_3s_3]^T . \qquad (10)
\end{aligned}
$$

Here $c_{\alpha,2} = \cos(\alpha + \theta_2)$, $c_{\beta,3} = \cos(\beta + \theta_3)$ and similarly for sines. For a fixed choice d of design parameters, we would anticipate the configuration space $C_d = F_d^{-1}(0)$ to have dimension 2, corresponding to its mobility.

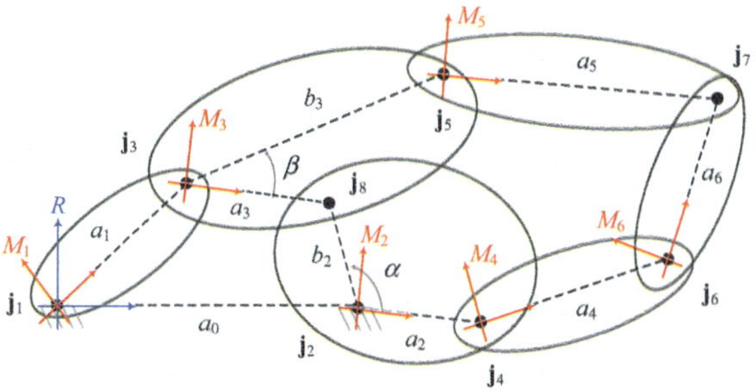

Fig. 2. Watt 7-bar mechanism

In order to determine whether this family of mechanisms is generic in the sense that its configuration space is a 2-dimensional manifold for almost all choices of design parameters, it is necessary to check whether the 16×29 Jacobian matrix of F (with respect to *all* parameters) drops rank. After some relevant row and column reductions, the matrix can be reduced to 3×16 as follows:

$$
\left[\begin{array}{cccccccccccccccc}
a_1c_1 & -a_2c_2 & b_3c_{\beta,3} & -a_4c_4 & a_5c_5 & -a_6c_6 & s_1 & s_{\beta,3} & b_3c_{\beta,3} & 0 & 0 & -s_2 & -s_4 & s_5 & 0 & -s_6 \\
0 & a_2s_2 - b_2s_{\alpha,2} & a_3s_3 - b_3s_{\beta,3} & a_4s_4 & -a_5s_5 & a_6s_6 & 0 & c_{\beta,3} & -b_3s_{\beta,3} & c_{\alpha,2} & -b_2s_{\alpha,2} & -c_2 & -c_4 & c_5 & -c_3 & -c_6 \\
-a_1c_1 & b_2c_{\alpha,2} & -a_3c_3 & 0 & 0 & 0 & -s_1 & 0 & 0 & s_{\alpha,2} & b_2c_{\alpha,2} & 0 & 0 & 0 & -s_3 & 0
\end{array}\right] \qquad (11)
$$

Given the assumptions on design parameters, this matrix is never rank deficient so that $F \pitchfork \{\mathbf{0}\}$, so that the total configuration space C is a submanifold of codimension 16.

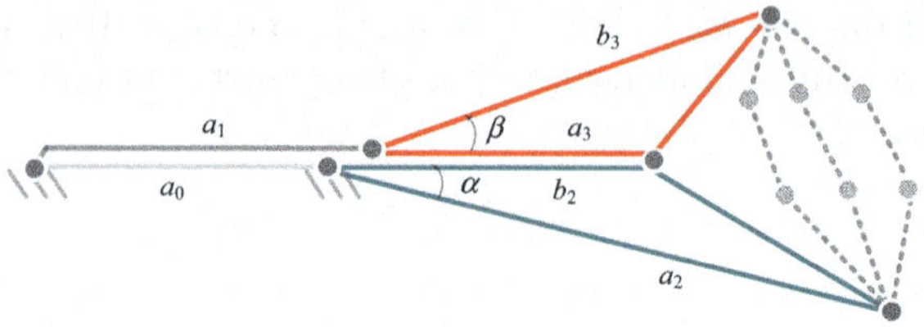

Fig. 3. Watt 7-bar $\mathcal{C}$-space singularity associated to the 4-bar at a flat configuration

Turning to the singularities, the Jacobian J_1F, after some relevant row/column reductions, reduces to a 4×6 matrix whose rank is the same:

$$\begin{bmatrix} a_1s_1 & -a_2s_2 & b_3s_{\beta,3} & -a_4s_4 & a_5s_5 & -a_6s_6 \\ -a_1c_1 & a_2c_2 & -b_3c_{\beta,3} & a_4c_4 & -a_5c_5 & a_6c_6 \\ -a_1s_1 & b_2s_{\alpha,2} & -a_3s_3 & 0 & 0 & 0 \\ a_1c_1 & -b_2c_{\alpha,2} & a_3c_3 & 0 & 0 & 0 \end{bmatrix} \tag{12}$$

Of the 15 4×4 minors, 12 are non-trivial and can be solved simultaneously using Mathematica or Maple to give distinct solutions determining the set $\Sigma_1^1 F$. From Eq. 8, it is expected that this set have codimension 3. However, one of the possible solutions is given by:

$$\theta_2 = \theta_1 - \alpha + n_2\pi, \qquad \theta_3 = \theta_1 + n_3\pi; \qquad n_2, n_3 \in \mathbb{Z} \tag{13}$$

defining submanifolds of codimension 2 in the space of pose parameters. Although there are other singular branches with various codimensions, this solution set relates simply to the base 4-bar linkage visible in Fig. 2. This tells us that there is a branch of singularities along which the Jacobian J_1F is not transverse to Σ^1. Nevertheless, since the conditions only relate to the 4-bar, we may still obtain the Grashof-type conditions:

$$a_0 \pm a_1 \pm b_2 \pm a_3 = 0. \tag{14}$$

as illustrated in Fig. 3. As mentioned, there are other singular branches, the most relevant one being of the form:

$$\theta_1 = g(\theta_2, \theta_3, \theta_4), \qquad \theta_5 = \theta_4 + n_5\pi, \qquad \theta_6 = \theta_4 + n_6\pi, \tag{15}$$

where g is specific function. Given this branch has the correct codimension, it is likely that it will give rise to a corresponding Grashof-type condition but the form of this has not yet been determined. One special case of Eq. 15 is given by adding conditions on the remaining pose parameters:

$$\theta_i = \theta_1 + \eta_i\pi, \quad i \in \{2,4,5,6\}, \quad \eta_i \in \mathbb{Z}; \qquad \theta_3 = \theta_1 - \beta + n\pi. \tag{16}$$

These correspond to the configuration where all the mobile links (i.e. all the x-axes of the moving frames) are parallel. Eliminating pose parameters leads to a collection of three Grashof-type conditions on the design parameters as follows:

$$a_0 \pm a_1 \pm a_2 \pm a_4 \pm a_5 \pm a_6 \pm b_3 = 0; \; a_0 \pm a_1 \pm b_2c_\alpha \pm a_3c_\beta = 0; \; b_2s_\alpha \pm a_3s_\beta = 0 \tag{17}$$

References

1. Amirinezhad, S.V., Donelan, P.: Kinematic constraint maps, C-space singularities and generalised Grashof conditions. In: 2017 IDETC/CIE Conference. ASME (2017)
2. Brockett, R.W.: Robotic manipulators and the product of exponentials formula. In: Mathematical Theory of Networks and Systems, pp. 120–129. Springer (1984)
3. Donelan, P.: Genericity conditions for serial manipulators. In: Lenarčič, J., Wenger, P. (eds.) Advances in Robot Kinematics: Analysis and Design, pp. 185–192. Springer, Dordrecht (2008)
4. Gibson, C.G., Newstead, P.E.: On the geometry of the planar 4-bar mechanism. Acta Applicandae Mathematicae **7**(2), 113–135 (1986)
5. Golubitsky, M., Guillemin, V.: Stable Mappings and Their Singularities. Graduate Texts in Mathematics. Springer, New York (2012)
6. Grashof, F.: Theoretische maschinenlehre, vol. 3. L. Voss, Hamburg (1890)
7. Guillemin, V., Pollack, A.: Differential Topology, vol. 370. American Mathematical Soc, Providence (2010)
8. Müller, A.: Generic mobility of rigid body mechanisms. Mech. Mach. Theor. **44**(6), 1240–1255 (2009)
9. Pai, D.K., Leu, M.C.: Generic singularities of robot manipulators. In: Proceedings of IEEE International Conference on Robotics and Automation, pp. 738–744. IEEE, May 1989
10. Tchoń, K., Muszynski, R.: Singular inverse kinematic problem for robotic manipulators: a normal form approach. IEEE Trans. Robot. Autom. **14**(1), 93–104 (1998)
11. Ting, K.L., Liu, Y.W.: Rotatability laws for N-bar kinematic chains and their proof. J. Mech. Des. **113**(1), 32–39 (1991)

Control and Dynamics

Lateral Stability of a 3-DOF Asymmetrical Spherical Parallel Manipulator with a Universal Joint Featuring Infinite Torsional Movement

Guanglei Wu[1(✉)] and Huiping Shen[2]

[1] School of Mechanical Engineering, Dalian University of Technology, Dalian 116024, China
gwu@dlut.edu.cn
[2] School of Mechanical Engineering, Changzhou University, Changzhou 213016, China
shp65@126.com

Abstract. In this paper, the dynamic lateral stability problem of an asymmetrical spherical parallel manipulator is studied by means of a monodromy matrix method. This manipulator adopts a universal joint as the ball-socket mechanism to support the mobile platform and to transmit the motion/torque between the input/output shaft. The linearized equations of motion of the mechanical system are established to analyze its stability according to the Floquet theory. The stable workspace region of the manipulator is identified and the critical rotating speed of the input shaft is detected.

Keywords: Spherical parallel manipulator · Universal joint
Dynamic stability · Infinite rotation

1 Introduction

Three degree-of-freedom (3-DOF) spherical parallel manipulators (SPMs) are dedicated for camera-orientating [6], minimally invasive surgical robots [8] and robotic joints [2] thanks to their large orientation workspace and high payload capacity. Besides, another potential application is that they can work as a tool head for complicated surface machining [15], where an unlimited torsion is needed in some common material processing such as milling or drilling. For this purpose, an asymmetrical SPM [15] as shown in Fig. 1 was proposed with a number of advantages compared to its symmetrical counterparts, such as enhanced positioning accuracy [13], infinite rotation [2], simple structure and low dynamic inertia [14], etc.. This manipulator adopts a universal (U) joint as the center of rotation supported by an input shaft, to generate an infinite input/output rotation transmission, which simplifies the manipulator architecture. On the other hand, the U joint suffers from one major problem, namely, it

J. Lenarcic and V. Parenti-Castelli (Eds.): ARK 2018, SPAR 8, pp. 233–241, 2019.
https://doi.org/10.1007/978-3-319-93188-3_27

transforms a constant input speed to a periodically fluctuating one, which may induce the vibrations and wear. This paper will investigate the dynamic stability problem for the SPM, focusing on the lateral stability.

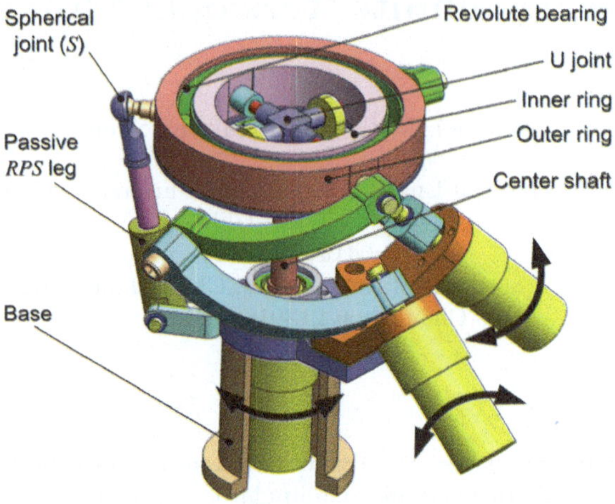

Fig. 1. CAD model of the 3-DOF asymmetrical SPM

Up to date, the stability problems of the U joint, concerning about the modeling and analysis, have been extensively studied [3,4,9–11]. Of the reported work, common approaches to analyze the linear/nonlinear dynamic model of the system include Floquet theory, Krylov–Bogoliubov method, Poincaré–Lyapunov method, etc.. As the relationship between the input and output shaft rotating speeds of the U joint is periodic, the Floquet theory will be an effective approach to analyze the stability problem, which will be adopted in this work. This paper investigates the dynamic stability problem of the proposed SPM with a U joint by means of a monodromy matrix method. To this end, a linear model consisting of input and output elements interconnected via a Hooke's joint is considered. The linearized equations of motion of the system are obtained, and the workspace region of instability is detected. Numerical study is carried out to assess the system stability.

2 Architecture of the SPM Wrist

Figure 1 depicts the asymmetrical SPM, which is composed of an outer and inner rings connected to each other with a revolute joint, the revolute joint being realized with a revolve bearing. The orientation of the outer ring is determined by two RRR limbs in parallel, respectively, and it is constrained by a fully passive RPS leg that is offset from the center of the mobile platform to eliminate the rotational motion around the vertical axis. Through a universal joint, the decoupled rotation of the inner ring is generated by the center shaft, which also supports the MP to improve the positioning accuracy.

The parameterization of the SPM is illustrated in Fig. 2. The reference coordinate system (x, y, z) is built as shown in Fig. 2, of which the origin is located at the center of rotation, namely, point O. The z axis points upward along the vertical direction and the x axis is parallel to the segment A_1A_2 and to the axis of the revolute joint in the passive leg. The local coordinate system (X, Y, Z) attached to the outer ring is built with the origin located at point O, of which Y axis passes through the geometric center of the spherical joint in the passive leg. The ith active leg consists of three revolute joints, whose axes are parallel to unit vectors $\mathbf{u}_i$, $\mathbf{v}_i$, and $\mathbf{w}_i$. Both of these two limbs have the same architecture, defined by α_1 and α_2 angles. The base platform is defined by angles α_0 and γ, while the design parameter of the mobile platform is β.

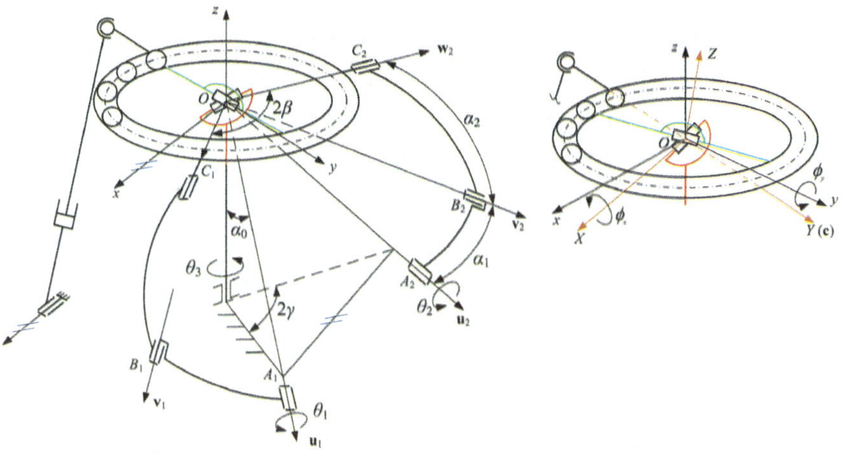

Fig. 2. Parameterizations of the asymmetrical SPM

3 Kinematics and Stiffness Matrix of the Manipulator

Under the prescribed coordinate system in Fig. 2, the unit vector $\mathbf{u}_i$ is derived as:

$$\mathbf{u}_i = \left[\varepsilon(i)\mathrm{S}\alpha_0\mathrm{S}\gamma \ \ \mathrm{S}\alpha_0\mathrm{C}\gamma \ \ -\mathrm{C}\alpha_0\right]^T, \quad \varepsilon(i) = \begin{cases} 1, \ i = 1 \\ -1, \ i = 2 \end{cases} \tag{1}$$

where S and C stand for the sine and cosine functions, respectively.

The unit vector $\mathbf{v}_i$ of the axis of the intermediate revolute joint in the ith leg is obtained in terms of the input joint angle θ_i following the angle-axis representation [1], namely,

$$\mathbf{v}_i = \left[a_i \ b_i \ c_i\right]^T = \mathbf{R}(\mathbf{u}_i, \theta_i)\mathbf{v}_i^*; \quad \mathbf{v}_i^* = \left[\varepsilon(i)\mathrm{S}\alpha_{01}\mathrm{S}\gamma \ \ \mathrm{S}\alpha_{01}\mathrm{C}\gamma \ \ -\mathrm{C}\alpha_{01}\right]^T \tag{2}$$

where $\mathbf{v}_i^*$ is the unit vector of the axis of the intermediate revolute joint in the ith leg in its home configuration and $\alpha_{01} = \alpha_0 + \alpha_1$.

The unit vector $\mathbf{w}_i$ of the top revolute joint in the ith leg, is a function of the MP orientation:

$$\mathbf{w}_i = \left[x_i \; y_i \; z_i\right]^T = \mathbf{Q}\mathbf{w}_i^*; \quad \mathbf{w}_i^* = \left[\varepsilon(i) \sin\beta \cos\beta \; 0\right]^T \tag{3}$$

where $\mathbf{w}_i^*$ is the unit vector of the axis of the top revolute joint of the ith leg when the mobile platform is located in its home configuration. Moreover, $\mathbf{Q} = \mathbf{R}_y(\phi_y)\mathbf{R}_x(\phi_x)$, $\{\phi_x, \phi_y\} \in (-\pi/2, \pi/2)$, represents the rotation matrix of the outer ring.

The inverse geometric problem of the $\underline{\text{R}}$RR leg is solved subject to the following kinematic constraints:

$$\mathbf{v}_i \cdot \mathbf{w}_i - \cos\alpha_2 = 0; \quad i = 1, 2 \tag{4}$$

whence, the input angle displacements are derived below [15]:

$$\theta_i = 2\tan^{-1}\frac{-I_i \pm \sqrt{I_i^2 + J_i^2 - K_i^2}}{K_i - J_i} \tag{5}$$

with

$$I_i = (x_i \mathrm{C}\gamma - \varepsilon(i) y_i \mathrm{S}\gamma)\mathrm{S}\alpha_1 \tag{6a}$$
$$J_i = (\varepsilon(i) x_i \mathrm{C}\alpha_0 \mathrm{S}\gamma + y_i \mathrm{C}\alpha_0 \mathrm{C}\gamma + z_i \mathrm{S}\alpha_0)\mathrm{S}\alpha_1 \tag{6b}$$
$$K_i = (\varepsilon(i) x_i \mathrm{S}\alpha_0 \mathrm{S}\gamma + y_i \mathrm{S}\alpha_0 \mathrm{C}\gamma - z_i \mathrm{C}\alpha_0)\mathrm{S}\alpha_1 - \mathrm{C}\alpha_2 \tag{6c}$$

Differentiating Eqs. (4) with respect to time leads to

$$\begin{bmatrix} A_{11} & A_{12} \\ A_{21} & A_{22} \end{bmatrix} \begin{bmatrix} \dot{\phi}_x \\ \dot{\phi}_y \end{bmatrix} = \begin{bmatrix} B_{11} & 0 \\ 0 & B_{22} \end{bmatrix} \begin{bmatrix} \dot{\theta}_1 \\ \dot{\theta}_2 \end{bmatrix} \quad \text{or} \quad \mathbf{A}\dot{\boldsymbol{\phi}} = \mathbf{B}\dot{\boldsymbol{\theta}} \tag{7}$$

with

$$A_{i1} = -\varepsilon(i) a_i \mathrm{C}\phi_x \mathrm{S}\phi_y \mathrm{C}\beta + \varepsilon(i) b_i \mathrm{S}\phi_x \mathrm{C}\beta - \varepsilon(i) c_i \mathrm{C}\phi_x \mathrm{C}\phi_y \mathrm{C}\beta \tag{8a}$$
$$A_{i2} = a_i(-\varepsilon(i)\mathrm{S}\phi_y \mathrm{S}\beta - \mathrm{S}\phi_x \mathrm{C}\phi_y \mathrm{C}\beta) - c_i(-\varepsilon(i)\mathrm{C}\phi_y \mathrm{S}\beta + \mathrm{S}\phi_x \mathrm{S}\phi_y \mathrm{C}\beta) \tag{8b}$$
$$B_{ii} = x_i \mathrm{S}\alpha_1(\mathrm{S}\theta_i \mathrm{S}\gamma \mathrm{C}\alpha_0 - \varepsilon(i)\mathrm{C}\theta_i \mathrm{C}\gamma) + y_i \mathrm{S}\alpha_1(\mathrm{S}\theta_i \mathrm{C}\gamma \mathrm{C}\alpha_0 + \varepsilon(i)\mathrm{C}\theta_i \mathrm{S}\gamma) - c_i \mathrm{S}\theta_i \mathrm{S}\alpha_0 \mathrm{S}\alpha_1 \tag{8c}$$

To this end, the kinematic Jacobian matrix of the manipulator is expressed as $\mathbf{J} = \mathbf{A}^{-1}\mathbf{B}$, subsequently, the stiffness matrix is derived as below

$$\mathbf{K} = k_\theta \mathbf{J}^{-T}\mathbf{J}^{-1} \tag{9}$$

where k_θ stands for the stiffness of the rotational actuated joint in the $\underline{\text{R}}$RR limb.

4 Equations of Motion–Rigid Dynamic Model

Splitting the outer ring and the two RRR limbs as well as the passive one, the remaining parts of the SPM can be equivalent to a U-joint mechanism, as displayed in Fig. 3. Considering the elastic deformations of the mobile platform, a moving coordinate frame (X', Y', Z') is achieved by a continuous rotation ξ_y followed by ξ_x about the frame (X, Y, Z). The angular velocity $\boldsymbol{\omega}'$ and acceleration $\dot{\boldsymbol{\omega}}'$ of the moving frame (X', Y', Z') with respect to the frame (X, Y, Z) is now expressed as [9]

$$\boldsymbol{\omega}' = \dot{\xi}_y \cos\xi_x \hat{\mathbf{y}}_2 - \dot{\xi}_y \sin\xi_x \hat{\mathbf{z}}_2 + \dot{\xi}_x \hat{\mathbf{x}}_2 \tag{10a}$$

$$\dot{\boldsymbol{\omega}}' = \ddot{\xi}_x \hat{\mathbf{x}}_2 + (\ddot{\xi}_y \cos\xi_x - \dot{\xi}_y \dot{\xi}_x \sin\xi_x)\hat{\mathbf{y}}_2 - (\ddot{\xi}_y \sin\xi_x + \dot{\xi}_y \dot{\xi}_x \cos\xi_x)\hat{\mathbf{z}}_2 \tag{10b}$$

where $\hat{\mathbf{x}}_2$, $\hat{\mathbf{y}}_2$ and $\hat{\mathbf{z}}_2$ denote unit vectors of X'–, Y'– and Z'–axis, respectively. Moreover, the speed of the driven shaft around the shaft axis is denoted by

$$\omega_O \equiv \omega_I \eta(t) = \omega_I \eta(t+\pi) = \frac{\omega_I \cdot \cos\sigma}{1 - \sin^2\sigma \sin^2\theta_3} \tag{11}$$

with

$$\theta_3 = \Omega_0 t + \theta_0 \tag{12}$$

where Ω_0 denote the constant velocity of the driving shaft and $\sigma = \cos^{-1}(\cos\phi_x \cos\phi_y)$ is the bend angle of the U joint.

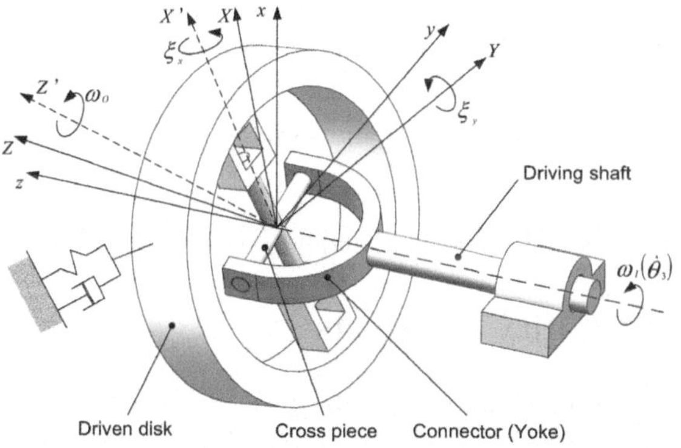

Fig. 3. U-joint mechanism

Here, the modified version of Euler's equations [7] are used, namely,

$$M_{X',O} = J_{X',O}\dot{\omega}_{X'} - (J_{Y',O} - J_{Z',O})\omega_{Y'}\dot{\omega}_{Z'} + J_{Z',O}\omega_{Y'}\omega \tag{13a}$$

$$M_{Y',O} = J_{Y',O}\dot{\omega}_{Y'} - (J_{Z',O} - J_{X',O})\omega_{X'}\dot{\omega}_{Z'} - J_{Z',O}\omega_{X'}\omega \tag{13b}$$

where $M_{a,b}$ is the moment about axis a with respect to reference point b, J's are mass moments of inertia with respect to point O, and subscripted ω and $\dot{\omega}$ stands for the corresponding elements in Eqs. (10a) and (10b).

Considering the stiffness of the mobile platform (springs in Fig. 3) and dampers (C_x, C_y), the moments that arise at the center of the U Joint can be expressed as

$$\mathbf{M}^* = \begin{bmatrix} M_x & M_y \end{bmatrix}^T = \mathbf{K}\boldsymbol{\xi} + \mathbf{C}\dot{\boldsymbol{\xi}} \tag{14}$$

with

$$\boldsymbol{\xi} = \begin{bmatrix} \xi_x & \xi_y \end{bmatrix}^T, \dot{\boldsymbol{\xi}} = \begin{bmatrix} \dot{\xi}_x & \dot{\xi}_y \end{bmatrix}^T, \mathbf{C} = \mathrm{diag}\begin{bmatrix} C_x & C_y \end{bmatrix} \tag{15}$$

Let $J_{X',O} = J_{Y',O} = J_{xy}$ and $J_{Z',O} = \rho J_{xy}$, substitution of Eqs. (13a), (13b) and (14) in Eqs. (10a) and (10b) leads to the following set of nonlinear ordinary differential equations in the system (x, y, z):

$$\begin{aligned} &J_{xy}\begin{bmatrix} 1 & 0 \\ 0 & 1 \end{bmatrix}\begin{bmatrix} \ddot{\xi}_x \\ \ddot{\xi}_y \end{bmatrix} + J_{xy}\mathrm{S}\xi_x\begin{bmatrix} 0 & (1-\rho)\mathrm{C}\xi_x \\ \rho & 0 \end{bmatrix}\begin{bmatrix} \dot{\xi}_x \\ \dot{\xi}_y \end{bmatrix}\dot{\xi}_y \\ &+\rho J_{xy}\omega_O\begin{bmatrix} 0 & -\mathrm{C}\xi_x \\ 1 & 0 \end{bmatrix}\begin{bmatrix} \dot{\xi}_x \\ \dot{\xi}_y \end{bmatrix} + \mathbf{C}\begin{bmatrix} \dot{\xi}_x \\ \dot{\xi}_y \end{bmatrix} + \mathbf{K}\begin{bmatrix} \xi_x \\ \xi_y \end{bmatrix} = \begin{bmatrix} 0 \\ 0 \end{bmatrix} \end{aligned} \tag{16}$$

Assuming that the deflections at are small, one can use small angle approximations, sequentially, Eq. (16) can be linearized as below:

$$\begin{aligned} &J_{xy}\begin{bmatrix} 1 & 0 \\ 0 & 1 \end{bmatrix}\begin{bmatrix} \ddot{\xi}_x \\ \ddot{\xi}_y \end{bmatrix} + \rho J_{xy}\omega_O\begin{bmatrix} 0 & -1 \\ 1 & 0 \end{bmatrix}\begin{bmatrix} \dot{\xi}_x \\ \dot{\xi}_y \end{bmatrix} + \mathbf{C}\begin{bmatrix} \dot{\xi}_x \\ \dot{\xi}_y \end{bmatrix} + \mathbf{K}\begin{bmatrix} \xi_x \\ \xi_y \end{bmatrix} = \begin{bmatrix} 0 \\ 0 \end{bmatrix} \\ &\text{or} \quad J_{xy}\mathbf{I}_2\ddot{\boldsymbol{\xi}} + \rho J_{xy}\Omega_0\eta_0(\tau)\mathbf{E}\dot{\boldsymbol{\xi}} + \mathbf{C}\dot{\boldsymbol{\xi}} + \mathbf{K}\boldsymbol{\xi} = \mathbf{0} \end{aligned} \tag{17}$$

with

$$\mathbf{E} = \begin{bmatrix} 0 & -1 \\ 1 & 0 \end{bmatrix}; \quad \eta_0(\tau) = \eta_0(\tau+\pi) = \frac{\cos\sigma}{1-\sin^2\sigma\sin^2\tau} \tag{18}$$

where $\ddot{\boldsymbol{\xi}} = \begin{bmatrix} \ddot{\xi}_x & \ddot{\xi}_y \end{bmatrix}^T$ and $\mathbf{I}$ stands for the identity matrix. Equation (17) can be rewritten as

$$\ddot{\boldsymbol{\xi}} + \mathbf{U}\dot{\boldsymbol{\xi}} + \mathbf{V}\boldsymbol{\xi} = \mathbf{0}; \quad \mathbf{U} = \rho\Omega_0\eta_0(\tau)\mathbf{E} + \mathbf{C}/J_{xy}, \ \mathbf{V} = \mathbf{K}/J_{xy} \tag{19}$$

5 Dynamic Stability Analysis

As Eq. (19) consists of a set of linear differential equations with π-periodic coefficients, a state-space representation of system (19) can be given as

$$\dot{\mathbf{x}}(t) = \mathbf{A}(t)\mathbf{x}(t); \quad \mathbf{x}(t) = \begin{bmatrix} \boldsymbol{\xi}^T & \dot{\boldsymbol{\xi}}^T \end{bmatrix}^T, \mathbf{A}(t) = \begin{bmatrix} \mathbf{0}_2 & \mathbf{I}_2 \\ -\mathbf{V} & -\mathbf{U} \end{bmatrix} \tag{20}$$

whence $\mathbf{A}(t)$ is a 4×4 π-periodic matrix. According to Floquet theory, a fundamental solutions' matrix of system (20) can be expressed as

$$\boldsymbol{\Phi}(\tau) = \mathbf{P}(\tau)e^{\tau\mathbf{R}} \tag{21}$$

where $\mathbf{P}(\tau)$ is a π-periodic matrix and $\mathbf{R}$ is a constant matrix which is related to another constant matrix $\mathbf{H}$, referred to as monodromy matrix, by $\mathbf{R} = \ln\mathbf{H}/\pi$. If the fundamental matrix is normalized so that $\mathbf{P}(0) = \mathbf{I}_4$, then $\mathbf{H} = \mathbf{P}(\pi)$.

The eigenvalues λ_i, $i = 1, 2, 3, 4$, of matrix $\mathbf{R}$, referred to as Floquet multipliers, govern the stability of the system. The system is asymptotically stable if and only if the real part of all the eigenvalues λ_i are non-positive [5]. In this study, the matrix $\mathbf{H}$ is obtained by numerically with the improved Runge Kutta Method [12] with step size 10^{-6}, and its eigenvalues are calculated to assess stability of the system. The monodromy matrix method is a simple and reliable method to determine the stability of parametrically excited systems.

6 Numerical Study on Stability

This section deals with the numerical stability analysis to detect the stable workspace region of the SPM, where the geometrical parameters are given by $\alpha_0 = \pi/3$, $\alpha_1 = 7\pi/18$, $\alpha_2 = \pi/2$, $\beta = \pi/3$, $\gamma = \pi/6$, and other parameters are $k_\theta = 10\,\text{Nm/rad}$, $J_{xy} = 0.008\,\text{kg}\cdot\text{m}^2$, $\rho = 2$, $C_x = C_y = 0.002$, respectively.

In accordance with the previous analysis, the stability of the asymmetrical SPM is analyzed. Figure 4 shows the stable region in blue of the SPM within its workspace with different rotating speed of the input (center) shaft, where σ represents the misalignment angle of the U joint. With the input speed $\Omega_0 = 2\pi$, namely, 60 rpm, the SPM can gave a large stable workspace region within the workspace $\sigma \in [0, 30°]$. It is obviously that the stable region dramatically decreases with the increasing rotating speed, from the comparison between Figs. 4(b) and (a), as it decreases 70% approximately. Moreover, when the rotating speed of the input shaft is equal to 15π, the manipulator is stable only if when the mobile platform is in its home configuration, i.e., $\phi_x = \phi_y = 0$. Therefore, the SPM is no longer stable when the input shaft speed is larger than the critical value 15π.

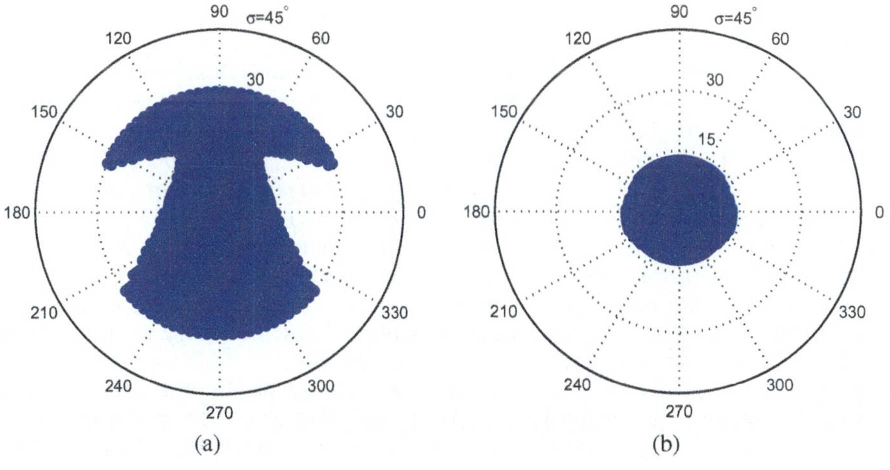

Fig. 4. The stable region of the SPM within the workspace: (a) $\Omega_0 = 2\pi$; (b) $\Omega_0 = 10\pi$.

7 Conclusion

This paper deals with the dynamic stability analysis of an asymmetrical spherical parallel manipulator with a universal joint. A linear model consisting of input shaft and output platform interconnected via a Hooke's joint is considered. The linearized equations of motion of the system are obtained, for which the stability problem is investigated by resorting to a monodromy matrix method. The method of stability analysis is numerically illustrated, wherein the stable workspace region is detected. A critical rotating speed of the input shaft is identified to avoid the instability for the manipulator. Future work includes the complete parametric stability analysis as well as the torsional stability of the U joint.

Acknowledgement. The support from the Doctoral Start-up Foundation of Liaoning Province (No. 20170520134), the Jiangsu Key Development Project (No. BE2015043) and the Chinese National Natural Science Foundation (No. 51475050) is acknowledged.

References

1. Angeles, J.: Fundamentals of Robotic Mechanical Systems: Theory, Methods, and Algorithms. Springer, New York (2007)
2. Asada, H., Granito, J.: Kinematic and static characterization of wrist joints and their optimal design. In: IEEE International Conference on Robotics and Automation, pp. 244–250 (1985)
3. Asokanthan, S.F., Hwang, M.C.: Torsional instabilities in a system incorporating a Hooke's joint. J. Vib. Acoust. **118**(3), 368–374 (1996)
4. Bulut, G., Parlar, Z.: Dynamic stability of a shaft system connected through a Hooke's joint. Mech. Mach. Theory **46**(11), 1689–1695 (2011)
5. Chicone, C.: Ordinary Differential Equations with Applications. Springer, New York (2006)
6. Gosselin, C., Hamel, J.: The Agile Eye: a high-performance three-degree-of-freedom camera-orienting device. In: IEEE International Conference Robotics and Automation, pp. 781–786 (1994)
7. Greenwood, D.: Principles of Dynamics. Prentice Hall, Englewood Cliffs (1988)
8. Li, T., Payandeh, S.: Design of spherical parallel mechanisms for application to laparoscopic surgery. Robotica **20**(2), 133–138 (2002)
9. Mazzei Jr., A.J., Argento, A., Scott, R.A.: Dynamic stability of a rotating shaft driven through a universal joint. J. Sound Vib. **222**(1), 19–47 (1999)
10. Ota, H., Kato, M.: Lateral vibrations of a rotating shaft driven by a universal joint : 1st report, generation of even multiple vibrations by secondary moment. Trans. Jpn. Soc. Mech. Eng. C **27**(231), 2002–2007 (2008)
11. Porter, B.: A theoretical analysis of the torsional oscillation of a system incorporating a Hooke's joint. ARCHIVE J. Mech. Eng. Sci. **3**(4), 324–329 (1961)
12. Szymkiewicz, R.: Numerical Solution of Ordinary Differential Equations. Academic Press, Cambridge (1971)

13. Wu, G., Bai, S., Kepler, J.: Mobile platform center shift in spherical parallel manipulators with flexible limbs. Mech. Mach. Theory **75**, 12–26 (2014)
14. Wu, G., Caro, S., Bai, S., Kepler, J.: Dynamic modeling and design optimization of a 3-DOF spherical parallel manipulator. Robot. Auto. Syst. **62**, 1377–1386 (2014)
15. Wu, G., Caro, S., Wang, J.: Design and transmission analysis of an asymmetrical spherical parallel manipulator. Mech. Mach. Theory **94**, 119–131 (2015)

On the Use of Instant Centers to Build Dynamic Models of Single-dof Planar Mechanisms

Raffaele Di Gregorio(✉)

Engineering Department, Ferrara University, 44122 Ferrara, Italy
raffaele.digregorio@unife.it

Abstract. Instant centers' (ICs') locations of single-degree-of-freedom (single-dof) planar mechanisms depend only on the mechanism configuration. In single-dof mechanisms, ICs' coordinates can be computed as a function of their generalized coordinate using a number of existing algorithms. Here, ICs are used to compute inertia forces and to build the dynamic model of these mechanisms by using two different dynamics formulations. The resulting dynamic models are novel and relate functional parameters (ICs locations) to the dynamic behavior and/or parameters of the mechanism. The last feature makes them useful tools for simultaneously imposing functional and performance requirements during design.

Keywords: Planar mechanisms · Dynamic model · Instant center

1 Introduction

Industrial applications diffusely adopt mechanical systems or subsystems that are single-dof planar mechanisms. Mathematical models of such mechanisms are useful for their both design and analysis. Kinematic models take functional aspects, whereas dynamic models take their performances. Therefore, dynamic models that involve significant kinematics entities are appealing for an integrated design or analysis.

ICs lend themselves to this task. In a previous paper [1], this author showed that ICs could explicitly appear in one dynamic model of single-dof planar mechanisms based on Eksergian's equation. Here, this aspect is further developed: ICs are used to compute inertia forces and to build the dynamic model of these mechanisms by using two other dynamics formulations. The resulting dynamic models are novel and tie functional parameters (ICs locations) to the dynamic performances of the mechanism. The last feature makes them useful for simultaneously imposing functional and performance requirements during design.

The paper is organized as follows. Section 2 recalls some background concepts, and Sect. 3 deduces the expressions of the resultant force and the resultant moment of the inertia forces by systematically using centrodes and their derivatives. Then, Sect. 4 builds the two dynamic models. Finally, Sect. 5 illustrates the critical points of the models through an example, and Sect. 6 draws the conclusions.

J. Lenarcic and V. Parenti-Castelli (Eds.): ARK 2018, SPAR 8, pp. 242–249, 2019.
https://doi.org/10.1007/978-3-319-93188-3_28

2 Background and Notations

Instant centers' (ICs') locations of single-dof planar mechanisms depend only on the mechanism configuration. In such mechanisms, ICs' coordinates as a function of their generalized coordinate (centrodes), say q, can be computed through a number of algorithms presented in the literature [2–8].

For an assigned value of q, the numerical or analytical solution of the closure equation system brings to compute the poses of all the mobile links. Links' poses straightly provide the coordinates of all the primary ICs (i.e., the ICs whose locations are immediately known by inspection of the mechanism configuration) and the parametric equations of all the straight lines the other (secondary) ICs lie on. Then, in the mechanisms that are not "indeterminate" (i.e., in the majority of the possible cases), the positions of all the secondary ICs are sequentially computed one by one as common intersection of two lines; whereas, in indeterminate mechanisms, this procedure is slightly modified since a subset of ICs must be simultaneously (i.e., not sequentially) computed (see [8] for details). These algorithms only involve the solution of simple linear systems.

ICs' coordinates allow the computation of all the velocity ratios (velocity coefficients (VCs)) [9]. In particular, let $\dot{\theta}_{ji}$ and $\dot{s}_{ji}$ be the signed magnitudes, respectively, of the relative angular velocity, in the case of instantaneous rotation, and of the relative translation velocity, in the case of instantaneous translation, between links j and i (see Fig. 1), the following formulas hold ($\mathbf{I}_{gh}$ denotes the position vector of the IC, I_{gh}, of the relative motion between links g and h):

$$\frac{\dot{\theta}_{ji}}{\dot{\theta}_{ki}} = \frac{(\mathbf{I}_{kj} - \mathbf{I}_{ki}) \cdot (\mathbf{I}_{kj} - \mathbf{I}_{ji})}{\left\|\mathbf{I}_{kj} - \mathbf{I}_{ji}\right\|^2}; \quad \frac{\dot{s}_{ji}}{\dot{\theta}_{ki}} = -(\mathbf{I}_{kj} - \mathbf{I}_{ki}) \cdot \mathbf{u}_{ji}; \quad \frac{\dot{s}_{ji}}{\dot{s}_{ri}} = \frac{(\mathbf{I}_{kj} - \mathbf{I}_{ki}) \cdot \mathbf{u}_{ji}}{(\mathbf{I}_{kr} - \mathbf{I}_{ki}) \cdot \mathbf{u}_{ri}} \tag{1}$$

where the rotation angle θ_{ji} and θ_{ki} are positive if counterclockwise (Fig. 1a); whereas, the unit vectors $\mathbf{u}_{ji}$ and $\mathbf{u}_{ri}$ are obtained from the unit vectors $\mathbf{t}_{ji}$ and $\mathbf{t}_{ri}$ (Fig. 1b), respectively, which give the positive directions of translation, through a

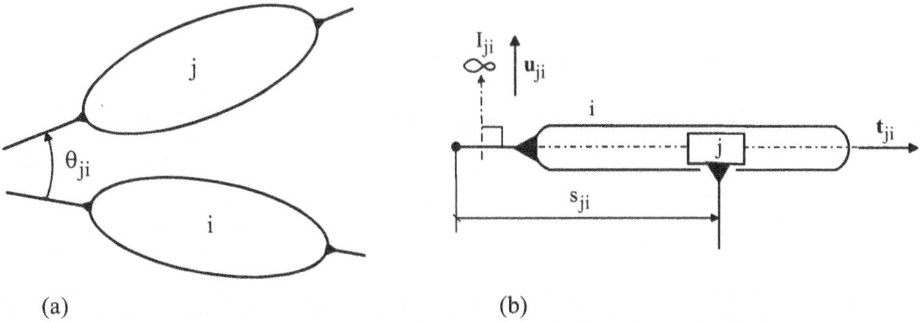

Fig. 1. Notations: (a) instantaneous rotation between links j and i (I_{ji} is a finite point of the motion plane; the motion is uniquely defined by I_{ji} and $\dot{\theta}_{ji}$), (b) instantaneous translation between links j and i (I_{ji} is the point at infinity of the lines perpendicular to the translation direction; the motion is uniquely defined by the translation velocity $\dot{s}_{ji}\mathbf{t}_{ji}$).

counterclockwise rotation of 90°. The 3rd equation is simply obtained as the ratio of two different forms of the 2nd equation, one with $\dot{s}_{ji}$ and the other with $\dot{s}_{ri}$.

The dot products appearing in Eqs. (1) are all between collinear planar vectors (see the Aronhold-Kennedy theorem [2]). VCs' expressions provided by Eqs. (1) give also the signs of the VCs. They reveal that, in single-dof planar mechanisms, VCs depend only on the mechanism configuration (i.e., on q) through the ICs that appear in their explicit expressions.

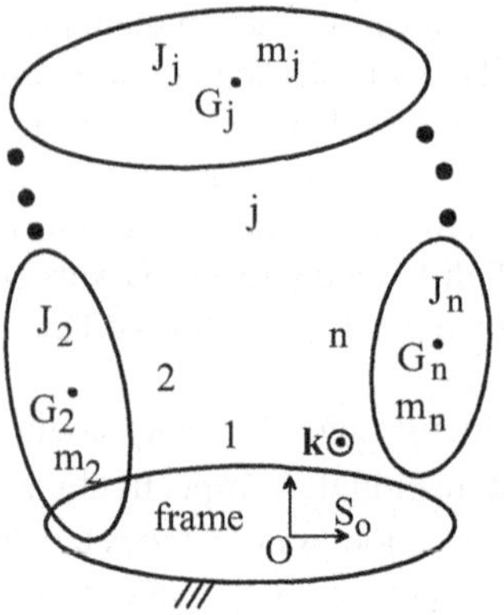

Fig. 2. Single-dof planar mechanism with n links.

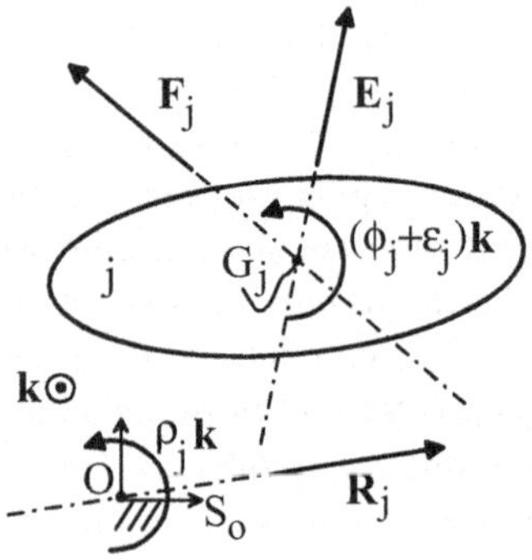

Fig. 3. Free-body diagram of the j-th link.

3 Inertia Forces and ICs

For the relative motion between the j-th link and the frame (link 1 of Fig. 2) of a single-dof planar mechanism with n links (see Fig. 1 with i = 1), the following relationships hold

$$v_j = \frac{\dot{\theta}_{j1}}{\dot{q}}, \quad \mathbf{a}_{I_{j1}|j} = -v_j\left(\mathbf{k} \times \frac{d\mathbf{I}_{j1}}{dq}\right)\dot{q}^2, \quad \ddot{\theta}_{j1} = v_j\,\ddot{q} + \frac{dv_j}{dq}\dot{q}^2 \tag{2}$$

where v_j is a VC whose explicit expression is given by one of formulas (1), $\mathbf{a}_{I_{j1}|j}$ is the acceleration the instant center I_{j1} would have, if it were a point fixed to the j-th link, $\mathbf{k}$ is a unit vector perpendicular to the plane of motion (see Fig. 2). Also, the acceleration, $\mathbf{a}_{G_j|j}$, of the center of mass, G_j, of the j-th link could be written as follows:

$$\mathbf{a}_{G_j|j} = \mathbf{a}_{I_{j1}|j} + \ddot{\theta}_{j1}\left[\mathbf{k} \times (\mathbf{G}_j - \mathbf{I}_{j1})\right] - \dot{\theta}_{j1}^2(\mathbf{G}_j - \mathbf{I}_{j1}) \tag{3}$$

The introduction of formulas (2) into Eq. (3) yields

$$\mathbf{a}_{G_j|j} = v_j\left[\mathbf{k} \times (\mathbf{G}_j - \mathbf{I}_{j1})\right]\ddot{q} + \left\{\frac{dv_j}{dq}\left[\mathbf{k} \times (\mathbf{G}_j - \mathbf{I}_{j1})\right] - v_j\left(\mathbf{k} \times \frac{d\mathbf{I}_{j1}}{dq}\right) - v_j^2\,(\mathbf{G}_j - \mathbf{I}_{j1})\right\}\dot{q}^2 \quad (4)$$

The resultant force, $\mathbf{F}_j$, and the resultant moment, $\phi_j\mathbf{k}$, about G_j of j-th link's inertia forces are $\mathbf{F}_j = -m_j\mathbf{a}_{G_j|j}$ and $\phi_j = -J_j\ddot{\theta}_{j1}$, where m_j and J_j are j-th link's mass and inertia moment about an axis passing through G_j and parallel to $\mathbf{k}$, respectively. Thus, by taking into account Eqs. (2) and (4), the following explicit expressions of these resultants can be deduced:

$$\phi_j = -J_j\left(v_j\,\ddot{q} + \frac{dv_j}{dq}\dot{q}^2\right) \quad (5)$$

$$\begin{aligned}\mathbf{F}_j = {} & -m_j v_j\left[\mathbf{k} \times (\mathbf{G}_j - \mathbf{I}_{j1})\right]\ddot{q} \\ & - m_j\left\{\frac{dv_j}{dq}\left[\mathbf{k} \times (\mathbf{G}_j - \mathbf{I}_{j1})\right] - v_j\left(\mathbf{k} \times \frac{d\mathbf{I}_{j1}}{dq}\right) - v_j^2(\mathbf{G}_j - \mathbf{I}_{j1})\right\}\dot{q}^2\end{aligned} \quad (6)$$

The coefficients of $\ddot{q}$ and $\dot{q}^2$ in Eqs. (5) and (6) contain only terms that depend on q through the position vectors $\mathbf{G}_j$ and $\mathbf{I}_{j1}$, and the velocity coefficient v_j. Equations (5) and (6) are the sought-after relationships that tie the inertia forces to the mechanism ICs since the VCs depend on the ICs through Eqs. (1).

4 Dynamic Models with ICs

Once the inertia forces have been tied to the ICs, all the dynamics formulations that directly use inertia forces can be adopted to build dynamic models based on the centrodes and their derivatives. All these models involve only a limited number of ICs. For instance, a single-dof planar mechanism (see Fig. 2) with n links requires the (n–1) ICs I_{j1} for j = 2,…, n, and the ICs necessary to determine the I_{j1}, but many of these ICs are primary ICs. Below, the dynamic models resulting from the Newton-Euler (N-E) formulation and from the use of the virtual work principle (D'Alembert's principle) are deduced.

4.1 Newton-Euler Formulation

N-E formulation builds a mechanism's dynamic model by writing the dynamic equilibrium equations of each mobile link of the mechanism. Figure 3 shows the free-body diagram of the j-th link. With reference to Fig. 3, $\mathbf{E}_j$ and $\varepsilon_j\mathbf{k}$ ($\mathbf{R}_j$ and $\rho_j\mathbf{k}$) are respectively the resultant force and the resultant moment about G_j (about O) of the external (reaction) force system applied to the j-th link. These notations yield the following equilibrium equations

$$\mathbf{E}_j + \mathbf{R}_j + \mathbf{F}_j = \mathbf{0} \quad j = 2, \ldots, n \tag{7}$$

$$\varepsilon_j + \rho_j + \phi_j + (\mathbf{E}_j + \mathbf{F}_j) \cdot [\mathbf{k} \times (\mathbf{G}_j - \mathbf{O})] = 0 \quad j = 2, \ldots, n \tag{8}$$

where ϕ_j and $\mathbf{F}_j$ are given by Eqs. (5) and (6), respectively, ε_j and $\mathbf{E}_j$ are known, and $\mathbf{R}_j$ and ρ_j are unknowns to be computed. Equation (8)[1] is the moment equilibrium about O (see Figs. (2) and (3)).

4.2 D'Alembert's Principle

D'Alembert's principle builds the dynamic model by equating to zero the generalized force, Q, associated to the generalized coordinate, q, of the single-dof mechanism. Only the active and inertia forces applied to the mechanism appear in the definition of Q, that is, $\mathbf{F}_j$, $\phi_j\mathbf{k}$, $\mathbf{E}_j$, $\varepsilon_j\mathbf{k}$ for j = 2,…,n, and the generalized torque, τ, which directly controls q.

Consequently, the dynamic model is

$$\tau + \sum_{j=2,n} \alpha_j \mu_j = 0 \tag{9}$$

where, if the j-th link rotates (Fig. 1(a)),

$$\alpha_j = \varepsilon_j + \phi_j + (\mathbf{E}_j + \mathbf{F}_j) \cdot [\mathbf{k} \times (\mathbf{G}_j - \mathbf{I}_{j1})], \quad \mu_j = \nu_j \tag{10}$$

whereas, if the j-th link translates (Fig. 1(b)),

$$\alpha_j = (\mathbf{E}_j + \mathbf{F}_j) \cdot \mathbf{t}_{j1}, \quad \mu_j = \frac{\dot{s}_{j1}}{\dot{q}} \tag{11}$$

The introduction of expressions (10) and (11) into Eq. (9) and, then, of expression (5) and (6) for ϕ_j and $\mathbf{F}_j$, respectively, into the resulting equation makes Eq. (9) a dynamic model that contains only terms depending on q through the position vectors $\mathbf{G}_j$ and $\mathbf{I}_{j1}$, and the velocity coefficients ν_j for j = 2,…,n.

4.3 Solution Algorithms

Dynamic models are mainly used to solve two dynamics problems: the "direct problem" (DP) and the "inverse problem" (IP). In single-dof mechanisms, the IP is the computation of the time history of the generalized torque, $\tau(t)$, once the time history of the generalized coordinate, q(t), is assigned. Vice versa, the DP is the computation of q (t) once $\tau(t)$ is known together with the initial motion conditions, q(0) and $\dot{q}(0)$.

---

[1] The use of the same reference point (point O) in Eqs. (8) for all the mobile links makes it possible to write the equilibrium equations of mechanism sub-systems by simply summing the equilibrium equations of the links that form the sub-systems. Indeed, with this choice of the reference point, this algebraic operation automatically cancel the internal reaction forces of the sub-system.

The solution algorithms both of these problems by using either Eqs. (7) and (8), or, as an alternative, Eq. (9) share the same type of kinematics analysis, which has to be implemented only once (even offline) if the mechanism geometry is kept unchanged. Such analysis, over the solution of the closure equations, involves the computation of ICs' position vectors and of the VCs appearing in those equations through simple algebraic manipulation or formulas (see Eqs. (1), and Refs. [1, 8] for details). The results of this analysis complete the set of input data necessary to compute all the coefficients of $\ddot{q}$ and $\dot{q}^2$ appearing in the proposed models. Thus, they make it possible to solve the so-obtained equations with ordinary analytical (when possible) and/or numerical techniques. Such techniques are fast and could be also implemented online for control purposes.

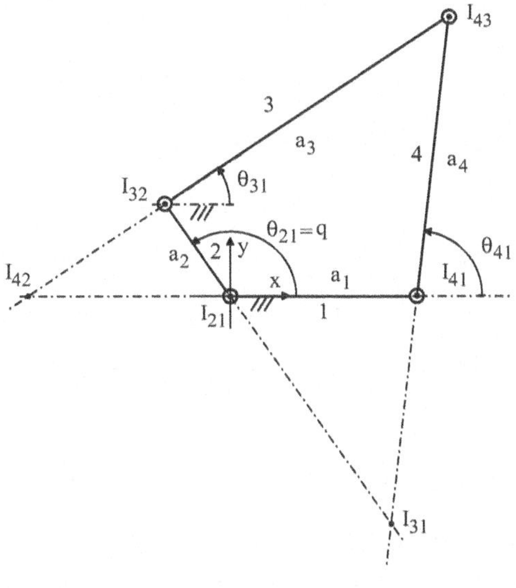

Fig. 4. Four-bar linkage: ICs and notations

5 Illustrative Example

Since the central point of the above-proposed models is the calculation of the ICs and VCs appearing in expressions (5) and (6) of inertia-forces' resultants, this computation for a four-bar linkage is presented here. Figure 4 shows the mechanism and the adopted notations. The solution, in closed form, of four-bar's closure equations is trivial; thus, without reporting such solution, the expressions of θ_{31} and θ_{41} as a function of θ_{21} ($\equiv$ q) are assumed known.

Links 2, 3, and 4 (see Fig. 4) are the mobile links. Their inertia forces' resultants (see Eqs. (5) and (6)) involve the instant centers I_{21}, I_{31}, and I_{41}, and the velocity coefficients v_2, v_3, and v_4. $\mathbf{I}_{21}$, $\mathbf{I}_{41}$, and v_2 are immediately known (see Fig. 4 and Eqs. (1) and (2)): $\mathbf{I}_{21} = (0,0)^T$, $\mathbf{I}_{41} = (a_1,0)^T$, and $v_2 = 1$. Also, the following relationships hold:

$$(\mathbf{I}_{32}-\mathbf{I}_{21}) = a_2(\cos\theta_{21},\sin\theta_{21})^T,\ (\mathbf{I}_{43}-\mathbf{I}_{41}) = a_4(\cos\theta_{41},\sin\theta_{41})^T,\ (\mathbf{I}_{43}-\mathbf{I}_{32}) = a_3(\cos\theta_{31},\sin\theta_{31})^T; \tag{12}$$

The position vector $\mathbf{I}_{31}$ can be computed by solving the vector equation (Fig. 4)

$$(\mathbf{I}_{43}-\mathbf{I}_{32}) = (\mathbf{I}_{43}-\mathbf{I}_{31}) - (\mathbf{I}_{32}-\mathbf{I}_{31}) \tag{13}$$

where

$$(\mathbf{I}_{32}-\mathbf{I}_{31}) = w_1(\mathbf{I}_{32}-\mathbf{I}_{21});\ (\mathbf{I}_{43}-\mathbf{I}_{31}) = w_2(\mathbf{I}_{43}-\mathbf{I}_{41}); \tag{14}$$

with w_1 and w_2 as scalar unknowns. Formulas (14), when compared with formulas (1) and (2) yield the following relationships:

$$v_3 = \frac{\dot\theta_{31}}{\dot\theta_{21}} \equiv w_1,\quad w_2 = \frac{\dot\theta_{41}}{\dot\theta_{31}},\quad v_4 = \frac{\dot\theta_{41}}{\dot\theta_{21}} \equiv w_1 w_2 \tag{15}$$

The solution of system (13) gives the explicit formulas

$$w_1 = -\frac{(\mathbf{I}_{43}-\mathbf{I}_{32})\cdot[\mathbf{k}\times(\mathbf{I}_{43}-\mathbf{I}_{41})]}{\mathbf{I}_{32}-\mathbf{I}_{21})\cdot[\mathbf{k}\times(\mathbf{I}_{43}-\mathbf{I}_{41})]} = \frac{a_3(\cos\theta_{31}\sin\theta_{41}-\sin\theta_{31}\cos\theta_{41})}{a_2(\sin\theta_{21}\cos\theta_{41}-\cos\theta_{21}\sin\theta_{41})} \tag{16}$$

$$w_2 = \frac{(\mathbf{I}_{43}-\mathbf{I}_{32})\cdot[\mathbf{k}\times(\mathbf{I}_{32}-\mathbf{I}_{21})]}{(\mathbf{I}_{43}-\mathbf{I}_{41})\cdot[\mathbf{k}\times(\mathbf{I}_{32}-\mathbf{I}_{21})]} = \frac{a_3(\sin\theta_{31}\cos\theta_{21}-\cos\theta_{31}\sin\theta_{21})}{a_4(\cos\theta_{21}\sin\theta_{41}-\sin\theta_{21}\cos\theta_{41})} \tag{17}$$

$$\mathbf{I}_{31} = a_2(1-w_1)\,(\cos\theta_{21},\,\sin\theta_{21})^T \tag{18}$$

The introduction of expressions (16) and (17) into Eqs. (15) and (18) together with the already known ICs and VCs (i.e., $\mathbf{I}_{21} = (0,0)^T$, $\mathbf{I}_{41} = (a_1,0)^T$, and $v_2 = 1$) provides the explicit expressions of all the ICs and VCs necessary to compute the resultants of the inertia forces for links 2, 3 and 4 with Eqs. (5) and (6). For the sake of brevity, such computation is not reported here[2].

[2] When calculating the explicit expressions of inertia-forces' resultants with Eqs. (5) and (6), since the computed formulas are, in general, of type $f(\theta_{21},\theta_{31},\theta_{41})$ with θ_{31} and θ_{41} depending only on θ_{21}, the following derivative rule can be used:

$$\frac{df}{d\theta_{21}} = \frac{\partial f}{\partial\theta_{21}} + v_3\frac{\partial f}{\partial\theta_{31}} + v_4\frac{\partial f}{\partial\theta_{41}}$$

6 Conclusions

The explicit expressions of inertia-forces' resultants as a function of suitable IC position vectors and VCs have been deduced for single-dof planar mechanisms. Such expressions have been directly employed to build these mechanisms' dynamic model through two dynamics formulations: N-E formulation and D'Alembert principle.

The two so-deduced models relate kinematics entities (centrodes and VCs), which enter into functional design, to the dynamic performances of these mechanisms. Thus, they are ideal tools for integrated design. In addition, they can be used to solve the two classic problems of dynamics analysis without losing numerical efficiency. Indeed, the peculiar kinematics analysis these models require is simple and somehow easier than others usually employed in dynamics analyses.

The two novel models, proposed here, together with the one proposed in a previous paper [1], which employs the Lagrangian formulation, complete the paramount of dynamic models for single-dof planar mechanisms that contain centrodes. Thus, the conclusion is that all the dynamics formulations can provide models for integrated design. Since each dynamics formulation stresses particular dynamic performances with respect to others, the availability of a complete set of dynamic models with centrodes allows a complete evaluation of the dynamic effects of changes on centrodes.

Acknowledgments. This work has been developed at the Laboratory of Advanced Mechanics (MECH-LAV) of Ferrara Technopole, supported by FAR2017 and FIR2016 funds and by Regione Emilia Romagna (District Councillorship for Productive Assets, Economic Development, Telematic Plan) POR-FESR 2007-2013, Attività I.1.1.

References

1. Di Gregorio, R.: A novel dynamic model for single degree-of-freedom planar mechanisms based on instant centers. ASME J. Mech. Robot. **8**(1), 011013-1–011013-8 (2016)
2. Barton, L.O.: Mechanism Analysis: Simplified Graphical and Analytical Techniques, 2nd edn. Marcel Dekker Inc., New York (1993)
3. Klein, A.W.: Kinematics of Machinery. McGraw-Hill Book Company Inc., New York (1917)
4. Foster, D.E., Pennock, G.R.: A graphical method to find the secondary instaneous centers of zero velocity for the double butterfly linkage. ASME J. Mech. Des. **125**(2), 268–274 (2003)
5. Pennock, G.R., Kinzel, E.C.: Path curvature of the single flier eight-bar linkage. ASME J. Mech. Des. **126**(3), 470–477 (2004)
6. Foster, D.E., Pennock, G.R.: Graphical methods to locate the secondary instantaneous centers of single-degree-of-freedom indeterminate linkages. ASME J. Mech. Des. **127**(2), 249–256 (2005)
7. Yan, H.-S., Hsu, M.-H.: An analytical method for locating instantaneous velocity centers. In: Proceedings of the 22nd ASME Biennial Mechanisms Conference, DE, vol. 47, pp. 353–359, Scottsdale, AZ, USA, 13–16 September 1992
8. Di Gregorio, R.: An algorithm for analytically calculating the positions of the secondary instant centers of indeterminate linkages. ASME J. Mech. Des. **130**(4), 042303-1–042303-9 (2008)
9. Paul, B.: Kinematics and Dynamics of Planar Machinery. Prentice-Hall Inc., Englewood Cliffs (1979)

Normal Forms of Robotic Systems with Affine Pfaffian Constraints: A Case Study

Krzysztof Tchoń(✉), Joanna Ratajczak, and Janusz Jakubiak

Wrocław University of Science and Technology, Wrocław, Poland
{krzysztof.tchon,joanna.ratajczak,janusz.jakubiak}@pwr.edu.pl

Abstract. We study robotic systems subject to Pfaffian constraints in the affine form. A control system representation of such a system is given in the form of a control-affine system. Normal forms under feedback are considered and used both as a description of configuration singularities of the robotic system as well as a tool for solving the motion planning problem. An example of a space manipulator is analysed in detail to illustrate the theoretical concepts.

Keywords: Affine constraints · Control-affine system
Normal forms · Singularities · Motion planning · Space manipulator

1 Introduction

It is well known that both in manipulator and mobile robots singular configurations impair the performance of their control algorithms. To cope with the problem of kinematic singularities of manipulators about 20 years ago we proclaimed the normal form approach [1]. The essence of this approach is to replace the kinematics by a simple map called the normal form. The normal form describes the kinematics in suitably chosen coordinates in the joint and in the task spaces. The normal forms not only allow one to characterize the kinematic singularities, but also to control the manipulator through and around singularities. Relying on the basic conceptual analogy between manipulator and mobile robots established within the Endogenous Configuration Space Approach (ECS) [2] we can adopt the normal form approach also to mobile robots. Singularities of the mobile robots are defined in terms of the singular controls. This paper is intended as a further step in this direction.

Specifically, we study robotic systems subject to affine Pfaffian constraints. Such systems are represented by control-affine systems. The concept of the normal form of a control system is well established in control theory, namely the normal form of a given system is an equivalent control system whose behaviour is easily understood and can be efficiently controlled. As the equivalence we use the feedback equivalence. By definition, singularities and control algorithms of a normal form are transparent, and due to the feedback equivalence they can

J. Lenarcic and V. Parenti-Castelli (Eds.): ARK 2018, SPAR 8, pp. 250–257, 2019.
https://doi.org/10.1007/978-3-319-93188-3_29

be transferred to the original system. These concepts are in the paper applied to a free-floating space manipulator subject to the conservation of the angular momentum. Using the normal forms discovered recently in [3] we provide a description of manipulator's singularities and obtain solutions of the motion planning problem.

2 Basic Concepts

2.1 Affine Pfaffian Constraints

The subject of our interest will be a robotic system characterized by generalized coordinates $q \in \mathbb{R}^n$ whose motion is subordinated to the phase constraints in the form

$$\mathcal{A}(q)\dot{q} = b(q). \tag{1}$$

Such a form will be referred to as the affine Pfaffian form. The matrix $\mathcal{A}(q)$ has dimension $l \times n$, and has everywhere the full row rank $l \leq n$.

The constraints (1) can be equivalently described by means of a control-affine system with $m = n - l$ inputs

$$\dot{q} = f(q) + G(q)u = f(q) + \sum_{i=1}^{m} g_i(q)u_i \tag{2}$$

defined in the following way. The drift vector field $f(q) = \mathcal{A}^{\#}(q)b(q)$, where $\mathcal{A}^{\#}(q)$ denotes any right inverse of $\mathcal{A}(q)$, and G is an $n \times m$ matrix whose columns g_i, $i = 1, 2, \ldots, m$, span the null space of $\mathcal{A}(q)$. The drift vector field and the control vector fields are assumed of the class C^{∞}. It should be noticed that the representation (2) is defined modulo a feedback transformation. Indeed, if we replace f by $\hat{f} = f + G\alpha$ and G by $\hat{G} = G\beta$, where $\alpha = \alpha(q) \in \mathbb{R}^m$ denotes an arbitrary vector and $\beta = \beta(q)$ is an $m \times m$ invertible matrix, then it is easily checked that the system $\dot{q} = \hat{f}(q) + \hat{G}(q)v$, with a control $v \in \mathbb{R}^m$, is another representation of the affine Pfaffian constraints (1). In this context, taking into account the observation from [4] that any two right inverses $\mathcal{A}^{\#1}$ and $\mathcal{A}^{\#2}$ of the matrix $\mathcal{A}$ are related as $\mathcal{A}^{\#1} = \mathcal{A}^{\#2} + GW$, where W is an $m \times m$ matrix, it may be justified to prefer in the definition of the drift vector field f the Moore-Penrose generalized inverse $\mathcal{A}^{+} = \mathcal{A}^T \left(\mathcal{A}\mathcal{A}^T\right)^{-1}$.

2.2 End-Point Map and Configuration Singularities

In the control-affine system (2) we let $T > 0$ denote a control time horizon. Then, the admissible control functions will be chosen the as Lebesgue square integrable functions of time defined on the interval $[0, T]$, with values in $\mathbb{R}^m$. These control functions belong to the Hilbert space $\mathcal{U} = L^2_m[0, T]$, equipped with the standard the inner product $< u(\cdot), \hat{u}(\cdot) >= \int_0^T u^T(t)\hat{u}(t)dt$. It will be assumed that the trajectory $q(t) = \varphi_{q_0,t}(u(\cdot))$ of the system (2), initialized at

$q(0) = q_0$ and steered by $u(\cdot)$, exists on $[0, T]$. With these assumptions we can define the end-point map $\mathcal{K}_{q_0,T} : \mathcal{U} \to \mathbb{R}^n$ of the system (2) as the value of its trajectory at the time instant T, i.e.

$$\mathcal{K}_{q_0,T}(u(\cdot)) = \varphi_{q_0,T}(u(\cdot)). \tag{3}$$

For the control-affine system the end-point map is continuously differentiable, with the derivative

$$J_{q_0,T}(u(\cdot)) = \mathsf{D}\mathcal{K}_{q_0,T}(u(\cdot))$$

that is called the Jacobian of system (2). For a given control function $v(\cdot) \in \mathcal{U}$ the Jacobian can be computed by the linear approximation to system (2) along the input–state trajectory $(u(t), q(t) = \varphi_{q_0,t}(u(\cdot)))$. It is well known that this linear approximation

$$\dot{\xi} = A(t)\xi + B(t)v \tag{4}$$

is determined by a pair of matrices $A(t) = \frac{\partial(f(q(t))+G(q(t))u(t))}{\partial q}$, $B(t) = G(q(t))$. Denoting by $\xi(t)$ the trajectory of the linear approximation (4) initialized at $\xi(0) = 0$, one derives the following formula for the Jacobian

$$J_{q_0,T}(u(\cdot))v(\cdot) = \xi(T) = \int_0^T \Phi(T,s)B(s)v(s)ds, \tag{5}$$

where $\frac{\partial \Phi(t,s)}{\partial t} = A(t)\Phi(t,s)$, $\Phi(s,s) = I_n$, is the fundamental matrix of (4).

For a fixed control function $u(\cdot)$ the Jacobian can be regarded as a linear map $J_{q_0,T}(u(\cdot)) : \mathcal{U} \longrightarrow \mathbb{R}^n$. Then, a control function $u(\cdot)$ is called regular (2), if the Jacobian map is surjective. Equivalently, this means that the Gram matrix

$$\mathcal{M}_{q_0,T}(u(\cdot)) = \int_0^T \Phi(T,t)B(t)B^T(t)\Phi^T(T,t)dt$$

has full rank $= n$. If the control function is not regular, i.e. the Gram matrix is rank deficient, $\operatorname{rank} \mathcal{M}_{q_0,T}(u(\cdot)) < n$, then $u(\cdot)$ is referred to as a configuration singularity of system (2). It is well known that the singular configurations are just the singular controls of the system (2), see [2].

The right inverse of the Jacobian is understood as any map $J^{\#}_{q_0,T}(u(\cdot)) : \mathbb{R}^n \longrightarrow \mathcal{U}$ such that $J_{q_0,T}(u(\cdot))J^{\#}_{q_0,T}(u(\cdot)) = I_n$. The most often encountered in the literature is the Jacobian pseudoinverse $J^{JP\#}_{q_0,T}$ that is defined at regular configurations as

$$\left(J^{JP\#}_{q_0,T}(u(\cdot))\eta\right)(t) = B^T(t)\Phi^T(T,t)\mathcal{M}^{-1}_{q_0,T}(u(\cdot))\eta. \tag{6}$$

2.3 Motion Planning

The motion planning problem in the control affine system (2) is stated in the following way: given an initial state q_0, a control horizon T, an initial control function $u_0(t)$, and a terminal state q_d, find a control function $u_d(\cdot)$ such that

$K_{q_0,T}(u_d(\cdot)) = q_d$. It is well known that the motion planning problem can be solved by a Jacobian algorithm based on the following differential equation

$$u'_\theta(\cdot) = -\gamma J^{\#}_{q_0,T}(u_\theta(\cdot))(K_{q_0,T}(u_\theta(\cdot)) - q_d), \quad u_{\theta=0}(\cdot) = u_0(\cdot), \tag{7}$$

where u' is the derivative with respect to the parameter $\theta \in \mathbb{R}$ and $J^{\#}_{q_0,T}$ denotes a right inverse of the Jacobian. Given a trajectory $u_\theta(\cdot)$ of this equation, and assuming that the necessary limit exists, the solution to the motion planning problem is obtained as the limit $u_d(t) = \lim_{\theta\to+\infty} u_\theta(t)$.

3 Feedback Equivalence and Normal Forms

Consider two control-affine systems

$$\sigma : \dot{x} = f(x) + G(x)u \quad \text{and} \quad \hat{\sigma} : \dot{\xi} = \hat{f}(\xi) + \hat{G}(\xi)v. \tag{8}$$

Hereabout $x, \xi \in \mathbb{R}^n$ and $u, v \in \mathbb{R}^m$ denote, respectively, the state variables and the control variables. The systems σ and $\hat{\sigma}$ are named feedback equivalent, if there exist a smooth change of coordinates $\xi = \varphi(x)$, and a smooth feedback $u = \alpha(x) + \beta(x)v$ containing an affine change of the control variable with invertible matrix $\beta(x)$, such that

$$D\varphi(x)(f(x) + G(x)\alpha(x)) = \hat{f}(\varphi(x)) \quad \text{and} \quad D\varphi(x)G(x)\beta(x) = \hat{G}(\varphi(x)).$$

In the case when the coordinate change $\varphi(x)$ exists only locally, we speak of the local feedback equivalence of these systems. By definition, the feedback equivalence is reflexive, symmetric and transitive, so in particular the feedback transformations are invertible.

A system $\hat{\sigma}$, feedback equivalent to the system σ whose behaviour is easily understood and controlled will be called a normal form of σ. If a normal form is known, having addressed any (analysis or synthesis) problem in the original system, we transform the system and the problem to the normal form, solve the problem in the normal form system, and finally, using invertibility of the feedback transformations, we get a solution of the problem for the original system.

In what follows we shall exploit the normal form approach in order to provide a description of configuration singularities, and to deliver a solution to a motion planning problem for a space manipulator.

4 Case Study

As an illustration of the concepts introduced in the previous sections we shall analyse a free-floating space manipulator designed in the Space Research Centre (SRC) of the Polish Academy of Sciences, called further the SRC manipulator [5]. The SRC manipulator is portrayed in Fig. 1. It is composed of a mobile base (a satellite) carrying on-board a 2DOF planar manipulator, and can move without the influence of the gravitation forces. The dynamic and geometric

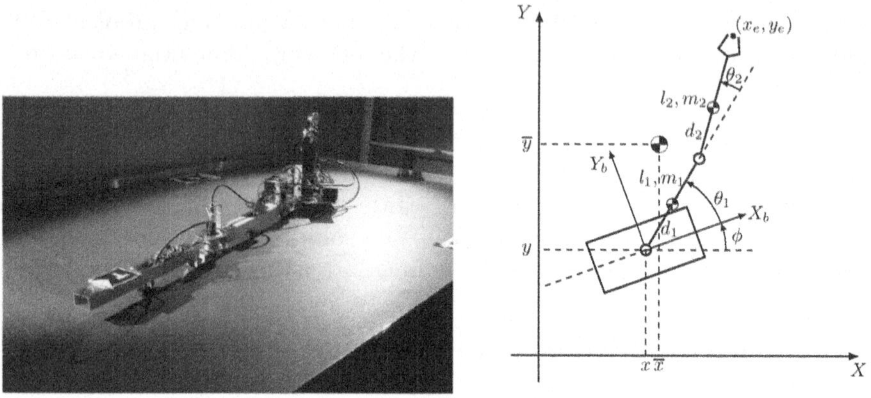

Fig. 1. SRC manipulator

parameters of the SRC manipulator can be found in [6]. Under assumption of the angular momentum conservation the SRC manipulator is subject to the affine Pfaffian constraints. Adopting the angular coordinates $(\theta_1, \theta_2, \phi) \in \mathbb{T}^3$ evolving on the 3-dimensional torus, the conservation of the angular momentum yields the affine Pfaffian constraints (1)

$$F(\theta_2)\dot{\phi} + G(\theta_2)\dot{\theta}_1 + H(\theta_2)\dot{\theta}_2 = p,$$

where $F(\theta_2) = I + B + C + 2D\cos\theta_2$, $G(\theta_2) = B + C + 2D\cos\theta_2$, $H(\theta_2) = C + D\cos\theta_2$, $p \neq 0$ denotes the constant angular momentum, and I, B, C and D are known mass and inertia parameters of the SRC manipulator. Equivalently, the dynamics of the SRC manipulator assume the following representation (2)

$$\dot{\theta}_1 = u_1,\ \dot{\theta}_2 = u_2,\ \dot{\phi} = a(\theta_2) + b(\theta_2)u_1 + c(\theta_2)u_2, \tag{9}$$

where $a(\theta_2) = \frac{p}{F(\theta_2)}$, $b(\theta_2) = -\frac{G(\theta_2)}{F(\theta_2)}$ and $c(\theta_2) = -\frac{H(\theta_2)}{F(\theta_2)}$. In [3] we have discovered two normal forms of system (9) under the feedback equivalence.

Theorem 1. *1. In a neighbourhood of a point $q \in \mathbb{T}^3$ for which $\theta_2 \neq 0, \pi$ system (9) is locally feedback equivalent to the normal form*

$$\dot{z}_1 = v_1,\ \dot{z}_2 = v_2,\ \dot{z}_3 = 1 + z_1 v_2. \tag{10}$$

2. The feedback establishing the equivalence of (9) to (10) is given by

$$z_1 = \cos\theta_2,\ z_2 = -\frac{D}{p}(2\phi + 2\theta_1 + \theta_2),$$
$$z_3 = \frac{1}{p}((I + B + C)\phi + (B + C)\theta_1 + C\theta_2), \quad u_2 = -\frac{1}{\sin\theta_2}v_1,$$
$$u_1 = -\frac{a(\theta_2)}{b(\theta_2) + 1} + \frac{c(\theta_2) + \frac{1}{2}}{(b(\theta_2) + 1)\sin\theta_2}v_1 - \frac{p}{2D(b(\theta_2) + 1)}v_2. \tag{11}$$

3. In a neighbourhood of a point $q \in \mathbb{T}^3$, such that $\theta_2 = 0, \pi$, system (9) is locally feedback equivalent to the normal form

$$\dot{z}_1 = v_1,\ \dot{z}_2 = v_2,\ \dot{z}_3 = 1 + z_1^2 v_2. \tag{12}$$

4. Around the point $\theta_2 = 0$ the feedback establishing the equivalence of (9) to (12) is given by

$$z_1 = \sqrt{2}\sin\frac{1}{2}\theta_2,\ z_2 = \frac{D}{p}(2\phi + 2\theta_1 + \theta_2),$$
$$z_3 = \frac{1}{p}((I+B+C+2D)\phi + (B+C+2D)\theta_1 + (C+D)\theta_2),\ u_2 = \frac{\sqrt{2}}{\cos\frac{\theta_2}{2}}v_1,$$
$$u_1 = -\frac{a(\theta_2)}{b(\theta_2)+1} - \frac{\sqrt{2}(c(\theta_2)+\frac{1}{2})}{(b(\theta_2)+1)\cos\frac{\theta_2}{2}}v_1 + \frac{p}{2D(b(\theta_2)+1)}v_2. \tag{13}$$

4.1 Configuration Singularities

An analysis of the configuration singularities of the normal forms (10) and (12) have been accomplished in the reference [3]. Using the property that singular controls are preserved by feedback, by inverting the feedback transformations (11) and (13) we have established the following characterization of configuration singularities of system (9).

Theorem 2. *Let $S = \{q = (\theta_1, \theta_2, \phi) \in \mathbb{T}^3 | \sin\theta_2 = 0\}$. Then:*

1. In the case $q \notin S$ ($z_1 \neq 0$) the configuration singularities of the normal form (10) are trivial, $v_1 = v_2 = 0$, so the configuration singularities of (9) are $u_1 = -\frac{a(\theta_2)}{b(\theta_2)+1} = const = \alpha$, and $u_2(t) = 0$. Along the singular curve $\theta_1(t) = \theta_1(0) + \alpha t$, $\theta_2(t) = \theta_2(0)$, and $\phi(t) = \phi(0) - \alpha t$. Also $\frac{d}{dt}(\phi(t) + \theta_1(t)) = 0$.
2. At the points $q \in S$ ($z_1 = 0$) the configuration singularities of the normal form (12) are $v_1 = 0$ and $v_2 = var$. This leads to the configuration singularities of (9) equal to $u_1 = var$ and $u_2 = 0$. The on-board manipulator stays in its singular position, while along the singular curve $a, b, c = const$, so $\theta_1(t) = var$, $\theta_2(t) = \theta_2(0)$ and $\phi(t) = \phi(0) + at + b(\theta_1(t) - \theta_1(0))$. It follows that $\frac{d}{dt}(\phi(t) + \theta_1(t)) = \frac{p}{2D}v_2$.

4.2 Motion Planning

In this section we shall address two motion planning problems in system (9). Problem 1 is defined by the initial state $q_0 = (0, 0, \frac{\pi}{6})$ and the terminal state $q_d = (\frac{\pi}{4}, -\frac{\pi}{4}, \frac{\pi}{2})$. Problem 2 corresponds to $q_0 = (0, 0, -\frac{\pi}{10})$, $q_d = (0, 0, \frac{\pi}{10})$; our intention was to force the system's trajectory to pass through the set S defined in Theorem 2. In both these problems we have set the control horizon to $T = 20$.

For comparative purposes these problems will be solved in three ways. First, we apply the Jacobian motion planning algorithm based on (6) and (7) directly to system (9). Then, we transform the motion planning problem by feedback to the normal form (10) or (12), solve the problem in the normal form using the same Jacobian motion planning algorithm, and transform the solution back to the original system (9). Finally, we again transform the problem to the normal forms, solve it by means of the sinusoid control algorithm, see [7] (Chap. 8), and transfer the solution back to the original system. These three solutions have been marked, respectively, as ECS+NH, ECS+NF and Sin+NF. In the applications of the Jacobian motion planning algorithms we have fixed the convergence rate as $\gamma = 0.02$, and used only the constant terms and the first order harmonics with the frequency $\omega = \frac{2\pi}{T}$ as well as suitably chosen initial values of the control coefficients. The sinusoid controls have been computed classically. Results of

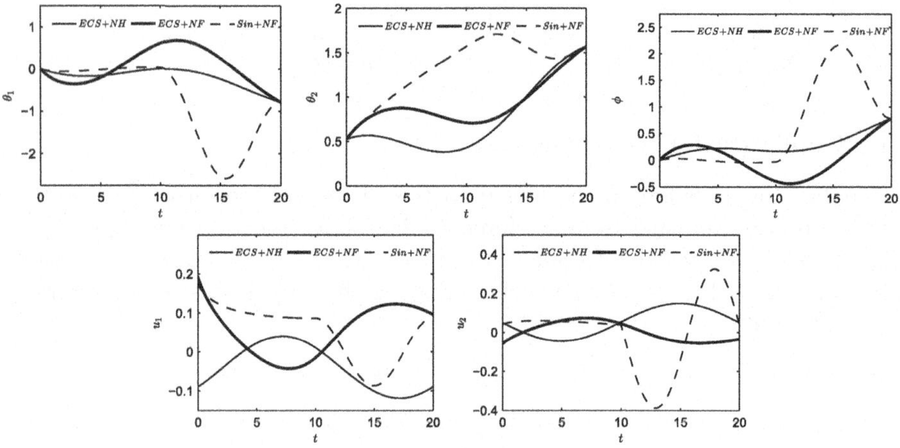

Fig. 2. Results for problem 1

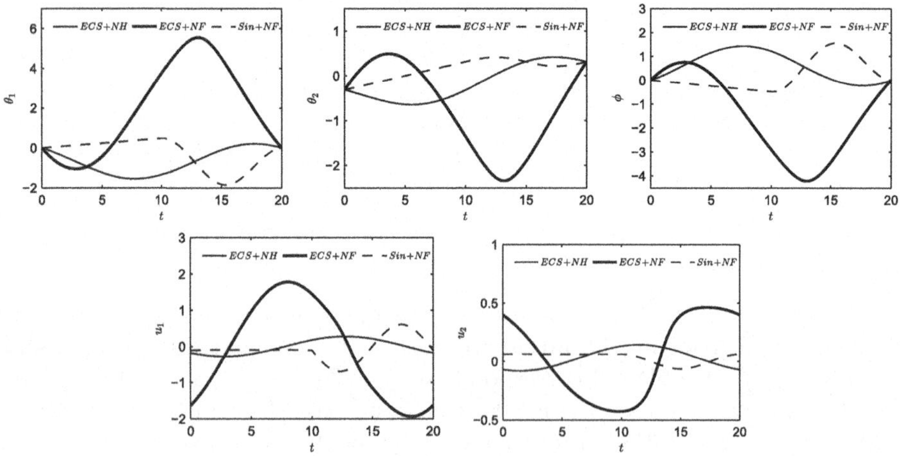

Fig. 3. Results for problem 2

computations are displayed below in Figs. 2 and 3. In the latter figure it is easily noticed that the normal form (12) can produce a trajectory passing through set S. This is also true for ECS-based algorithms which break down only at configuration singularities that are not related with S. We have observed that on a processor 3.4 GHz equipped with 8 GB RAM the computation times are in the range: ECS+NH – 7 s, CS+NF – 4.5 s and Sin+NF – 0.1 s, what could be expected since the sinusoid control algorithm is not iterative whatsoever.

5 Conclusions

In an attempt at extending the normal form approach from holonomic manipulators to mobile robots we have shown an application of normal forms of a space manipulator to describing its singularities and to solving some motion planning problems.

Acknowledgement. Author 2 was assisted by the National Science Centre Poland, grant decision No. 2015/17/B/ST7/03995. Authors 1 and 3 were supported by the Wrocław University of Science and Technology under a research project number 0401/0143/17.

References

1. Tchoń, K., Muszyński, R.: Singular inverse kinematic problem for robotic manipulators: a normal form approach. IEEE Trans. Robot. Autom. **14**, 93–104 (1998)
2. Tchoń, K.: Endogenous configuration space approach: an intersection of robotics and control theory. In: van de Wouw, N., Lefeber, E., Lopez Arteaga, I. (eds.) Nonlinear Systems. LNCIS, vol. 470, pp. 209–234. Springer, Cham (2017). https://doi.org/10.1007/978-3-319-30357-4_10
3. Tchoń, K., Respondek, W., Ratajczak, J.: Normal forms and configuration singularities of a space manipulator. J. Intell. Robot. Syst. (2018). https://doi.org/10.1007/s10846-018-0883-8
4. Roberts, R.G., Maciejewski, A.A.: Nearest optimal repeatable control strategies for kinematically redundant manipulators. IEEE Trans. Autom. Control **8**, 327–337 (1992)
5. Rybus, T., Seweryn, K.: Planar air-bearing microgravity simulators: review of applications, existing solutions and design parameters. Acta Astronaut. **120**, 239–259 (2016)
6. Rybus, T.: Control of a satellite manipulator during the interception maneuver and the motion stabilization. Ph.D. dissertation. Wrocław University of Science and Technology (2016)
7. Murray, R.M., Li, Z., Sastry, S.S.: A Mathematical Introduction to Robotic Manipulation. CRC Press, Boca Raton (1994)

Experimental Identification of Stress-Strain Material Models of UHMWPE Fiber Cables for Improving Cable Tension Control Strategies

Philipp Tempel(✉), Felix Trautwein, and Andreas Pott

Institute for Control Engineering of Machine Tools and Manufacturing Units ISW, University of Stuttgart, Stuttgart, Germany
{philipp.tempel,felix.trautwein}@isw.uni-stuttgart.de,
andreas.pott@ipa.fraunhofer.de

Abstract. Ultra-high-molecular-weight polyethylene fibers like Dyneema or Spectra are employed in a vast variety of cable-driven parallel robots. The stress-strain dynamics of such cables are highly non-linear with time-varying mechanical parameters, resulting in involved modeling and control of robot dynamics. To improve controllability of cable robots, the cable stress-strain dynamics need to be known and explicitly considered feedforward or closed-loop control. A model can only be deemed suitable, if its inherent dynamics is confirmed through experiments and if it allows for reasonable parameter estimation. We present results of experimental identification of stress-strain dynamics of UHMWPE cables made of Dyneema in different stages of operation. Due to the internal material mechanics, four stages can be identified: the tensing and relaxing transition as well as plateaus coming from either. The implications of verified and parametrized stress-strain models for cable robot tension control strategies is expedited.

Keywords: Cable-driven parallel robots · Cable strain
UHMWPE fibers · Dyneema

1 Introduction

Cable-driven parallel robots (herein *cable robots*) are parallel kinematics with their rigid links replaced by elastic and flexible cables. Using this setup enables such systems to achieve high dynamics, high payloads, and large workspaces by reducing the actuated inertia to a minimum since only the mobile platform and the comparatively light-weight cables need to be accelerated. However, this merit is also the design's inherent flaw: cables can only exert tensile forces and their force transmission behavior from winch to mobile platform is different to their rigid counterparts like prismatic joints. Most notably, lightweight cables are made from laying several hundred thin fibers into strains which in turn are

J. Lenarcic and V. Parenti-Castelli (Eds.): ARK 2018, SPAR 8, pp. 258–265, 2019.
https://doi.org/10.1007/978-3-319-93188-3_30

woven into the cable. It is no surprise that a single 6 mm diameter cable consists of thousands of such fibers. This mechanical construction of these fiber cables inadvertently influences the stress-strain dynamics yielding, on a macroscopic level, an elastoviscoplastic system with the viscoelastics being the short-term dominant effect.

Ultimate goal in operating a cable robot is to position its planar or spatial mobile platform in space through change of length of the driving cables. Additionally, the static or dynamic external wrench on the platform must be compensated for which requires cable tension control. Conventionally, fiber cables in cable robots are considered linearly elastic as spring or spring-damper system for simplification of the kinematics and dynamics [1,5]. Different elastohysteretic cable models for cable-driven serial and parallel robots of small scale were developed e.g., [7,8], while dynamics and stability of suspended cable robots with viscoelastic stress-strain models is also investigated e.g., [6]. Recently, Piao et al. presented a cable model based on Burgers material model extended by another Kelvin-Voigt material element [9]. The model's three elastic and two viscoelastic parameters were estimated using polynomial fit of the instantaneous stress response dynamics during loading and unloading i.e., without considering any dynamic excitation input.

While in biomechanics, experimental validations of material models have been presented e.g., [2,12], contributions to validation of existing material models for fiber cables used in cable robots are limited and have almost exclusively focused on small diameter cables i.e., $D \leq 2$mm. For larger diameter fiber cables, elaborate validation and parameter estimation is still given short shrift, which also includes understanding long-term effects of stress relaxation i.e., creep. Since the operational state of a cable robot alters between motion and stationary points, stress relaxation after tensing impacts stiffness at these stationary points. Simple control of cable tension may be employed to compensate for creep, yet a model-based feed-forward creep-compensation requires less actuating and sensing capabilities, since stress and strain are inherently mechanically linked. Independent of the actual implementation of cable robot control and tension control, a model with only few parameters suitable for describing the stress-strain dynamics will lead to significant impact on tension control strategies and thus robot accuracy.

This paper focuses on experimentally validating stress-strain material models for use in cable robots given operationally characteristic stress trajectories. In particular, we probe Liros D-Pro SK78 6 mm (used in cable robots of the IPAnema family [10]) and compare it against Liros D-Pro XTR SK99 6 mm, supposedly having higher stiffness thus having smaller dynamic strain. This contribution is structured as follows: Sect. 2 introduces the models under investigation and their constitutive stress-strain equations. In Sect. 3, we present the experimentally obtained data, while Sect. 4 gives qualitative and quantitative insight into the obtained results, and discusses the effect of the investigated behavior on tension control strategies. Lastly, Sect. 5 closes with concluding remarks and presentation of future steps.

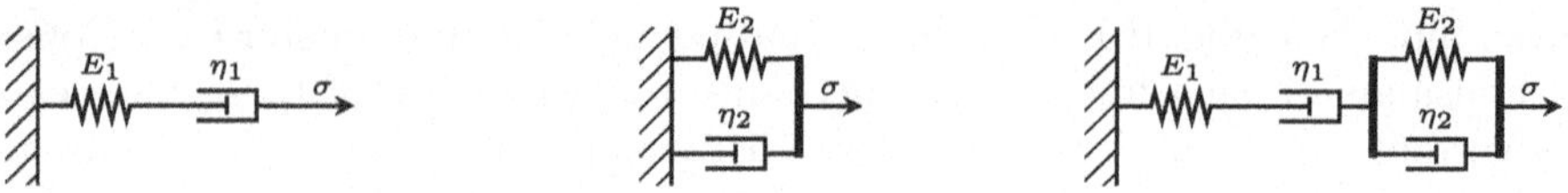

(a) Maxwell material (b) Kelvin-Voigt material (c) Burgers material

Fig. 1. Possible viscoelastic material models for stress-strain dynamics of Liros D-Pro6 mm and Liros D-Pro XTR6 mm.

2 Material Models for Stress-Strain Dynamics

Models describing purely elastic or viscoelastic or plastoelastic materials are composed of linear springs and dampers [14]. A viscoelastic material can be modeled as either Maxwell material (cf. Fig. 1(a)) or Kelvin-Voigt material (cf. Fig. 1(b)); the serial combination of both these models is called Burgers model (cf. Fig. 1(c)).[1] More detailed models can be implemented as combination of Mawell/Kelvin-Voigt material with additional serial or parallel springs or dampers. Each material model describes a certain characteristic stress-strain dynamics as inherent material property, which will be briefly covered in the following.

Maxwell Material describes viscoelastic properties with the constitutive equation

$$\varepsilon = \frac{1}{E_1}\dot{\sigma} + \frac{1}{\eta_1}\sigma, \tag{1a}$$

and its characteristic solution

$$\varepsilon(t) = \sigma\left(\frac{1}{E_1} + \frac{1}{\eta_1}t\right), \tag{1b}$$

showing a linear response behavior for the stress given a step strain input with offset $\varepsilon(0) = \sigma/E_1$. Therefore, the Maxwell model is not suitable as linear creep strain at constant stress does not conform with experiments.

Kelvin-Voigt Material describes viscoelastic behavior that is affected directly by the retarded dynamics of the linear damper yielding the constitutive equation

$$\eta_2\dot{\varepsilon} + E_2\varepsilon = \sigma, \tag{2a}$$

with its characteristic solution

$$\varepsilon(t) = \frac{\sigma}{E_2}\left(1 - \exp\left(-\frac{E_2}{\eta_2}t\right)\right), \tag{2b}$$

showing a first-order lag behavior with a settling strain $\varepsilon\,(t \to \infty) = \sigma/E_1$. This response behavior does not match experimental data, as it does not capture an instantaneous strain response (cf. Fig. 3(a)).

[1] We follow mechanical notation with ε being the material's total strain and σ the applied stress.

Burgers Model consists of Maxwell material and Kelvin-Voigt material in series (cf. Fig. 1(c)) yielding the constitutive equation

$$\frac{\eta_1\eta_2}{E_2}\ddot{\varepsilon} + \eta_1\dot{\varepsilon} = \frac{\eta_1\eta_2}{E_1E_2}\ddot{\sigma} + \left(\frac{\eta_1}{E_1} + \frac{\eta_1}{E_2} + \frac{\eta_2}{E_2}\right)\dot{\sigma} + \sigma \tag{3a}$$

with the characteristic solution

$$\varepsilon(t) = \frac{\sigma}{E_1}\left(1 + \frac{E_1}{\eta_1}t + \frac{E_1}{E_2}\left(1 - \exp\left(-\frac{E_2}{\eta_2}t\right)\right)\right), \tag{3b}$$

showing an inclined asymptote for strain increases with decreasing strain rate. Burgers model matches experimental measurements best as it captures both infinite increase of the strain as well as instantaneous strain response.

Table 1. Mechanical properties of cables Liros D-Pro and Liros D-Pro XTR.

Cable	Diameter (mm)	Breaking Load (daN)	Weight (g/m)
Liros D-Pro	6	4300	23
Liros D-Pro XTR	6	5600	23

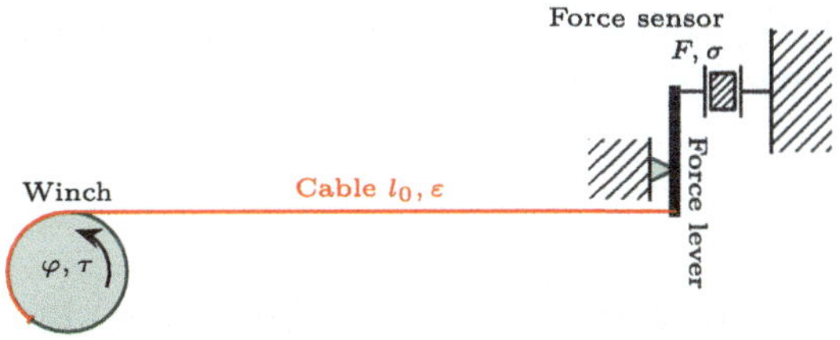

Fig. 2. Sketch of the experimental setup with the winch at the cable proximal end and a force sensor at the cable distal end.

3 Experimental Identification

We perform experiments with two different types of cables: Liros D-Pro 6 mm and Liros D-Pro XTR 6 mm (cf. Table 1 for the mechanical properties according to the manufacturer). While Liros D-Pro is a well known cable used in several different cable robots e.g., [10,11,13], Liros D-Pro XTR has not been presented in any cable robot application. The manufacturer promotes the cable as having small working stretch. Especially high payload applications with large strain-to-breaking load factor may benefit from Liros D-Pro XTR.

The stress-strain cable test bench for experimental identification, based on [3,4] and shown in Fig. 2, is comprised of a continuous axis servo drive attached

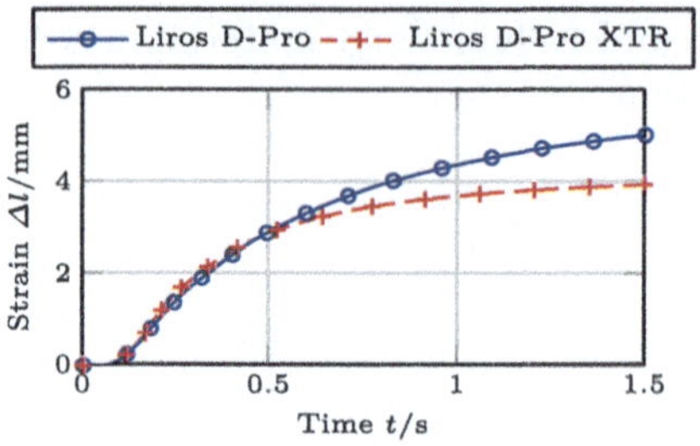

(a) Average strain response trajectories for step stress input exeriment to 450 N at $t_{\text{step}} = 0.10\,\text{s}$.

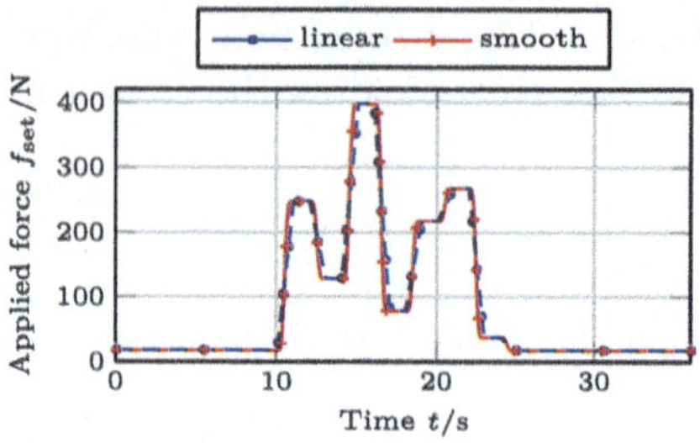

(b) Linear and differentially smooth stress input trajectories for parameter estimation and validation.

Fig. 3. Stress step response of Liros D-Pro and Liros D-Pro XTR (a); experimental trajectory used for validation of estimated parameters (b).

to the cable drum of diameter 190 mm changing the strained cable length. The cable distal end is located at a horizontal distance of 2.78 m where it is connected via a lever to a force sensor. Given the force sensor reading and the cable's linear strain from the drive's angular position, the input/output to the stress-strain models are available.

At first, we obtain the parameter values from the stress step-response shown in Fig. 3(a), performing a nonlinear least square curve fitting over the parameter set $\boldsymbol{\xi} = \{E_1, E_2, \eta_1, \eta_2\}$. These parameters are then used as initial guess for estimation of the dynamics-based parameters with an input as shown in Fig. 3(b).

4 Results

Stress-strain dynamics of both Liros D-Pro and Liros D-Pro XTR cables can be described well using the Burgers model, even under real world assumptions of stress inputs (see Figs. 4 and 5). Over a strain of 4 mm for Liros D-Pro and 5 mm for Liros D-Pro XTR, the error between simulative strain response and experimentally acquired is less than 0.5 mm and 0.4 mm, respectively, or equivalently 12% and 8%, respectively. Of special interest are the areas of constant stress and after the cable stress is removed. The strain settling behavior for both cables is matched very well and the error between simulation and experiment vanishes over time. Lastly we can also show that Liros D-Pro XTR is of higher stiffness matching the data sheet properties well (cf. Table 1). This can be very advantageous for that the cables elongate less over the same range of stress than Liros D-Pro.

Given the current estimation procedure is not considering any drive dynamics i.e., the cable linear strain and its time-derivative are directly related to the angular position and velocity of the servo drive, the estimated mechanical parameters still provide a close match between model prediction and experimentally observed dynamics. Stress relaxation at the end of motion is represented

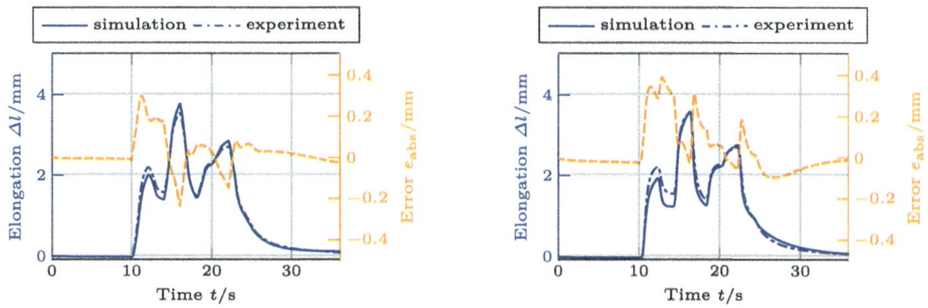

(a) Identified and experimental strain of Liros D-Pro for linear stress transition.

(b) Identified and experimental strain of Liros D-Pro for smooth stress transition.

Fig. 4. Simulated and experimental strain of Liros D-Pro for linear stress transition trajectory (a) and smooth stress transition trajectory (b).

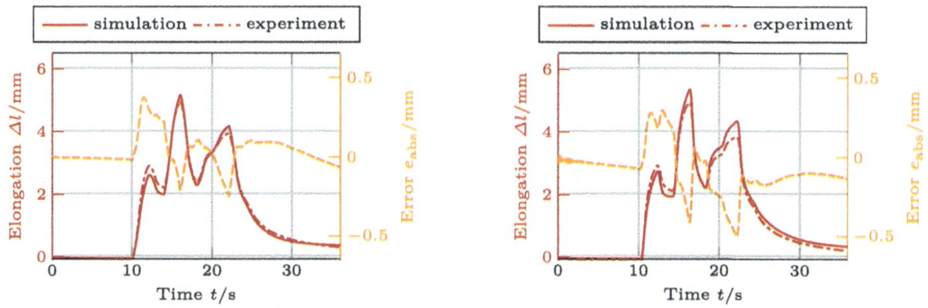

(a) Identified and experimental strain of Liros D-Pro XTR for linear stress transition.

(b) Identified and experimental strain of Liros D-Pro XTR for smooth stress transition.

Fig. 5. Simulated and experimental strain of Liros D-Pro XTR for linear stress transition trajectory (a) and smooth stress transition trajectory (b).

well in the simulative validation. Predicting this behavior is of high interest as such validated and parametrized models can then be used for closed-loop controller design which goes beyond the scope of a simple Maxwell material model. Additionally, the estimated model parameters, as stated in Table 2, are within a range that is to be expected from literature and the manufacturer's data sheets.

Deviation between simulative data and experimental data lies in both physical phenomena as well as inconsistencies of the initial conditions and environmental constraints. While the model assumptions assume both ends of the cable to be rigidly fixed to a ground and an ideally moving object, respectively, the experimental rig setup has neither. On the distal end i.e., on the force sensor, the cable is also wound around a cylinder. While several windings around both the actuating drum and the sensing drum allow for a free cable end, stick-slip effects may still be observed—this is not considered in the parameter estimation model. Additional consideration or compensation of gravitational forces on the cable may further improve the estimation results since the experimental setup is subject to a unified load on the cable (cf. Fig. 2). Especially when transitioning from a high-tension to a low-tension state, this unified load introduces transversal waves—which can be observed visually during experiments—affecting the measured cable tension.

Table 2. Estimated Burgers model parameters based on step response and dynamic trajectory response.

Parameter	Liros D-Pro		Liros D-Pro XTR	
	Step-Response	Motion	Step-Response	Motion
E_1 [MPa]	76,686.5	14,928.4	102,706.0	10,883.8
E_2 [MPa]	539.2	21,664.5	397.1	13,451.4
η_1 [MPa/s]	53,754.6	441,482.1	20,069.0	718,093.1
η_2 [MPa/s]	257.8	66,862.7	264.6	37,734.1

5 Conclusion and Perspective

Burgers model for viscoelastic materials describes the stress-strain dynamics of both Liros D-Pro and Liros D-Pro XTR well during changes in stress but also captures strain relaxation after stress decline. Applying Burgers model to the stress-strain dynamics of both cables leads to better matching of simulation and experimental observation. We can also observe that a smoother stress transition trajectory provides decreased cable strain as the contribution of instantaneous strain response can be reduced drastically. With better understanding and parameter identification of cable stress-strain models, cable robot tension control and accuracy can be greatly improved.

To further assess applicability of the models and also consider the fact of changing nominal length as well as cable-pulley interaction, the experiments need to be more diversified. We will begin by altering the nominal cable length between winch and force sensor and investigating the effect of this parameter on the scale of the model's characteristic parameters. Additionally, we may also add pulleys to the setup to model the loss of stress over the pulley due to frictional forces. After deeper assessment of the model parameters, an offline feedforward control scheme will be derived to operate a 6 degree of freedom 8 cable robot.

Acknowledgements. The authors Ph. T. and A. P. would like to thank the German Research Foundation (DFG) for financial support of the project within the Cluster of Excellence in Simulation Technology (EXC 310/2) at the University of Stuttgart.

References

1. Bruckmann, T.: Auslegung und Betrieb redundanter paralleler Seilroboter. PhD Thesis, Universität Duisburg-Essen, Duisburg-Essen, Germany (2010)
2. Budday, S., Sommer, G., Birkl, C., Langkammer, C., Haybaeck, J., Kohnert, J., Bauer, M., Paulsen, F., Steinmann, P., Kuhl, E., Holzapfel, G.A.: Mechanical characterization of human brain tissue. Acta biomaterialia **48**, 319–340 (2017)
3. Czichos, H., Saito, T., Smith, L.: Springer Handbook of Materials Measurement Methods. Springer, Heidelberg (2006)
4. Findley, W.N., Lai, J.S., Onaran, K.: Creep and relaxation of nonlinear viscoelastic materials: With an Introduction to Linear Viscoelasticity. Dover and London: Constable, New York (1989, 1976)
5. Kawamura, S., Choe, W., Tanaka, S., Pandian, S.R.: Development of an ultrahigh speed robot FALCON using wire drive system. In: ICRA IEEE International Conference on Robotics and Automation, pp. 215–220 (1995)
6. Korayem, M.H., Yousefzadeh, M., Beyranvand, B.: Dynamics and control of a 6-dof cable-driven parallel robot with visco-elastic cables in presence of measurement noise. J. Intell. Robot. Syst. **88**(1), 73–95 (2017)
7. Miermeister, P., Kraus, W., Lan, T., Pott, A.: An elastic cable model for cable-driven parallel robots including hysteresis effects. In: Pott, A., Bruckmann, T. (eds.) Cable-Driven Parallel Robots. MMS, vol. 32, pp. 17–28. Springer, Cham (2015)
8. Miyasaka, M., Haghighipanah, M., Li, Y., Hannaford, B.: Hysteresis model of longitudinally loaded cable for cable driven robots and identification of the parameters. In: ICRA IEEE International Conference on Robotics and Automation, pp. 4051–4057 (2016)
9. Piao, J., Jin, X., Jung, J., Choi, E., Park, J.O., Kim, C.S.: Open-loop position control of a polymer cable-driven parallel robot via a viscoelastic cable model for high payload workspaces. Adv. Mech. Eng, 9(12) (2017)
10. Pott, A., Mütherich, H., Kraus, W., Schmidt, V., Miermeister, P., Verl, A.: IPAnema: a family of cable-driven parallel robots for industrial applications. In: Bruckmann, T., Pott, A. (eds.) Cable-Driven Parallel Robots, Mechanisms and Machine Sciences, vol. 12, pp. 119–134. Springer, Berlin Heidelberg, Berlin, Heidelberg (2013)
11. Reichert, C., Müller, K., Bruckmann, T.: Robust internal force-based impedance control for cable-driven parallel robots. In: Pott, A., Bruckmann, T. (eds.) Cable-Driven Parallel Robots, Mechanisms and Machine Sciences, pp. 131–143. Springer, Cham (2015)
12. Romanyk, D.L., Melenka, G.W., Carey, J.P.: Modeling stress-relaxation behavior of the periodontal ligament during the initial phase of orthodontic treatment. J. Biomech. Eng. **135**(9), 091007 (2013)
13. Schmidt, V., Kraus, W., Ho, W.Y., Seon, J., Pott, A., Park, J.O., Verl, A.: Extending dynamic trajectories of cable-driven parallel robots as a novel robotic roller coaster. In: ISR/Robotik 41st International Symposium on Robotics, pp. 1–7 (2014)
14. Skrzypek, J.J., Ganczarski, A.W.: Constitutive equations for isotropic and anisotropic linear viscoelastic materials. In: Skrzypek, J.J., Ganczarski, A.W. (eds.) Mechanics of Anisotropic Materials, Engineering Materials, pp. 57–85. Springer, Cham (2015)

Modelling

A General Discretization-Based Approach for the Kinetostatic Analysis of Closed-Loop Rigid/Flexible Hybrid Mechanisms

Genliang Chen(✉), Zhuang Zhang, Zhengtao Chen, and Hao Wang

The State Key Laboratory of Mechanical System and Vibration, Shanghai Jiao Tong University, Shanghai 200240, China
{leungchan,z.zhang,cztllg,wanghao}@sjtu.edu.cn

Abstract. This paper presents a general approach for the kinetostatic modeling and analysis of closed-loop mechanisms consisting of both rigid and flexible links. Based on the principal axes decomposition of structural compliance matrix, the flexible links are approximated by hyperredundant linkages with rigid bodies connected by passive elastic joints. Hence, the kinetostatic analysis of rigid/flexible hybrid mechanisms can be formulated as static equilibrium problems of the approximate linkages which are regarded as hyperredundant underactuated multibody systems. By taking advantage of mechanism kinematics/statics, the large deflection problems of flexible links can be efficiently solved using the presented discretization-based approach. In order to demonstrate the effectiveness of the proposed method, a hybrid planar four-bar mechanism is provided as an example. And FEA simulations have been conducted to verify the correctness of obtained results.

Keywords: Kinetostatic analysis · Rigid/flexible hybrid mechanisms
Discretization-based approach

1 Introduction

Compliant mechanisms [1] have been widely used because of their inherent advantages, such as compactness in structure, wear and backlash reduction, and conservation of energy [2,3]. Due to the flexible structure, large deflection problems usually occur in compliant mechanisms. In these situations, solving the large deflection problems of flexible links effectively and efficiently becomes one of the most fundamental issues in the design, analysis and control of these mechanisms.

To overcome this problem, many researchers have made their contributions during the past few years and several approaches were proposed to solve the large deflection problems. Based on the Bernoulli-Euler beam assumption, classical elliptic integral [4,5] was used to derive the closed-form solution to the second-order, nonlinear differential equations. Howell and Midha [6] introduced

J. Lenarcic and V. Parenti-Castelli (Eds.): ARK 2018, SPAR 8, pp. 269–276, 2019.
https://doi.org/10.1007/978-3-319-93188-3_31

the pseudo-rigid-body model to the approximate path coordinates of flexible beams, which has been successfully applied to the synthesis and analysis of compliant mechanisms [7,8]. Additionally, discretization-based techniques were also employed to solve the large deflection problems, including Frame finite element method [9], chain algorithm [10], beam-constraint model [11] and so on.

In this paper, a general discretization-based approach is proposed for the kinetostatics modeling and simulation of mechanisms consisting of both rigid and flexible links. Based on the principal axes decomposition of structural compliance matrices [12], a particular type of elements is established to approximate the force-deflection behavior of discretized small segments via passive linkages with elastic joints. Then, the kinetostatics analysis can be transformed to the static equilibrium simulation of hyper-redundant underactuated mechanisms of active, passive, and elastic joints. The main advantage of the proposed method comes from the use of linkage kinematics/statics, rather than elastic mechanics. Thus, closed-form solutions to motion/force transmission properties can be developed in a straightforward manner. And the static equilibrium configurations of the approximation mechanism can be identified efficiently by means of gradient-based searching algorithms.

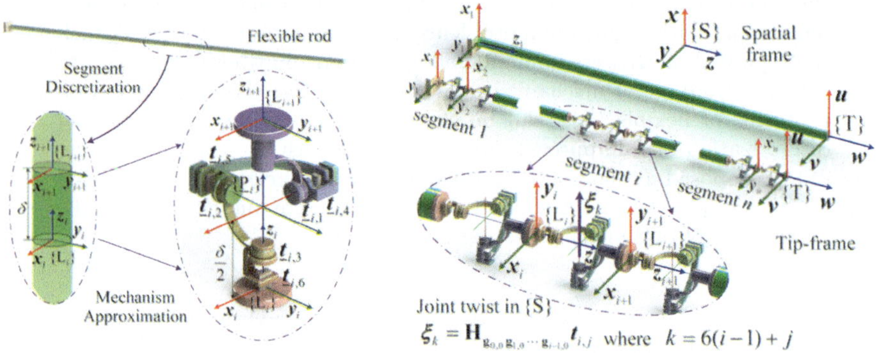

Fig. 1. Discretization-based approximation of flexible links.

2 Linkage Approximation of Flexible Links

In this section, the linkage approximation method for the flexible links in hybrid mechanisms is presented as the fundamental of the proposed approach.

2.1 Stiffness Modeling of Segment Elements

As shown in Fig. 1, in the proposed approach, the flexible links are discretized into small segments. And the force-deflection behavior of these segment elements is approximated by passive elastic linkages based on the principal axes decomposition of their structural compliance matrices [12].

For each discretized segment, the approximation linkage can be determined according to the joint twists and compliance, which are obtained by the following rank-one decomposition of compliance matrix as

$$\mathbf{C}_i = \begin{bmatrix} \frac{\delta}{EI_{xx}} & 0 & 0 & 0 & -\frac{\delta^2}{2EI_{xx}} & 0 \\ 0 & \frac{\delta}{EI_{yy}} & 0 & \frac{\delta^2}{2EI_{yy}} & 0 & 0 \\ 0 & 0 & \frac{\delta}{GI_{zz}} & 0 & 0 & 0 \\ 0 & \frac{\delta^2}{2EI_{yy}} & 0 & \frac{\delta^3}{3EI_{yy}} & 0 & 0 \\ -\frac{\delta^2}{2EI_{xx}} & 0 & 0 & 0 & \frac{\delta^3}{3EI_{xx}} & 0 \\ 0 & 0 & 0 & 0 & 0 & \frac{\delta}{EA} \end{bmatrix} = \sum_{j=1}^{6} c_{i,j} \boldsymbol{t}_{i,j} \boldsymbol{t}_{i,j}^T \tag{1}$$

where $\mathbf{C}_i$ is the structural compliance matrix of the discretized segment. δ and A are the segment's length and area of cross section. I_{xx}, I_{yy}, and I_{zz} are the three principal area moments. E and G are the elastic and shear modulus, respectively. While the joint twists and compliance, namely $\boldsymbol{t}_{i,j}$ and $c_{i,j}$, can be given by

$$\begin{cases} \boldsymbol{t}_1 = \begin{bmatrix} \boldsymbol{e}_1 \\ \boldsymbol{0} \end{bmatrix}, & \boldsymbol{t}_2 = \begin{bmatrix} \boldsymbol{e}_2 \\ \boldsymbol{0} \end{bmatrix}, & \boldsymbol{t}_3 = \begin{bmatrix} \boldsymbol{e}_3 \\ \boldsymbol{0} \end{bmatrix}, & \boldsymbol{t}_4 = \begin{bmatrix} \boldsymbol{0} \\ \boldsymbol{e}_1 \end{bmatrix}, & \boldsymbol{t}_5 = \begin{bmatrix} \boldsymbol{0} \\ \boldsymbol{e}_2 \end{bmatrix}, & \boldsymbol{t}_6 = \begin{bmatrix} \boldsymbol{0} \\ \boldsymbol{e}_3 \end{bmatrix} \\ c_1 = \frac{\delta}{EI_{xx}} & c_2 = \frac{\delta}{EI_{yy}} & c_3 = \frac{\delta}{GI_{zz}} & c_4 = \frac{\delta^3}{3EI_{yy}} & c_5 = \frac{\delta^3}{3EI_{xx}} & c_6 = \frac{\delta}{EA} \end{cases} \tag{2}$$

where $\boldsymbol{e}_1 = [1, 0, 0]^T$, $\boldsymbol{e}_2 = [0, 1, 0]^T$ and $\boldsymbol{e}_3 = [0, 0, 1]^T$ are unit vectors.

According to (1), the force-deflection behavior of a segment element can be approximately replaced by a spatial passive linkage of three prismatic and three revolute elastic joints. As illustrated in Fig. 1, the three decomposed revolute joints are mutually perpendicular to each other and concurrently pass through the segment's center of compliance, so are those prismatic ones. Meanwhile, the joint compliance can be identified readily from the geometrical parameters and mechanical property of flexible links.

Thus, the deflection of segment can be determined by the static equilibrium configuration satisfying

$$\begin{cases} \mathbf{g}_i(\boldsymbol{\theta}_i) = \exp(\hat{\mathbf{t}}_{i,1}\theta_{i,1}) \cdots \exp(\hat{\mathbf{t}}_{i,6}\theta_{i,6})\, \mathbf{g}_{i,0} \\ \boldsymbol{\tau}_i = \mathbf{K}_i \boldsymbol{\theta}_i = \mathbf{J}_i^T(\boldsymbol{\theta}_i) \boldsymbol{F}_i \end{cases} \tag{3}$$

where $\hat{\mathbf{t}}_{i,j} \in se(3), j = 1, \cdots, 6$ denotes the standard representation of the joint twists. Here, the joint variables $\boldsymbol{\theta}_i = [\theta_{i,1}, \cdots, \theta_{i,6}]^T$ relate to the deflections of the elastic joints. $\mathbf{g}_{i,0} \in SE(3)$ is the initial pose of the local frame $\{\mathrm{L_{i+1}}\}$ with respect to $\{\mathrm{L_i}\}$ in an undeformed configuration. While the 6×6 diagonal matrix $\mathbf{K}_i = \mathrm{diag}(\frac{1}{c_{i,1}}, \cdots, \frac{1}{c_{i,6}})$ relates to the stiffness of elastic joints. Hence, $\boldsymbol{\tau}_i$ represents the joints' forces/torques due to elastic deflection. $\mathbf{J}_i$ is the Jacobian matrix of approximation linkage, which is a function of joint deflections. And $\boldsymbol{F}_i$ is the external wrench which can be considered, to some extent, as stress exerted on the discretized segment.

Therefore, the equilibrium configuration of segments can be achieved as the reaction forces/torques due to external load are balanced by the elastic deflection of passive joints. It should be noted that, although the proposed approximation is derived based on the linearized force-deflection relation of segments, namely the structural compliance matrices, nonlinear effect can be taken into account by means of the linkage's forward displacement analysis, rather than its first-order approximation.

2.2 Force-Deflection Analysis of Flexible Links

By assembling the segments' approximation linkages together, a hyper-redundant rigid-body linkage with passive elastic joints, as shown in Fig. 1, can be obtained for a flexible link. Then, the force-deflection characterization of a flexible link can be accomplished via the static equilibrium analysis of the corresponding passive hyper-redundant elastic linkage.

Assembling (3) for all segments, the deflection of linkage can be obtained as

$$\begin{cases} \mathbf{g}_{st}(\boldsymbol{\theta}) = \mathbf{g}_0 \prod_{i=1}^{n} \mathbf{g}_i(\boldsymbol{\theta}_i) = \prod_{k=1}^{6n} (\exp(\hat{\boldsymbol{\xi}}_k \theta_k))\, \mathbf{g}_{st,0} \\ \boldsymbol{\tau} = \mathbf{K}\boldsymbol{\theta} = \mathbf{J}^T(\boldsymbol{\theta})\, \boldsymbol{F} \end{cases} \tag{4}$$

where $\boldsymbol{\theta} = [\theta_1, \cdots, \theta_N]^T \in \mathbb{R}^{N\times 1}$ denotes the joint variables of the whole hyper-redundant linkage. $\boldsymbol{\xi}_k = \mathrm{Ad}\left(\mathbf{g}_{0,0} \cdots \mathbf{g}_{i-1,0}\right) \boldsymbol{t}_{i,j}$ are the joint twists, transformed from their local frames to the spatial one {S}. $\mathbf{g}_{st,0} = \mathbf{g}_{0,0}\, \mathbf{g}_{1,0} \cdots \mathbf{g}_{n,0}$ relates to the initial pose of {T} with respect to {S}. The subscript $k = 6(i-1)+j$, where $i = 1, \cdots, n, j = 1, \cdots, 6$. And $N = 6n$ is the total number of joints. The overall joint stiffness matrix $\mathbf{K}$ is defined in a similar way as the segment one. And $\boldsymbol{F}$ corresponds to the external wrench exerted at the tip frame {T}.

In the proposed formulation, the product-of-exponential (POE) formula is employed to establish the forward kinematics. Thus, the Jacobian matrix can be derived in a concise and compact form as

$$\mathbf{J}_\theta = \left(\frac{\partial \dot{\mathbf{g}}_{st}}{\partial \boldsymbol{\theta}} \mathbf{g}_{st}^{-1}\right)^\vee = [\, \boldsymbol{\xi}_1',\ \boldsymbol{\xi}_2',\ \cdots,\ \boldsymbol{\xi}_N' \,] \in \mathbb{R}^{6\times N} \tag{5}$$

where the components $\boldsymbol{\xi}_k' = \mathrm{Ad}(\exp(\hat{\boldsymbol{\xi}}_1\theta_1) \cdots \exp(\hat{\boldsymbol{\xi}}_{k-1}\theta_{k-1}))\boldsymbol{\xi}_k$ relate to the joint twists in the current configuration and represented with respect to {S}.

Therefore, the linkage's static equilibrium analysis (4) can be modeled as an unconstrained optimization problem as

$$\min \quad \boldsymbol{c}(\boldsymbol{\theta}, \boldsymbol{F}) = \begin{bmatrix} \log(\mathbf{g}_t^{-1}\mathbf{g}_{st}(\boldsymbol{\theta}))^\vee \\ \mathbf{K}\boldsymbol{\theta} - \mathbf{J}^T\boldsymbol{F} \end{bmatrix} \in \mathbb{R}^{(6n+6)\times 1} \tag{6}$$

where $\mathbf{g}_t \in SE(3)$ denotes the target pose for the tip-frame. $(\log(\mathbf{g}_t^{-1}\mathbf{g}(\boldsymbol{\theta})))^\vee \in \mathbb{R}^{6\times 1}$ corresponds to the twist deviation of current pose from the target one.

And the gradient of the objective function (6) with respect to the variables $\boldsymbol{\theta}$ and $\boldsymbol{F}$ can be derived readily as

$$\nabla = \begin{bmatrix} \frac{\partial \boldsymbol{c}}{\partial \boldsymbol{\theta}}, \frac{\partial \boldsymbol{c}}{\partial \boldsymbol{F}} \end{bmatrix} = \begin{bmatrix} \mathbf{J} & \mathbf{0} \\ \mathbf{K} + \mathbf{K}_\mathbf{J} & -\mathbf{J}^T \end{bmatrix} \in \mathbb{R}^{(N+6)\times(N+6)} \tag{7}$$

where the configuration-dependent stiffness item $\mathbf{K_J}$ is given by

$$\mathbf{K_J} = -\begin{bmatrix} \frac{F^T \partial \xi'_1}{\partial \theta_1} & \cdots & \frac{F^T \partial \xi'_1}{\partial \theta_N} \\ \vdots & \ddots & \vdots \\ \frac{F_i^T \partial t'_{i,6}}{\partial \theta_{i,1}} & \cdots & \frac{F_i^T \partial t'_{i,6}}{\partial \theta_{i,6}} \end{bmatrix} = \begin{bmatrix} 0 & 0 & \cdots & 0 \\ \boldsymbol{F}^T[\boldsymbol{\xi}'_1, \boldsymbol{\xi}'_2] & 0 & \ddots & \vdots \\ \vdots & \ddots & \ddots & 0 \\ \boldsymbol{F}^T[\boldsymbol{\xi}'_1, \boldsymbol{\xi}'_N] & \cdots & \boldsymbol{F}^T[\boldsymbol{\xi}'_{N-1}, \boldsymbol{\xi}'_N] & 0 \end{bmatrix} \tag{8}$$

where $[\boldsymbol{\xi}'_i, \boldsymbol{\xi}'_j]$ represents the Lie bracket, also known as cross product, of the joint twist $\boldsymbol{\xi}'_i$ and $\boldsymbol{\xi}'_j$. Please refer to [13] for more details.

3 Kinetostatic Modeling of Rigid/Flexible Mechanisms

In this section, a planar four-bar linkage is provided as an example to extend the proposed method to the kinetostatic analysis of mechanisms consisting of both rigid and flexible links.

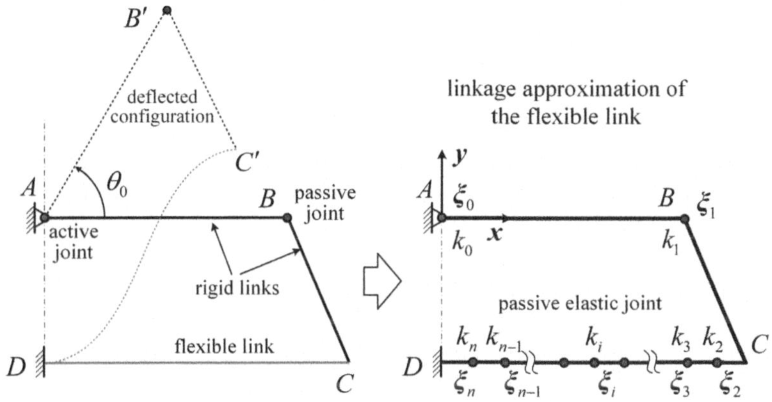

Fig. 2. Planar four-bar mechanism with a flexible link.

As shown in Fig. 2, the studied four-bar mechanism consists of two rigid links and flexible one. The crank AB is actuated by an active joint and connected to the coupler BC through a passive one. At the other end, the flexible rocker CD is attached to the coupler and ground, respectively.

Using the approximation method presented in Sec. 2, the flexible link CD can be discretized into small segments whose structural compliance is replaced by passive elastic joints, as shown in Fig. 2. It should be noted that, in planar cases, only the revolute joint perpendicular to the plane and the prismatic ones within the plane will affect the deflection of the flexible link. As a matter of fact, they correspond to the planar bending and shearing deformations, respectively. Moreover, the effect of shearing is much less than the bending one, so that only the revolute joints that perpendicular to the plane are taken into account in the studied example.

As a result, the studied planar four-bar mechanism can be considered as an under-actuated linkage constituting active, fully passive, and elastic joints, as illustrated in Fig. 2. Note that the first two joints are actuated and passive, respectively. It yields $k_0 = \infty$ and $k_1 = 0$. On the other hand, the joints decomposed from the flexible link are elastic ones. Hence, they have finite-values stiffness which can be determined according to (2).

Using the POE formula, the kinematics closure equation can be expressed as

$$\exp(\hat{\boldsymbol{\xi}}_0\theta_0)\exp(\hat{\boldsymbol{\xi}}_1\theta_1)\exp(\hat{\boldsymbol{\xi}}_2\theta_2)\cdots\exp(\hat{\boldsymbol{\xi}}_n\theta_n) = \mathbf{I}_4 \tag{9}$$

where $\mathbf{I}_4$ denotes the fourth-order identify matrix. $\hat{\boldsymbol{\xi}}_i \in se(3), i = 0, 1, \cdots, n$ are the joint twists at the mechanism's undeformed configuration.

Since the first joint is active, the kinematic equation can be rewritten as

$$\mathbf{g}(\boldsymbol{\theta}) = \exp(\hat{\boldsymbol{\xi}}_1\theta_1)\exp(\hat{\boldsymbol{\xi}}_2\theta_2)\cdots\exp(\hat{\boldsymbol{\xi}}_n\theta_n) = \exp(-\hat{\boldsymbol{\xi}}_0\theta_0) \tag{10}$$

where $\boldsymbol{\theta} = [\theta_1, \cdots, \theta_n]^T \in \mathbb{R}^{n\times 1}$.

Then, the static equilibrium condition can be represented as same as the one in (4), where the external load $\boldsymbol{F}$ corresponds to the reaction force between the flexible link and the fixed ground. Another difference is that the first element of the diagonal stiffness matrix is zero since the corresponding joint is fully passive. Thus, the kinetostatic analysis of mechanism can be modeled as an optimization problem similar to (6) and solved by gradient-based searching algorithms as (7).

The main benefit of the proposed method is that the kinetostatics analysis of rigid/flexible mechanism can be achieved in the same way as the large deflection problems of flexible links. The only difference is to set the stiffness of active and passive joints to infinity and zero, respectively. Moreover, since the gradient of objective function can be derived in a concise form, the corresponding simulation can be accomplished by means of efficient searching algorithms.

4 Numerical Simulation and Verification

In this section, numerical simulations are conducted on the studied planar flexible four-bar mechanism using the proposed approach. Meanwhile, FEM results are also provided to verify the correctness and efficiency.

Table 1. Link lengths of the four-bar linkage.

Parameters	l_{AD}	l_{AB}	l_{BC}	l_{CD}
Values	0.60 m	1.00 m	0.65 m	1.25 m
Parameters	E	I_{xx}	δ	k_i
Values	72 GPa	4.91×10^{-14} m^4	0.025 m	0.414 Nm/rad

The link lengths of the studied four-bar mechanism are given in Table 1, in which the mechanical property of the flexible link is also provided. The length of discretized segments is set to $\delta = 0.025\,\mathrm{m}$. In other words, 50 segment elements, as well as 25 elastic joints, are decomposed in the studied four-bar mechanism. Therefore, the kinetostatics analysis has been transformed to the determination of equilibrium configuration of an underactuated hyper-redundant linkage with one active, one passive and 50 elastic joints.

Using the gradient-based searching algorithm, the update theme of joint variables and reaction force/torque can be represented as

$$\begin{bmatrix} \boldsymbol{\theta} \\ \boldsymbol{F} \end{bmatrix}_{k+1} = \begin{bmatrix} \boldsymbol{\theta} \\ \boldsymbol{F} \end{bmatrix}_{k} - \boldsymbol{\nabla}_k^{-1} \boldsymbol{c}_k \tag{11}$$

where the gradient $\boldsymbol{\nabla}_k$ can be obtained in the same way as (7). And the residual of the objective function $\boldsymbol{c}_k$ is defined in (6). Here, the subscript k denotes the step of iterative searching algorithm.

The searching algorithm (11) will be iteratively repeated until the objective function $\boldsymbol{c}$ approaches zero and the variables $\boldsymbol{\theta}$ and $\boldsymbol{F}$ stably converge to some values. In the numerical simulation, the full-cycle rotation of the active joint is equally divided into 30 driving inputs which are calculated successively. And the corresponding equilibrium configurations are illustrated in Fig. 3.

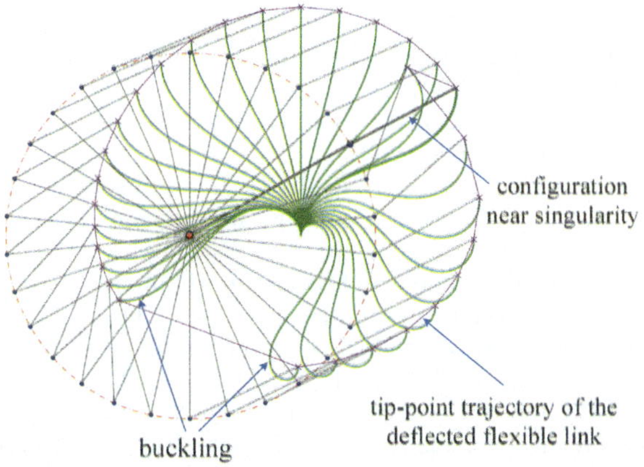

Fig. 3. Numerical results of kinetostatics of the studied four-bar mechanism.

From the results, it is known that the proposed discretization-based method do work for the kinetostatics analysis of mechanisms which constitute flexible links undergoing large deflections. Analogous to the rigid four-bar linkages, the flexible one also suffers from singularity, as indicated in the figure, near which different solutions can be obtained to the same deflection problems. Moreover, due to the buckling effect, the solutions to the kinetostatics model jump between different stable configurations even when the deviation of actuations is extremely small. This can be considered as a specific type of singularity that particularly arises in mechanisms constituting large-deflection flexible links, which will be intensively in our future work. The correctness of the numerical results have been verified via the commercial FEA software Abaqus/Standard.

5 Conclusions

In this paper, a general approach is presented for the kinetostatics modeling and simulation of closed-loop mechanisms with both rigid and flexible links. Based on the principal axes decomposition of discretized segments' structural compliance, those mechanisms can be modeled as underactuated hyper-redundant linkages, which constitute active, passive and elastic joints. A planar four-bar mechanism is studied as an example and the simulation results have verified the effectiveness and efficiency of the proposed method. Since the principal axes decomposition of stiffness matrices was originally developed for spatial robot compliance, the proposed method can be readily extended to spatial mechanisms having complexer topology structures, such as the continuum parallel manipulators.

References

1. Howell, L.: Compliant Mechanisms. John Wiley & Sons, New York (2001)
2. Saxena, R., Saxena, A.: Design of electrothermally compliant MEMS with hexagonal cells using local temperature and stress constraints. J. Mech. Des. **131**(5), 051006 (2009)
3. Asbeck, A., Cutkosky, M.: Designing compliant spine mechanisms for climbing. J. Mech. Robot. **4**(3), 031007 (2012)
4. Awtar, S., Slocum, A., Sevincer, E.: Characteristics of beam-based flexure modules. J. Mech. Des. **129**(6), 625–639 (2007)
5. Zhang, A., Chen, G.: A comprehensive elliptic integral solution to the large deflection problems of thin beams in compliant mechanisms. J. Mech. Robot. **5**(2), 021006 (2013)
6. Howell, L., Midha, A.: Parametric deflection approximations for end-loaded, large-deflection beams in compliant mechanisms. J. Mech. Des. **117**(1), 156–165 (1995)
7. Su, H.: A pseudo-rigid-body 3R model for determining large deflection of cantilever beams subject to tip loads. J. Mech. Robot. **1**(2), 021008 (2009)
8. Zhu, S., Yu, Y.: Pseudo-rigid-body model for the flexural beam with an inflection point in compliant mechanisms. J. Mech. Robot. **9**(3), 031005 (2017)
9. Saxena, A., Ananthasuresh, G.: Topology synthesis of compliant mechanisms for nonlinear force-deflection and curved path specifications. J. Mech. Des. **123**(1), 33–42 (2001)
10. Her, I., Methodology for compliant mechanisms design. Ph.D. thesis, Purdue University (1986)
11. Chen, G., Ma, F.: Kinetostatic modeling of fully compliant bistable mechanisms using Timoshenko beam constraint model. J. Mech. Des. **137**(2), 022301 (2015)
12. Chen, G., Wang, H., Lin, Z., Lai, X.: The principal axes decomposition of spatial stiffness matrices. IEEE Trans. Robot. **31**(1), 191–207 (2015)
13. Chen, G., Zhang, Z., Wang, H.: A general approach to the large deflection problems of spatial flexible rods using principal axes decomposition of compliance matrices. J. Mech. Robot. **10**(3), 031012 (2018)

A Pure-Inertia Method for Dynamic Balancing of Symmetric Planar Mechanisms

Jan J. de Jong[1,3](✉), Yuanqing Wu[2], Marco Carricato[2], and Just L. Herder[1,3]

[1] University of Twente, Enschede, The Netherlands
j.j.dejong@utwente.nl
[2] University of Bologna, Bologna, Italy
{yuanqing.wu,marco.carricato}@unibo.it
[3] Delft University of Technology, Delft, The Netherlands
j.l.herder@tudelft.nl

Abstract. In this paper, we present a novel method for the dynamic balance of planar mechanisms, by transforming a mechanism into a dynamically equivalent form where all links have zero mass but non-zero moment of inertia. The dynamic balance of such pure-inertia systems is shown to be governed by mirror symmetry that cancels out the system's total angular momentum. Our method not only covers well-known dynamically balanced 1-DOF mechanisms, such as the slider-crank and four-bar linkages, but also leads to the discovery of a novel dynamically balanced 2-DOF planar mechanism.

Keywords: Dynamic balance · Dynamic equivalence
Equimomental systems · Pure-inertia method · Symmetric space

1 Introduction

For robots and machines operating at high speeds, dynamic balance is desired to eliminate varying reaction forces and moments (also termed shaking forces and moments), which are known to be a major source of wear, noise and accuracy degradation [1]. The necessary and sufficient condition for dynamic balancing is that both the linear and angular momenta of the mechanism are constant. A mechanism is force balanced when only the linear momentum is constant, which in practice comes down to having a fixed center of mass (COM) of the system.

In principle, dynamic balance can be found by symbolically solving the kinematic and balancing equations for unknown kinematic and inertial parameters. A solution is in general not guaranteed, or otherwise very difficult to find, due to high algebraic complexity [1,2]. Alternatively, one may construct dynamically balanced mechanisms from primitive functional modules which, either are dynamically balanced [3] or have the dynamic balance conditions that are easy to obtain symbolically [4–6]. The dynamic balance equations can be simplified by

J. Lenarcic and V. Parenti-Castelli (Eds.): ARK 2018, SPAR 8, pp. 277–284, 2019.
https://doi.org/10.1007/978-3-319-93188-3_32

reducing the number of parameters through dynamic equivalence [7]. Two mechanisms with equal kinematics, but different mass distributions, are said to be dynamically equivalent or equimomental [8] if their linear and angular momenta are equal when undergoing a common arbitrary motion. Consequently, the two mechanisms will always have the same shaking forces and moments even though their masses, COMs, and moments of inertia are different.

A fruitful source of dynamically balanced mechanisms is the class of linkages, such as the pantograph [6], the slider-crank [4], and the four-bar [1], whose geometric structures exhibit a mirror symmetry about a plane moving with half magnitude (hence also half velocity) of that of the end-effector for full-cycle motion. Recently, the second and third author of this paper made an addendum to this class of mechanisms by systematically investigating plane and line symmetry of a class of parallel mechanisms known as symmetric subspace motion generators [9,10].

The main contribution of this paper is twofold. Firstly, we propose a dynamic balance method via dynamic equivalence to massless systems with only moments of inertia. This approach, called pure-inertia method (PIM), provides a unified understanding of several existing dynamically balanced mechanisms that were previously synthesized by solving complex algebraic equations. Secondly, the PIM is applied to a novel 2-DOF planar symmetric mechanism proposed in [10]. Numerical simulation confirms that the mechanism is dynamically balanced.

2 Pure-Inertia Method

The pure-inertia method relies on the dynamic equivalence of all links in a mechanism to a zero mass with non-zero moment of inertia. A pure inertia may be interpreted as dynamically equivalent to an infinitely large and thin ring.

The principle of dynamic equivalence is illustrated in Fig. 1. A range of dynamically equivalent links can by found via *point mass redistribution* [7]. The initial inertial parameters comprise mass m_o, COM $\boldsymbol{c}_o$, and moment of inertia g_o (Fig. 1(a)), whereas for the dynamic equivalent set we have: m_f, $\boldsymbol{c}_f$, and g_f (Fig. 1(b)). If we start with a link hinged at $\boldsymbol{o}$, we may transfer a point mass of

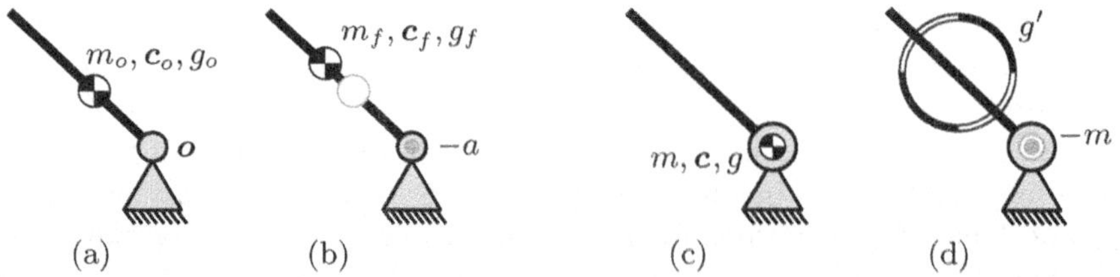

Fig. 1. Dynamic equivalence via mass redistribution. (a) A simple pendulum; (b) a dynamic equivalent of (a) via adding or subtracting point mass at the hinge; (c) a pendulum with COM located at the hinge; (d) a dynamic equivalent of (c) with pure moment of inertia.

arbitrary value a - located at the hinge - from the link to the base and vice versa, without changing the link's momentum and shaking forces and moments. In the same manner, the link may exchange a point mass with any body i hinged to it by a revolute joint at a location denoted by $\boldsymbol{o}_i$. This is a valid dynamic equivalent operation, since at a revolute joint the linear velocities of both connecting bodies are equal. Therefore changing the attachment of the a point mass from one body to the other does not affect the mechanism's momentum. We have the following equations:

$$m_f = m_o + \sum_{i=1}^{n} a_i, \quad m_f \boldsymbol{c}_f = m_o \boldsymbol{c}_o + \sum_{i=1}^{n} a_i \boldsymbol{o}_i, \quad g_f = g_o - m_f \|\boldsymbol{c}_f\|^2 + \sum_{i=1}^{n} a_i \|\boldsymbol{o}_i\|^2 \tag{1}$$

To ensure dynamic equivalence, the point mass added to body i should be subtracted from body j, which we refer to as *mass continuity*. If both $m_f = 0$ and $\boldsymbol{c}_f = \boldsymbol{0}$ (Fig. 1(c)), i.e., if $m_o = -\sum_{i=1}^{n} a_i$ and $m_o \boldsymbol{c}_o = -\sum_{i=1}^{n} a_i \boldsymbol{o}_i$, the dynamics of the body is completely determined by a pure moment of inertia (Fig. 1(d)). This pure moment of inertia - or for brevity; pure inertia - shall be denoted with a prime: g'. When this holds for all bodies in the mechanism, the mechanism is said to be a pure-inertia mechanism.

A pure-inertia body is not physically feasible, but it can be transformed back into a feasible one by applying the reverse process of *point mass recomposition* [7], i.e., by adding point masses at the revolute joints. For the purpose of this paper, we shall consider bodies connected with at most two revolute joints, and therefore all vectors in Eq. 1 can be treated as scalars. By assuming that one of the joints is located at the origin and the other at a distance l, the following relations hold for the mass, COM and moment of inertia:

$$m = a_1 + a_2, \qquad mc = a_2 l, \qquad g = g' + a_2 \left(1 - \frac{a_2}{a_1 + a_2}\right) l^2 \tag{2}$$

The links generated from dynamically equivalent pure-inertia bodies, remain physically feasible if the equivalent mass m and moment of inertia g satisfy the following positivity constraints:

$$m > 0 \Leftrightarrow a_1 > -a_2, \qquad g > 0 \Leftrightarrow \frac{g'}{l^2} > a_2 \left(\frac{a_2}{a_1 + a_2} - 1\right) \tag{3}$$

which implies that a pure inertia, denoted g', is allowed to have a negative value. The limits on a_1 and a_2 are carried over to adjacent bodies due to mass continuity. See [7] for more details about the mass redistribution method. It should be noted that a pure-inertia mechanism has a fixed COM with respect to the base and is therefore necessarily force balanced. However, the reverse does not hold in general.

2.1 Dynamic Balance

In the literature, force balance conditions are usually inspected prior to moment balancing. In comparison, for a pure-inertia mechanism, force balance conditions are automatically satisfied and therefore only the angular momentum, denoted ξ, needs to be canceled out to achieve dynamic balance:

$$\xi = \sum_i g_i' \omega_i = 0 \tag{4}$$

where the moments of inertia g_i' are pose independent. The pose dependence of the angular velocities ω_i, on the other hand, imposes conditions on the geometry of the mechanism [1].

2.2 Dynamic Balance of the Slider-Crank Linkage Using the PIM

We shall now illustrate the PIM by considering the dynamic balance of a slider-crank linkage, as shown in Fig. 2. According to [4], a slider-crank linkage can be dynamically balanced under symmetric kinematic conditions ($l_1 = l_2$, $l_3 = 0$), if the total COM is located at the base revolute joint for all configurations, and links' inertial parameters satisfy:

$$c_2 = 0, \qquad m_1 c_1 = -m_2 l_1, \qquad g_1 = g_2 - m_2 \left(\frac{m_2}{m_1} + 1 \right) l_1^2 \tag{5}$$

It is apparent that the symmetry conditions constrain link 1 and 2 to move with equal and opposite angular velocities: $\omega_1 = -\omega_2$. In reference to Eq. 4, the dynamic balance condition for the slider-crank linkage reduces to:

$$g_2' = g_1' \tag{6}$$

We shall show that Eq. 6 is exactly equivalent to the conditions given in [4]. Via the mass recomposition - refer to Eq. 2 - we have:

$$\begin{aligned} m_1 &= a_{10} - a_{21}, & m_1 c_1 &= -a_{21} l_1, & g_1 &= g_1' - a_{21} \left(1 + \tfrac{a_{21}}{a_{10} - a_{21}}\right) l_1^2 \\ m_2 &= a_{21}, & m_2 c_2 &= 0, & g_2 &= g_1' \end{aligned} \tag{7}$$

We remark that the dynamics of the slider can be easily included in the above process. Because it generates no angular momentum (it only translates), it is dynamically equivalent to a point mass which can be included in the dynamic balance conditions via point mass recomposition, as in Eq. 1.

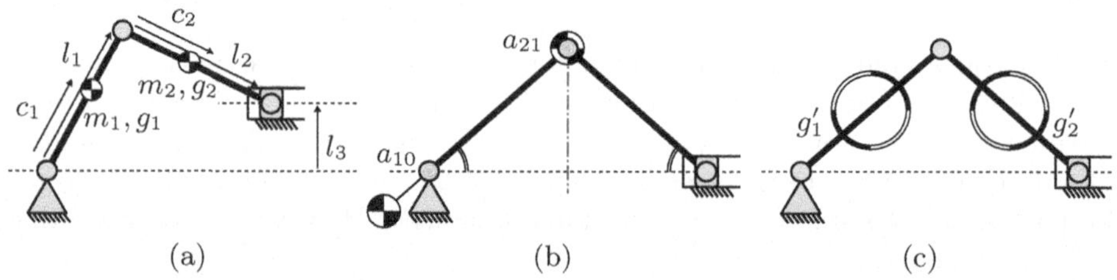

Fig. 2. Dynamic balance conditions for a slider-crank mechanism. (a) Kinematic and inertial parameters. (b) Conventional dynamic balance conditions for a symmetric slider-crank. The symmetry line is represented by the green dash-dotted line. (c) The pure-inertia equivalent of the dynamically balanced slider-crank

2.3 Dynamic Balance of the Four-Bar Linkage Using the PIM

Gosselin and Ricard [1] showed that two types of symmetric four-bar linkages can be dynamically balanced. The first one is the kite type, shown in Fig. 3. It is symmetric about a plane passing through joint q_2 and q_4 (or equivalently q_1 and q_3). The second type, the anti-parallelogram, is symmetric about a plane which bisects the angle formed by the lines through body l_1 and l_3.

We shall focus on the first type here (the same procedure can be applied to the anti-parallelogram). The kinematic symmetry conditions are: $l_1 = l_2$, and $l_0 = l_3$ [3], whereas the dynamic balance conditions are:

$$\begin{aligned} & & g_1 &= g_{c1} + m_1 c_1 (l_1 - c_1) \\ m_2 c_2 &= m_1 c_1 + m_2 l_1, & g_2 &= g_{c1} + m_2 c_2 (l_1 - c_2) \\ m_3 c_3 &= \tfrac{l_3}{l_1}\left(c_2 m_2 + l_1 m_3\right), & g_3 &= -g_{c1} - m_3 c_3 (l_3 - c_3) \end{aligned} \tag{8}$$

in which g_{c1} is a useful collection of inertial parameters introduced in [3]. The corresponding pure-inertia mechanism can be generated by considering the following mass redistribution:

$$\begin{aligned} a_{21} &= a_{10} - m_1, \qquad a_{32} = a_{21} - m_2, \qquad a_{03} = a_{32} - m_3 \\ m_1' c_1' &= m_2' c_2' = m_1 (c_1 - l_1) + a_{10} l_1, \qquad m_3' c_3' = \frac{l_3}{l_1} m_1' c_1' \end{aligned} \tag{9}$$

The link masses can then be canceled out by setting $a_{10} = -m_1/l_1(c_1 - l_1)$, thus resulting in the following pure-inertia dynamic balance conditions:

$$g_1' = g_2' = -g_3' = g_{c1} \tag{10}$$

which offers a novel interpretation of g_{c1}. It should be noted that these conditions can also be derived from the relation between the angular velocities $\omega_1 + \omega_2 = \omega_3$. Based on these pure-inertia conditions, we can find a range of dynamically equivalent mechanisms, as long as the selection of inertial parameters g_1', a_{10}, a_{21}, a_{32}, and a_{03} respect the positivity constraint $m_i > 0$ and $g_i > 0$.

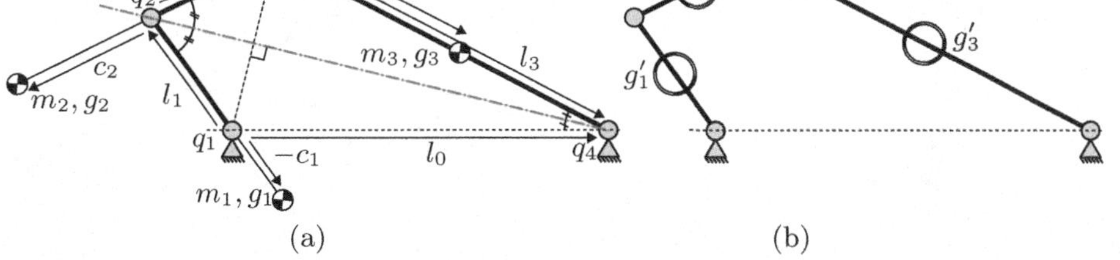

Fig. 3. (a) The dynamically balanced four-bar linkage. (b) A pure-inertia dynamic equivalent of the linkage shown in (a). The green dash-dotted line shows the symmetric plane, bisecting the kite through joint q_2 and q_4.

3 Dynamic Balance of a Novel 2-DOF Symmetric Mechanism Using the PIM

A novel 2-DOF symmetric planar mechanism, capable of generating finite rotations about any axis on its symmetric plane, was recently proposed in [10]. A variant is illustrated in Fig. 4. It comprises two RRR chains acting in-parallel on the end-effector and an angle-bisecting device (gray links in Fig. 4) to ensure a symmetric motion. The angle-bisecting device consists of a pantograph, attached via two revolute joints to two collinear slider joints on the symmetric plane. The symmetric plane bisects both "elbow" angles of the two RRR chains for full-cycle motion. Due to the *half-angle property* [11], the symmetric plane also bisects the angle between the base and the end-effector.

We shall apply the PIM to dynamically balance this novel mechanism in two steps. In the first step, the mass and moment of inertia of the angle-bisecting device are ignored (but assuming that the geometric constraint is still effective). We set base joints (q_1 and q_3) as the input variables. Because of the symmetry conditions, the angular velocity $\omega_{i,j}$ of body i with respect to the actuator $j = \{1, 3\}$ can be written as:

$$\begin{aligned} \omega_{1,1} + \omega_{2,1} &= \omega_{4,1} = \omega_{5,1} \\ \omega_{3,3} + \omega_{4,3} &= \omega_{2,3} = \omega_{5,3} \end{aligned} \tag{11}$$

Both relations are in fact equivalent to those of the kite-type four-bar linkage: when one of the base joints is locked, the platform and its adjacent bodies move as a rigid body, thus originating a kite-shaped four-bar linkage. Consequently, dynamic balance conditions can be derived from Eq. 10 as:

$$g_1' = g_2' = -g_4' - g_5', \qquad g_3' = g_4' = -g_2' - g_5' \tag{12}$$

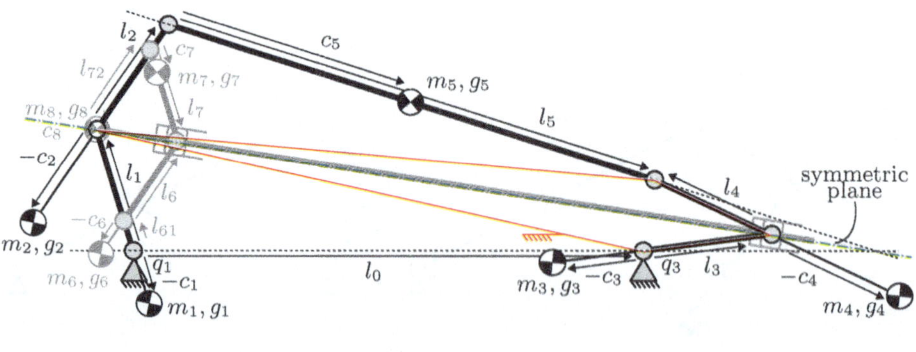

Fig. 4. The dynamically balanced 2-DOF planar symmetric mechanism (a variant of the M_{2A}-type symmetric mechanism proposed in [10]). The angle bisecting pantograph-slide on the symmetry plane is shown in gray. When fixing one of the base joints, in this case q_1, the mechanism acts as a symmetric four-bar mechanism (red).

Table 1. Geometric and inertial parameters for the dynamically balanced 2-DOF symmetric mechanism used in the simulation.

Length $[m]$		PI $[kgm^2]$		Point mass $[kg]$		Mass $[kg]$		COM $[m]$		Inertia $[kgm^2]$	
l_1	1.0	g'_1	1.00	a_{10}	−1.93	m_1	1.3	c_1	−0.43	g_1	0.17
l_2	1.0	g'_2	1.00	a_{21}	−0.570	m_2	0.30	c_2	−0.89	g_2	0.50
l_3	1.0	g'_3	1.01	a_{30}	−1.18	m_3	0.69	c_3	−0.71	g_3	0.17
l_4	1.0	g'_4	1.01	a_{43}	−0.490	m_4	0.23	c_4	−1.1	g_4	0.46
l_5	4.0	g'_5	−2.03	a_{52}	−0.260	m_5	0.52	c_5	2.0	g_5	55×10^{-3}
				a_{54}	−0.260						
l_{61}	0.25	g'_6	10.0×10^{-3}	a_{61}	-10.0×10^{-3}	m_6	7.0×10^{-3}	c_6	−0.32	g_6	7.6×10^{-3}
l_6	0.75	g'_7	10.0×10^{-3}	a_{72}	-10.0×10^{-3}	m_7	13×10^{-3}	c_7	−0.17	g_7	11×10^{-3}
l_{71}	0.25	g'_8	10.0×10^{-3}	a_{76}	-3.00×10^{-3}	m_8	10×10^{-3}	c_8	0.0	g_8	10×10^{-3}
l_7	0.75			a_{81}	-10.0×10^{-3}						

In the second step, the entire mechanism is dynamically balanced. First, note that links l_6 and l_7 of the pantograph (the angle-bisecting device in Fig. 4) have the same angular velocity as that of links l_2 and l_1, respectively. Their moments of inertia can then be easily lumped together in the dynamic balance conditions derived in the first step (Eq. 12). Secondly, due to the half-angle property, the angular velocity of the connecting link l_8, is exactly half the angular velocity of the end-effector. Thus we have:

$$\omega_6 = \omega_2, \qquad \omega_7 = \omega_1, \qquad \omega_8 = 0.5\omega_5 \tag{13}$$

The dynamic balance conditions for the entire mechanism are then given by:

$$g'_1 + g'_7 = g'_2 + g'_6, \qquad g'_3 = g'_4, \qquad g'_5 + 0.5g'_8 = -g'_2 - g'_4 \tag{14}$$

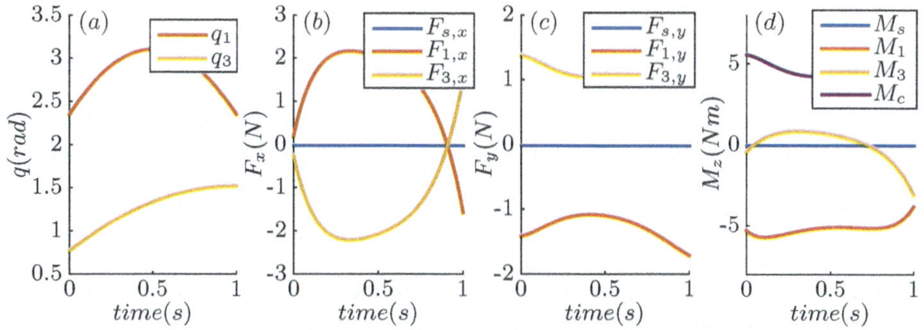

Fig. 5. Simulation results of the novel 2-DOF manipulator. The figure shows the input angles (a), shaking forces in x and y direction (b, and c respectively) and the shaking moment (d). The shaking forces (F_s) are the sum of reaction forces on base joints (F_1, F_3). The shaking moment (M_s) is the sum of motor torques (M_1, M_3) and the couple (M_c) induced by base joint forces.

A simulation of the dynamically balanced 2-DOF mechanism was performed to validate the PIM. The numerical values chosen for the lengths, COM, masses and moments of inertia can be found in Table 1. The simulation results, as shown in Fig. 5, confirm that the sum of the shaking forces and moments are zero for arbitrary motion.

4 Conclusion

All dynamically balanced mechanisms discussed in this paper share the same property of kinematic symmetry and dynamical equivalence to pure-inertia mechanisms. The pure-inertia method provides a simplified approach for achieving dynamic balance for two reasons: (i) Force balance conditions are automatically satisfied; (ii) in the planar case, moments of inertia are pose invariant and can be treated as scalars, reducing the dynamic balance conditions to linear kinematic relations between angular velocities. These relations, in the class of mechanisms shown here, are linear due to kinematic symmetry conditions. Based on this new understanding, a new dynamically balanced 2-DOF mechanism was presented.

References

1. Ricard, R., Gosselin, C.M.: On the development of reactionless parallel manipulators. In: Proceedings of ASME DETC CIE, vol. 1, pp. 1–10 (2000)
2. Moore, B., Schicho, J., Gosselin, C.M.: Determination of the complete set of shaking force and shaking moment balanced planar four-bar linkages. Mech. Mach. Theory **44**(7), 1338–1347 (2009)
3. Wu, Y., Gosselin, C.M.: Synthesis of reactionless spatial 3-DoF and 6-DoF mechanisms without separate counter-rotations. Int. Jo. Robot. Res. **23**(6), 625–642 (2004)
4. Briot, S., Bonev, I.A., Gosselin, C.M., Arakelian, V.: Complete shaking force and shaking moment balancing of planar parallel manipulators with prismatic pairs. Proc. Inst. Mech. Eng. **223**(1), 43–52 (2009)
5. Van der Wijk, V., Herder, J.L.: Dynamic balancing of a single crank-double slider mechanism with symmetrically moving couplers. In: Proceedings of the 3rd IFToMM EUCOMES, vol. 5, pp. 413–420 (2010)
6. Van der Wijk, V., Herder, J.L.: Synthesis method for linkages with center of mass at invariant link pointpantograph based mechanisms. Mech. Mach. Theory **48**, 15–28 (2012)
7. de Jong, J., van Dijk, J., Herder, J.: On the dynamic equivalence of planar mechanisms, an inertia decomposition method. In: Proceedings of the 6th IFToMM EUCOMES, vol. 43, pp. 51–59 (2017)
8. Routh, E.J.: An Elementary Treatise on the Dynamics of a System of Rigid Bodies. Macmillan, London (1877)
9. Loos, O.: Symmetric Spaces. Benjamin, New York (1969)
10. Wu, Y.Q., Carricato, M.: Symmetric subspace motion generators. IEEE Trans. Robot. (2018, in press). https://doi.org/10.1109/TRO.2018.2813377
11. Wu, Y.Q., Löwe, H., Carricato, M., Li, Z.: Inversion symmetry of the euclidean group: theory and application to robot kinematics. IEEE Trans. Rob. **32**(2), 312–326 (2016)

Stiffness and Deformation of Mechanisms with Locally Flexible Bodies: A General Method Using Expanded Passive Joints

Gonzalo Moreno[1,2(✉)], Julio C. Frantz[1], Lauro C. Nicolazzi[1], Rodrigo S. Vieira[1], and Daniel Martins[1]

[1] Federal University of Santa Catarina, Florianopolis, SC 88040-900, Brazil
jcfrantz@gmail.com, daniel.martins@ufsc.br, {lauro,rvieira}@grante.ufsc.br
[2] University of Pamplona, 543050 Pamplona, Colombia
gmoren@hotmail.com

Abstract. This paper presents a methodology that allows the stiffness and deformation analysis of locally flexible bodies using passive joints. The proposed model takes into account the influence of the external and internal loadings on the axial and shear deformation of the mechanical components depending of the mechanism configuration, this technique is denominated as the expanded passive joints method (*EPJM*). The Davies method is used to obtain the static of the mechanism, and the simulation example is presented to demonstrate the ability of this technique.

Keywords: Stiffness · Deformation · Theory of buckling Passive joints · Davies method

1 Introduction

Stiffness and deformation analysis are the most important indicators in performance evaluation of mechanisms systems. These parameters are important for designing mechanical components and control systems where the performance of the mechanism in certain position(s) is important. Prismatic joints "*P*" are frequently used in the description of the stiffness and the elastic deformations of links; this technique is also known as the Virtual Joint Method (*VJM*), which is based on the extension of the traditional rigid model by adding virtual joints (localized passive prismatic joint) [1, 2] for the static analysis of mechanisms. This method is based on the stiffness matrix, which uses the inverse of the Jacobian matrix, but in several cases cited in the literature getting the inverse of the Jacobian matrix is generally hard task, and it takes a long computational time [3, 4].

Besides, when it is analyzed the buckling of columns, the bar under the action of a force can have two kinds of deformation: a substantial axial deformation and a smaller but not negligible shear deformation; for the statics solution of the mechanism these deformations can be associated to a helical compression spring and a

J. Lenarcic and V. Parenti-Castelli (Eds.): ARK 2018, SPAR 8, pp. 285–292, 2019.
https://doi.org/10.1007/978-3-319-93188-3_33

torsion spring at the base, respectively [5,6]. Making an analogy between stability of columns and the static analysis of flexible bodies, the technique mentioned above is applied to the modelling and analysis of mechanisms with locally flexible bodies; and the formalism described by Davies [7] was used as the primary mathematical tool to analyze the mechanisms statically, this method was selected since it allows the static model for the mechanism to be obtained in a straightforward manner and it is also easily adapted using this approach [8,9]. In our previous work [10], the technique mentioned above was used together with mechanism theory to developed a simplified kinematic chain of the rigid suspension, which allowed to calculate its influence in the stability of heavy vehicles.

Based in the *VJM* method together with the theory of buckling of columns, the authors present the Extended Passive Joints Method (*EPJM*) for the static analysis of mechanisms with locally flexible bodies. The paper is organized as follows. Section 2 introduces mechanism model, which is analyzed and discussed within a case study, and finally, conclusions are drawn in Sect. 3.

2 Deformation Model

Elastic and plastic deformations of different components of mechanisms can be caused by the internal or external forces that act on the mechanism, or by changes in temperature. For the representation of the axial deformation in mechanical components, passive prismatic joints are used frequently [1], but some mechanical components have a relatively small shear deformation that other methods commonly ignore. This deformation, generally small, is however crucial for the solution of mechanisms with flexible bodies, such as: suspensions, tires, hydraulic and pneumatic systems [8].

Taking into account these aspects and making an adaptation of the theory of buckling of columns [5,11], the following model was developed: using a column where the buckling mode adopted is one end fixed and one end free (Fig. 1(a)), the bar under the action of the force F has an axial deformation and a small shear deformation (Fig. 1(b)); these deformations allow two degrees of freedom (2 DoF) of the bar (displacement in the longitudinal y-axis and a roll rotation around the z-axis). Besides, these deformations can be represented by a prismatic joint with helical compression spring and a revolute joint with torsion spring at the base respectively [5], as shown in Fig. 1(c), where k_1 and k_2 are the stiffness of the compression spring and the torsional spring respectively ($k_2 >> k_1$) [5], and θ is the rotation angle of the bar. Figures 2(a) and (b) show the proposed kinematic chain of the bar, which is composed of three links (n = *3*) identified by the letters A (the base) and B-C (helical compression spring), and the two joints (j = *2*) identified by numbers as follow: a revolute joint "*R*"(1), and a prismatic joint "*P*" (2).

Figure 2(c) shows the forces that act on the joints of the mechanism: F_n and F_p are the normal and perpendicular loads in the passive prismatic joint respectively, M is the moment that act on the prismatic joint, and T is the torque that act on the revolute joint with torsional spring.

On the other hand, in a continuous body, a strain field results from a displacement field induced by applied forces or to changes in the temperature field

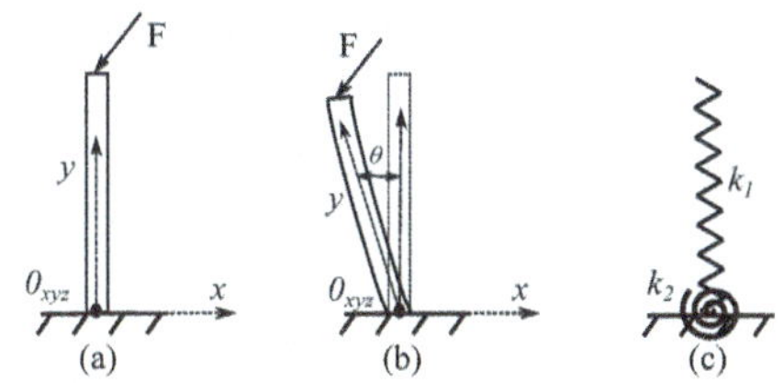

Fig. 1. Deformation model of a buckling column.

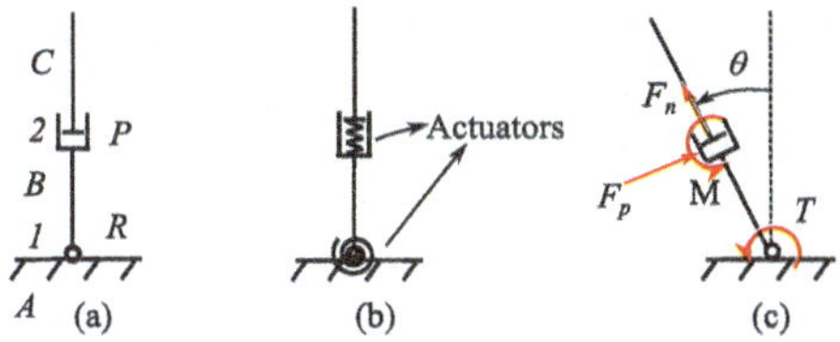

Fig. 2. Proposed kinematic chain of a buckling column with $M = 2$.

inside the body. In a mechanism, the total axial deformation of structural members (δ_{Total}) is influenced by the deformation induced by forces (δ_F) and deformation induced by the changes in temperature (δ_T), as shown in Eq. (1).

$$\delta_{Total} = \delta_F + \delta_T \tag{1}$$

The main mechanisms of deformation by force and temperature are briefly summarized in Table 1, where all notations are adapted to those used in this paper; δ is the elastic deformation, δ_S is the spring deformation, δ_{LS} is the leaf spring deformation, δ_t is the rubber axial deformation, δ_T is the thermal deformation, L is the original length of the bar, A is the cross-sectional area of the bar, E is the modulus of elasticity of the material, $\triangle F$ is the algebraic change in the initial load, F^{start} is the initial load, k_S is the equivalent spring stiffness, L_s is the length of the leaf spring, N is the number of the leaves, B

Table 1. Summary of the related works and expressions for the deformation of mechanical members.

Element	Deformation by	Model
Links [13]	Force	$\delta = -\frac{F_n L_b}{A_b E}$
Spring [14]	Force	$\delta_S = \frac{\triangle F}{k_s} = \frac{F^{start} - F_n}{k_s}$
Leaf spring [15,16]	Force	$\delta_{LS} = \frac{3 \triangle F l^3}{8 E_s N B T_l^3} = \frac{\triangle F}{k_{Ls}}$
Rubber - Tires [17,18]	Force	$\delta_t = \frac{3 \triangle F}{k_t + a_c} = \frac{\triangle F}{k_T}$
Links [19]	Temperature	$\delta_T = \alpha \triangle T L$
Torsional spring [5,20,21]	Torque	$\theta = \frac{\tau}{k_{st}}$

is the width of the leaf, T_{Ls} is the thickness of the leaf, E_s is the modulus of elasticity for a multiple leaf, k_{Ls} is the equivalent stiffness of the suspension, k_t is the radial stiffness of the rubber, a is the regression coefficient, k_T is the equivalent rubber stiffness, $\triangle T^o$ is the algebraic change in the temperature, α is the linear coefficient of thermal expansion, τ is the torque, and k_{ts} is the spring torsional coefficient.

The mechanism shown in Fig. 3 is used to lower and retract the landing gear on small airplanes. If on the tire-ground interface acting two forces (P_x - traction or brake force, and P_y - normal force); the problem is to determine the torque required to drive the motor and to maintain the position of the mechanism, taking into account that when the traction force is applied, the shear deformation of the tire changes the point of application of the forces.

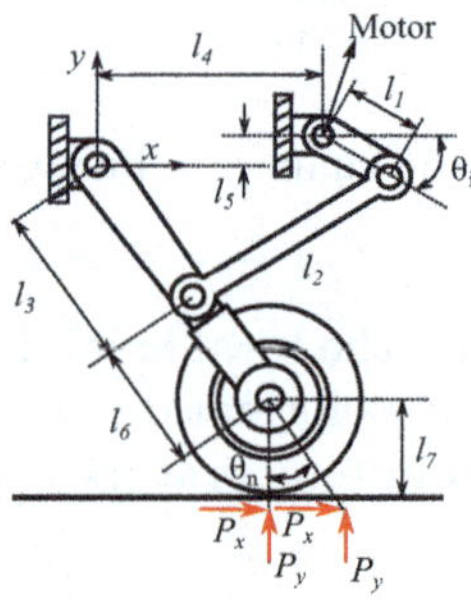

Fig. 3. Mechanism example 2.

Using the *EPJM* method, the mechanism can be represented by the following kinematic chains: Fig. 4(a) shows the model proposed by Myszka [22] without tire deformation; Figs. 4(b) and 4(c) show the proposed models with axial deformation of the tire, and with axial and shear deformations of the tire respectively.

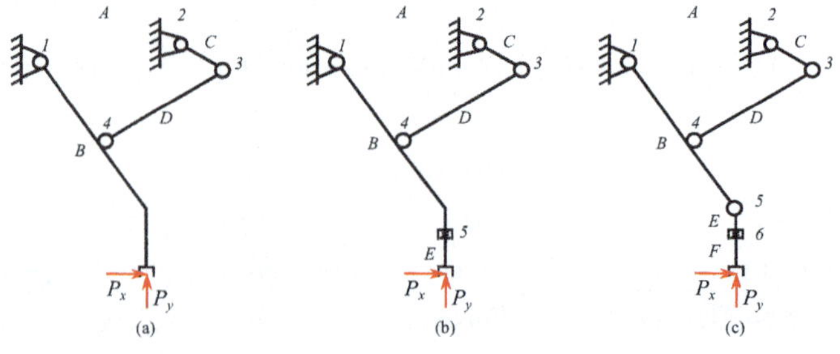

Fig. 4. Proposed kinematic chains of the mechanism.

The kinematic chain of the Fig. 4(c) is composed of six links identified by letters A (the base), B(link and damping system), C and D (links), and E and F (tire); and the six joints are identified by numbers as follow: four revolute joints "R"(1, 2, 3, 4 and 5), and a prismatic joint "P" (6). The developed mechanism of Fig. 4(c) has 3-DoF, and it requires three actuators to control its movements. The mechanism has an active actuator in the joint 1 (motor), a passive actuator in the joint 6 (axial deformation of the tire), and a passive actuator in the joint 5 (torsion spring - shear deformation of the tire); these actuators control the movement of the mechanism as shown in Fig. 5.

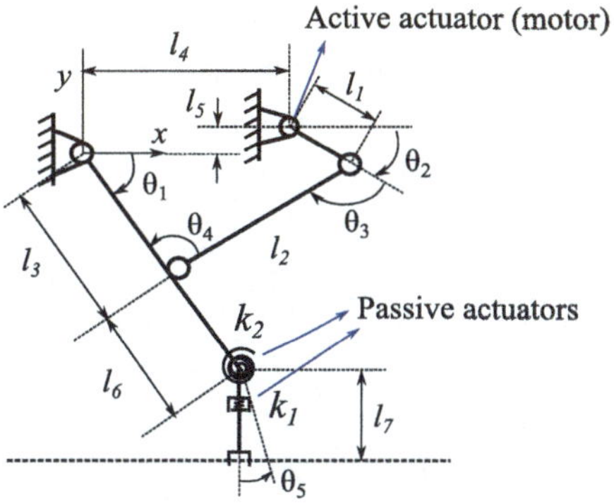

Fig. 5. Parameters of the mechanism.

Using graph theory, the mechanism of the Fig. 4(c) can be represented by the graph shown in the Fig. 6(a); where the vertices correspond to the links and the edges correspond to the constraints of each joint [12] as well as the external forces present in the mechanism. Figure 6(b) represents the cut-set action graph, and using the cut-set law [8], the statics of the mechanism can be defined, as exemplified in Eq. (2).

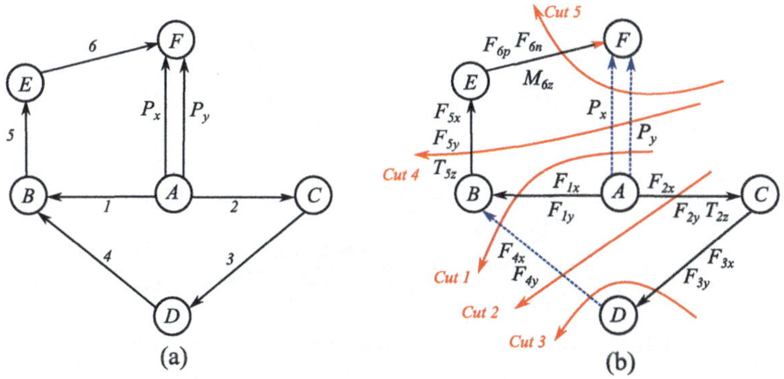

Fig. 6. (a) Direct coupling graph. (b) Cut-set action graph.

$$\left[\widehat{A}_n\right]_{15\times 17}[\Psi]^T_{17\times 1} = [0]_{15\times 1} \tag{2}$$

Solving the static system using the Gauss-Jordan elimination method, all secondary variables of the system being function of the two primary variables of the mechanism (P_x and P_y), Table 2 and Fig. 7 shows the parameters used for the example and the results obtained:

Table 2. Parameters of the example 2.

	Parameters	Test 1	Test 2	Test 3	Units
		Fig. 4(a)	Fig. 4(b)	Fig. 4(c)	
Inputs	l_1	0.54	0.54	0.54	m
	l_2	0.91	0.91	0.91	m
	l_3	0.71	0.71	0.71	m
	l_4	0.61	0.61	0.61	m
	l_5	0.23	0.23	0.23	m
	l_6	0.61	0.61	0.61	m
	l_7	0.54	0.54	0.54	m
	θ_2	30	30	30	^{o}
	k_1	-	160	160	$kN.m^{-1}$
	k_2	-	-	12	$kN.m.rad^{-1}$
	P_x	3	3	3	kN
	P_y	9.8	9.8	9.8	kN
Outputs	δ_{l7}	0	0.0612	0.0612	m
	T_{3z}	8.352	8.221	8.835	kN.m
	θ_5	0	0	10.96	^{o}

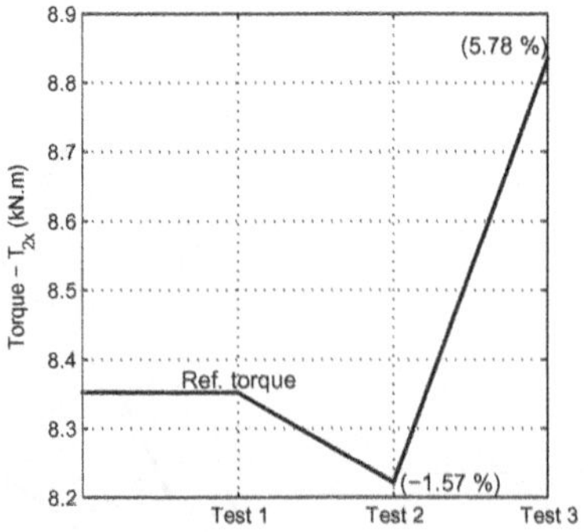

Fig. 7. Torque required to drive the motor link.

Table 2 and Fig. 7 shows that:

- when the mechanism takes into account the axial deformations of the tire (Test 2, case b), the torque of the motor has a decrease about 1.57 %, using the Test 1 as a reference
- when the mechanism takes into account the axial and shear deformations of the tire (Test 3, case c), the torque of the motor has an increase about 5.78 %, using the Test 1 as a reference.

In summary, the local flexibilities (both axial and shear) are not negligible and should be taken into account in the static analysis of mechanisms.

3 Conclusions

The Extended Passive Joints Method (*EPJM*), proposed in the work allows the modelling of mechanical component where the linear deformation and the shear deformation plays an important role on the stiffness analysis of the mechanisms.

The developed technique can be applied in different types of mechanisms and mechanical structures (as shown in the developed example), this example shows that the differences in the results obtained are important; these results enables a better design and selection of mechanical components.

The example shows possible effects of design that may arise when is taking into account the axial and shear deformations of flexible bodies in mechanisms.

The *EPJM* method can be extended to other mechanism where the axial and shear deformations play an important role to design and selection of mechanical components.

References

1. Pashkevich, A., Klimchik, A., Chablat, D.: Enhanced stiffness modeling of manipulators with passive joints. J. Mech. Mach. Theory **46**, 662–679 (2011). https://doi.org/10.1016/j.mechmachtheory.2010.12.008
2. Rauh, J., Schiehlen, W.: A unified approach to modelling of flexible robot arms. In: RoManSy 6, pp. 99–106. Springer, Boston (1987)
3. Zhun-Qing, H., Zhi-Guo, F., Hai-Rong, F.: Study of singularity robust inverse of Jacobian matrix for manipulator. In: Proceedings of the International Conference on Machine Learning and Cybernetics, vol. 1, pp. 406-410 (2002). https://doi.org/10.1109/ICMLC.2002.1176785
4. Sariyildiz, E., Ucak, K., Oke, G., Temeltas, H.: A trajectory tracking application of redundant planar robot arm via support vector machines. In: Proceedings of the Second International Conference on Adaptive and Intelligent Systems, ICAIS 2011, pp. 192–202. Springer, Klagenfurt (2011). ISBN 978-3-642-23856-7
5. Popov, E.: Engineering Mechanics of Solids. Pearson (1998). ISBN-10:0137261598
6. Adman, R., Saidani, M.: Elastic buckling of columns with end restraint effects. J. Construct. Steel Res. **87**, 1–5 (2014). https://doi.org/10.1016/j.jcsr.2013.03.022
7. Davies, T.H.: Mechanical networks-III wrenches on circuit screws. J. Mech. Mach. Theory **18**(2), 107–112 (1983)

8. Moreno, G., Nicolazzi, L., Vieira, R.S., Martins, D.: Suspension and tyres: the stability of heavy vehicles. Int. J. Heavy Veh. Syst. **24**(4), 305–326 (2017). https://doi.org/10.1504/IJHVS.2017.10007635. ISSN online: 1741–5152. ISSN print: 1744–232X
9. Tsai, L.W.: Robot Analysis - The Mechanics of Serial and Parallel Manipulators. Wiley, New York (1999). ISBN 0-471-32593-7
10. Moreno, G., Nicolazzi, L., Vieira, R.S., Martins, D.: Stability of long combination vehicles. Int. J. Heavy Veh. Syst. **25**(1), 113–131 (2018). https://doi.org/10.1504/IJHVS.2018.10011111. ISSN online: 1741–5152. ISSN print: 1744–232X
11. Wood, R.H.: Effective lengths of columns in multi-storey buildings Part II. J. Struct. Eng. **52**(8), 295–302 (1974)
12. Tsai, L.W.: Mechanism Design: Enumeration of Kinematic Structures According to Function. CRC Press, Boca Raton (2001). ISBN 0849309018
13. Huang, Y., Qu, S., Hwang, K.C., Li, M., Gao, H.: A conventional theory of mechanism-based strain gradient plasticity. Int. J. Plast. **20**(4), 753–782 (2004)
14. Bakhshesh, M., Bakhshesh, M.: Optimization of steel helical spring by composite spring. Int. J. Multidiscip. Sci. Eng. **3**(6) (2012)
15. Rill, G.: Vehicle modeling by subsystems. J. Braz. Soc. Mech. Sci. Eng. **28**(4), 430–442 (2006)
16. Dhoshi, N., Ingole, N., Gulhane, U.: Analysis and modification of leaf spring of tractor trailer using analytical and finite element method. Int. J. Mod. Eng. Res. **1**(2), 719–722 (2011)
17. Taylor, R., Bashford, L., Schrock, M.: Method for measuring vertical tire stiffness. Am. Soc. Agric. Eng. **42**(6), 1415–1419 (2000)
18. Rill, G.: Road Vehicle Dynamics: Fundamentals and Modeling. CRC Press, Boca Raton (2011). ISBN 978-1-4398-3898-3
19. Fang, W., Tsai, H., Lo, C.: Determining thermal expansion coefficients of thin films using micromachined cantilevers. J. Sens. Actuators A Phys. **77**(1), 21–27 (1999)
20. Farley, C., Morgenroth, D.: Leg stiffness primarily depends on ankle stiffness during human hopping. J. Biomech. **32**(3), 267–273 (1999)
21. Wang, J., Howard, I.: The torsional stiffness of involute spur gears. Proc. Inst. Mech. Eng. Part C J. Mech. Eng. Sci. **218**(1), 131–142 (2004)
22. Myszka, D.H.: Machines & Mechanisms, 4th edn, pp. 357–363. Prentice Hall, Upper Saddle River (2012). ISBN-10:0-13-215780-2

Kinematic Characteristics of Parallel Continuum Mechanisms

Oscar Altuzarra(✉), Diego Caballero, Quichen Zhang, and Francisco J. Campa

University of the Basque Country UPV/EHU, 48013 Bilbao, Spain
{oscar.altuzarra,diego.caballero,fran.campa}@ehu.es,
qzhang001@ikasle.ehu.eus
http://www.ehu.eus/compmech/oscar-altuzarra/

Abstract. Parallel Continuum Mechanisms are manipulators whose end-effector is connected with several compliant legs. Continuum manipulators of such kind have been considered historically serial devices because most of them have a long slender form. Nevertheless, these architectures contain aspects resembling their rigid-link parallel analogs. In this paper, authors go over the parallelisms between both kinds of devices regarding several kinematic aspects. We will show that in the field of Parallel Continuum Mechanisms there are alter egos to concepts of rigid-link manipulators such as: assembly modes and working modes, or singularities. However, there are also distinct features that will be highlighted.

Keywords: Parallel continuum mechanisms · Compliant · Buckling Singularity · Instability

1 Introduction

Compliant mechanisms, where a notch in the material plays the role of a kinematic pair, are mechanical systems that match very closely classical rigid-link mechanisms. Parallel structures in Compliant mechanisms for MEMS applications are not rare [1]. Kinematically they have analogous properties to their rigid counterparts.

Continuum mechanisms are long slender devices, subjected to large deformations by design to accomplish a task. Bending is distributed along the elements rather than lumped at a point. They have become quite popular in surgical applications to provide access to difficult spaces with obstacles. However, many continuum designs contain aspects resembling parallel architectures, such as serial assemblies of intermediate disks through flexible elements or springs, actuated in parallel by cables that arrive to different points of a final end-effector [2].

More recently, such parallel analogy has raised some attention and further development. Parallel Continuum Manipulators are being considered as a type of mechanisms on their own [3]. In these, a single rigid element is joined to a fixed system through flexible elements that introduce actuation. A typical case

J. Lenarcic and V. Parenti-Castelli (Eds.): ARK 2018, SPAR 8, pp. 293–301, 2019.
https://doi.org/10.1007/978-3-319-93188-3_34

is the hexapod where pneumatic limbs are replaced with flexible rods whose free ends are moved with actuators placed on a fixed frame [4].

Kinematic analysis of these latter mechanisms have strong similarities to their rigid-link counterparts. This paper is intended to show similarities and differences regarding some kinematic features such as: solution methods, multiplicity of solutions, and singularities.

2 Solving Methods for Position Problems

Position problems in rigid-link parallel mechanisms are ruled by a system of non-linear equations obtained from loop-closure equations and constraint equations. These are stated solely from geometric relationships, i.e. forces applied have no effect on the solution. Finding all solutions requires algebraic solving methods such as elimination, Gröbner basis or homotopy, and it is very often a difficult task that may require finally numerical methods.

In cable-driven parallel robots the problem is even more complex because the system is larger as first static equations have to be considered, and deformation in cables may play a role. When cables are considered with their mass and elasticity, catenary equations must be used to relate length of cables to forces exerted on the platform. As a result, a set of equations that are not algebraic is generated and we will have to resort to numerical methods [5].

In some sense, Parallel Continuum Manipulators are very close in nature to these latter robots. Each leg connected to the end-effector is ruled by a set of equations where shape and forces are related by differential equations that need to be integrated. In addition, a set of boundary conditions due to the geometric attachment of legs to the end-effector plus the equations of equilibrium of forces and moments in the platform must be accomplished.

A classical method for the analysis of deformation of a slender flexible element subjected to arbitrary loads is the Cosserat rod theory. In the case of neglecting extension and transverse shear strain, this simplifies to the so called Kirchhoff model. Three aspects are combined here: kinematics definition of deformed rod, material constitutive equations, and static equilibrium equations.

The deformed shape of the rod is defined by a parametric curve in space $\mathbf{p}(s)$ as a function of arc-length s, and a rotation matrix $\mathbf{R}(s)$ to define cross-section orientation. Derivatives of these with respect to arc-length in local frame, define local stretch and shear deformation, i.e. $\Delta\mathbf{v}$, and curvature $\mathbf{u}$, through:

$$\begin{aligned} \dot{\mathbf{p}} &= \mathbf{R}\mathbf{v} \\ \dot{\mathbf{R}} &= \mathbf{R}\mathbf{U} = \mathbf{R}\hat{\mathbf{u}} \end{aligned} \tag{1}$$

The material constitutive equations relate deformations to internal forces $\mathbf{n}(s)$ and moments $\mathbf{m}(s)$. A linear constitutive law is defined in local frame using stiffness matrices as:

$$\begin{aligned} \mathbf{n} &= \mathbf{K}_{SE}\Delta\mathbf{v} \\ \mathbf{m} &= \mathbf{K}_{BT}\Delta\mathbf{u} \end{aligned} \tag{2}$$

Finally, static equilibrium equations on each differential section of the rod subjected to applied distributed loads, $\mathbf{f}(s)$ and $\mathbf{l}(s)$, produces a set of differential equations:

$$\begin{aligned} \dot{\mathbf{n}} + \mathbf{f} &= \mathbf{0} \\ \dot{\mathbf{m}} + \dot{\mathbf{p}} \times \mathbf{n} + \mathbf{l} &= \mathbf{0} \end{aligned} \tag{3}$$

The resulting set of differential equations constitute a boundary value problem that relates loads and boundary conditions to the shape of the rod. Solving this problem numerically can be approached with the use of a shooting method to process the boundary value problem as an initial value problem from guessed values, using a Runge-Kutta method of fourth order, subjected to a root finding method.

When several rods are assembled to an end-effector we get a larger set of differential equations as we must introduce the geometrical conditions and force equilibrium of the assembly. Parallel Continuum manipulators can be solved like this, for example, the Gough platform controlled by a set of extensible flexible legs in Fig. 1, as reported in [6].

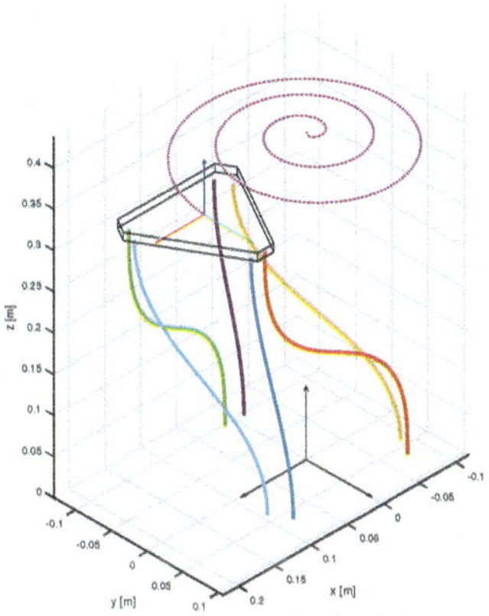

Fig. 1. Inverse kinematic problem in a Gough platform with flexible legs

Kinematic features are analyzed easier in planar mechanisms. We take a two DOF planar mechanism with a punctual end-effector $\mathcal{P}$ (see Fig. 2). Each flexible leg is clamped at one extreme to an actuator on a frame, i.e. a controlled angle θ_j will move the mechanism, while the other extreme is articulated to tip-point $\mathcal{P}$. We will denote this mechanism by $2\underline{Cl}FR$, that stands for clamped, flex, and revolute. Forward Kinematics (FK) will get the output position of $\mathcal{P}$ for known values of the input angles and load assuring the force equilibrium. Therefore, here arises the first difference with those rigid-link counterparts, the position problem is load-dependent (see Fig. 2 right). Such problems may have multiple solutions. For a real-time FK, recorded values of encoders and torque in the

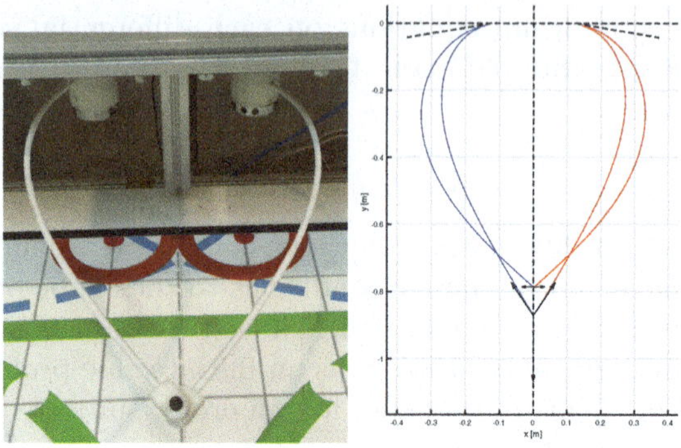

Fig. 2. $2\underline{Cl}FR$, and deformed shapes when loaded and unloaded.

motors, i.e. θ_j and bending moments at clamped ends, can produce the only feasible solution along with the unknown load.

When analyzing rods in plane deformations subjected only to a load at tip $\mathbf{F}_{\mathcal{P}}$ (expressed in terms of modulus R and orientation ψ), the set of differential equations above can be further developed arriving to an expression of the deformed shape in terms of elliptic integrals, $F(k,\phi_i)$ and $E(k,\phi_i)$, ruled by the elliptic modulus k, and a variable ϕ_i in a range $[\phi_1\ \phi_2]$ to express location on arc-length [7]:

$$\begin{aligned} x(s) &= -\sqrt{\tfrac{EI}{R}}\cos\psi\,[2E(k,\phi_i)-2E(k,\phi_1)-F(k,\phi_i)+F(k,\phi_1)] \\ &+ \qquad -\sqrt{\tfrac{EI}{R}}2k\sin\psi\,[\cos\phi_i-\cos\phi_1] \\ y(s) &= -\sqrt{\tfrac{EI}{R}}\sin\psi\,[2E(k,\phi_i)-2E(k,\phi_1)-F(k,\phi_i)+F(k,\phi_1)] \\ &+ \qquad +\sqrt{\tfrac{EI}{R}}2k\cos\psi\,[\cos\phi_i-\cos\phi_1] \end{aligned} \tag{4}$$

Here, $x(s)$ and $y(s)$ are coordinates in a mobile local frame attached to the actuator, i.e. oriented with θ_j.

Planar continuum mechanisms formed this way can be solved with a set of equations whose solution can be better controlled to avoid missing solutions, a common issue in general numerical methods of integration. Also, buckling mode, i.e. a deformed shape with different number of inflection points, can be specified easily allowing a better discrimination of multiple solutions. Nevertheless, boundary conditions on the assembled mechanism must be considered. For the FK, given θ_1 and θ_2, the set of equations is:

$$\begin{aligned} \mathbf{p}_1(L_1)-\mathbf{p}_2(L_2) &= \mathbf{0} \\ \mathbf{n}_1+\mathbf{n}_2+\mathbf{F}_{\mathcal{P}} &= \mathbf{0} \end{aligned} \tag{5}$$

where unknowns are k_1, k_2, ψ_1, ψ_2. For the Inverse Kinematics (IK), equations are:

$$\begin{aligned} \mathbf{p}_1(L_1) - \mathbf{p}_{\mathcal{P}} &= \mathbf{0} \\ \mathbf{p}_2(L_2) - \mathbf{p}_{\mathcal{P}} &= \mathbf{0} \\ \mathbf{n}_1 + \mathbf{n}_2 + \mathbf{F}_{\mathcal{P}} &= \mathbf{0} \end{aligned} \tag{6}$$

where θ_1, θ_2 are also unknowns.

Hence, in this kind of closed-loop mechanisms, the FK is simpler than the IK.

3 Multiplicity of Solutions of the IK and FK

The multiplicity of solutions in the IK and FK, defining working modes and assembly modes in rigid-link parallel manipulators, is also present in continuum manipulators. In the rigid-link $2\underline{R}RR$ mechanism, alter ego to the $2\underline{Cl}FR$, solutions are found by the different assemblies of dyads. In flexible mechanisms the force balance at $\mathcal{P}$ plays a central role. Possible geometrical solutions for each of the flexure legs have to accomplish with a force balance of the whole. Hence, a difference arises here, solutions of the IK cannot be obtained by simple combination of geometrical solutions obtained for each leg.

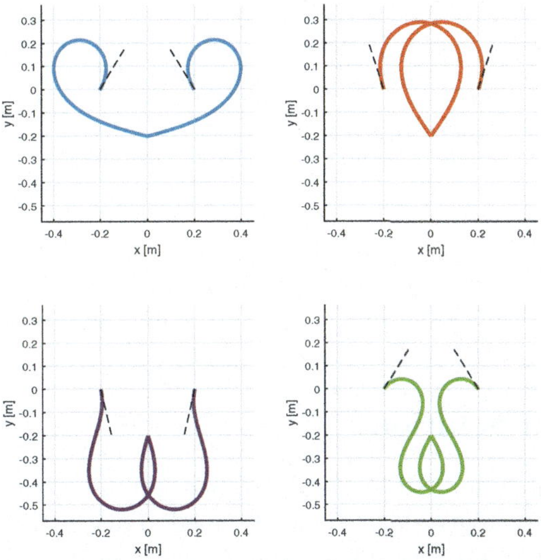

Fig. 3. Solutions of the IK problem.

In Fig. 3 there are several solutions of the IK problem for a given position of $\mathcal{P}$ and no load, obtained with symmetrical inputs. The two upper solutions have a buckling mode with no inflection point, the solutions below have one. Dotted lines at clamped ends indicate input angles. Note that it is not possible to combine the different solutions for each leg because the force balance must be accomplish. Only symmetrical solutions, where force at tip-end of each leg is horizontal and balanced, are possible.

In Fig. 4 we have plotted some solutions for the FK problem in the event of given symmetrical inputs and no load.

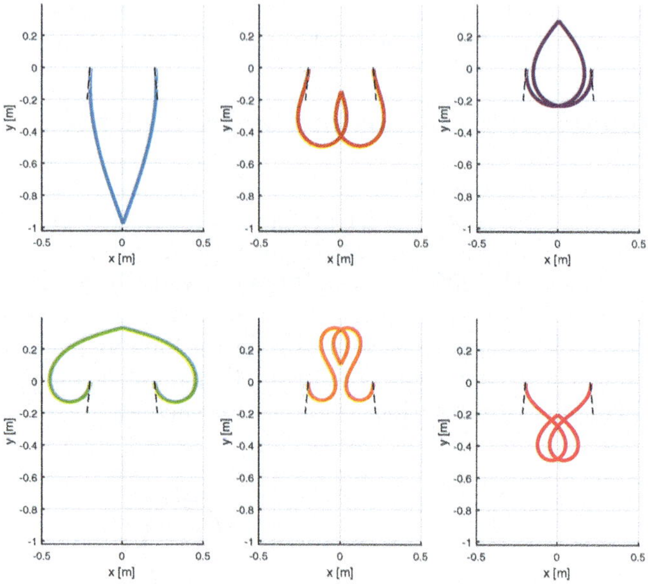

Fig. 4. Solutions of the FK problem.

It must be pointed out that an exhaustive solving of the full FK or IK position problem is not straightforward. There is not an initial estimation of the number of solutions for each buckling mode, or up to which buckling mode we will get feasible solutions.

4 Stability of Solutions

Another important issue to be taken into account in such mechanisms is the stability of the mathematical solutions obtained. This is the same phenomenon that rules the well-known bistable compliant mechanisms used in some MEMS [8]. In the parallel continuum mechanisms under analysis, stability should be checked in view of the elastic energy of the mechanism. This elastic energy can be obtained numerically from the deformed shapes of the elements. Then, we can find the stiffness matrix of the mechanism and the Hessian to decide whether the pose is stable or not.

$$\mathbf{F}_{\mathcal{P}} = \mathbf{K} \cdot \Delta \mathbf{p}_{\mathcal{P}} \quad (7)$$

For example, in Fig. 5 we have plotted the elastic energy on the vicinity of each IK pose shown in Fig. 3. Upper solutions are stable ones with a minimum of the elastic energy function, while solutions below are unstable due to the saddle point of the function.

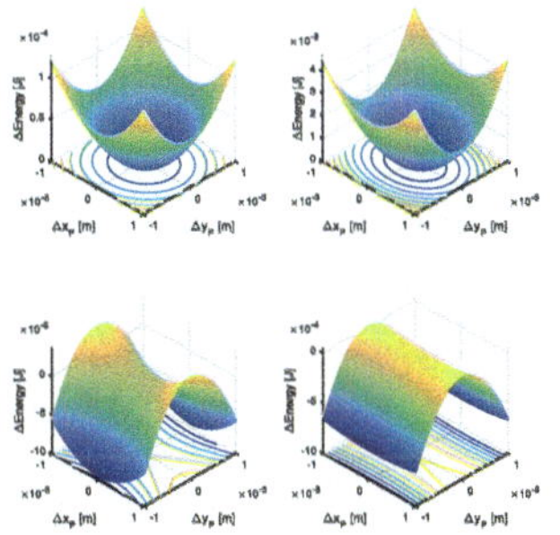

Fig. 5. Elastic energy of the mechanism for the vicinity of IK solutions

In Fig. 6, the elastic energy on FK poses in Fig. 4 indicate stability of each solution. Solutions in the middle are unstable: upper one due to saddle point, and the one below due to a maximum.

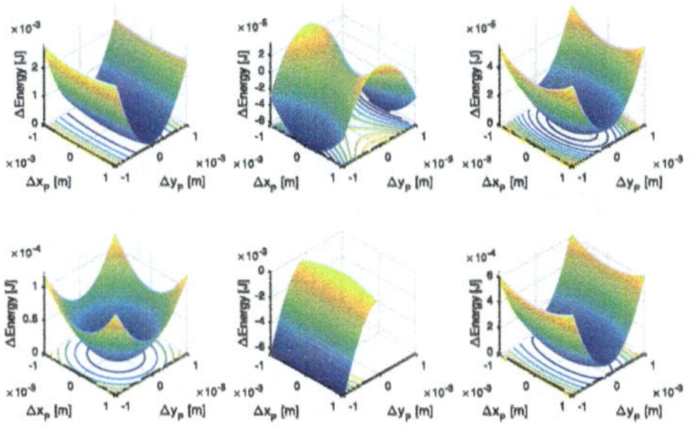

Fig. 6. Elastic energy of the mechanism for FK solutions on the vicinity of poses

5 Singularities

Singular poses occur when the Jacobians relating input to output velocities become singular. In rigid-link parallel manipulators, algebraic finding of these equations provide a set of linear equations that define the Jacobians analytically, and hence geometric conditions for singular poses. In continuum parallel manipulators this is not the case, Jacobians are found numerically and singular poses are found by iterative searching upon evaluation of:

$$\begin{aligned} \Delta \mathbf{p}_{\mathcal{P}} &= \mathbf{J}_{IK} \cdot \Delta \boldsymbol{\theta} \\ \Delta \boldsymbol{\theta} &= \mathbf{J}_{FK} \cdot \Delta \mathbf{p}_{\mathcal{P}} \end{aligned} \tag{8}$$

In Fig. 7 there are two poses of singularities of the IK and FK. This concept is not to be mistaken for instabilities.

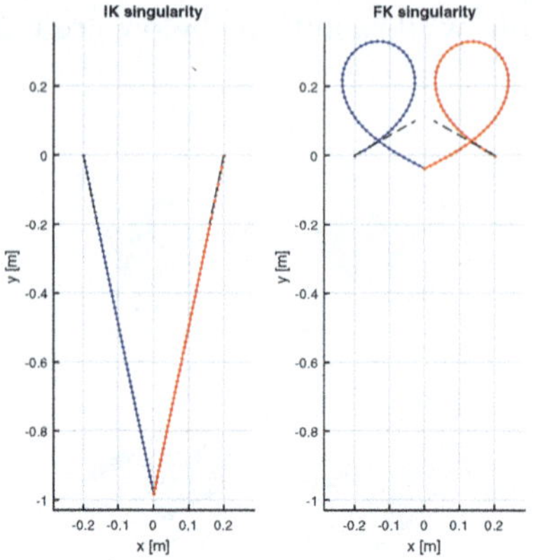

Fig. 7. Singularity poses

6 Conclusions

Parallel Continuum manipulators have kinematic features analogous to those of rigid-links but there are also important differences mentioned along the paper. Position problems are numerically solved and subject to solution missing. FK problem is easier than IK problem, and both are load-dependent. Apart from singular poses, there are also unstable solutions, equally undesirable.

Acknowledgements. Authors wish to acknowledge financial support from Spanish Government (DPI2015-64450-R), MEC Ph.D. grant (FPU15/04422) and Regional Government of the Basque Country (Project IT949-16).

References

1. Teo, Y.J., Yang, G., Chen, I-M.: Compliant manipulators. In: Handbook of Manufacturing Engineering and Technology, pp. 2229–2300. Springer, London (2014). https://doi.org/10.1007/978-1-4471-4976-7_102-1
2. Remirez, A.A., Rucker, D.C., Webster, R.J.: Continuum Devices and Control Methods Thereof. U.S. Patent US 2016/0016319 A1 (2016)
3. Bryson, C.E., Rucker, D.C.: Toward parallel continuum manipulators. In: Proceedings of the IEEE International Conference on Robotics and Automation, May 2014, pp. 778–785 (2014). https://doi.org/10.1109/ICRA.2014.6906943
4. Zhu, Z., Cui, H., Pochiraju, K.: Flexible Parallel Manipulator for Nano-, Meso- or Macro-positioning with Multi-degrees of Freedom. U.S. Patent US 2008/0257096 A1 (2008)
5. Merlet, J.P., Alexandre-dit-Sandretto, J.: The forward kinematics of cable-driven parallel robots with sagging cables. In: Cable-Driven Parallel Robots, vol. 32, pp. 3–15. Springer, Cham (2015). https://doi.org/10.1007/978-3-319-09489-2_1

6. Till, J., Bryson, C.E., Chung, S., Orekhov, A., Rucker, D.C.: Efficient computation of multiple coupled Cosserat rod models for real-time simulation and control of parallel continuum manipulators. In: Proceedings of the IEEE International Conference on Robotics and Automation, May 2015, pp. 5067–5074 (2015). https://doi.org/10.1109/ICRA.2015.7139904
7. Zhang, A., Chen, G.: A comprehensive elliptic integral solution to the large deflection problems of thin beams in compliant mechanisms. J. Mech. Robot. **5**(2), 021006 (2013). https://doi.org/10.1115/1.4023558
8. Chen, G., Ma, F.: Kinetostatic modeling of fully compliant bistable mechanisms using Timoshenko beam constraint model. J. Mech. Des. **137**, 022301 (2015). https://doi.org/10.1115/1.4029024

Spatial Orientations of Principal Vector Planes for Inherent Dynamic Balancing

Volkert van der Wijk(✉)

Delft University of Technology, Delft, The Netherlands
v.vanderwijk@tudelft.nl

Abstract. The spatial orientations of the planes defined by principal vectors of the pantograph, the closed 4-principal vector linkage, and the grand 4R four-bar based inherently balanced linkage architecture are investigated in this paper. These principal vector planes form the basis for the synthesis of planar and spatial inherently balanced linkages, i.e. linkages derived with balance as starting point in the design process of which the center of mass is in an invariant point in one of the links. It is shown that a pantograph consists of a single plane of principal vectors, the 4-principal vector linkage consists of three planes of principal vectors, and the grand 4R four-bar based inherently balanced linkage architecture consists of ten planes of principal vectors with four parallel sets. It is shown how the principal vector planes can be used for synthesis by example of a spatial inherently balanced double Bennett linkage.

Keywords: Inherent balance · Principal vectors · Center of mass

1 Introduction

Principal vectors have shown to be key elements in the synthesis and design of planar inherently balanced linkages, i.e. linkages derived with balance as starting point in the design process of which the center of mass is in an invariant point in one of the links [4]. Instead of balancing a given mechanism, generally leading to unpractical solutions with relatively large mass and inertia [7] and complexity [2,3], especially for multi degree-of-freedom mechanisms, the inherent dynamic balance approach based on principal vectors where balance is the starting point in the design process has shown to have high potential for finding kinematic solutions with excellent balance characteristics [8]. With this approach already a variety of new balanced linkages have been found.

The inherent balance approach based on principal vectors has already been well studied for planar linkages [4]. The goal of this paper is to make the first steps in thoroughly studying the principal vectors in three-dimensional space. In three-dimensional space the principal vectors define planes with different orientations. Understanding the structure of these principal vector planes is needed for the synthesis of spatial inherently balanced linkages of which the kinematics are fully determined by the motion of the principal vector planes. With the knowledge of

J. Lenarcic and V. Parenti-Castelli (Eds.): ARK 2018, SPAR 8, pp. 302–309, 2019.
https://doi.org/10.1007/978-3-319-93188-3_35

the principal vector planes it will become possible to find new spatial balance solutions.

In the next section first the spatial orientations of the principal vectors of the pantograph are shown. Then the principal vector planes of the 4-principal vector linkage is investigated followed by the study of the principal vector planes of the grand 4R four-bar based inherently balanced linkage architecture from [5]. At the end it is shown how from the model of the principal vector planes balance solutions are found by example of the spatial inherently balanced double Bennett linkage from [6].

All figures are drawn to scale to give realistic impressions. Only mass symmetric links are considered.

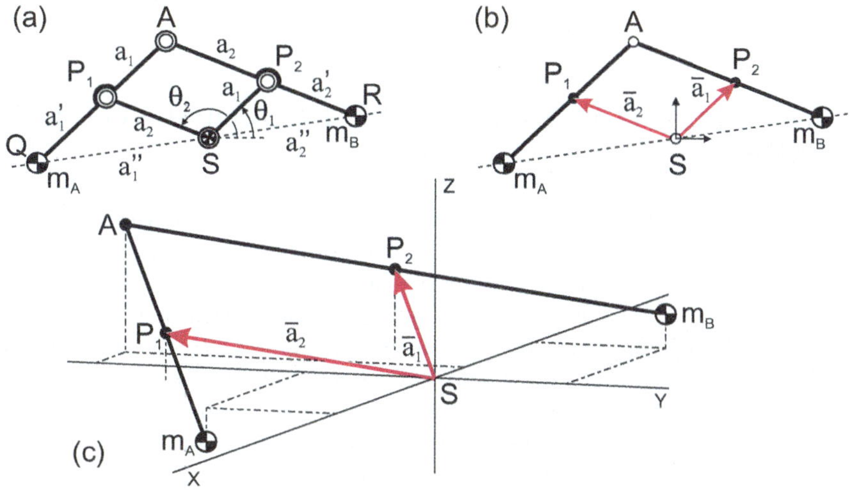

Fig. 1. (a) Pantograph linkage with the common CoM of m_A and m_B in S for all motion [4]; (b) Principal vectors $\bar{a}_1$ and $\bar{a}_2$ defining the pantograph with principal (link) points P_1 and P_2; (c) Spatial orientation of the principal vectors which span a single plane.

2 Principal Vector Plane of the Pantograph

The planar pantograph linkage from [4], shown in Fig. 1a, consists of the four links SP_1, P_1A, AP_2, and P_2S which are arranged as parallelogram with link lengths a_1 and a_2. The pantograph has two degrees of freedom (DoF) which are indicated by θ_1 and θ_2. For any motion, the common CoM of mass m_A in Q and mass m_B in R is stationary in joint S for the conditions $m_A a_1' = m_B a_1$ and $m_A a_2 = m_B a_2'$. Q is a point in link P_1A at a distance a_1' from P_1, as illustrated, and R is a point in link AP_2 at a distance a_2' from P_2 as illustrated. If S is made a fixed point in the base, then the common CoM is stationary for any motion and the pantograph has become a shaking force balanced linkage.

The design of the balanced pantograph is determined by the two principal vectors $\bar{a}_1$ and $\bar{a}_2$ as illustrated in Fig. 1b. These vectors determine the location of the principal points P_1 and P_2 with respect to joint A with a_1 and a_2 being the principal dimensions, which are the lengths of the principal vectors. Since principal vectors have a constant length, they can be transformed into links such as links SP_1 and SP_2 in Fig. 1a.

To extend the theory to the three dimensional space, Fig. 1c shows how the principal vectors can be depicted. With solely two principal vectors the result is quite similar to the planar case. Also here the two principal vectors span a single principal vector plane but the plane now has a spatial orientation. Since the complete mechanism is within the same plane, the spatial orientation will not change the synthesis of balance solutions with respect to the planar situation. When more than two principal vectors are involved then the spatial orientation will affect the synthesis of balance solutions as will be shown in the next sections.

3 Principal Vector Planes of the 4-Principal Vector Linkage

Figure 2a shows the four principal vectors $\bar{a}_1$, $\bar{a}_{21}$, $\bar{a}_{23}$, and $\bar{a}_3$ which define the position of common CoM in S of an open chain of three links (l_1, l_2, and l_3) and also of a closed chain of four links, four-bar linkage $A_0A_1A_2A_3$ [4,5]. The vectors are aligned with the links, forming three parallelograms with the principal points P_1, P_2, and P_3 in links l_1, l_2, and l_3, respectively, similar to the pantograph in Fig. 1. Also here, because of their constant lengths, the principal vectors can be transformed into links with, for the planar case, revolute pairs in all the joints S, B_1, B_2, P_1, P_2, P_3. The resulting linkage is referred to as the *4-principal vector linkage*, i.e. the inherently balanced linkage architecture based on four principal vectors with links along all principal vectors.

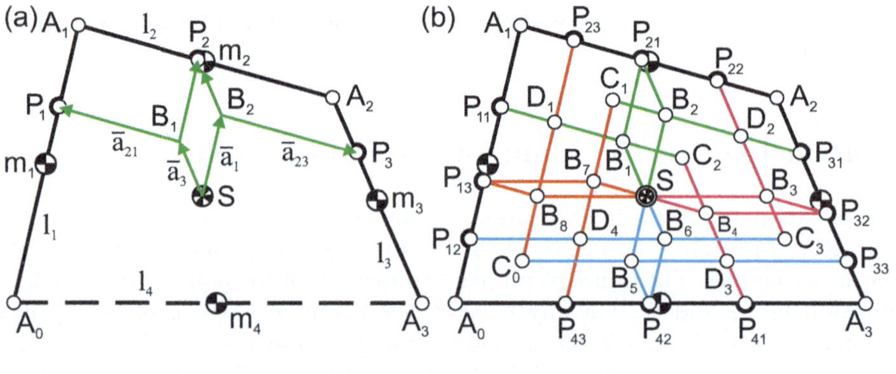

Fig. 2. (a) The four principal vectors $\bar{a}_1$, $\bar{a}_{21}$, $\bar{a}_{23}$, and $\bar{a}_3$ define the common CoM in S of an open chain of three links or of a closed chain of four links [4]; (b) Grand 4R four-bar based inherently balanced linkage architecture [5].

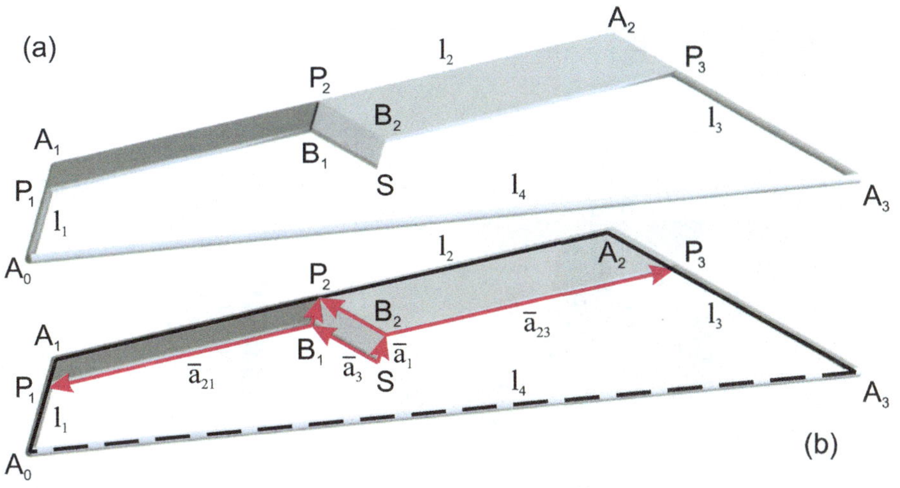

Fig. 3. (a) Spatial orientations of the three principal vector planes for a specific pose, maintaining inherent balance with the common CoM in S; (b) Illustration of the four principal vectors describing the spatial orientations of the principal vector planes.

For investigation of the spatial motions of the 4-principal vector linkage for inherent balance the type of joints is not (yet) of importance. The spatial motions are solely determined by the principal vector planes, which are illustrated in Fig. 3a. For clarity the shape of the planes is depicted equal to the enclosure by the principal vectors. It is observed that each of the three parallelograms in Fig. 2a is a plane with own orientation. In Fig. 3b also the principal vectors are included, showing that the orientation of each plane is defined by two principal vectors. Plane $P_1A_1P_2B_1$ is defined by principal vectors $\overline{a}_1$ and $\overline{a}_{21}$ and also includes links A_0A_1 and A_1A_2. Plane $P_2A_2P_3B_2$ is defined by principal vectors $\overline{a}_3$ and $\overline{a}_{23}$ and also includes the links A_1A_2 and A_2A_3. The third plane $SB_1P_2B_2$ is defined by principal vectors $\overline{a}_1$ and $\overline{a}_3$. The three planes intersect in point P_2.

In [1] a linkage can be found which can be seen as a special case of the model in Fig. 3a without link A_0A_3, applied as a statically balanced spatial manipulator with two springs. That linkage solution was designed having solely revolute pairs with the orientation of plane $SB_1P_2B_2$ for all motion orthogonal to both the other planes and also orthogonal to the line A_1A_2. This means that the rotational axes of links A_0A_1 and A_2A_3 then are aligned with line A_1A_2.

4 Principal Vector Planes of the Grand 4R Four-Bar Based Inherently Balanced Linkage Architecture

Figure 2b shows the grand 4R four-bar based inherently balanced linkage architecture that was derived in [5]. It is the combination of the four possible principal

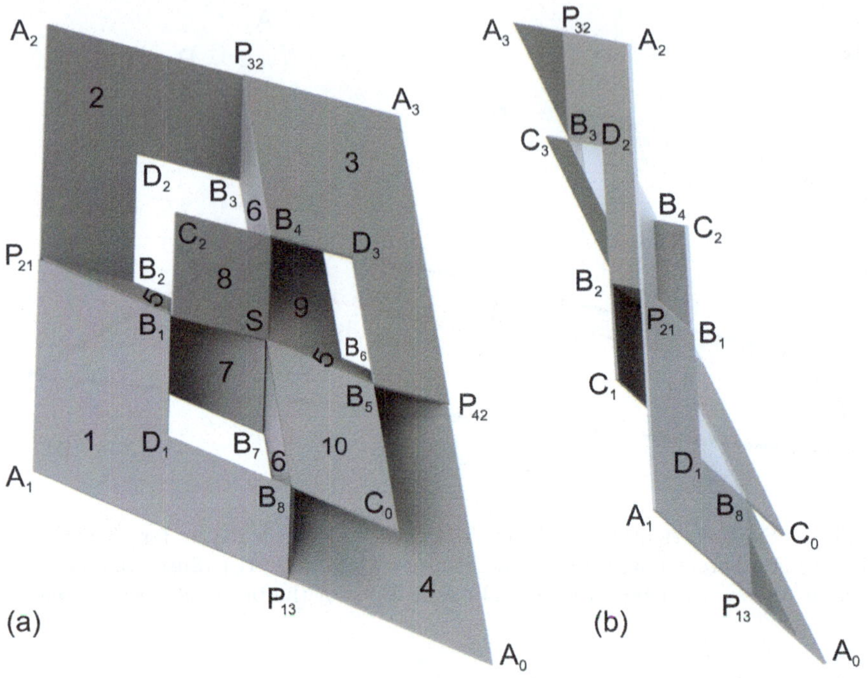

Fig. 4. (a) Front view and (b) side view of the spatial orientations of the principal vector planes of the grand 4R four-bar based inherently balanced linkage architecture, consisting of ten planes of which four times two planes are in parallel.

vector linkages of a closed 4R four-bar linkage. In green the principal vector linkage according to Fig. 2a is shown, in red, blue, and purple the other three possible principal vector linkage of the same four-bar are shown, which all connect together in S and in joints C_i and D_i. Together they form an advanced inherently balanced linkage architecture from which a variety of planar inherently balanced mechanism solutions can be found.

Also this grand architecture can be moved spatially, disregarding the type of joints. The sixteen principal vectors define ten planes of which the orientations are illustrated in Fig. 4. In each of the four corners there is a (L-shaped) plane which is based on four principal vectors each: planes 1, 2, 3, and 4. This is redundant since a plane is sufficiently defined by solely two vectors. These four planes are connected by the cross over planes 5 and 6, which are the plane through P_{21}, B_1, B_2, S, B_5, B_6 and P_{42} and the plane through P_{13}, B_7, B_8, S, B_3, B_4 and P_{32}, respectively. It is interesting to note that these two sets of points are both within a single plane. Also both of these planes are based on four principal vectors each and hence are defined redundantly.

Each of the four smaller central planes 7, 8, 9, and 10 is in parallel with a corner plane, planes 1, 2, 3, and 4, respectively, as shown also in Fig. 4b. These four central planes together with planes 5 and 6 all intersect in point S.

When the linkage architecture is moved spatially, the planes rotate relative to each other and their shapes enclosed by the principal vectors change. However the structure of the orientations of the planes remains equal to Fig. 4. There is only one situation where the complete structure reduces to one single plane, which is, of course, for motion into the planar situation as illustrated in Fig. 2b.

The six principal vector planes 1, 2, 3, 4, 5, and 6 are defined redundantly with each four principal vectors. This is due to the choice to have mass symmetric links. In general the CoM of a link may not be on the line through the joints. In the general case with each link CoM off the line through the joints, each of the six redundantly defined planes will turn out as two separate planes. This means that for the general case the model of the grand 4R four-bar based inherently balanced linkage architecture will have sixteen principal vector planes in total, equal to the number of principal vectors and each defined solely by two principal vectors.

5 Synthesis of Spatial Inherently Balanced Linkages

Understanding the characteristics of the principal vector planes is the first step in the synthesis of spatial inherently balanced linkages, similarly to the synthesis of

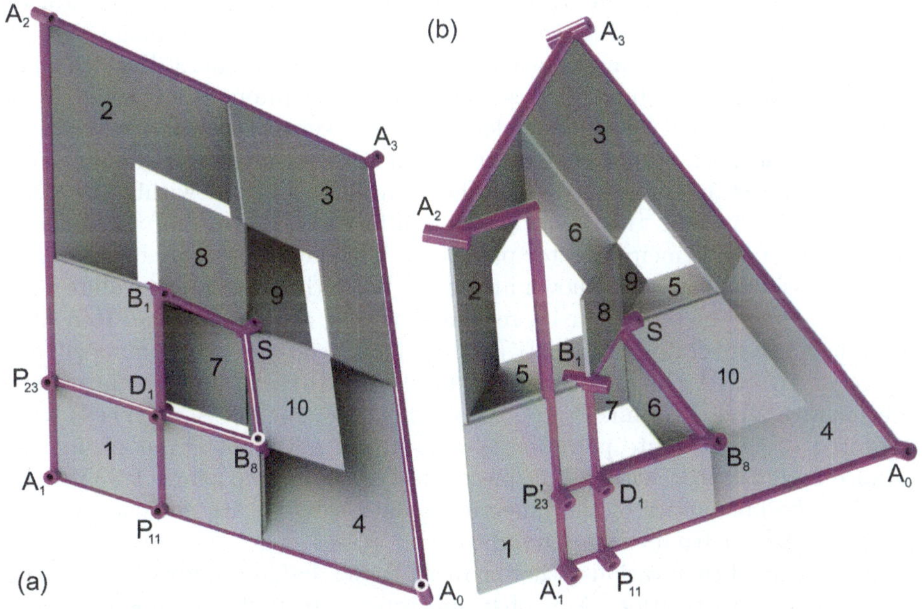

Fig. 5. (a) Illustration of the spatial inherently balanced double Bennett linkage [6] which can be found also from the model of the principal vector planes. The revolute pairs of A_1, P_{11}, P_{23}, and D_1 are in plane 1, of A_2 is in plane 2, of A_3 is in plane 3, of A_0 is in plane 4, of B_1 is in plane 8, of B_8 is in plane 10, and of S is in plane 9; (b) Possible change in the design of a link for which inherent balance is maintained.

planar inherently balanced linkages where understanding of the principal vectors is the first step in the synthesis process. The second step then is to find linkages which move according the principal vector planes. These solutions have inherent balance characteristics, which means that with a chosen combination of mass distributions of the links balance is obtained in the final step. As a third step before this final step it is possible to adapt the design of the linkage if needed by using the principal vector planes as guidelines with which the inherent balance characteristics are kept.

In [6] the first spatial inherently balanced linkage was presented, a mechanism consisting of two similar Bennett linkages, one inside the other with a parallelogram in between. It was derived in an intuitive way, but it can be explained by using the model of the principal vector planes. Figure 5a shows the adapted version of the model in Fig. 4 with the double Bennett linkage illustrated within the vector planes. The model was adapted by including the Bennett conditions, making links A_0A_1 and A_2A_3 and links A_1A_2 and A_3A_0 of equal lengths. The linkage consists of eight links in total, with D_1 the joint of the links $P_{11}B_1$ and $P_{23}B_8$. It can be seen that all the revolute pairs have their rotational freedom within one of the principal vector planes. The revolute pairs in A_1, P_{11}, P_{23}, and D_1 rotate within plane 1, the revolute pairs in A_2 rotate within plane 2, the revolute pairs in A_3 rotate within plane 3, the revolute pairs in A_0 rotate within plane 4, the revolute pairs in B_1 rotate within plane 8, the revolute pairs in B_8 rotate within plane 10, and the revolute pairs in S rotate within plane 9.

The design of the linkage solution as shown in Fig. 5a can still be changed if this is desired. Changing the design for spatial solutions is expected to be more complicated than for planar linkage solutions since it is important that the kinematic properties do not change. In planar linkages revolute pairs and links can relatively easily be moved to another (parallel) location within the plane but for spatial linkages this is not as straightforward. Because of the different orientations of the principal vector planes the kinematics are more complicated.

One possible adaption is shown in Fig. 5b. Since there are four revolute pairs in plane 1, some of them can be relocated within the same plane. Here it is shown how the revolute pairs in A_1 and P_{23} have been moved to points A_1' and P_{23}', respectively. With this change the link A_1A_2 is translated within plane 1 and remains parallel to link $P_{11}B_1$. As a result the new link $A_1'P_{23}'$ needs to be connected to the revolute pairs in A_2 with a link part that is out of line, for instance as illustrated with a link piece that is also within plane 1. With this adaption the kinematics of the inherent balance solution are not changed and therefore inherent balance is maintained.

It is expected that various other changes in the design are possible, but this needs further investigation. Also of future topic is to find new inherently balanced linkage solutions from the model of the principal vector planes. One main challenge will be to find the right type and combination of joints that constrain the kinematics properly. The motivation to search for solutions is strong: it is hypothesized that all possible inherently balanced linkages must be inside the model of the principal vector planes. This is since the principal vectors form the basis of all balanced linkages.

6 Conclusions

For the first time the spatial orientations of principal vector planes have been investigated. It was shown that the pantograph consists of a single principal vector plane, that the closed 4-principal vector linkage consists of three principal vector planes with their own orientations based on two principal vectors each, and that the grand 4R four-bar based inherently balanced linkage architecture consists of ten principal vector planes with four parallel sets. As an illustration how the model of the principal vector planes can be used for the synthesis of spatial inherently balanced linkages it was shown how the spatial inherently balanced double Bennett linkage is found. Also an example was shown of a possible modification of the found solution for which the inherent balance characteristics are maintained.

Acknowledgement. This publication was financially supported by the Netherlands Organization for Scientific Research (NWO, 15146).

References

1. Agrawal, S.K., Fattah, A.: Gravity-balancing of spatial robotic manipulators. Mech. Mach. Theory **39**, 1331–1344 (2004)
2. Briot, S., Bonev, I.A., Gosselin, C.M., Arakelian, V.: Complete shaking force and shaking moment balancing of planar parallel manipulators with prismatic pairs. Multi-body Dyn. **223**(K), 43–52 (2009)
3. Gosselin, C.M., Vollmer, F., Côté, G., Wu, Y.: Synthesis and design of reactionless three-degree-of-freedom parallel mechanisms. IEEE Trans. Robot. Autom. **20**(2), 191–199 (2004)
4. Van der Wijk, V.: Methodology for analysis and synthesis of inherently force and moment-balanced mechanisms - theory and applications. Dissertation, University of Twente (2014). https://doi.org/10.3990/1.9789036536301
5. Van der Wijk, V.: On the grand 4R four-bar based inherently balanced linkage architecture. In: Wenger, P., Flores, P. (eds.) New Trends in Mechanism and Machine Science 2016. MMS, vol. 43, pp. 473–480. Springer (2017)
6. Van der Wijk, V.: Inherently balanced double Bennett linkage. In: Zeghloul, S., et al. (eds.) Computational Kinematics 2017. MMS, vol. 50, pp. 3–10. Springer (2018)
7. Van der Wijk, V., Herder, J.L., Demeulenaere, B.: Comparison of various dynamic balancing principles regarding additional mass and additional inertia. Mech. Robot. **1**(4), 1006 (2009)
8. Van der Wijk, V., Krut, S., Pierrot, F., Herder, J.L.: Design and experimental evaluation of a dynamically balanced redundant planar 4-RRR parallel manipulator. Int. J. Robot. Res. **32**(6), 744–759 (2013)

Static Modeling of Sagging Cables with Flexural Rigidity and Shear Forces

Hussein Hussein(✉), Marc Gouttefarde, and François Pierrot

LIRMM, Université de Montpellier, CNRS, Montpellier, France
{hussein.hussein,marc.gouttefarde,Francois.pierrot}@lirmm.fr

Abstract. Cables are considered perfectly flexible in the classic modeling without flexural rigidity and with purely tangential cable forces. In this paper, a static modeling of cables is presented where flexural rigidity and shear forces are considered. The calculation details are presented with and without considering the axial extensibility of cables for three problems where one of the parameters including cable length, end point positions and forces is determined by knowing the two others. The effects of considering flexural rigidity and shear forces are then analyzed on a simple example of a cable-driven parallel robot consisting of a point mass attached to two cables.

Keywords: Cable modeling · Flexural rigidity and shear forces
Cable-driven parallel robots

1 Introduction

In several previous works, e.g. [3,6,8,14], the static analysis of large-dimension cable-driven parallel robots has been studied using the well-known elastic catenary cable modeling [7] in order to account for the cable mass and elasticity. The resulting kinetostatic modeling consists of a set of nonlinear equations involving the cable lengths, the forces applied by the cables on the robot mobile platform, and the mobile platform pose. Due to their nonlinear nature, numerical methods must be considered to solve the corresponding inverse and direct kinetostatic problems [8,11,12]. These methods might be relatively slow because of the involved computations which may be impractical in real-time environments. Hence, simplifications of this kinetostatic modeling have been considered in [5,10,13] where assumptions on the cable model are made, e.g. considering that the cable elasticity has a little influence on the cable shape. In fact, the elastic catenary cable model [7] is also based on some assumptions, one of them being to consider that the cable has a negligible flexural rigidity.

The contribution of the present paper is thus a formulation of the static modeling of cables of non-negligible mass, where flexural rigidity and shear forces are taken into account. Three problems are considered in the modeling where one of the parameters including cable length, end point positions and forces is

J. Lenarcic and V. Parenti-Castelli (Eds.): ARK 2018, SPAR 8, pp. 310–318, 2019.
https://doi.org/10.1007/978-3-319-93188-3_36

determined by knowing the two others. A simple application to a cable-driven parallel robot consisting of a point mass attached to two cables is also presented. In contrast to other works such as [1,2,9], the cable modeling in the present paper is an extension of the elastic catenary cable modeling [7] and is notably based on the Euler-Bernoulli bending moment expression for large deformations. This modeling aims at investigating the effects of considering bending moments on the accuracy of end point positioning, and on the determination of cable length and forces, in order to be used for cable-driven parallel robots.

The paper is organized as follows. In Sect. 2, the static modeling of an elastic cable with consideration of flexural rigidity and shear forces is presented. Based on this modeling, Sect. 3 discusses three problems where the cable length, end point positions or forces is to be determined when the two others are known. Finally, in Sect. 4, an example application of a point mass attached to two cables is presented.

2 Cable Model

The static modeling of an elastic cable with consideration of flexural rigidity and shear forces is introduced in this section. Figure 1 shows a schematic of a cable of length L hanged between two points A and B which are not necessarily at the same level. The forces and moments applied on the cable and on a segment of the cable of length p are shown in Fig. 1.

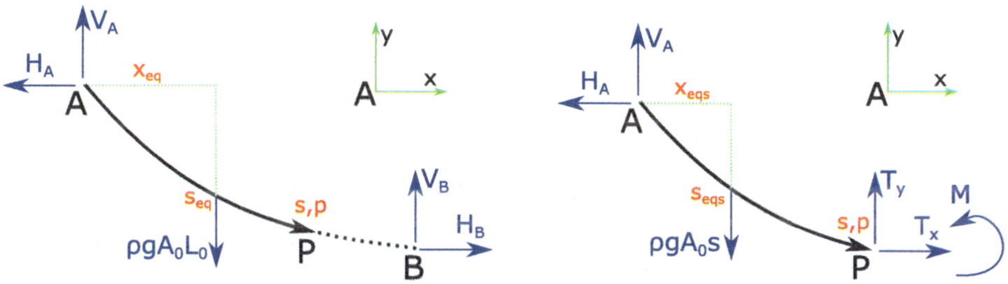

Fig. 1. Schematic of the cable with the different notations.

No moment is considered to be applied on the cable end points which is generally the case in cable robot applications. The axes x and y are the horizontal and vertical axes, respectively. The point A is taken as the origin of the x-y frame. The cable profile is denoted by w (y coordinate). The strained length of the cable segment between point A and point P is denoted by p while its unstrained length is denoted by s. The coordinates of point P are (x, w). The relationship between x, w and p is defined by the following geometric constraint:

$$dp^2 = dx^2 + dw^2 \tag{1}$$

The force and moment static equilibrium equations of the cable segment AP are:

$$\begin{cases} T_x = H_A \\ T_z = \rho g A_0 s - V_A \\ M = V_A x + H_A w - \rho g A_0 s\,(x - x_{eqs}) \end{cases} \tag{2}$$

where g is the gravity acceleration, ρ the cable density, A_0 the unstrained cable cross-section area, and H_A and V_A are the horizontal and vertical cable force components at point A, respectively.

The force components (H_B, V_B) at point B can be determined using the first two equations in (2), with $s = L$, once the cable length and force components at point A are determined. Thus, only the force components at point A, H_A and V_A, are considered in the sequel. Note that $\rho g A_0 s$ is the weight of the cable segment AP. It is written using the unstrained length s since the known parameters are generally the unstrained cable length L and the unstrained cross-section area A_0.

As illustrated in Fig. 1, x_{eqs} in (2) is the position along x of the center of mass of the cable segment $[0, s]$. x_{eqs} is given by:

$$x_{eqs} = \frac{1}{s}\int_0^s x\, ds \tag{3}$$

Considering axial elasticity, s and p are related by Hooke's law as follows:

$$\frac{dp}{ds} = 1 + \frac{T_\alpha}{EA_0} \tag{4}$$

where E is the Young's modulus and T_α is the tangential component of the cable force defined as:

$$T_\alpha = T_x \cos\alpha + T_z \sin\alpha = H_A \cos\alpha + (\rho g A s - V_A)\sin\alpha \tag{5}$$

where α is the slope angle of the cable along its length.

The classic sagging cable model (elastic catenary) used in civil engineering since the 1930s, takes the cable mass and axial elasticity into account. It is based on the equations presented above and will be referred to as the "classic model" in the sequel. The mathematical details of this model are presented e.g. in [7,8]. The cable in the classic model is assumed to be perfectly flexible, devoid of flexural rigidity (bending moments are neglected). The action of any part of the cable upon its neighbor is purely tangential. The static cable profile is then defined by a set of non-linear equations:

$$\begin{cases} x = \frac{H_A}{EA_0} s + \frac{H_A}{\rho g A_0} \ln \frac{\sqrt{H_A^2 + V_A^2} + V_A}{\sqrt{H_A^2 + (V_A - \rho g A_0 s)^2} + V_A - \rho g A_0 s} \\ w = \frac{s}{EA_0}\left(\frac{\rho g A_0 s}{2} - V_A\right) + \frac{1}{\rho g A_0}\left(\sqrt{H_A^2 + (V_A - \rho g A_0 s)^2} - \sqrt{H_A^2 + V_A^2}\right) \end{cases} \tag{6}$$

The modeling presented in this paper does not consider the assumptions of the classic model and investigates the effects of the bending moments on the statics of the cable. The Euler-Bernoulli law states that the bending moment M is proportional to the change in the curvature produced by the action of the load [4]. The bending moment expression according to this law is:

$$M = EI \frac{\frac{d^2 w}{dx^2}}{\left(1 + \left(\frac{dw}{dx}\right)^2\right)^{\frac{3}{2}}} \tag{7}$$

where I is the second moment of area of the cable cross-section. For simplification, the following variable substitution can be made:

$$\begin{cases} dx &= dp \cos\alpha \\ dw &= dp \sin\alpha \end{cases} \tag{8}$$

Considering the new variable α, and substituting (7) and (3) in the third equation of (2), the governing equation of the cable is obtained as:

$$EI \frac{d\alpha}{dp} = V_A x + H_A w - \rho g A_0 s x + \rho g A_0 \int_0^s x \, ds \tag{9}$$

where the following equation has been used:

$$\frac{d\alpha}{dp} = \frac{\frac{d^2 w}{dx^2}}{\left(1 + \left(\frac{dw}{dx}\right)^2\right)^{\frac{3}{2}}} \tag{10}$$

The differentiation in (9) must be made with respect to the unstrained length s when the boundary conditions are expressed in s. The relation between dp and ds is obtained from (4). This relation is dependent of the tangential cable tension T_α.

The cable force being assumed to be tangent to the cable profile in the classic model, the right-hand side of (9) can be shown to be zero [7]. Then, according to (10), $\frac{d^2 w}{dx^2} = 0$ which means that the cable is a straight line and, thus, without sagging. It does not correspond to the real case since the cable mass is considered in the modeling. To be rigorous, the bending moment should thus be taken into account and shear forces must also be considered since the cable force is not necessarily tangent to the cable profile.

Let us note that the moment expression with small deformations, $M = EI\frac{d^2 w}{dx^2}$ in (7), cannot be used in the case of inclined cables since the cable slope $\frac{dw}{dx}$ cannot be neglected. One solution for taut cables is to use an inclined orthogonal frame (x', y') where x' is parallel to the line AB. In this case, the governing equation obtained from the equilibrium is simplified with the small deformation expression and can be solved using Fourier series. This solution is not shown in this paper since the more general case with large deformation expression of the moment is treated.

3 Modeling Problems and Solutions

Generally, in cable robot applications, the position of the end point A is assumed to be known, and the main parameters to be calculated are the cable unstrained length L, the position of the cable end point B, and the force components at the cable end points. Three main problems can then be defined. In the first one, the length and forces are known and the end point position must be determined. In the second one, the length and end point position are known and the forces have to be calculated. In the third one, the end point position and forces are known and the unknown is the cable length. With the cable modeling introduced in Sect. 2, a solution method is presented for each of these problems in the following subsections.

3.1 First Problem: Knowing Length and Forces and Searching for Position

The differential equation in (9) governs the static behavior of the cable. Because of the nonlinearity of the equation, explicit expressions for x and w cannot be obtained, contrary to the case of the classic model.

Hence, an ODE must be considered and the first, second, and third problems defined above are boundary value problems that can be solved using numerical methods. The solvers 'bvp4c' or 'bvp5c', which are finite difference codes available in Matlab, have been used. Accordingly, a system of differential equations of the first order must be written and the number of boundary conditions must be equal to the number of equations in this system.

In order to solve the first problem, the ODE (9) is differentiated with respect to s and written as a system of five first order differential equations ($\boldsymbol{y}' = f_1(s, \boldsymbol{y})$) with five variables ($\boldsymbol{y} = [x, w, \alpha, \frac{d\alpha}{ds}, \int_0^s x ds]$). When the axial elasticity (extensibility) is not considered, the system of first order differential equations is simplified by considering that s and p are equal and ds/dp is equal to 1. Five boundary conditions are introduced in order to solve the problem, which are related to the position of the cable end point A ($x(0) = 0, w(0) = 0$), to the value of $\int_0^s x\, ds$ at A ($\int_0^s x\, ds(0) = 0$), and to the moments at A and B ($M(0) = M(L) = 0$).

3.2 Second Problem: Knowing Length and Position and Searching for Forces

When searching for the forces in the case of the second problem, the force components H_A and V_A are unknown in the governing ODE (9) while two additional boundary conditions related to the position of the cable end point B are added.

The solvers 'bvp4c' and 'bvp5c' allow solving ODE with additional unknown parameters by adding more boundary conditions. Taking advantage of this possibility, using the same system of first order differential equations defined for the first problem, the second problem can be solved by considering H_A and V_A as unknowns. Seven boundary conditions are introduced in this case in order to solve the problem. These conditions are related to the position of the cable end points A ($x(0) = 0, w(0) = 0$) and B ($x(L) = x_B, w(L) = y_B$), to the value of $\int_0^s x\ ds$ at A ($\int_0^s x\ ds(0) = 0$), and to the moments at A and B ($M(0) = M(L) = 0$).

3.3 Third Problem: Knowing Position and Forces and Searching for Length

In the case of the third problem, the length is unknown which requires to change the differentiation variable in the governing ODE. In the first two cases, differentiation with respect to s is possible since the bounds on s ($s \in [0, L]$) are well defined, which is not the case in the third problem.

The other possible differentiation variables are x and w. Differentiation with respect to x is considered since x is a monotonic function of the cable length s or p, which is not the case for w. Therefore, another variable substitution is made as follows:

$$\begin{cases} dx = dp \cdot \sqrt{1-u^2} \\ dw = dp \cdot u \end{cases} \tag{11}$$

where u is defined by (11), the latter being consistent with (1).

Considering the new variable substitution with u, the ODE in (9) becomes:

$$EI\frac{du}{dx} = V_A x + H_A w - \rho g A_0 \int_0^s s\ dx \tag{12}$$

Differentiating with respect to x one time, the following ODE is obtained:

$$\frac{d^2u}{dx^2} = V_A + \frac{H_A}{EI}\frac{u}{\sqrt{1-u^2}} - \frac{\rho g A_0}{EI}s \tag{13}$$

For a given position of B, note that the cable tension components H_A and V_A are dependent on each other. Consequently, only one force component H_A or V_A can be defined. This is obvious in the case of the inextensible cable model since the differentiation of (13) with respect to x eliminates V_A so that the problem can be solved without considering it.

Moreover, considering that the cable tension $T_A = \sqrt{H_A^2 + V_A^2}$ can be measured using a force sensor, it is considered that T_A is known in the third problem.

Besides, knowing that H_A is always positive, it can be replaced by $H_A = \sqrt{T_A^2 - V_A^2}$ in the equations. The governing ODE (13) becomes:

$$\frac{d^2u}{dx^2} = V_A + \frac{\sqrt{T_A^2 - V_A^2}}{EI}\frac{u}{\sqrt{1-u^2}} - \frac{\rho g A_0}{EI}s \tag{14}$$

Thereby, the system of first order equations defining the third problem ($\boldsymbol{y}' = f_2(x, \boldsymbol{y})$) has five variables ($\boldsymbol{y} = [w, s, \int_0^x s dx, u, \frac{du}{dx}]$), one unknown V_A and six boundary conditions. These conditions are related to the position of the cable end points A ($w(0) = 0, s(0) = 0$) and B ($w(L) = 0$), to the value of $\int_0^x s\ dx$ at A ($\int_0^x s\ dx(0) = 0$), and to the moments at A and B ($M(0) = M(L) = 0$).

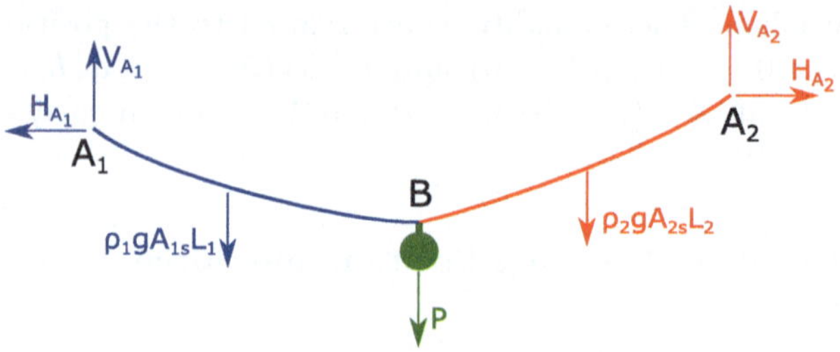

Fig. 2. Schematic of two cables attached to a point mass B.

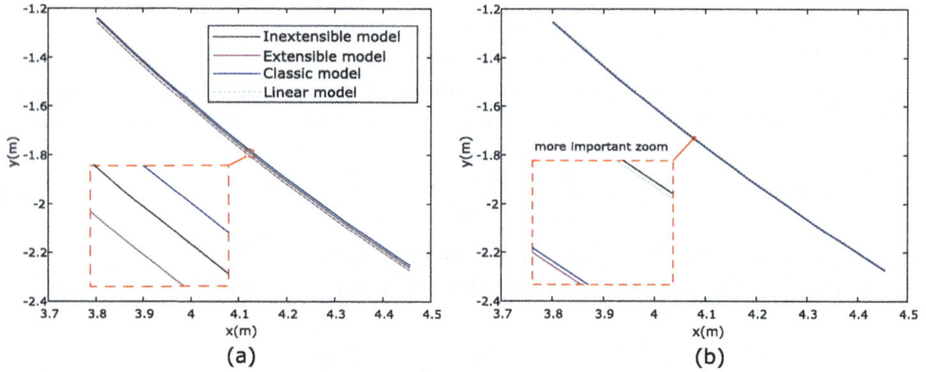

Fig. 3. Trajectories obtained with the different models for a weight P = 0.5 kN (a) and P = 10 KN (b).

4 Application: Two Cables Attached to a Point Mass

In this section, the extensible and inextensible cable models introduced in the previous sections (referred to as the new models in the sequel) are compared with the classic cable model (elastic catenary). To this end, as a simple example of a cable-driven parallel robot, the trajectory of a point mass attached to two cables is considered (Fig. 2). Changing the length of one cable, the trajectory of the point mass is calculated with the new and classic models as well as with a linear model where no sagging is considered.

The cables have a radius of 1 cm, density 8050 kg/m^3 and Young's modulus 170 GPa. The coordinates of A_1 and A_2 are defined as $(0;0)m$ and $(10;2)m$, respectively. The unstrained length L_2 of the second cable is fixed at 7 m while the length L_1 of the first cable is varied between 4 m and 5 m. Figure 3 shows the trajectory obtained with the different models and with two weights $P = 0.5$ kN and P = 10 kN of the point mass.

The trajectory calculated with the linear model is independent of the weight P. Figure 3 shows that the trajectory with the new extensible model is close to the linear model for a moderate weight P. The rationale of this rather surprising result requires further investigation. In addition, for a moderate weight P, the trajectories obtained with the new inextensible model and with the classic model are farther away. Besides, we notice that the trajectories get closer to the linear model when the weight is increased. This can be expected since the effect of the cable sagging is lower for higher tensions. In effect, the trajectories for a heavy weight become much closer to each other Fig. 3b.

Besides, as shown in the zoom in Fig. 3b, the trajectory of the new inextensible model becomes much closer to the trajectory of the linear model. However, the trajectories of the new extensible model and classic model are much closer to each other at a lower level along the vertical y axis which is due to the effect of the axial elongation of the cables.

Note that the calculation of the trajectory with the new models is not possible using directly 'bvp4c' or 'bvp5c' solvers since the governing equations of the two cables are coupled by the static equilibrium of the point mass and these solvers do not solve multi-region boundary value problems. The trajectories shown in Fig. 3 were calculated at each point of a discretization of the trajectory by sweeping the unknown parameters so as to satisfy the equilibrium equations. The corresponding calculations are time consuming. One future work on the modeling presented in this paper is thus to reduce this computation time, which is required if a cable-driven parallel robot driven by six or more cables is considered. It should be possible either by using previously calculated data in online calculations or by finding an algorithm to solve directly the coupled multi-region boundary value problem.

5 Conclusion

A static modeling of cables was presented in this paper. It takes into account the cable flexural rigidity which is not considered in the usual elastic catenary model. The shear forces are thereby not neglected and the cable force is not necessarily tangent to the sagging cable profile. The calculation details were presented with and without considering extensibility for three main problems where one of the parameters among cable length, end point positions and forces is determined by knowing the two others. A comparison with the elastic catenary cable model was made in a simple application of a cable-driven parallel robot consisting of a point mass attached to two cables. Future works include comparisons to experimental data and computation efficiency improvements in the case of an application to cable-driven parallel robots.

Acknowledgements. The research leading to these results has received funding from the European Union's H2020 Programme (H2020/2014-2020) under grant agreement No. 732513.

References

1. Boyer, F., De Nayer, G., Leroyer, A., Visonneau, M.: Geometrically exact kirchhoff beam theory: application to cable dynamics. J. Comput. Nonlinear Dyn. **6**(4), 14 (2011). 041004
2. Caballero, A., Poser, M.: Local bending stresses in stay cables with an elastic guide. Struct. Eng. Int. **20**(3), 254–259 (2010)
3. Du, J., Bao, H., Duan, X., Cui, C.: Jacobian analysis of a long-span cable-driven manipulator and its application to forward solution. Mech. Mach. Theor. **45**, 1227–1238 (2010)
4. Fertis, D.G.: Basic theories and principles of nonlinear beam deformations. In: Nonlinear Structural Engineering: With Unique Theories and Methods to Solve Effectively Complex Nonlinear Problems, pp. 1–61 (2006)
5. Gouttefarde, M., Collard, J.F., Riehl, N., Baradat, C.: Simplified static analysis of large-dimension parallel cable-driven robots. In: Proceedings of IEEE International Conference on Robotics and Automation (ICRA), pp. 2299–2305 (2012)
6. Gouttefarde, M., Nguyen, D.Q., Baradat, C.: Kinetostatic analysis of cable-driven parallel robots with consideration of sagging and pulleys. In: Lenarčič, J., Khatib, O. (eds) Advances in Robot Kinematics, pp. 213–221. Springer, Cham (2014)
7. Irvine, M.: Cable Structures. MIT Press, Cambridge (1981)
8. Kozak, K., Zhou, Q., Wang, J.: Static analysis of cable-driven manipulators with non-negligible cable mass. IEEE Trans. Robot. **22**(3), 425–433 (2006)
9. Lang, H., Linn, J., Arnold, M.: Multi-body dynamics simulation of geometrically exact cosserat rods. Multibody Syst. Dyn. **25**(3), 285–312 (2011)
10. Li, H., Zhang, X., Yao, R., Sun, J., Pan, G., Zhu, W.: Optimal force distribution based on slack rope model in the incompletely constrained cable-driven parallel mechanism of FAST telescope. In: Bruckmann, T., Pott, A. (eds.) Cable-Driven Parallel Robots, pp. 87–102. Springer, Heidelberg (2013)
11. Merlet, J.P.: The kinematics of cable-driven parallel robots with sagging cables: preliminary results. In: Proceedings of IEEE International Conference on Robotics and Automation (ICRA), Seattle, Washington, pp. 1593–1598 (2015)
12. Merlet, J.P.: A new generic approach for the inverse kinematics of cable-driven parallel robots with 6 deformable cables. In: Advances in Robot Kinematics (ARK) (2016)
13. Nguyen, D.Q., Gouttefarde, M., Company, O., Pierrot, F.: On the simplifications of cable model in static analysis of large-dimension cable-driven parallel robots. In: Proceedings of IEEE/RSJ Interantional Conference on Intelligent Robots and Systems (IROS), Tokyo, Japan, pp. 928–934 (2013)
14. Ottaviano, E., Castelli, G.: A study on the effects of cable mass and elasticity in cable-based parallel manipulator. In: Proceedings of the 18th CISM-IFToMM Symposium on Robot Design, Dynamics and Control, pp. 149–156. Springer, Udine (2010)

From Differential Geometry of Curves to Helical Kinematics of Continuum Robots Using Exponential Mapping

Stanislao Grazioso[1(✉)], Giuseppe Di Gironimo[1], and Bruno Siciliano[2]

[1] Dipartimento di Ingegneria Industriale,
Università degli Studi di Napoli Federico II, P.le Tecchio 80, 80125 Napoli, Italy
{stanislao.grazioso,giuseppe.digironimo}@unina.it
[2] Dipartimento di Ingegneria Elettrica e Tecnologie dell'Informazione,
Università degli Studi di Napoli Federico II, Via Claudio 21, 80125 Napoli, Italy
bruno.siciliano@unina.it

Abstract. Kinematic modeling of continuum robots is challenging due to the large deflections that these systems usually undergone. In this paper, we derive the kinematics of a continuum robot from the evolution of a three-dimensional curve in space. We obtain the spatial configuration of a continuum robot in terms of exponential coordinates based on Lie group theory. This kinematic framework turns out to handle robotic helical shapes, i.e. spatial configurations with constant curvature and torsion of the arm.

Keywords: Differential geometry · Continuum robotics · Kinematics

1 Introduction

The great capabilities for locomotion and manipulation exhibited by biological structures as elephant's trunks, snakes and octopus tentacle have inspired researchers towards recreating their robotics counterpart. A continuum (or continuous backbone) robot can be defined as a continuously bending, infinite degrees-of-freedom robot with an elastic structure [1]. The use of continuum robots in practical applications requires accurate mathematical models which can predict the robot's shape and motion. The inherent compliance of such mechanical systems suggests to use classical elasticity theories for long slender objects to describe their nonlinear behavior [2]. Despite recent improvements which have led to a faster computation of the governing equations for the robot's shape [3,4], we are still far to use these methods in real-time. Thus, over the years, a lot of effort has been done in deriving simplified and more analytically tractable models. A promising approach, mostly limited to kinematic analysis, begins by describing the desired three-dimensional curve in space, then fitting the physical robot to the theoretical analytical curve [5].

Using this approach, the constant curvature model have been developed [6]. It represent the robot as a series of mutually tangent circular arcs, which can be

J. Lenarcic and V. Parenti-Castelli (Eds.): ARK 2018, SPAR 8, pp. 319–326, 2019.
https://doi.org/10.1007/978-3-319-93188-3_37

described by only three parameters (radius of curvature, angle of the arc, bending plane). The constant curvature approximation has been successfully applied to many continuum robots [7,8]. However, this approach does not match the effective biological counterpart of novel designs, as the Festo's Bionic Handling Assistant (BHA), whose backbone curve has a continuously changing curvature [9]. The current variable curvature framework models the backbone of a single segment of the continuum manipulator using a piecewise constant curvature assumption. The major limit of current variable curvature kinematics framework is that they consider only bending and neglect torsion [10]. Hence, a variable curvature kinematic framework which can handle all the components of deformation is highly desirable.

In this paper, we derive a kinematic model for continuum manipulators using differential geometry techniques [11] and exponential coordinate representation [12], to provide an intuitive and effective description of the robot's configuration. The variable deformation framework is then integrated analitically under the hypothesis of constant curvature and torsion of the backbone. The shape of the robot turns out to possibly deform as a series of mutually tangent helixes, instead of circular arcs as in common variable curvature frameworks. We start by describing the evolution of a three-dimensional curve in space using the Serret-Frenet formulas. Then, we fit a continuum robot to the analytical curve through the exponential mapping. Finally, we give two examples and we demonstrate that this approach includes the constant curvature kinematics framework.

2 Geometric Description of a Spatial Curve

In classical differential geometry [11], a material abscissa $s \in [0,1]$ is usually used to parametrize a three-dimensional curve Γ. The position vector of a point $P \in \Gamma$ can be defined as $\boldsymbol{u}(s) = [u_1 \;\; u_2 \;\; u_3]^T$. The unit tangent vector is $\boldsymbol{t}(s) = \mathrm{d}\boldsymbol{u}(s)/\mathrm{d}\,s$, the unit normal vector $\boldsymbol{n}(s)$ is such that $\boldsymbol{t} \cdot \boldsymbol{n} = \boldsymbol{0}$, and the unit bi-normal vector $\boldsymbol{b}(s)$ is defined as $\boldsymbol{b} = \boldsymbol{t} \times \boldsymbol{n}$. The collection of these three unit vectors constitutes a local triad $\mathbf{R}(s) = [\boldsymbol{t}(s) \;\; \boldsymbol{n}(s) \;\; \boldsymbol{b}(s)]$ along the curve. The evolution of the curve in space is conveniently described by the formulas of Serret-Frenet [11], which take the form

$$\boldsymbol{t}'(s) = \kappa(s)\,\boldsymbol{n}(s) \tag{1}$$

$$\boldsymbol{n}'(s) = -\kappa(s)\boldsymbol{t}(s) + \tau(s)\boldsymbol{b}(s) \tag{2}$$

$$\boldsymbol{b}'(s) = -\tau(s)\boldsymbol{n}(s) \tag{3}$$

where $(\cdot)'$ denotes the derivative with respect to s, while $\kappa(s)$ and $\tau(s)$ indicate the curvature and the torsion of the curve. The vectors $\boldsymbol{t}'(s)$, $\boldsymbol{n}'(s)$, $\boldsymbol{b}'(s)$ can be collected in a 3×3 matrix $\mathbf{R}'(s) = \left[\boldsymbol{t}'(s) \;\; \boldsymbol{n}'(s) \;\; \boldsymbol{b}'(s)\right]$ such that the evolution of the curve in space can be written in the compact form (see e.g., Fig. 1)

$$\mathbf{H}'(s) = \mathcal{H}(\mathbf{R}'(s), \boldsymbol{u}'(s)) = \begin{bmatrix} \mathbf{R}'(s) & \boldsymbol{u}'(s) \\ \mathbf{0}_{1\times 3} & 1 \end{bmatrix} \in SE(3) \tag{4}$$

where $SE(3)$ is the Special Euclidean group of the 4×4 homogeneous matrices.

Let us introduce a 4×4 matrix $\widetilde{\boldsymbol{f}}(s)$ as

$$\widetilde{\boldsymbol{f}}(s) = \begin{bmatrix} 0 & -f_{\omega_3}(s) & f_{\omega_2}(s) & f_{u_1}(s) \\ f_{\omega_3}(s) & 0 & -f_{\omega_1}(s) & f_{u_2}(s) \\ -f_{\omega_2}(s) & f_{\omega_1}(s) & 0 & f_{u_3}(s) \\ 0 & 0 & 0 & 0 \end{bmatrix} \in \mathfrak{se}(3) \tag{5}$$

where $\mathfrak{se}(3)$ is the Lie algebra associated to $SE(3)$. In screw theory, the elements of $\mathfrak{se}(3)$ are called twists. In this case, $\widetilde{\boldsymbol{f}}(s)$ takes the meaning of deformation twist of the curve. Notice that the Lie algebra $\mathfrak{se}(3)$ is isomorphic to $\mathbb{R}^6$, with $\boldsymbol{f}(s) = [\boldsymbol{f}_u^T(s)\ \boldsymbol{f}_\omega^T(s)]^T \in \mathbb{R}^6$.

With this in mind, Eq. 4 can be rewritten as

$$\mathbf{H}'(s) = \mathbf{H}(s)\widetilde{\boldsymbol{f}}(s) \tag{6}$$

stating that the spatial differential kinematics of the curve is simply expressed by the product of the deformation twist with the 4×4 homogeneous matrix

$$\mathbf{H}(s) = \mathcal{H}(\mathbf{R}(s), \boldsymbol{u}(s)) = \begin{bmatrix} \mathbf{R}(s) & \boldsymbol{u}(s) \\ \mathbf{0}_{1\times 3} & 1 \end{bmatrix} \in SE(3) \tag{7}$$

which describe the configuration of the curve in three-dimensional space. Therefore, the current configuration of the curve in this variable deformation framework is computed by integrating (6) over the length of the curve, once that its deformation is known. This is not a trivial problem, since the three-dimensional curve evolves on $SE(3)$, which is a non-linear and non-commutative space. Indeed, closed-form solutions to (6) are usually unknown; general methods must employ a numerical integration scheme. Particularly appealing are the geometric integration schemes which do not involve parametrization of rotation, as the method of Crouch and Grossman [13] and the Munthe-Kaas [14]. In the next section we will provide an analytical solution to (6), which is appealing for describing the kinematics of a particular class of continuous backbone robots.

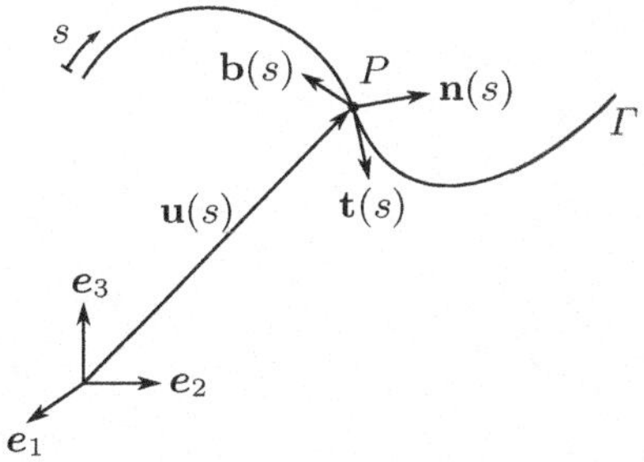

Fig. 1. Geometric description of a spatial curve.

3 Kinematic Modelling

Our kinematic approach is to fit the physical manipulator to the analytically desirable three-dimensional curve, whose evolution in space is described by (6). In this section we will derive a consistent helical kinematics framework which accounts for constant curvature and torsion of the robot's shape. Throughout this paper we will assume that the robot is inextensible and no shear effects are considered. With this hypothesis, the position and orientation parts of the deformation vector read

$$\boldsymbol{f}_u = \begin{bmatrix} 1 \\ 0 \\ 0 \end{bmatrix} \qquad \boldsymbol{f}_\omega = \begin{bmatrix} \tau \\ 0 \\ \kappa \end{bmatrix} \tag{8}$$

where we are considering initially straight configurations of the robot's shape aligned with $\boldsymbol{e}_1$ axis.

3.1 Forward Kinematics

Under the assumption of constant curvature and torsion, $\widetilde{\boldsymbol{f}}$ does not depend on s. Thus, the solution of (6) exists in closed form and it is given by

$$\mathbf{H}(s) = \mathbf{H}_0 \exp_{SE(3)}(s\boldsymbol{f}) \tag{9}$$

where $s \in [0, L]$, being L the length of the arm, $\mathbf{H}_0$, the configuration at $s = 0$, is a constant of integration, and $\exp_{SE(3)}(\cdot)$ is the exponential mapping on $SE(3)$, which maps an element of the Lie algebra $\widetilde{\boldsymbol{f}} \in \mathfrak{se}(3)$ into an element of the Lie group $\mathbf{H} \in SE(3)$. Formally, it is defined by

$$\exp_{SE(3)}(\boldsymbol{f}) = \begin{bmatrix} \exp_{SO(3)}(\boldsymbol{f}_\omega) & \mathbf{T}^T_{SO(3)}(\boldsymbol{f}_\omega)\boldsymbol{f}_u \\ \mathbf{0}_{1\times 3} & 1 \end{bmatrix} \tag{10}$$

where $\exp_{SO(3)}(\cdot)$ is the exponential map on the special Orthogonal group $SO(3)$, the group of the rotation matrices. This is given by

$$\exp_{SO(3)}(\boldsymbol{f}_\omega) = \mathbf{I}_{3\times 3} + \alpha(\boldsymbol{f}_\omega)\widetilde{\boldsymbol{f}}_\omega + \frac{\beta(\boldsymbol{f}_\omega)}{2}\widetilde{\boldsymbol{f}}^2_\omega \tag{11}$$

and it is known as Rodrigues' formula. In (11), we have

$$\alpha(\boldsymbol{f}_\omega) = \frac{\sin(\|\boldsymbol{f}_\omega\|)}{\|\boldsymbol{f}_\omega\|} \qquad \beta(\boldsymbol{f}_\omega) = 2\frac{1-\cos(\|\boldsymbol{f}_\omega\|)}{\|\boldsymbol{f}_\omega\|^2} \tag{12}$$

Indeed, $\mathbf{T}^T_{SO(3)}(\cdot)$ is the transpose of the tangent operator on $SO(3)$, which is defined from the derivative of the exponential map as

$$\mathbf{T}_{SO(3)}(\boldsymbol{f}_\omega) = \mathbf{I}_{3\times 3} - \frac{\beta(\boldsymbol{f}_\omega)}{2}\widetilde{\boldsymbol{f}}_\omega + \frac{1-\alpha(\boldsymbol{f}_\omega)}{\|\boldsymbol{f}_\omega\|^2}\widetilde{\boldsymbol{f}}^2_\omega \tag{13}$$

By separating the rotation and position part of (9), we obtain

$$\mathbf{R}(s) = \mathbf{R}_0 \exp_{SO(3)} (s\boldsymbol{f}_\omega) \tag{14}$$

$$\boldsymbol{u}(s) = \boldsymbol{u}_0 + \mathbf{R}_0 \mathbf{T}^T_{SO(3)} (s\boldsymbol{f}_\omega)\, s\boldsymbol{f}_u \tag{15}$$

where $\mathbf{R}_0$, the rotation matrix at $s = 0$, and $\boldsymbol{u}_0$, the position vector at $s = 0$, are constant of integration. Since we are considering arms aligned with $\boldsymbol{e}_1$ and initially straight, we have $\mathbf{R}_0 = \mathbf{I}_{3\times3}$ and $\boldsymbol{u}_0 = \mathbf{0}_{3\times1}$.

As a matter of fact, the exponential mapping introduces a local parametrization which allows describing the configuration of a curve, which belongs to $SE(3)$, i.e. a non-commutative and non-linear space, with an element belonging to a linear space, namely the Lie algebra $\mathfrak{se}(3)$.

3.2 Inverse Kinematics

The inverse kinematics maps the configuration of the continuum robot to the deformation vector. In our hypothesis, the inversion of (9) exists in closed form and it is given by

$$\boldsymbol{f} = \log_{SE(3)} \left(\mathbf{H}_0^{-1}\mathbf{H}_L\right) \tag{16}$$

where $\mathbf{H}_L$ is the configuration at $s = L$, and $\log_{SE(3)}(\cdot)$ is the logarithmic mapping on $SE(3)$, which maps back an element of the Lie group into an element of its corresponding Lie algebra. It is defined by

$$\log_{SE(3)}(\mathbf{H}) = \begin{bmatrix} \widetilde{\boldsymbol{f}}_\omega & \mathbf{T}^{-T}_{SO(3)}(\boldsymbol{f}_\omega)\boldsymbol{u} \\ \mathbf{0}_{1\times3} & 1 \end{bmatrix} \tag{17}$$

where we have indicated $\mathbf{H} = \left(\mathbf{H}_0^{-1}\mathbf{H}_L\right)$, $\widetilde{\boldsymbol{f}}_\omega = \log_{SO(3)}(\mathbf{R})$ and

$$\log_{SO(3)}(\mathbf{R}) = \frac{\theta}{2\sin\theta}(\mathbf{R} - \mathbf{R}^T) \tag{18}$$

with

$$\theta = \operatorname{acos}\left(\frac{1}{2}(\operatorname{trace}(\mathbf{R}) - 1)\right), \qquad \theta < \pi \tag{19}$$

Indeed, $\mathbf{T}^{-T}_{SO(3)}(\cdot)$ is the transpose of the inverse of the tangent operator on $SO(3)$, which is defined as

$$\mathbf{T}^{-1}_{SO(3)}(\boldsymbol{f}_\omega) = \mathbf{I}_{3\times3} + \frac{1}{2}\widetilde{\boldsymbol{f}}_\omega + \frac{1-\gamma(\boldsymbol{f}_\omega)}{\|\boldsymbol{f}_\omega\|^2}\widetilde{\boldsymbol{f}}^2_\omega \tag{20}$$

with

$$\gamma(\boldsymbol{f}_\omega) = \frac{\|\boldsymbol{f}_\omega\|}{2}\cot\left(\frac{\|\boldsymbol{f}_\omega\|}{2}\right) \tag{21}$$

4 Test Cases

In this section we consider two test examples, one relating a continuum arm with constant curvature, one relating a continuum arm with constant curvature and torsion.

4.1 Continuum Arm with Constant Curvature

Let $\boldsymbol{f}_\omega = [0 \;\; 0 \;\; \kappa]^T$. By developing (9), it is easy to demonstrate that the configuration of the robotic arm is given by

$$\mathbf{H}(s) = \begin{bmatrix} \cos(s\kappa) & -\sin(s\kappa) & 0 & \frac{1}{\kappa}\sin(s\kappa) \\ \sin(s\kappa) & \cos(s\kappa) & 0 & \frac{1}{\kappa}(1-\cos(s\kappa)) \\ 0 & 0 & 1 & 0 \\ 0 & 0 & 0 & 1 \end{bmatrix} \tag{22}$$

which corresponds to the robot's shape obtained with constant curvature kinematics framework [6,15]. A circular desired end point trajectory is shown in Fig. 2. The parameters used in this example are: $L = 1$m; $\kappa = \pi/10 : \pi/10 : \pi \text{m}^{-1}$.

It is easy to verify that, computing $\mathbf{H}_0 = \mathbf{H}(s=0)$ and $\mathbf{H}_L = \mathbf{H}(s=L)$ from (22), the use of (17) gives back $\boldsymbol{f}_\omega = [0 \;\; 0 \;\; \kappa]^T$.

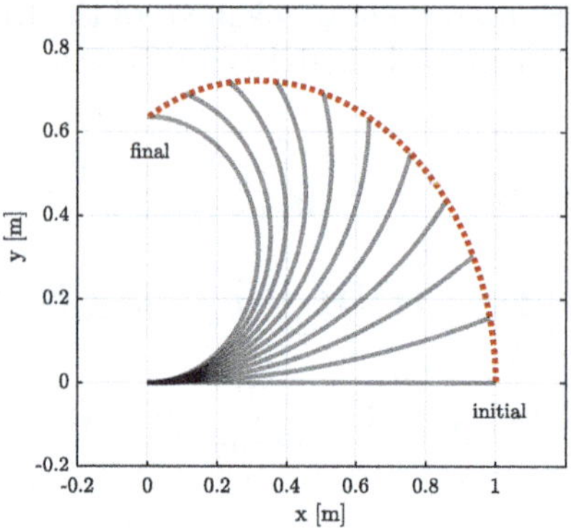

Fig. 2. Whole arm shape configuration under a desired circular end point trajectory.

4.2 Continuum Arm with Constant Curvature and Torsion

Let $\boldsymbol{f}_\omega = [\tau \;\; 0 \;\; \kappa]^T$. By developing (9), the configuration of the robotic arm is given by

$$\mathbf{H}(s) = \begin{bmatrix} 1-(1-\cos(s\kappa_g))\frac{\kappa^2}{\kappa_g^2} & -\sin(s\kappa_g)\frac{\kappa}{\kappa_g} & (1-\cos(s\kappa_g))\frac{\kappa\tau}{\kappa_g^2} & s+(\sin(s\kappa_g)-s\kappa_g)\frac{\kappa^2}{\kappa_g^3} \\ \sin(s\kappa_g)\frac{\kappa}{\kappa_g} & \cos(s\kappa_g) & -\sin(s\kappa_g)\frac{\tau}{\kappa_g} & (1-\cos(s\kappa_g))\frac{\kappa}{\kappa_g^2} \\ (1-\cos(s\kappa_g))\frac{\kappa\tau}{\kappa_g^2} & \sin(s\kappa_g)\frac{\tau}{\kappa_g} & 1-(1-\cos(s\kappa_g))\frac{\tau^2}{\kappa_g^2} & (s\kappa_g-\sin(s\kappa_g))\frac{\kappa\tau}{\kappa_g^3} \\ 0 & 0 & 0 & 1 \end{bmatrix} \tag{23}$$

where $\kappa_g = \sqrt{\kappa^2 + \tau^2}$ is the Gaussian curvature of the arm. Equation (23) correspond to an helical-shaped configuration of the whole arm. Figure 3 illustrate a typical screw motion. The left hand figure shows the motion corresponding to a fixed torsion $\tau = 3\,\mathrm{m}^{-1}$ and varying curvature $\kappa = 0 : \pi/2 : 2\pi\,\mathrm{m}^{-1}$. Indeed, the right hand figure shows the motion corresponding to a fixed curvature $\kappa = 3\,\mathrm{m}^{-1}$ and varying torsion $\tau = 0 : \pi/2 : 2\pi\,\mathrm{m}^{-1}$.

It is easy to verify that, again, computing $\mathbf{H}_0 = \mathbf{H}(s = 0)$ and $\mathbf{H}_L = \mathbf{H}(s = L)$ from (23), the use of (17) gives back $\boldsymbol{f}_\omega = [\tau \;\; 0 \;\; \kappa]^T$.

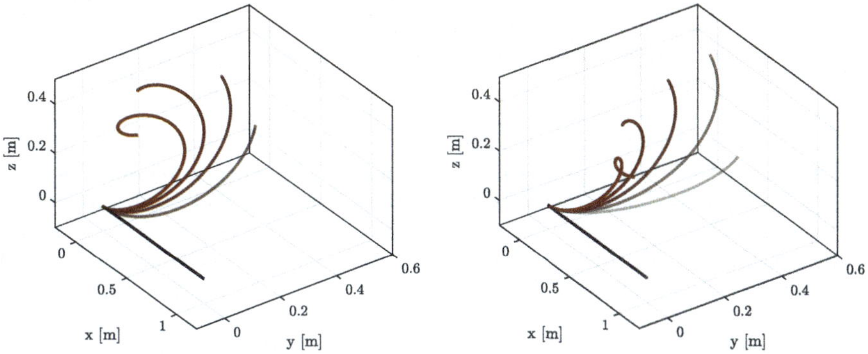

Fig. 3. Whole arm screw motion of a manipulator with constant curvature and torsion.

5 Conclusions

A variable deformation kinematic framework based on differential geometry has been presented for spatial three-dimensional curves. In order to derive analytical solutions, we have integrated the spatial differential kinematic model by using a constant deformation assumption. In particular, we have developed the analytical solutions for constant curvature and torsion robots, whose backbone configuration in space resembles the shape of an helix, i.e. a spatial curve with constant curvature and torsion.

The kinematic modeling of compound continuum robots will arise naturally from the framework. Future works of the authors will investigate closed-loop inverse kinematics algorithms to iteratively solve the inverse kinematics of single-segment and multiple-segments continuum robots.

Acknowledgments. This work was partially supported by the FlexARM project, which has received funding from the European Commission's Euratom Research and Training Programme 2014–2018 under the EUROfusion Engineering Grant EEG-2015/21 *"Design of Control Systems for Remote Handling of Large Components"* and partially by the RoDyMan project, which has received funding from the European Research Council under Advanced Grant 320992.

References

1. Walker, I.D., Choset, H., Chirikjian, G.S.: Snake-like and continuum robots. In: Springer Handbook of Robotics, pp. 481–498. Springer, Cham (2016)
2. Simo, J.C., Vu-Quoc, L.: A three-dimensional finite-strain rod model. part II: Computational aspects. Comput. Methods Appl. Mech. Eng. **58**(1), 79–116 (1986)
3. Grazioso, S., Sonneville, V., Di Gironimo, G., Bauchau, O., Siciliano, B.: A nonlinear finite element formalism for modelling flexible and soft manipulators. In: 2016 IEEE International Conference on Simulation, Modeling, and Programming for Autonomous Robots, pp. 185–190. IEEE (2016)
4. Grazioso, S., Di Gironimo, G., Siciliano, B.: A geometrically exact model for soft robots: the finite element deformation space formulation. Soft Robotics
5. Andersson, S.B.: Discretization of a continuous curve. IEEE Trans. Rob. **24**(2), 456–461 (2008)
6. Webster III, R.J., Jones, B.A.: Design and kinematic modeling of constant curvature continuum robots: a review. Int. J. Rob. Res. **29**(13), 1661–1683 (2010)
7. Hannan, M.W., Walker, I.D.: Kinematics and the implementation of an elephant's trunk manipulator and other continuum style robots. J. Field Rob. **20**(2), 45–63 (2003)
8. Jones, B.A., Walker, I.D.: Kinematics for multisection continuum robots. IEEE Trans. Rob. **22**(1), 43–55 (2006)
9. Grzesiak, A., Becker, R., Verl, A.: The bionic handling assistant: a success story of additive manufacturing. Assem. Autom. **31**(4), 329–333 (2011)
10. Mahl, T., Hildebrandt, A., Sawodny, O.: A variable curvature continuum kinematics for kinematic control of the bionic handling assistant. IEEE Trans. Rob. **30**(4), 935–949 (2014)
11. Struik, D.J.: Lectures on classical differential geometry. Courier Corporation, North Chelmsford (2012)
12. Lynch, K.M., Park, F.C.: Modern Robotics: Mechanics, Planning, and Control. Cambridge University Press, Cambridge (2017)
13. Crouch, P.E., Grossman, R.: Numerical integration of ordinary differential equations on manifolds. J. Nonlinear Sci. **3**(1), 1–33 (1993)
14. Munthe-Kaas, H.: Runge-Kutta methods on lie groups. BIT Numer. Math. **38**(1), 92–111 (1998)
15. Grazioso, S., Di Gironimo, G., Siciliano, B.: Analytic solutions for the static equilibrium configurations of externally loaded cantilever soft robotic arms. In: 2018 IEEE International Conference on Soft Robotics. IEEE (2018)

A Ligament Model Based on Fibre Mapping for Multibody Simulations

Fabrizio Nardini[1], Nicola Sancisi[2(✉)], and Vincenzo Parenti-Castelli[2]

[1] NCS Lab Srl., Via Pola esterna 4/12, 41012 Carpi, Italy
fabrizio.nardini2@unibo.it
[2] DIN-Department of Industrial Engineering, University of Bologna, Viale del Risorgimento 2, 40136 Bologna, Italy
{nicola.sancisi, vincenzo.parenti}@unibo.it

Abstract. A ligament model is proposed to define the wrench of a ligament on the bones during multibody simulations. The model is based on the thin-plate spline mapping to adapt reference data on the anatomy of a specific subject and to define a correspondence between the points on the two insertions of the ligament on the bones, thus simulating the ligament fibrous structure. Tests performed on a ligament show that the model is faster, more accurate and easier to be implemented than a standard multi-bundle model. The accuracy can be also easily adjusted to the application in order to further reduce computational time.

Keywords: Multibody · Ligament model · Thin-plate spline mapping

1 Introduction

Musculoskeletal multibody models allow the kinematic, kinetostatic and dynamic analysis of a skeletal system made of different bony segments. These segments are generally considered as rigid bodies connected by joints, which represent the different articulations of the skeletal system; muscles generate the activation forces that equilibrate the system under given loading conditions. In this context the joints are often modelled either by lower pairs (such as revolute or spherical joints) or by planar or spatial linkages [1]. The latter are parallel mechanisms that, through a more detailed representation of the joint structures (i.e., ligaments and contacts), allow a better description of the three-dimensional joint motion [2, 3] and the determination of the forces applied by each considered joint structure. However, these models are rigid and cannot represent possible changes in the joint configuration due to the effect of elasticity and loads. Joint models with compliant elements were also proposed for multibody analyses and, despite the greater computational complexity, their application is increasing to improve the description of the joint behaviour under loads [4, 5].

Compliant ligaments are often modelled by a simplified approach, for instance representing the ligament with a single or multiple fibres (multi-bundle model, or MBM) [6, 7], considering or not the fibre-bone and fibre-fibre interactions [8–10]. Model [10] was actually proposed for the muscles but it could be applied to any fibrous body. The main limitation of a MBM is that results depend on the positioning of fibres, and only partial information are reported in the literature to optimize this choice.

J. Lenarcic and V. Parenti-Castelli (Eds.): ARK 2018, SPAR 8, pp. 327–334, 2019.
https://doi.org/10.1007/978-3-319-93188-3_38

Moreover, a high number of fibres is generally needed to obtain accurate results, requiring long computational time and awkward model definitions. Finite element models were also proposed [11, 12] but they are too computationally expensive for multibody analyses.

A new ligament model is proposed in this study. The model is based on thin-plate spline mapping, through which the resultant force and moment of each ligament on bones is represented by a continuous function over the ligament insertion. This approach has several advantages: it allows a fast model definition based on reference data from experiments or from the literature; it does not require the definition of arbitrary fibres; it is faster and more accurate than a MBM; its accuracy can be scaled depending on the application. The whole procedure is described and a computational example is shown.

2 Ligament Model

The main idea is to define mappings that transform one dataset into another by continuous functions, satisfying a perfect or approximate correspondence between some elements of the dataset (namely, the landmarks). Mappings are used here for two purposes:

1. To register and warp generic reference data (e.g., from the literature or previous experiments) of ligament insertions on the anatomical geometry of the specific subject;
2. To create a one-to-one correspondence between each point in one ligament insertion with one point in the other insertion, simulating the ligament fibre connections.

If two-dimensional data are considered, these mappings are functions $\Phi : \mathbb{R}^2 \rightarrow \mathbb{R}^2$ that transform all points $P(x, y)$ in one plane, into corresponding points $Q(u, v)$ in another plane, so that n given landmarks $P_i(x_i, y_i)$ $(i = 1, \ldots, n)$ in one dataset are mapped either exactly or with a given approximation level into corresponding landmarks $Q_i(u_i, v_i)$ in the other dataset.

2.1 Thin-Plate Spline Mapping

Among all possible solutions, the thin-plate spline mapping is chosen here [13–15]:

$$\begin{aligned} u(x,y) &= \alpha_0 + \alpha_1 x + \alpha_2 y + \sum\nolimits_{i=1}^{n} a_i g_i(x,y) \\ v(x,y) &= \beta_0 + \beta_1 x + \beta_2 y + \sum\nolimits_{i=1}^{n} b_i g_i(x,y) \end{aligned} \tag{1}$$

where n is the number of landmarks, $\alpha_k, \beta_k, a_i, b_i$ $(k = 0, 1, 2;\ i = 1, \ldots, n)$ are coefficients to be determined and $g_i(x,y)$ is a particular radial basis function, i.e. a univariate function of the distance from the landmark $P_i(x_i, y_i)$:

$$\begin{aligned} g_i(x,y) &= r_i^2 \log(r_i) \\ r_i &= \|(x - x_i, y - y_i)\| \end{aligned} \tag{2}$$

It could be shown that this choice for $g_i(x,y)$ reduces to a minimum the overall warp of the mapping [13, 14], since it is connected to the minimum bending energy of a thin plate subjected to point constraints. More generally, however, different radial basis functions (with a more global or local warping effect of the mapping) could be chosen depending on the application [15].

The coefficients α_j, β_j, a_i, b_i could be obtained by imposing the correspondence between landmarks $Q_i(u_i, v_i)$ and $P_i(x_i, y_i)$ in Eq. (1). Since these coefficients appear linearly in Eq. (1), the problem is reduced to the solution of a linear system. The full mathematical derivation is developed elsewhere [13], thus only the final formulas are reported here. Given the following definitions:

$$\mathbf{G} = (g_{ij}) = \begin{bmatrix} 0 & g_{12} & \cdots & g_{1n} \\ g_{21} & 0 & \cdots & g_{2n} \\ \vdots & \vdots & \ddots & \vdots \\ g_{n1} & g_{n2} & \cdots & 0 \end{bmatrix}_{n\times n} \qquad \mathbf{P} = \begin{bmatrix} 1 & x_1 & y_1 \\ 1 & x_2 & y_2 \\ \vdots & \vdots & \vdots \\ 1 & x_n & y_n \end{bmatrix}_{n\times 3} \tag{3}$$

where $g_{ij} = g_i(x_j, y_j)$ $(i,j = 1,\ldots,n)$, and given:

$$\mathbf{L} = \begin{bmatrix} \mathbf{G} & \mathbf{P} \\ \mathbf{P}^{\mathrm{T}} & \mathbf{0} \end{bmatrix}_{(n+3)\times(n+3)} \qquad \mathbf{Q} = \begin{bmatrix} u_1 & v_1 \\ u_2 & v_2 \\ \vdots & \vdots \\ u_n & v_n \\ 0 & 0 \\ 0 & 0 \\ 0 & 0 \end{bmatrix}_{(n+3)\times 2} \tag{4}$$

where $\mathbf{0}$ is a 3×3 matrix of zeros, mapping coefficients can be obtained from:

$$\begin{bmatrix} a_1 & b_1 \\ a_2 & b_2 \\ \vdots & \vdots \\ a_n & b_n \\ \alpha_0 & \beta_0 \\ \alpha_1 & \beta_1 \\ \alpha_2 & \beta_2 \end{bmatrix} = \mathbf{L}^{-1}\mathbf{Q} \tag{5}$$

Substituting coefficients (5) in (1) returns the thin-plate spline mapping that guarantees the perfect correspondence between landmarks Q_i and P_i and the minimum warp. The landmark correspondence could be relaxed by additional parameters λ_i that weight the thin-plate bending energy (i.e., the overall warp) with the landmark correspondence violation [15]. In this case, mapping coefficients could be obtained as in Eq. (5), where the minor **G** in matrix **L** is substituted by:

$$\mathbf{G}_\lambda = \begin{bmatrix} \lambda_1 & g_{12} & \cdots & g_{1n} \\ g_{21} & \lambda_2 & \cdots & g_{2n} \\ \vdots & \vdots & \ddots & \vdots \\ g_{n1} & g_{n2} & \cdots & \lambda_n \end{bmatrix} \tag{6}$$

In particular, if $\lambda_i = 0$, $\forall i = 1, \ldots, n$, then $\mathbf{G}_\lambda = \mathbf{G}$ and the correspondence between landmarks is perfectly respected; if $\lambda_i \to +\infty$, $\forall i = 1, \ldots, n$, then no warp is introduced, i.e., the mapping Φ is a rigid rototranslation that registers the points P_i on Q_i in order to minimize the sum of their squared distances.

2.2 Reference Data Registration

Each ligament is composed of several bundles, which, however, cannot be easily recognized in vivo by standard measurements. More likely, specific experiments can be performed in vitro to find these bundles and, as a consequence, their insertions on the bones. Some studies in the literature also report these anatomical bundle maps [6, 7]. The thin-plate spline mapping is thus used at a first step to register generic reference data on the anatomy of a specific subject.

The procedure is reported in Fig. 1 for the anterior cruciate ligament (ACL) of the knee, but it can be extended to any other ligament. With reference to the femur, the subject-specific insertion (red area in Fig. 1) is obtained by medical imaging. This is then projected on a best-fitting plane in order to obtain its planar approximation. The reference data include the external contour of the insertion and the centroids of the bundles (black dots in Fig. 1). A local coordinate system is defined both for the subject (S_{sf}) and for the reference (S_{rf}) femoral insertion. The coordinates of the centroids of the bundles are measured in S_{rf}, together with some landmarks on the insertion external contour. Corresponding landmarks on the insertion external contour are measured in S_{sf}. The transformation Φ_F is defined by means of Eq. (1), mapping the generic point $P(x, y)$ in S_{rf} to the point $Q(u, v)$ in S_{sf}. The landmarks on the external contour are used to create the correspondence and solve Eq. (5); due to the uncertainties of this correspondence, the matrix in Eq. (6) is used in this case.

The same procedure is repeated for the tibial insertion, thus defining the transformation Φ_T between the local coordinate system of the reference (S_{rt}) and the subject (S_{st}) insertion. The coordinates of the bundle centroids are then mapped by Φ_F and Φ_T from the coordinate systems of the reference data to the coordinate systems of the subject insertions. This can be done by substitution of the centroid coordinates into Eq. (1).

2.3 Insertion-to-Insertion Correspondence

A similar procedure is used to obtain a one-to-one correspondence between the points in S_{st} and S_{sf}. In particular, Φ is a thin-plate mapping that transforms a generic point $P(x, y)$ in S_{st} to the point $Q(u, v)$ in S_{sf}. It is obtained by imposing that the centroids in

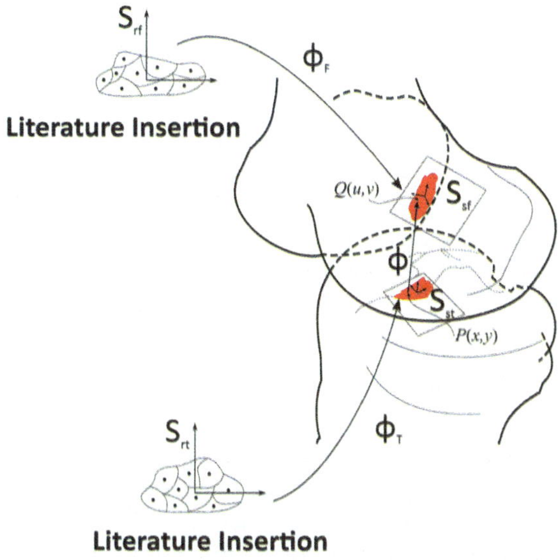

Fig. 1. Mapping procedure: reference data registration (Φ_F and Φ_T) and insertion-to-insertion correspondence (Φ).

S_{st} perfectly corresponds to those in S_{sf} by solving Eq. (5). The use of Eq. (1)–(2) also guarantees the minimum warp for the transformation.

2.4 Ligament Wrench

Standard multibody simulations require the determination of the wrench applied by each ligament to the bones, in order to solve the kinetostatic and dynamic analysis of the system. The wrench can be found by integrating the contribution of each infinitesimal bundle (i.e., each fibre) to the resultant force and torque over the insertion area.

In particular, with reference to Fig. 1, the mapping Φ associates a point $Q(u, v)$ in S_{sf} to a point $P(x, y)$ in S_{st}. Since the pose (i.e., position and orientation) of these local coordinate systems on the bones is known from the subject anatomy, straightforward coordinate transformations make it possible to represent the position vectors of Q and P (namely, $\mathbf{q}$ and $\mathbf{p}$) with respect to a common global reference system S_g for each relative pose of the femur and tibia.

The force $\boldsymbol{\tau}$ and torque $\boldsymbol{\mu}_O$ (with respect to a pole O) per unit area applied by each fibre on the tibia takes the form:

$$\begin{aligned} \boldsymbol{\tau} &= f(c_k, \varepsilon)\hat{\mathbf{s}} = f(c_k, l, l_0)\hat{\mathbf{s}} \\ \boldsymbol{\mu}_O &= \mathbf{o} \times \boldsymbol{\tau} \end{aligned} \tag{7}$$

where $\varepsilon = (l - l_0) / l_0$ is the fibre strain; $l = \|\mathbf{p} - \mathbf{q}\|$ is the fibre length; l_0 is the fibre free length; f is the constitutive relation between the strain ε and $\boldsymbol{\tau}$, defined by some stiffness parameters c_k; $\hat{\mathbf{s}} = (\mathbf{p} - \mathbf{q}) / l$ is the unit vector of each fibre; $\mathbf{o}$ is the moment arm of $\boldsymbol{\tau}$ with respect to O. It is worth noting that both l_0 and c_k are related to the constitutive relation of the ligament and can be considered either constant or function of $\mathbf{p}$. Finally, since $\mathbf{q}$ is also function of $\mathbf{p}$ through the mapping Φ, and since $\mathbf{p}$ is a function of the coordinates x, y of P in S_{st}:

$$\begin{aligned} \tau &= \tau(\mathbf{p}) = \tau(x, y) \\ \boldsymbol{\mu}_O &= \boldsymbol{\mu}_O(\mathbf{p}) = \boldsymbol{\mu}_O(x, y) \end{aligned} \tag{8}$$

Equation (7) are thus continuous functions of the coordinates x, y: the result of integration of these equations over the tibial insertion area is the wrench of the ligament on the tibia.

3 Numerical Example

The new model (hereinafter called the thin-plate spline model, or TPSM) was tested with data measured in vivo on a volunteer in a previous study [3]. The geometry of the ACL at a given knee configuration was obtained from these data. A reference map of the ACL bundles was taken from the literature [6], which reports both the contour and the centroids of the ACL bundles at the tibial and femoral insertions.

The full procedure reported in Sect. 2 was then applied. The resulting one-to-one correspondence is reported in Fig. 2, where corresponding points are connected by linear segments. The constitutive relation of fibres was obtained from the literature [7]. Since the scope of this example was not a detailed ligament parameter identification, the free length l_0 of each fibre was chosen constant over the insertion area, and fixed at a value for which fibres could be both in a tensioned and slack state. This was done in order to test the capability of the TPSM to handle these two states of the fibres. Equation (7) were numerically integrated. A first result was obtained using a high number of points to obtain the *exact* result of the integration. Then the number of points was reduced to 500 and the difference with respect to the exact result was quantified.

A standard MBM was also applied [6]. The fibres representing the ligament bundles were chosen in order to cover the whole insertion. Fibre free lengths l_0 were the same as in the TPSM. Fibre stiffness parameters were also the same. However, since in Eq. (8) stiffness parameters are computed per unit area, the parameters for each fibre were obtained by multiplying them for the area assigned to each bundle, namely the ratio between the tibial insertion total area and the number of bundles. In this way, the whole stiffness does not change between the TPSM and the MBM. The number of fibres of the MBM was increased until the accuracy (i.e., the difference with respect to the exact result of the integration) was similar to that of the TPSM with 500 integration points.

With respect to the MBM, the TPSM resulted more accurate for the same number of points/fibres. Indeed, the results obtained with the TPSM with 500 integration points showed a difference that was less than 2% with respect to the exact result. Changing integration points did not affect these results significantly. The MBM results were more sensitive to the fibre distribution, and in general were less accurate: depending on the distribution, similar but lower accuracy (difference up to 10%) could be obtained by the MBM with a number of fibres between 5000 and 7000 in the tests performed in this study. As a result, the computational time of the TPSM was much lower than the

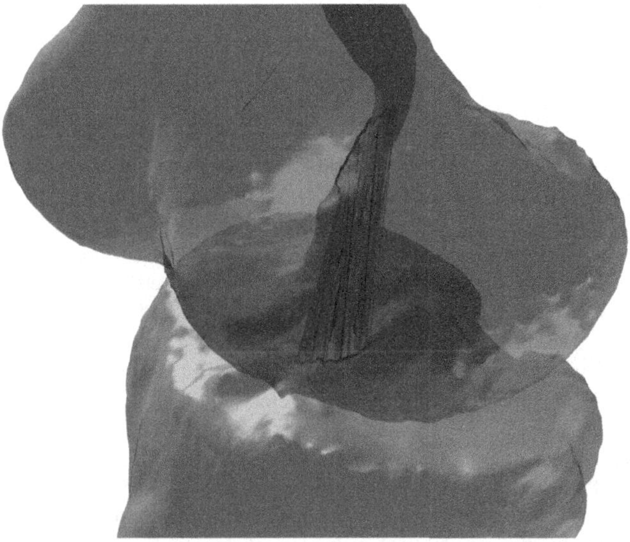

Fig. 2. Representation of the one-to-one correspondence between insertion points. The black lines represent the landmark correspondence.

MBM, for a similar accuracy. The computational time could be reduced further in the TPSM by lowering the number of integration points to be less than 500, still obtaining a good accuracy.

4 Conclusions

The new ligament model based on the thin-plate spline mapping (TPSM) proved to be an efficient alternative to standard multi-bundle models (MBM). The procedure presented here can be easily implemented and solved, requiring the solution of a linear problem. It is worth noting that the TPSM could be seen as a generalization of the standard MBM, but with some advantages. Firstly, the fibres are intrinsically infinite in the TPSM, representing a fibre distribution rather than a discrete description like in the MBM. Secondly, the fibre distribution, i.e. the mapping, can be defined in a rigorous but simple way, thus reducing the arbitrariness of the MBM. These characteristics make the TPSM more robust (i.e., less sensitive), more accurate and faster than the standard MBM, as showed with a numerical example. The present model implementation implicitly considers each fibre as a straight line. However, the proposed procedure could be generalized, and the same mapping of the TPSM could be used to connect corresponding points on the ligament insertions with other strategies [9, 10] that can take into account fibre-bone and fibre-fibre interactions.

References

1. Leardini, A., Belvedere, C., Nardini, F., Sancisi, N., Conconi, M., Parenti-Castelli, V.: Kinematic models of lower limb joints for musculo-skeletal modelling and optimization in gait analysis. J. Biomech. **62**, 77–86 (2017)
2. Sancisi, N., Parenti-Castelli, V.: A new kinematic model of the passive motion of the knee inclusive of the patella. J. Mech. Robot. **3**(4), 041003 (2011)
3. Nardini, F., Sancisi, N., Belvedere, C., Conconi, M., Leardini, A., Parenti-Castelli, V.: Definition of a subject-specific model of the knee in vivo. Gait Posture **49**, S6 (2016)
4. Sintini, I., Sancisi, N., Parenti-Castelli, V.: Comparison between anatomical and approximate surfaces in a 3D kinetostatic model of the knee for the study of the unloaded and loaded joint motion. Meccanica **53**, 7–20 (2018)
5. Marra, M.A., Vanheule, V., Fluit, R., Koopman, B.H., Rasmussen, J., Verdonschot, N., Andersen, M.S.: A subject-specific musculoskeletal modeling framework to predict in vivo mechanics of total knee arthroplasty. J. Biomech. Eng. **137**(2), 020904 (2015)
6. Mommersteeg, T.J., Kooloos, J.G., Blankevoort, L., Kauer, J.M., Huiskes, R., Roeling, F.Q.: The fibre bundle anatomy of human cruciate ligaments. J. Anat. **187**(Pt 2), 461 (1995)
7. Mommersteeg, T.J.A., Blankevoort, L., Huiskes, R., Kooloos, J.G.M., Kauer, J.M.G.: Characterization of the mechanical behavior of human knee ligaments: a numerical-experimental approach. J. Biomech. **29**(2), 151–160 (1996)
8. Hefzy, M.S., Grood, E.S.: An analytical technique for modeling knee joint stiffness—part II: ligamentous geometric nonlinearities. J. Biomech. Eng. **105**(2), 145–153 (1983)
9. Franci, R., Parenti-Castelli, V.: A new tool to investigate the interactions between elastic fibers and rigid bodies. In: Proceedings of 12th IFToMM World Congress, Besançon, 18–21 June 2007
10. Scholz, A., Sherman, M., Stavness, I., Delp, S., Kecskeméthy, A.: A fast multi-obstacle muscle wrapping method using natural geodesic variations. Multibody Syst. Dyn. **36**(2), 195–219 (2016)
11. Hirokawa, S., Tsuruno, R.: Three-dimensional deformation and stress distribution in an analytical/computational model of the anterior cruciate ligament. J. Biomech. **33**(9), 1069–1077 (2000)
12. Gardiner, J.C., Weiss, J.A.: Subject-specific finite element analysis of the human medial collateral ligament during valgus knee loading. J. Orthop. Res. **21**(6), 1098–1106 (2003)
13. Bookstein, F.L.: Principal warps: thin-plate splines and the decomposition of deformations. IEEE Trans. Pattern Anal. Mach. Intell. **11**(6), 567–585 (1989)
14. Bookstein, F.L.: Thin-plate splines and the atlas problem for biomedical images. In: Biennial International Conference on Information Processing in Medical Imaging, pp. 326–342. Springer, Heidelberg (1991)
15. Arad, N., Dyn, N., Reisfeld, D., Yeshurun, Y.: Image warping by radial basis functions: application to facial expressions. CVGIP Graph. Models Image Process. **56**(2), 161–172 (1994)

Generalized Path Following Constraints with Spatial Curves for Roller Coaster Applications

Jorge Ambrosio(✉), Pedro Antunes(✉), and Mário Viegas(✉)

IDMEC, Instituto Superior Técnico, University of Lisbon, Lisbon, Portugal
{jorge.ambrosio,pedro.antunes,
mario.viegas}@tecnico.ulisboa.pt

Abstract. A roller coaster ride consists of a vehicle negotiating a track characterized by a sequence of spatial curves and straight lines. During the track negotiation, the occupants are subjected to accelerations that depend, not only on the car speed variation, but also on the instantaneous curvature of the track. The design of the roller coaster geometry requires reliable computational tools to simulate roller coaster rides. An important ingredient to simulate the exposure of the occupant to accelerations is the modelling of the vehicle-track interaction. In this work, an approach to model the car-track interaction is proposed, being two new path motion constraints implemented for the purpose. These constraints allow to prescribe the path of each wheelset along each one of the track rails. Two paths are generated based on the roller coaster geometry, representing the geometry of the rails of the track. A multibody models to represent the roller coaster vehicle is developed and analyzed, serving as application examples of the tools developed in this work.

Keywords: Path motion kinematic constraint · Multibody dynamics
Roller coaster vehicle

1 Introduction

Roller coaster s are, apparently, relatively simple mechanical systems when compared to modern railways or cars. Since most of the roller coaster s represent unique designs, extensive testing of real world prototypes is not possible from an economical point of view. To avoid expensive testing before the final installation of the roller coaster track, reliable computer aided design tools are required.

The design of a safe roller coaster system requires the knowledge of the human tolerances to injury [1, 2]. The roller coaster ride cannot be boring, it must be exciting and stimulating, but ensuring that the riders are not injured. The physiological excitement of the passenger is achieved through the roller coaster track, which assures the human body is subjected to accelerations in different directions, within the human tolerance thresholds [1, 2], but not any further. Therefore, the basis for the analysis is the ability to model and analyze the roller coaster dynamics.

The first step, for the dynamic analysis, is to model the vehicle-track interaction. Pombo and Ambrósio [3] proposed a methodology for the accurate description of the

J. Lenarcic and V. Parenti-Castelli (Eds.): ARK 2018, SPAR 8, pp. 335–343, 2019.
https://doi.org/10.1007/978-3-319-93188-3_39

track centerline, in the general case of a fully three-dimensional track geometry. Within the framework of the description of multibody systems using Cartesian coordinates, as presented by Nikravesh [4], Pombo presented a methodology for representing a spatial curve constraint in multibody systems, using a Lagrange formulation of multibody dynamics [3]. The geometric description of the curve for a roller coaster track starts with the definition of a moving referential, in which the tangent, normal and binormal vectors define an orthogonal frame. Both Frenet and Darboux frames are candidates to play the role of the required moving referential [5, 6]. Both have singularities in general spatial curve geometries, as discussed by Tändl and Kecskemethy [7, 8]. The work of Pombo uses the fundamental theory of differential geometry and kinematics of motion along spatial curves. The roller coaster position coordinates, the Frenet frame vectors and their parametric derivatives are required for the definition of path motion kinematic constraints. These calculations are made in a pre-processor, before the simulation, being the derived quantities available to the code in the form of a multidimensional lookup table, which is interpolated online. For general roller coaster features, cubic splines are used when describing the spatial trajectory, as shown by Ambrósio et al. [9].

The second element of the system is the vehicle, which is modelled here using the multibody dynamics approach. As in the literature no roller coaster vehicle models could be identified a new vehicle model is proposed and demonstrated in this work. Its dynamic analysis in a realistic track is developed and the possibility of equipping the vehicle with a tilting secondary suspension to minimize for the non-compensated acceleration that affects the rider [10] is explored.

2 Path Following Kinematic Constraint

Let a curve be described using an nth order spline segments, interpolating a set of control points, be defined as

$$\mathbf{g}(u) = \begin{Bmatrix} x(u) \\ y(u) \\ z(u) \end{Bmatrix} = a_0 + a_1 u + a_2 u^2 + a_3 u^3 + \ldots + a_n u^n \tag{1}$$

where $\mathbf{g}(u)$ is the vector locating a point on the curve, u is the local parametric variable and a_i are unknown algebraic coefficients that must be calculated using points with known coordinates. Although Eq. (1) is generic for any polynomial interpolation, in this work only cubic polynomials are considered.

2.1 Curve Moving Frame

There are different available frames definitions that can be used to represent general curves, of which the Frenet frame, shown in Fig. 1(a), is used here, since it provides an appropriate curve referential at every point with non-vanishing ouverture. The definition of the Frenet frame starts with the identification of the osculating plane, at a given point P on a curve, i.e., the plane of closest contact to the curve in the vicinity of P [5].

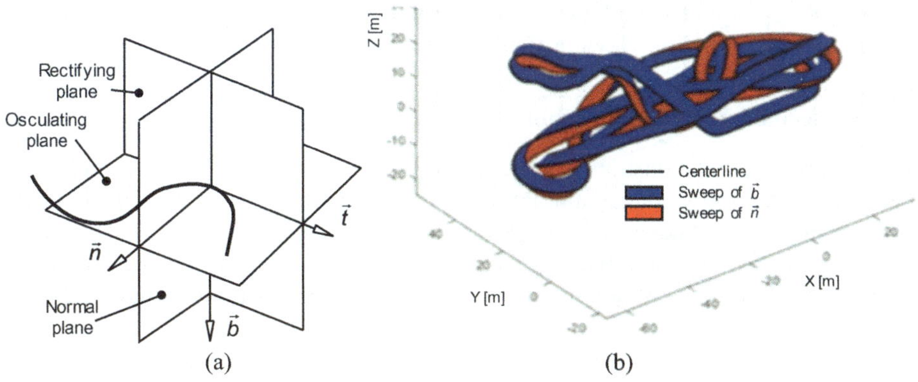

Fig. 1. Description of spatial curves: (a) Frenet frame of a given curve; (b) sweep of the normal and binormal vectors of the centerline of a roller coaster.

The tangent vector **t** and the principal normal vector **n** are defined in the osculating plane. The binormal vector **b** is defined as being normal to the other two vectors, as shown in Fig. 1(a). These vectors are defined in the intersection of the normal, rectifying and osculating planes, point P, and can be written as [3]

$$\mathbf{t} = \frac{\mathbf{g}^u}{\|\mathbf{g}^u\|}; \quad \mathbf{n} = \frac{\mathbf{k}}{\|\mathbf{k}\|}; \quad \mathbf{b} = \tilde{\mathbf{t}}\mathbf{n} \tag{2}$$

where $\tilde{\mathbf{t}}\mathbf{n}$ means a cross product and the auxiliary vector **k** is given by

$$\mathbf{k} = \mathbf{g}^{uu} - \frac{\mathbf{g}^{uu^T}\mathbf{g}^u}{\|\mathbf{g}^u\|^2}\mathbf{g}^u \tag{3}$$

in which $\mathbf{g}^u$ and $\mathbf{g}^{uu}$ are, respectively, the first and second derivatives of the parametric curve $\mathbf{g}(u)$ with respect to the parametric variable u.

The curve parameter u does not ensure that the polynomial exhibits a constant velocity, i.e., the arc length corresponding to a similar increment of the parameter is different in different parts of the curve. The implementation of a prescribed point constraint requires that the piecewise polynomial parameter u is replaced by a curve arc-length parameter L with respect to which the interpolating polynomial has a constant velocity. Consider the parametric variable u^P, corresponding to a point P, located on the kth polynomial segment to which a curve length L_k^P measured from the kth segment origin, is associated. The parameter u^P is such that [3]

$$\int_0^{u^P} \sqrt{g_k^{u^T} g_k^u}\, du - L_k^P = 0 \tag{4}$$

In terms of its computer implementation, the non-linear Eq. (4) is solved in the program pre-processor, using Newton-Raphson method [4].

2.2 Prescribed Motion Constraints

The prescribed point constraint is the basis for the definition of the vehicle-track interaction in the roller coaster multibody system. The wheelsets of the vehicle model are enforced to move along the rails of the track by kinematic constraints. The wheel-rail contact forces, related to the Lagrange multipliers associated to this constraint, are evaluated in the process of solving the multibody system equations of motion [4].

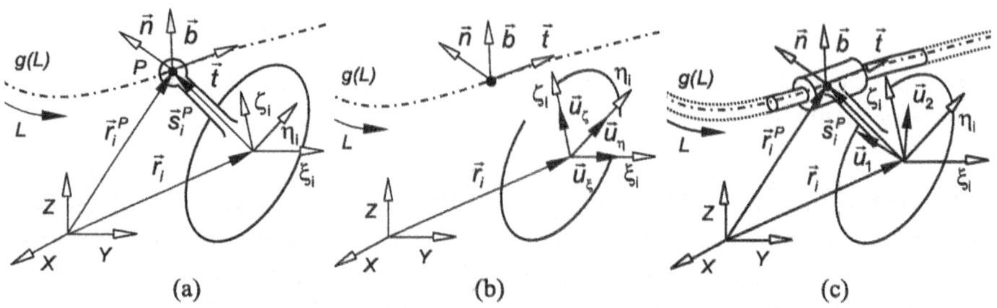

Fig. 2. Prescribed motion constraints: (a) point motion constraint; (b) local frame alignment constraint; (c) cylindrical path following constraint.

Point Motion Constraint. The prescribed point constraint enforces that a point *P*, located on rigid body *i*, follows a specified path, as depicted in Fig. 2(a). The path is defined by a parametric curve $\mathbf{g}(L)$, in which *L* represents the length travelled by the point along the curve from the origin. The point motion constraint is defined by [3, 4].

$$\mathbf{\Phi}^{(pmc,3)} \equiv \mathbf{r}_i^P - \mathbf{g}(L) = \mathbf{0} \tag{5}$$

where $\mathbf{r}_i^P$ is the position of point *P* expressed in the global coordinate system (*X, Y, Z*).

The velocity equation for the point motion constraint is the derivative of Eq. (5)

$$\dot{\mathbf{\Phi}}^{(pmc,3)} \equiv \mathbf{\Phi}_{\mathbf{q}}^{(pmc,3)}\dot{\mathbf{q}} = \mathbf{0} \tag{6}$$

where $\dot{\mathbf{q}} = \left\{ \dot{\mathbf{r}} \quad \omega' \quad \dot{L} \right\}^T$ and the Jacobian matrix is

$$\mathbf{\Phi}_{\mathbf{q}}^{(pmc,3)} = \begin{bmatrix} \mathbf{I} & -\tilde{\mathbf{s}}_i^R \mathbf{A}_i & -\frac{d\mathbf{g}}{dL} \end{bmatrix} \tag{7}$$

where $\mathbf{A}_i$ is the transformation matrix from the body fixed to the inertia frame. The acceleration equation is the time derivative of Eq. (6), being written as

$$\ddot{\mathbf{\Phi}}^{(pmc,3)} \equiv \mathbf{\Phi}_{\mathbf{q}}^{(pmc,3)}\ddot{\mathbf{q}} = \gamma^{(pmc,3)} \tag{8}$$

where the right hand side (r.h.s) of the acceleration equation is

$$\gamma^{(pmc,3)} = -\tilde{\omega}_i\tilde{\omega}_i\mathbf{A}_i\mathbf{s}_i'^P + \frac{d^2\mathbf{g}}{dL^2}\dot{L}^2 \tag{9}$$

Cylindrical Path Following Constraint. A local frame alignment constraint [3], as implied in Fig. 2(b), is too restrictive for the roller coaster model envisaged in this work and, therefore, is not used here. The cylindrical path following constraint, shown in Fig. 2(b), not only includes the prescribed point motion constraint, but also restrains orientations of body i, only allowing it to roll about the path tangent vector. Vectors $\mathbf{u}_l$ and $\mathbf{u}_2$ are fixed to rigid body i and they are not collinear with each other or with vector $\mathbf{t}$. The relative orientation between the body fixed vectors and the curve Frenet frame is such that,

$$\mathbf{\Phi}^{(lfac,2)} = \begin{Bmatrix} \mathbf{t}^T\mathbf{A}_i\mathbf{u}_1' \\ \mathbf{t}^T\mathbf{A}_i\mathbf{u}_2' \end{Bmatrix} - \begin{Bmatrix} a \\ b \end{Bmatrix} = \mathbf{0} \tag{10}$$

where a, b and c are the cosine of the angles between the Frenet and local frame axis for the initial conditions. Therefore, the cylindrical path following constraint is

$$\mathbf{\Phi}^{(cpfc,5)} = \begin{Bmatrix} \mathbf{\Phi}^{(pmc,3)^T} & \mathbf{\Phi}^{(lfac,2)^T} \end{Bmatrix}^T = \mathbf{0} \tag{11}$$

The Jacobian matrix and the r.h.s. of the acceleration constraint equation are obtained by simply appending the individual contributions of the constraints as

$$\mathbf{\Phi}_{\mathbf{q}}^{(cpfc,5)} = \begin{bmatrix} \mathbf{\Phi}_{\mathbf{q}}^{(pmc,3)^T} & \mathbf{\Phi}_{\mathbf{q}}^{(lfac,2)^T} \end{bmatrix}^T; \quad \gamma^{(cfac,5)} = \begin{Bmatrix} \gamma^{(pmc,3)^T} & \gamma^{(lfac,2)^T} \end{Bmatrix}^T \tag{12}$$

in which the contributions of the alignment condition are

$$\begin{aligned} \mathbf{\Phi}_{\mathbf{q}}^{(lfac,2)} &= \begin{bmatrix} \mathbf{0}^T & -\mathbf{t}^T\mathbf{A}_i\tilde{\mathbf{u}}_1'\left(\frac{d\mathbf{t}}{dL}\right)^T & \mathbf{A}_i\mathbf{u}_1' \\ \mathbf{0}^T & -\mathbf{t}^T\mathbf{A}_i\tilde{\mathbf{u}}_2'\left(\frac{d\mathbf{t}}{dL}\right)^T & \mathbf{A}_i\mathbf{u}_2' \end{bmatrix}; \\ \gamma^{(lfac,2)} &= -\begin{Bmatrix} \left[2\dot{L}\left(\frac{d\mathbf{t}}{dL}\right)^T\mathbf{A}_i\tilde{\omega}_i' + \mathbf{t}^T\mathbf{A}_i\tilde{\omega}_i'\tilde{\omega}_i' + \dot{L}^2\left(\frac{d^2\mathbf{t}}{dL^2}\right)^T\mathbf{A}_i\right]\mathbf{u}_1 \\ \left[2\dot{L}\left(\frac{d\mathbf{t}}{dL}\right)^T\mathbf{A}_i\tilde{\omega}_i' + \mathbf{t}^T\mathbf{A}_i\tilde{\omega}_i'\tilde{\omega}_i' + \dot{L}^2\left(\frac{d^2\mathbf{t}}{dL^2}\right)^T\mathbf{A}_i\right]\mathbf{u}_2 \end{Bmatrix} \end{aligned} \tag{13}$$

It must be noticed that the online parametrization of the spatial curve, implied by Eq. (4), and the calculation of its parametric derivatives, used in Eqs. (5) through (13), is computationally expensive. Instead, a lookup table with the precomputed values for the curve and parametric derivatives, as a function of the curve arc-length, L, with a small length step, ΔL, is made available for the online evaluation of the prescribed path following constraints, using the procedure proposed in Pombo and Ambrosio [3].

3 Multibody Equations of Motion

A multibody system is, generally, composed by several types of rigid and flexible, bodies, connected by mechanical joints and acted upon by passive or active force elements. Let position and orientations of the n_b bodies of a multibody system, denoted by $\mathbf{q} = \left[\mathbf{q}_1^T \ldots \mathbf{q}_{n_b}^T\right]^T$ with the position $\mathbf{r}_i$ and orientation $\mathbf{p}_i$ of each individual body i, be collected in vector $\mathbf{q}_i = \left[\mathbf{r}_i^T \quad \mathbf{p}_i^T\right]^T$. The equations of motion for a multibody system are [4]

$$\begin{bmatrix} \mathbf{M} & \mathbf{\Phi}_\mathbf{q}^T \\ \mathbf{\Phi}_\mathbf{q} & \mathbf{0} \end{bmatrix} \begin{bmatrix} \ddot{\mathbf{q}} \\ -\boldsymbol{\lambda} \end{bmatrix} = \begin{bmatrix} \mathbf{g} \\ \boldsymbol{\gamma} - 2\alpha\dot{\mathbf{\Phi}} - \beta^2\mathbf{\Phi} \end{bmatrix} \tag{14}$$

where $\mathbf{\Phi}_\mathbf{q}$ is the Jacobian matrix of the kinematic constraints, $\mathbf{M}$ is the system mass matrix, $\boldsymbol{\lambda}$ is the vector of unknown Lagrange multipliers, $\ddot{\mathbf{q}}$ is the acceleration vector, $\mathbf{g}$ the vector of applied forces and $\boldsymbol{\gamma}$ the vector with the right-hand-side of the kinematic acceleration equations. The equations of motion of the multibody system, given by Eq. (14) have to be solved and integrated in time. This numerical process is prone to the accumulation of numerical errors because the position and velocity constraint equations are not used explicitly. The Baumgarte constraint violation method, implied in the r.h.s. of the second equation, is used to maintain the constraint violations inside workable limits [4, 11].

4 Application to Roller Coaster Track and Vehicle Models

The methodology proposed in this work is applied to the dynamic analysis of a roller coaster vehicle while riding a realistic track, i.e., spatial curves with general geometries but excluding inflections points. The topology of the vehicle is described in Fig. 3,

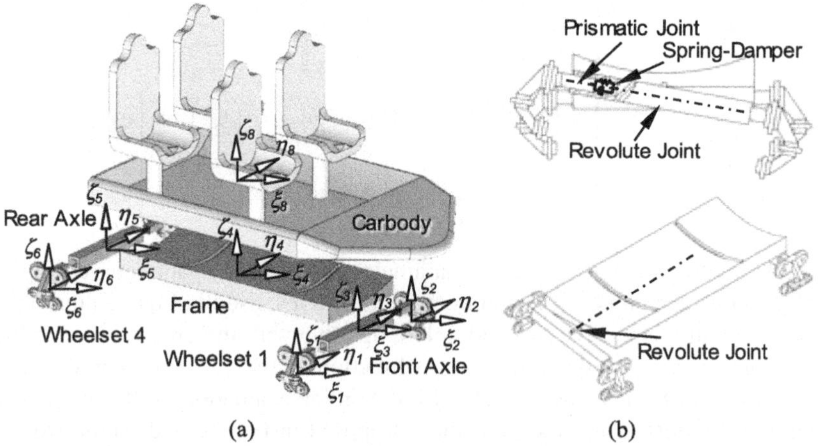

Fig. 3. Roller coaster vehicle: (a) exploded view; (b) details of the vehicle primary suspension.

Table 1. Physical properties and initial conditions for the vehicle multibody system.

ID	Rigid bodies	Mass (Kg)	Inertia ($Kg\ m^2$) ($I_{\xi\xi}$, $I_{\eta\eta}$, $I_{\zeta\zeta}$)	Initial positions/orientations (m) (x, y, z)/(e_1, e_2, e_3)
1	Wheelset FrontLft	10.1	(0.141, 0.134, 0.933)	(−2.100, 0.500, 0.000)/(0.000, 0.000, 1.000)
2	Wheelset FrontRht	10.1	(0.141, 0.134, 0.933)	(−2.100, −0.500, 0.000)/(0.000, 0.000, 0.000)
3	Front Axle	51.5	(0.098, 2.34, 2.34)	(−2.100, 0.000, 0.125)/(0.000, 0.000, −0.707)
4	Frame	245	(13.0706, 77.9, 90.9)	(−1.100, 0.000, 0.185)/(0.000, 0.000, 1.000)
5	Rear Axle	51.5	(0.098, 2.34, 2.34)	(−0.100, 0.000, 0.125)/(0.000, 0.000, −0.707)
6	Wheelset RearLft	10.1	(0.141, 0.134, 0.933)	(−0.100, 0.500, 0.000)/(0.000, 0.000, 1.000)
7	Wheelset RearRht	10.1	(0.141, 0.134, 0.933)	(−0.100, −0.500, 0.000)/(0.000, 0.000, 0.000)
8	Carbody	280	(30.8, 127. 9, 145.8)	(−1.175, 0.000, 0.480)/(0.000, 0.000, 1.000)

Table 2. Definition of the kinematic joints for the vehicle multibody system.

ID	Kinematic constraint	Bodies i	Bodies j	Attachment points local coordinates (m) $s_i^{\prime P}$	$s_i^{\prime Q}$	$s_j^{\prime P}$	$s_j^{\prime Q}$
1	Prescribed point	1	Rail	(0, 0, 0)	–	–	–
2	Prescribed point	2	Rail	(0, 0, 0)	–	–	–
3	Prescribed point	6	Rail	(0, 0, 0)	–	–	–
4	Prescribed point	7	Rail	(0, 0, 0)	–	–	–
5	Prismatic	3	1	(0.100, 0, 0)	(0.200, 0, 0)	(0, 0.100, 0.125)	(0, 0.200, 0.125)
6	Prismatic	3	2	(−0.100, 0, 0)	(−0.200, 0, 0)	(0, 0.100, 0.125)	(0, 0.200, 0.125)
7	Prismatic	5	6	(−0.100, 0, 0)	(−0.200, 0, 0)	(0, 0.100, 0.125)	(0, 0.200, 0.125)
8	Prismatic	5	7	(0.100, 0, 0)	(0.200, 0, 0)	(0, 0.100,0.125)	(0, 0.200, 0.125)
9	Revolute	4	5	(−1.00, 0, 0)	(−1.100, 0, 0)	(0, 0, 0.060)	(0, 0.100, 0.060)
10	Revolute	4	8	(1.050, 0, 1.050)	(1.075, 0, 1.050)	(0.975, 0, 0.755)	(0.950, 0, 0.755)
11	Rigid	3	4	(0, 0, 0)	–	(0, 0, 0)	–

while the inertia properties and initial positions of the rigid bodies of the model are presented in Table 1. The data to define the kinematic constraints is shown in Table 2.

Two models of the vehicle are considered here, one with a secondary suspension that allows for the tilting of the car body with respect to the frame, and a second one in which these two bodies are rigidly attached, by substituting the revolute joint with ID = 10, in Table 2, by a rigid joint. The models of the vehicles in the roller coaster tracks, at the start of the analysis, are shown in Fig. 4.

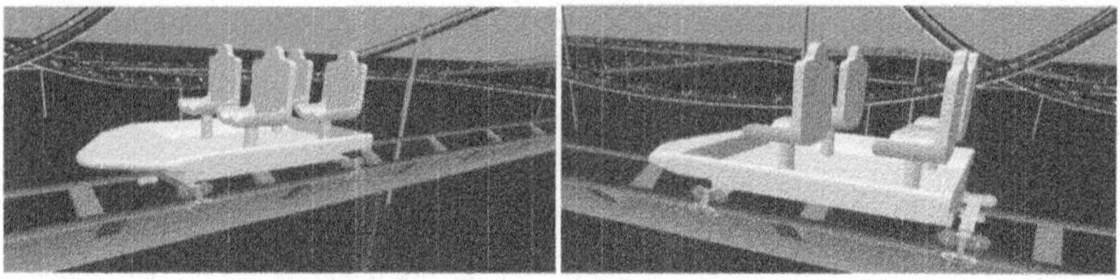

Fig. 4. Roller coaster vehicle on the track.

Among other dynamic analysis of the roller coaster results of interest, those associated to the rider comfort are depicted in Fig. 5 with the non-compensated acceleration (NCA), directly associated to the lateral accelerations experienced by the rider, placed in the front left seat, and the car body roll motion with respect to the horizontal plane.

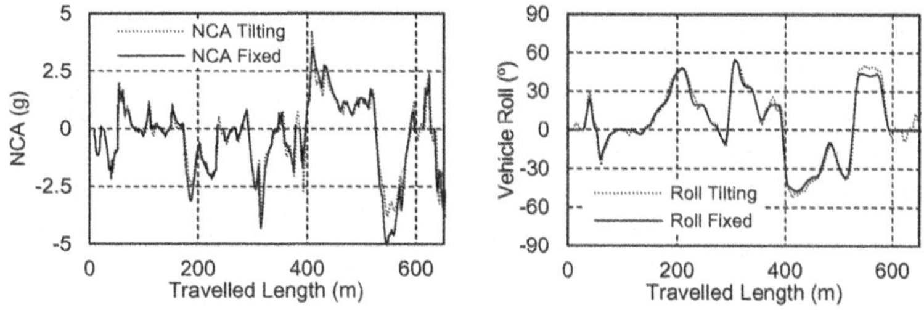

Fig. 5. Non-compensated acceleration and roll of the carbody for the vehicle models with and without a tilting system.

These results allow to identify that the tilting mechanism has the potential to lower the NCA for longer curves. It is also observed that, for the speed at which the vehicle operates, there is a cant deficiency in some curves. However, it is also observed that the current settings of the tilting system of vehicle lead to undesired oscillations when negotiating short curves and curve transitions, suggesting the need for adjusting the suspension properties to the actual characteristics of the roller coaster track.

References

1. Fraser, T.M.: Human response to sustained acceleration: a literature review. In: Scientific and Technical Information Division, National Aeronautics and Space Administration, United States Government Printing Office, Albuquerque, New Mexico (1966)
2. Schmitt, K.-U., Zürich, P., Muser, M., Walz, F.: Trauma Biomechanics: Introduction to Accidental Injury. Springer, Heidelberg (2013)
3. Pombo, J., Ambrósio, J.: General spatial curve joint for rail guided vehicles: kinematics and dynamics. Multibody Syst. Dyn. **9**(3), 237–264 (2003)

4. Nikravesh, P.E.: Computer-Aided Analysis of Mechanical Systems. Prentice-Hall, Englewood Cliffs (1988)
5. Frenet, F.: Sur les courbes à double courbure. Journal des mathématiques pures et appliquées **17**, 437–447 (1852)
6. Lagrange, J.L., Serret, J.A., Darboux, G.: Oeuvres [de Lagrange]: Publiées par les soins de JA Serret. Gauthier-Villars, Paris (1889)
7. Tändl, M., Kecskemethy, A.: Singularity-free trajectory tracking with Frenet frames. In: Husty, M., Schroecker, H.-P. (eds.) Proceedings of the 1st Conference EuCoMeS. Innsbruck University Press, Obergurgl (2006)
8. Tändl, M.: Dynamic Simulation and Design of Roller Coaster Motion. VDI Verlag, Düsseldorf (2009)
9. Ambrósio, J., Antunes, P., Pombo, J.: On the requirements of interpolating polynomials for path motion constraints. In: Kecskeméthy, A., Geu Flores, F. (eds.) Interdisciplinary Applications of Kinematics: Proceedings of the International Conference, pp. 179–197. Springer, Dordrecht (2015)
10. Magalhães, H., Madeira, J., Ambrósio, J., Pombo, J.: Railway vehicle performance optimization using virtual homologation. Veh. Syst. Dyn. **54**(9), 1177–1207 (2016)
11. Ambrósio, J., Neto, A.: Stabilization methods for the integration of DAE in the presence of redundant constraints. Multibody Syst. Dyn. **10**(1), 81–105 (2003)

Design

Line-Symmetric Motion Generators

Yuanqing Wu(✉) and Marco Carricato

University of Bologna, 40136 Bologna, Italy
{yuanqing.wu,marco.carricato}@unibo.it

Abstract. A class of line-symmetric motion generators is synthesized here for the first time. Line-symmetric motions comprise motions of a axial-symmetric object from an initial to a target configuration, by screwing along the common perpendicular of the initial and the goal axis, without incurring self spin or sliding of the object about and along its own axis. These motions have application in several robot mechanism synthesis for line-symmetric manipulation tasks, e.g., minimal invasive surgery robots and haptic interfaces.

Keywords: Line-symmetric motion · Symmetric space
Type synthesis

1 Introduction

Line-symmetric motions refer to a special class of one-parameter motions generated by half-turns of a given body about a line ℓ undergoing a one-parameter motion [3], as illustrated in Fig. 1. The space of all line-symmetric motions sharing a common initial line ℓ_0 is a 4-dimensional submanifold M_4 of the Euclidean group SE(3), and is a *symmetric subspace* of SE(3)[7]. From a practical viewpoint, M_4 entails motions of an object with line symmetry from an initial to a goal configuration, by screwing along the common perpendicular of the initial and the goal axis, without incurring self spin or sliding of the object about and along its own axis. This kind of motions have applications in robotic motion planning and design for line-symmetric manipulation tasks (e.g., minimal invasive surgery robots [2], haptic interfaces [5]).

In this paper, we focus on the mechanism synthesis problem, namely, on the synthesis of 4-DoF mechanisms capable of generating arbitrary line-symmetric motions, or in other words, mechanisms whose motion manifolds are open submanifolds of M_4. As far as we are aware, no such mechanisms exist in the literature. It is proved in [6] that M_4 cannot be generated by 4-DoF serial mechanism comprising Reuleaux lower pairs, because M_4 does not have a product-of-exponentials representation. Therefore, this paper probably introduces for the first time a class of *line-symmetric motion generators* (LSMGs). It should be clear that the motion manifold of a LSMG is not the entirety of M_4, but only an open submanifold of it (due to workspace boundary and singularities).

J. Lenarcic and V. Parenti-Castelli (Eds.): ARK 2018, SPAR 8, pp. 347–355, 2019.
https://doi.org/10.1007/978-3-319-93188-3_40

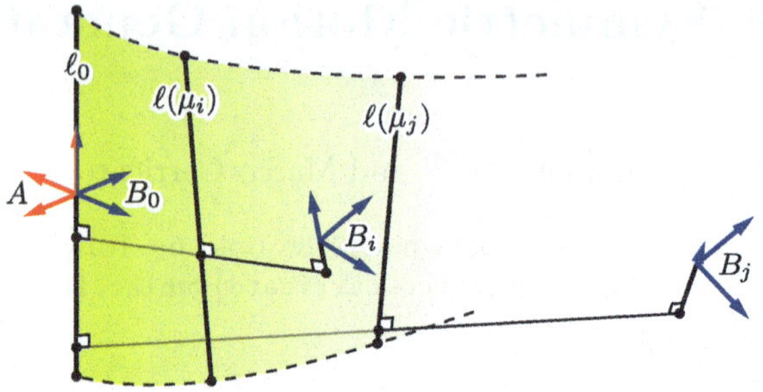

Fig. 1. Line-symmetric motion about a moving line $\ell(\mu)$ parameterized by $\mu \in \mathbb{R}^+$. The half-turned image B_0 of frame A about ℓ_0 serves as the reference frame.

The paper is organized as follows. Section 2 gives a brief review of the motion manifold M_4. Section 3 proposes a class of LSMGs by using a symmetric subspace motion synthesis procedure recently proposed in [6]. Finally, we conclude this paper with a discussion about possible followup work.

2 Line-Symmetric Motion Manifold

We shall denote quaternion units as $\mathbf{i}, \mathbf{j}, \mathbf{k}$, the dual number as ε, and identify elements of the Lie algebra $\mathfrak{se}(3)$ of SE(3) with dual quaternion vectors [4]. A *half-turn* ℓ is a rotation about a line for a magnitude of π. A line-symmetric motion parameterized by $\mu \in \mathbb{R}$ is given by:

$$\mathbf{g}(\mu) = \ell(\mu)\ell_0^{-1} = \ell(\mu)\ell_0, \qquad \ell_0 = \ell(0) \tag{1}$$

where it is apparent that $\ell_0^{-1} = \ell_0$ (Fig. 1). As illustrated in Fig. 1, $\mathbf{g}(\mu)$ is the screw motion along the common perpendicular of ℓ_0 and $\ell(\mu)$ with magnitudes of rotation and translation specified by twice the angle and distance between ℓ_0 and $\ell(\mu)$. In other words, all finite screw axes of the line-symmetric motion manifold M_4 perpendicularly intersect the *characteristic line* ℓ_0 and their corresponding Plücker coordinates span the following 4D twist subspace:

$$\mathfrak{m}_4 = \{\boldsymbol{\xi} = (a_1\mathbf{i} + a_2\mathbf{j}) + \varepsilon(b_1\mathbf{i} + b_2\mathbf{j}) \mid a_1, a_2, b_1, b_2 \in \mathbb{R}\} \tag{2}$$

where ℓ_0 is aligned with the $\mathbf{z}$-axis of the reference frame B_0, and $\mathrm{M}_4 = \exp \mathfrak{m}_4$. $\mathfrak{m}_4$ defines the following derived Lie subalgebra:

$$\mathfrak{h}_4 := [\mathfrak{m}_4, \mathfrak{m}_4] = \{a_3\mathbf{k} + \varepsilon b_3\mathbf{k} \mid a_3, b_3 \in \mathbb{R}\} \tag{3}$$

where the commutator $[\cdot,\cdot]$ is defined by $[\boldsymbol{\xi}_1, \boldsymbol{\xi}_2] := \boldsymbol{\xi}_1 \times \boldsymbol{\xi}_2, \forall \boldsymbol{\xi}_1, \boldsymbol{\xi}_2 \in \mathfrak{se}(3)$. $\mathfrak{h}_4$ is the cylindrical algebra along the $\mathbf{z}$-axis, which satisfies $\mathfrak{m}_4 \cap \mathfrak{h}_4 = \{0\}$ and $\mathfrak{m}_4 \oplus \mathfrak{h}_4 = \mathfrak{se}(3)$.

3 Type Synthesis of Line Symmetric Motion Generators

3.1 Generating M_4 Using Symmetric Chains

In [7], we introduced the concept of *symmetric twist chains* (SCs) for the generation of symmetric subspaces of SE(3). In particular, a $\mathfrak{m}_4$-SC is an 8-tuple of unit twists, denoted as $(\boldsymbol{\xi}_1^+, \boldsymbol{\xi}_2^+, \boldsymbol{\xi}_3^+, \boldsymbol{\xi}_4^+, \boldsymbol{\xi}_4^-, \boldsymbol{\xi}_3^-, \boldsymbol{\xi}_2^-, \boldsymbol{\xi}_1^-)$, which comprises four nesting *symmetric pairs* (SP) $(\boldsymbol{\xi}_i^+, \boldsymbol{\xi}_i^-)$ defined by:

$$\begin{cases} \boldsymbol{\xi}_i^+ := \boldsymbol{\xi}_i + \boldsymbol{\eta}_i \\ \boldsymbol{\xi}_i^- := \boldsymbol{\xi}_i - \boldsymbol{\eta}_i \end{cases} \qquad \boldsymbol{\xi}_i \in \mathfrak{m}_4, \boldsymbol{\eta}_i \in \mathfrak{h}_4, i = 1, \ldots, 4 \tag{4}$$

and satisfies one of the following equivalent conditions (see [7] for details):

$$\text{span}\{\boldsymbol{\xi}_1, \boldsymbol{\xi}_2, \boldsymbol{\xi}_3, \boldsymbol{\xi}_4\} = \mathfrak{m}_4 \tag{5a}$$

$$\text{span}\{\boldsymbol{\xi}_1^+, \boldsymbol{\xi}_2^+, \boldsymbol{\xi}_3^+, \boldsymbol{\xi}_4^+\} \oplus \mathfrak{h}_4 = \mathfrak{se}(3) \tag{5b}$$

$$\text{span}\{\boldsymbol{\xi}_4^-, \boldsymbol{\xi}_3^-, \boldsymbol{\xi}_2^-, \boldsymbol{\xi}_1^-\} \oplus \mathfrak{h}_4 = \mathfrak{se}(3) \tag{5c}$$

Geometrically, each $\boldsymbol{\xi}_i^+$ is always line-symmetric to $\boldsymbol{\xi}_i^-$ about the characteristic line ℓ_0. For example, as illustrated in Fig. 2, $(\boldsymbol{\xi}_i^+, \boldsymbol{\xi}_i^-)$ is generated from $\boldsymbol{\xi}_i \in \mathfrak{m}_4$ ($\boldsymbol{\xi}_i$ has finite pitch p_i) by first applying a pair of rotations about $\boldsymbol{\delta}_i \in \mathfrak{m}_4$ and then a pair of translations along $\boldsymbol{\delta}_i'$, both with equal and opposite magnitudes. When $\boldsymbol{\xi}_j \in \mathfrak{m}_4$ has infinite pitch, $(\boldsymbol{\xi}_j^+, \boldsymbol{\xi}_j^-)$ is generated by applying a pair of rotations about $\boldsymbol{\delta}_j \in \mathfrak{m}_4$ with equal and opposite magnitudes.

Denote the joint variable of $\boldsymbol{\xi}_i^\pm$ by $\theta_i^\pm$. A $\mathfrak{m}_4$-SC generates M_4 under the *symmetric movement condition* (SMC):

$$\theta_i^+ = \theta_i^-, \qquad i = 1, 2, 3, 4 \tag{6}$$

For the sake of simplicity, we shall consider synthesis of $\mathfrak{m}_4$-SCs with only revolute joints, which amounts to specifying a quadruple of 0-pitch screws $\mathcal{M}^- := (\boldsymbol{\xi}_4^-, \boldsymbol{\xi}_3^-, \boldsymbol{\xi}_2^-, \boldsymbol{\xi}_1^-)$ satisfying (5c):

$$S^- := \text{span}\{\boldsymbol{\xi}_4^-, \boldsymbol{\xi}_3^-, \boldsymbol{\xi}_2^-, \boldsymbol{\xi}_1^-\}, \qquad S^- \oplus \mathfrak{h}_4 = \mathfrak{se}(3) \tag{7}$$

In general, the axes of all 0-pitch screws of a 4-dimensional twist space (a 4-*system*, in brief) form a two-parameter family of lines known as a *linear congruence*. Following Hunt's classification of general and special 4-systems [1], we have the following cases[1]:

(G) The collection of all 0-pitch screws of a *general* 4-*system* comprises a linear congruence, which may:

[1] Case G refers to the general 4-system, whereas cases I, II, III, IV and V refer to special 4-systems of types 1 through 5. A subscript 0, 1 or 2 indicates the presence of, respectively, 0, 1 or 2 real directrices.

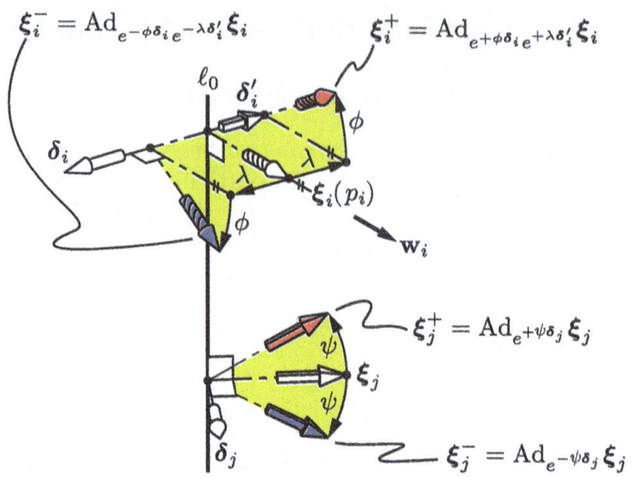

Fig. 2. Examples of SPs of $\mathfrak{m}_4$. Round, prism and threaded arrows represent, respectively, 0-pitch, infinite-pitch and finite-nonzero-pitch screws.

(G_2) have all its lines meet *two* real skew lines called *directrices* (*hyperbolic net*) (Fig. 3(a)); the reciprocal wrench subspace $(S^-)^\perp$ is spanned by 0-pitch screws ζ_1 and ζ_2 aligned with the directrices;
G_1) have only *one* real directrix (*parabolic net*); the reciprocal wrench subspace $(S^-)^\perp$ is spanned by a finite nonzero-pitch screw ζ_1 and a 0-pitch screw ζ_2 aligned with the directrix (Fig. 3(b));
(G_0) have no real but two complex directrices (*elliptic net*); the congruence is a one-parameter family of reguli on ∞^1 coaxial hyperboloids (Fig. 3(c)).

(I) The 0-pitch screws of a *first special* 4-*system* may:
(I_0) comprise an axial-symmetric elliptic net if its reciprocal wrench system is formed by a pencil of finite-nonzero-pitch screws (Fig. 3(c));
(I_2) meet two intersecting real directrices, resulting in a degenerate congruence comprising a field of lines in the directrices' plane, as well as a bundle of lines with center at the intersection of the directrices (Fig. 3(d)).

(II) The 0-pitch screws of a *second special* 4-*system* form a degenerate congruence comprising:
(II_1) a one-parameter family of pencils of parallel lines on as many parallel planes, if its reciprocal wrench system is formed by a parallel pencil of finite-nonzero-pitch screws (Fig. 4(a)).
(II_2) a parallel bundle formed by all lines parallel to a real directrix and a field formed by all lines on a plane containing the real directrix (Fig. 4(b)).

(III) The 0-pitch screws of a *third special* 4-*system* (the Schönflies algebra) form a parallel bundle formed by all lines parallel to a given direction; there are no four linearly independent 0-pitch screws.

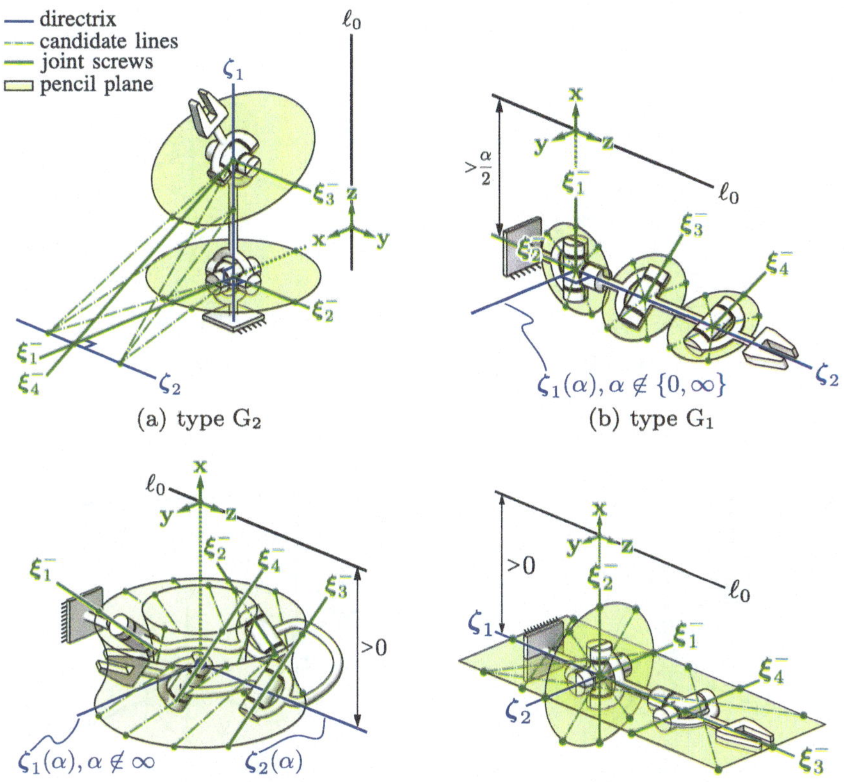

(a) type G_2 (b) type G_1 (c) type I_0 (G_0) (d) type I_2

Fig. 3. Congruences formed by the axes of the 0-pitch screws in a general (a, b) and a first special (c, d) 4-system, with their associated kinematic chains. In the absence of one or two real directrices, $\zeta_1(\alpha)$ and $\zeta_2(\beta)$ are basis screws of the reciprocal two-system, with α and β denoting the screw pitches. When α or β is zero, the axis of the corresponding screw is a real directrix.

(IV) The 0-pitch screws of a *fourth special* 4-*system* form a one-parameter family of pencils of concurrent lines on as many parallel planes, with their centers lying on a real directrix (Fig. 4(c)).
(V) The 0-pitch screws of a *fifth special* 4-*system* lie on all lines intersecting perpendicularly a real directrix (Fig. 4(c)).

The aforementioned cases, except for the third special 4-system, lead to distinct $\mathfrak{m}_4$-SC types, which are summarized in Table 1.

3.2 Type Synthesis of Line-Symmetric Motion Generators

When SMC is not imposed, $\mathfrak{m}_4$-SCs are essentially 8-DoF kinematic chains, and therefore cannot directly form a parallel mechanism (PM) with a 4D motion space. The SMC is imposed by adding internal interconnections within the PM,

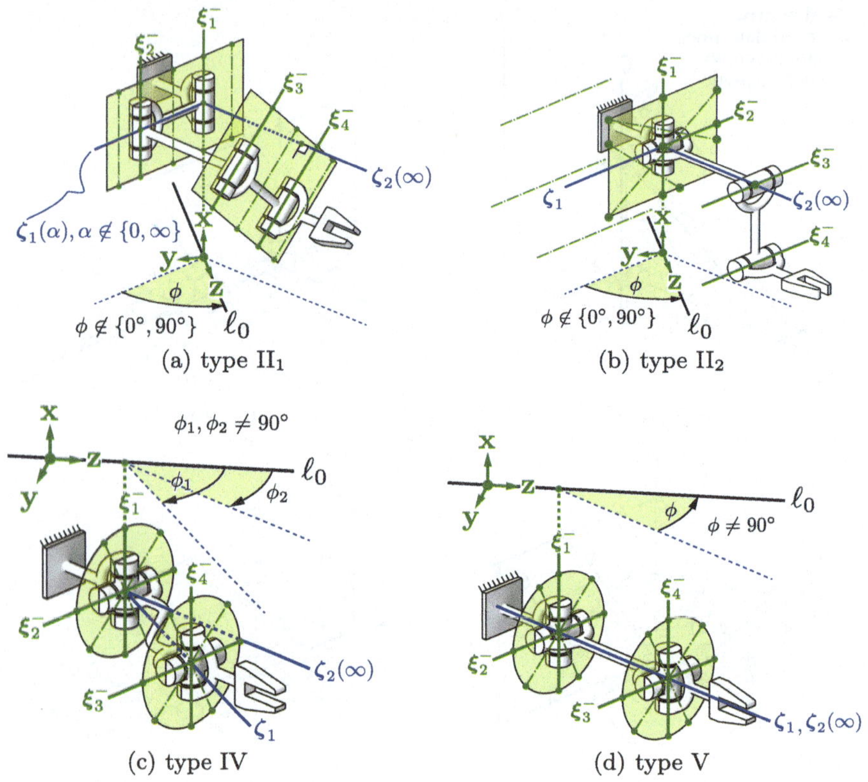

Fig. 4. Congruences formed by the axes of the 0-pitch screws in special 4-systems of type 2 (a), type 3 (b), type 4 (c) and type 5 (d), with their associated kinematic chains.

which generate additional constraints. Given a PM comprising l $\mathfrak{m}_4$-SCs, denoted $\mathcal{M}_i := (\boldsymbol{\xi}_{i1}^+, \ldots, \boldsymbol{\xi}_{i4}^+, \boldsymbol{\xi}_{i4}^-, \ldots, \boldsymbol{\xi}_{i1}^-)$, $i = 1, \ldots, l$, we proposed in [6] the following procedure to synthesize a LSMG:

Step (1) Synthesize, for $i = 1, \ldots, l$, a chain of four revolute joints represented by the quadruple $\mathcal{M}_i^- := (\boldsymbol{\xi}_{i4}^-, \boldsymbol{\xi}_{i3}^-, \boldsymbol{\xi}_{i2}^-, \boldsymbol{\xi}_{i1}^-)$ satisfying (5c), namely

$$S_i^- := \mathrm{span}\{\boldsymbol{\xi}_{i4}^-, \boldsymbol{\xi}_{i3}^-, \boldsymbol{\xi}_{i2}^-, \boldsymbol{\xi}_{i1}^-\} \qquad S_i^- \oplus \mathfrak{h}_4 = \mathfrak{se}(3)$$

Step (2) Complete the $\mathfrak{m}_4$-SC using line symmetry, as shown in Fig. 2.
Step (3) Verify the *constraint synthesis condition* (CSC):

$$\textstyle\sum_{i=1}^{l} (S_i^-)^\perp = \mathfrak{se}(3)^* \tag{8}$$

where $(S_i^-)^\perp$ denotes the reciprocal wrench subspace of S_i^- and $\mathfrak{se}(3)^*$ is the 6D wrench space.
Step (4) Connect the innermost links (i.e., the link joining $\boldsymbol{\xi}_{i4}^-$ to $\boldsymbol{\xi}_{i4}^+$) of the $\mathcal{M}_i$'s to each other, by way of cylindrical joints along ℓ_0 (i.e., the $\mathbf{z}$-axis). □

Table 1. Four-systems and their associated $\mathfrak{m}_4$-SCs comprising revolute ($\mathcal{R}$), universal ($\mathcal{U}$) and spherical ($\mathcal{S}$) joints.

	congruence of 0-pitch screws	$\mathcal{M}^-$	condition (5c)
G_2	2 pencils with centers lying on the same directrix	$\mathcal{UU}$	the cylindroid spanned by ζ_1 and ζ_2 contains no axis that perpendicularly intersects ℓ_0
G_1	3 skew lines intersecting a 4th one at right-angle	$\mathcal{URR}$	
G_0	3 lines lying on a regulus of a hyperboloid, and a 4th one outside it	$4\mathcal{R}$	
I_0		$4\mathcal{R}$	
I_2	a bundle and a line not passing through its center (or equivalently two non-coplanar pencils sharing a line)	$\mathcal{RS}(\mathcal{UU})$	
II_1	2 pairs of parallel lines lying in 2 parallel planes	$4\mathcal{R}$	neither ζ_1 nor ζ_2 is perp. to ℓ_0
II_2	a pencil and 2 parallel lines parallel to the pencil plane	$\mathcal{URR}$	
IV,V	2 pencils in 2 parallel planes	$\mathcal{UU}$	

Since Step (1) is already considered in the last subsection and Step (2), (4) are trivial, we are only left with Step (3). When $(S_i^-)^\perp$ may be spanned by two 0-pitch screws ζ_{i1}, ζ_{i2}, $i = 1, \ldots, l$ (which is the case for type G_2, I_2, II_2, IV and V), denoted , Step (3) reduces to the problem of verifying (8) with a set of generators $(\zeta_{11}, \zeta_{12}, \ldots, \zeta_{l1}, \zeta_{l2})$ comprising 0-pitch screws:

$$\operatorname{span}\{\zeta_{11}, \zeta_{12}, \ldots, \zeta_{l1}, \zeta_{l2}\} = \mathfrak{se}(3)^* \tag{9}$$

Thus, at least three $\mathfrak{m}_4$-SCs are required for the synthesis of LSMGs. If the $\mathfrak{m}_4$-SCs in a LSMG are of the same kind, we have the following LSMG types:

1. **type G_2** (Fig. 5) and **type I_2** (Fig. 6) When ζ_{i1}, ζ_{i2} is a pair of real finite lines, for $i = 1, 2, 3$, the three $\mathfrak{m}_4$-SCs can be arranged so that, at the home configuration ζ_{11}, ζ_{21} and ζ_{31} span a bundle of parallel lines (parallel to the **z**-axis), whereas ζ_{12}, ζ_{22} and ζ_{32} span a field in a plane (**xy**-plane) perpendicular to the parallel bundle.
2. **type II_2, IV** and **V**. In this case, ζ_{i2} $(i = 1, 2, 3)$ can be conveniently chosen so as a line at infinity, namely as an infinite-pitch screw. The three $\mathfrak{m}_4$-SCs can be arranged so that ζ_{12}, ζ_{22}, ζ_{32} span a bundle of infinite-pitch screws, and ζ_{11}, ζ_{21}, ζ_{31} (after a convenient translation to become concurrent) span a bundle of 0-pitch screws. Figures of type II_2, IV, V LSMGs are not reported due to space limitations.

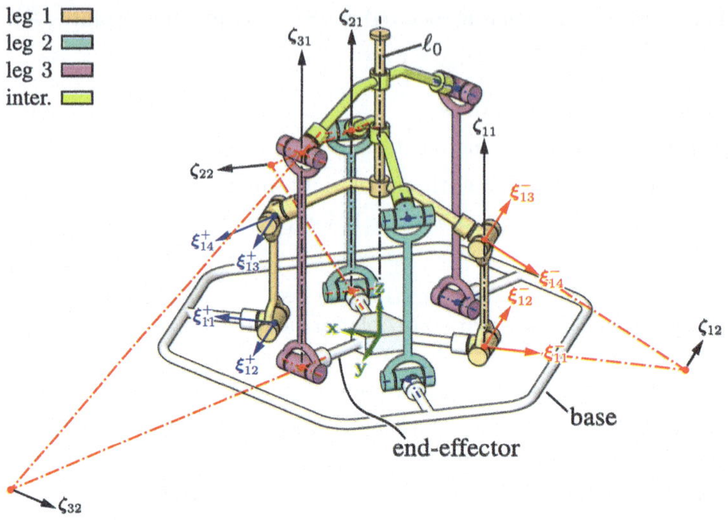

Fig. 5. Line symmetric motion generator of type G_2. For clarity, only the joint screws of the first leg are shown.

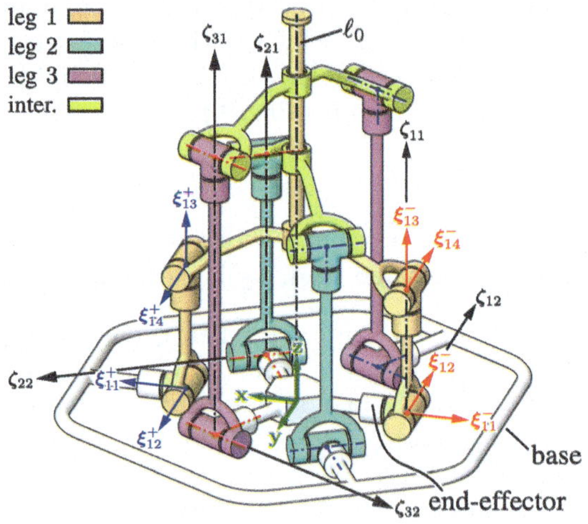

Fig. 6. Line symmetric motion generator of type I_2. For clarity, only the joint screws of the first leg are shown.

4 Conclusions

We presented for the first time a class of 4-DoF mechanisms capable of generating arbitrary line-symmetric motions, based on both symmetric space theory and classic screw theory. The synthesis conditions for the LSMGs are given, which can be used to systematically generate a whole spectrum of novel mechanisms, even besides those that are reported here as examples.

References

1. Hunt, K.H.: Kinematic geometry of mechanisms. Oxford University Press, Oxford (1978)
2. Kuo, C.H., Dai, J.S.: Robotics for minimally invasive surgery: a historical review from the perspective of kinematics. In: International symposium on history of machines and mechanisms, pp. 337–354. Springer, Dordrecht (2009)
3. Selig, J., Husty, M.: Half-turns and line symmetric motions. Mech. Mach. Theory **46**(2), 156–167 (2011)
4. Selig, J.M.: Geometric fundamentals of robotics. Springer, New York (2004)
5. Stocco, L.J., Salcudean, S.E., Sassani, F.: Optimal kinematic design of a haptic pen. IEEE/ASME Trans. Mechatron. **6**(3), 210–220 (2001)
6. Wu, Y., Carricato, M.: Symmetric subspace motion generators. IEEE Trans. Robot. **34**(3), 716–735 (2018)
7. Wu, Y., Löwe, H., Carricato, M., Li, Z.: Inversion symmetry of the Euclidean group: theory and application to robot kinematics. IEEE Trans. Robot. **32**(2), 312–326 (2016)

Kinematic Synthesis of Planar Multi-limb Mechanisms for Multi-directional Interaction with Bodies in the Environment

Nina Robson[1] and Gim Song Soh[2(✉)]

[1] Department of Mechanical Engineering, California State University, Fullerton, USA
nrobson@fullerton.edu

[2] Engineering Product Development, Singapore University of Technology and Design, Singapore, Singapore
sohgimsong@sutd.edu.sg

Abstract. The paper considers the synthesis of planar mechanisms to manipulate a body in multiple desired directions in the vicinity of a contact pose, a critical focus not fully captured in existing design methodologies. The mechanism consists of multiple planar Revolute-Prismatic-Revolute (RPR) limbs in contact with the body. Under equilibrium grasp, each of the desired directions impose normal forces and curvature constraints in the vicinity of the limb-body contact. We discuss how to transform these constraints into velocity and acceleration requirements of the body, and formulate the synthesis equations of planar RPR serial chains to achieve each desired manipulation task. The simultaneous design of two planar RPR limbs able to manipulate an object in two different directions in the vicinity of a contact pose illustrates the feasibility of the proposed design approach.

Keywords: Kinematic synthesis · Differential kinematics
Manipulation

1 Introduction

Design of human interactive systems for robotics applications have recently gain significant attention [1]. This involves a variety of design considerations in human factors such as physical (physiological and bio-mechanical), cognitive, social or emotional. Typically, the physical interface between human and machine needs to be selected from anthropomorphic or non-anthropomorphic architectures [2], each with its pros and cons. For instance, anthropomorphic architecture have the advantage of a simpler design through the matching of limb shapes and the kinematic joint degrees-of-freedom (DOF) of the human operator. However, it is unable to mimimic the complex compositional articulations of multiple anatomical structures within a human joint. This poses a significant challenge

J. Lenarcic and V. Parenti-Castelli (Eds.): ARK 2018, SPAR 8, pp. 356–363, 2019.
https://doi.org/10.1007/978-3-319-93188-3_41

for the design of these devices [3] for safety and transparent operation. Currently, the design of these devices relies on kinematic synthesis of each individual fingers to achieve the required task and limited efforts have been applied to approach the design problem from a higher level task intent. Thus, a methodology for simultaneous design of multiple fingers at a whole hand level still does not exist. For a good review, see [4].

Rimon and Burdick [5,6] generalized the study of grasping constraint of a rigid body using the fingers of a mechanical hand by considering the configuration space of movement of the body relative to obstacles formed by the fingers. These work introduced the concept of second order mobility of a constrained body, and showed how the relative curvature of the surfaces in contact act to limit movement and can be used to reduce the number of fingers/fixtures used in planar grasping tasks to a minimum of two. Inspired from their work, Robson and McCarthy [7] considered how second order effects can be used to synthesize planar linkages that guide a rigid body such that it satisfies obstacle constraints. The authors focused on first and second order problem which ensures the rigid body maintains the velocity/force and curvature constraint specifications. Subsequently, this work was extended to deal with the simultaneous design of multi-fingered hands for manipulation tasks using first order kinematic specifications [8]. Recently, Robson and Soh [9] and Tan et al. [10] presented techniques for the design of planar passive multi-loop wearable devices capable of maintaining relative curvature task specification in one direction using first and second order kinematic task constraints.

In this paper, we focus on the manipulation problem and seek to generalize the above mentioned efforts to obtain a planar synthesis theory that incorporates multiple first and second order kinematic task specifications within a contact location that can output a variety of designs able to maintain the relative curvature task specifications for body manipulation in more than one direction in the vicinity of a contact pose. This extends the methodology in [8] by relaxing the constraints on the limb-length, expanding the design capability to two different directions in the vicinity of a contact pose and introducing more freedom for the designer to customize the resulting linkage device for human interactive applications.

2 Differential Kinematics of Body-Linkage Manipulation

When articulated limbs interact with objects or the environment itself, the constraints imposed by the relative motion of the object/environment need to be taken into account for stability. For example, a roll-slide manipulation of a body by the tip(s) of a kinematic chain(s) in two different directions in the vicinity of a body-linkage contact, can be defined by two sub-poses denoted by u and u' (see Fig. 1). The geometry of the kinematic chains cause the contact points to travel on the surface of the body as the configurations of the linkages change, moving from u to u' during manipulation. The rolling motion of the fingertip on the body requires that the velocities of the contact points on the fingers and

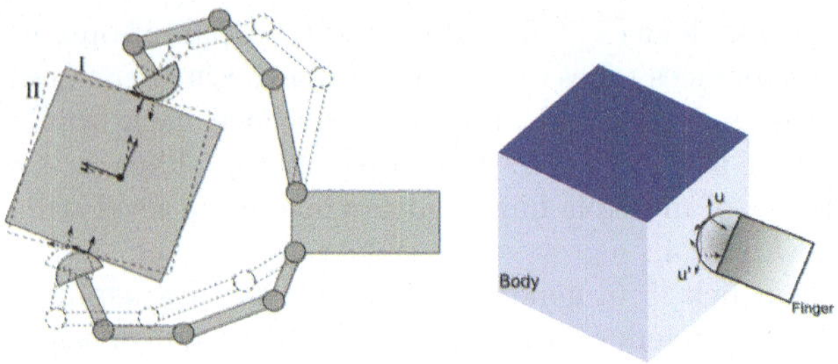

Fig. 1. Left: The planar rolling problem in clockwise and counterclockwise direction: the kinematic chains are designed and their motion synchronized so that the body is moved from configuration I to configuration II and back. Right: Rolling occurs in each of the two desired directions within the contact pose, defined between the two sub-poses u and u', respectively. Notice that u and u' may not be parallel as the contact surface may not be even.

the body to remain the same, denoting a higher pair. This means for the planar case as shown, its general mobility is seven. Thus, incorporating such linkage-body interaction constraints into the synthesis problem involve the application of differential geometry.

2.1 The Contact Velocity and Acceleration Constraints

Consider a body M guided by two objects in the environment with contact at A and B as shown in Fig. 2. If the contact is frictionless, then the contact forces $\mathbf{F}_A$ and $\mathbf{F}_B$ act normal to the contact plane, and are perpendicular to the direction of the velocity vectors $\dot{\mathbf{A}}$ and $\dot{\mathbf{B}}$. For quasi-static equilibrium grasp up to the first order, this requires the body and objects to have common velocity at the respective contact points. This is achieved by imposing velocity constraints on the body M such that the body rotates about its velocity pole V with an angular velocity $\dot{\phi}$. The velocity pole can be easily obtained as the intersection of the line of action of these two forces defined by

$$\begin{aligned} L_A: &\quad (\mathbf{x} - \mathbf{A}) \cdot [J]\mathbf{F}_A = 0, \\ L_B: &\quad (\mathbf{x} - \mathbf{B}) \cdot [J]\mathbf{F}_B = 0, \end{aligned} \tag{1}$$

where $\mathbf{x} = (x, y)$ are variable point coordinates and the cross product operation to perform a $90°$ rotation in the plane $[J] = \begin{bmatrix} 0 & -1 \\ 1 & 0 \end{bmatrix}$.

Similarly, for quasi-static equilibrium grasp up to the second order, it is solely dependent of the relative magnitude of the distance of the velocity pole from the point of contact ρ, and the radii of curvature r of the body or finger depending if the velocity pole is located on the body or finger side respectively. If $\rho = r$, we get a 1st order roll-slide motion that is also a 2nd order roll-slide motion. If $\rho > r$, the body separation occurs and if $\rho < r$, body penetration occurs.

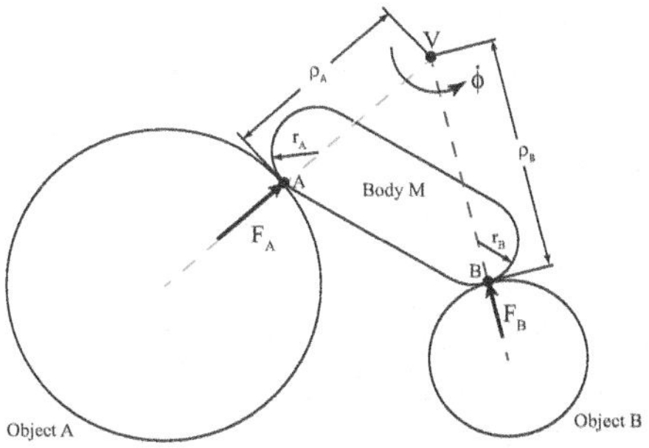

Fig. 2. The body is guided by two objects for each manipulation direction $n = 1, 2$ in the vicinity of the contact, such that the trajectories of $\mathbf{A}(t)_n$ and $\mathbf{B}(t)_n$ have known radii of curvature, R_{A_n} and R_{B_n} in the vicinity of a specified position.

A given contact point trajectory $\mathbf{P}$ and its curvature are related by [11]

$$\kappa = \frac{([J]\dot{\mathbf{P}}) \cdot \ddot{\mathbf{P}}}{|\dot{\mathbf{P}}|^3} \tag{2}$$

Hence, given the object curvature, Eq. (2) allows us to determine the required acceleration at the contact point and subsequently the required acceleration constraint at body M. For details, see [7].

3 Kinematic Synthesis of Two Planar Non-anthropomorphic Limbs for a Bi-directional Body Manipulation

In this section we present the simultaneous kinematic synthesis of two RPR mechanical limbs, able to interact with the body by roll-slide along the body's curvature within a contact location (defined by two different sub-poses) while keeping contact with the body at all times (see Fig. 3).

3.1 The Kinematics of Planar RPR Chain

The kinematic equations of a planar RPR chain equate the 3×3 homogeneous transformation $[T]$ between the end-effector and the base frame to the sequence of local coordinate transformations around the joint axes and along the links of the serial chain,

$$[T] = [G_1][Z(\theta_1)][X(R)][Z(\theta_2)][H_1] = [D_1]. \tag{3}$$

The parameters θ_i define the movement of each joint and R is the length of the sliding link, which is the sum of its slide s and its initial length R_0 as shown in Fig. 3. The transformation $[G_1]$ defines the position of the base G_1 relative to

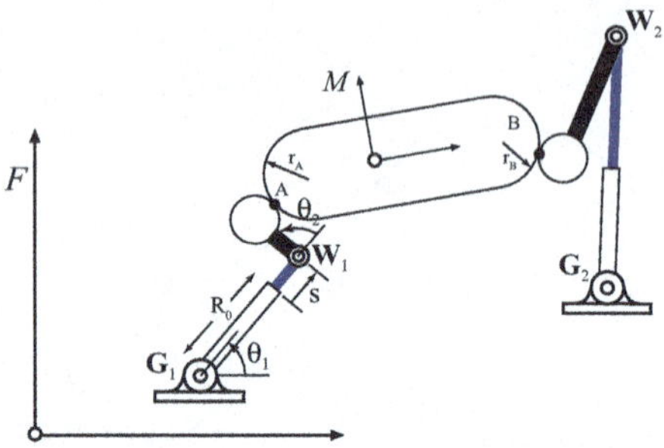

Fig. 3. Body M manipulated by two RPR kinematic chains, such that the trajectories of $\mathbf{A}(t)$ and $\mathbf{B}(t)$ have known radii of curvature, r_A and r_B in the vicinity of the contact, defined by two sub-poses.

the fixed frame, and $[H_1]$ locates the task frame M relative to the position of the moving pivot W_1.

The kinematics equation that relates the velocity task $[\Omega]$ to its angular and linear velocities around the joint axes and along its links is;

$$[\Omega] = \begin{bmatrix} \omega \ \mathbf{v} \\ 0 \ 0 \end{bmatrix} = [\dot{D}][D]^{-1}, \tag{4}$$

where $\omega = \begin{bmatrix} 0 & -(\dot{\theta}_1 + \dot{\theta}_2) \\ \dot{\theta}_1 + \dot{\theta}_2 & 0 \end{bmatrix}$ and $\mathbf{v} = -\omega\mathbf{d} + \dot{\mathbf{d}}$ denote the end-effector angular and linear velocity vector respectively.

Similarly, the kinematics equation for the acceleration task $[\Lambda]$ equates the joint angular and linear accelerations:

$$[\Lambda] = \begin{bmatrix} \alpha \ \mathbf{a} \\ 0 \ 0 \end{bmatrix} = [\ddot{D}][D]^{-1}, \tag{5}$$

where $\alpha = \begin{bmatrix} -(\dot{\theta}_1 + \dot{\theta}_2)^2 & -(\ddot{\theta}_1 + \ddot{\theta}_2) \\ \ddot{\theta}_1 + \ddot{\theta}_2 & -(\dot{\theta}_1 + \dot{\theta}_2)^2 \end{bmatrix}$ and $\mathbf{a} = -\omega^2\mathbf{d} - \dot{\omega}\mathbf{d} + \ddot{\mathbf{d}}$ denote the end-effector angular and linear acceleration vector respectively.

3.2 The Design Equations

Assume that constraint forces and contact curvatures within the contact location, defined by two different sub-poses compatible with two different body manipulation directions are specified. This means that the two sub-poses, as well as two velocities and two accelerations of M are defined in the vicinity of the contact. The geometry of the RPR chain satisfies the constraint equation

$$(\mathbf{W} - \mathbf{G}) \cdot (\mathbf{W} - \mathbf{G}) = R^2. \tag{6}$$

where $\mathbf{W} = (x, y, 1)$ denotes its moving pivot and $\mathbf{G} = (u, v, 1)$ its ground pivot in the fixed frame F. Two derivatives of this equation yield the additional constraint equations for the velocities and accelerations specifications such that

$$\dot{\mathbf{W}} \cdot (\mathbf{W} - \mathbf{G}) = R\dot{R},$$
$$\ddot{\mathbf{W}} \cdot (\mathbf{W} - \mathbf{G}) + \dot{\mathbf{W}} \cdot \dot{\mathbf{W}} = \dot{R}^2 + R\ddot{R}. \tag{7}$$

Note that in the last two equations the changing link length R is related to the known initial link length R_0, the stiffness coefficient k, the slide distance s, the slide velocity $\dot{s}$ and the slide acceleration $\ddot{s}$. These are free design parameters and any three of which the designer needs to specify for the successful synthesis of RPR chain. The design parameters satisfy the following:

$$R = R_0 + ks,$$
$$\dot{R} = k\dot{s},$$
$$\ddot{R} = k\ddot{s}. \tag{8}$$

Using Eqs. (6) to (8), six design equations can be formulated for the synthesis of the RPR chain by determining the trajectory of the moving frame M such that

$$\mathscr{P}_1: \quad 0 = (\mathbf{W} - \mathbf{G}) \cdot (\mathbf{W} - \mathbf{G}) - R_0^2, \tag{9}$$
$$\mathscr{V}_1: \quad 0 = [\Omega_1]\mathbf{W} \cdot (\mathbf{W} - \mathbf{G}), \tag{10}$$
$$\mathscr{A}_1: \quad 0 = [\Lambda_1]\mathbf{W} \cdot (\mathbf{W} - \mathbf{G}) + [\Omega_1]\mathbf{W} \cdot [\Omega_1]\mathbf{W}, \tag{11}$$
$$\mathscr{P}_2: \quad 0 = ([T_{12}]\mathbf{W} - \mathbf{G}) \cdot ([T_{12}]\mathbf{W} - \mathbf{G}) - R^2, \tag{12}$$
$$\mathscr{V}_2: \quad 0 = [\Omega_2]\mathbf{W} \cdot (\mathbf{W} - \mathbf{G}) - R\ddot{R}, \tag{13}$$
$$\mathscr{A}_2: \quad 0 = [\Lambda_2]\mathbf{W} \cdot (\mathbf{W} - \mathbf{G}) + [\Omega_2]\mathbf{W} \cdot [\Omega_2]\mathbf{W} - \dot{R}^2 - R\ddot{R}, \tag{14}$$

where $[T_{12}] = [T_2][T_1]^{-1}$ denotes the relative displacement of the body as it moves from M_1 to M_2, while $[\Omega_i]$ and $[\Lambda_i]$, $i = 1, 2$, correspond to the body's velocity and acceleration matrices in each direction respectively.

4 Simultaneous Design of Two Non-anthropomorphic Fingers for Body Manipulation in Two Directions

We start by obtaining the required manipulation movement $[T_1(t)]$ and $[T_2(t)]$, that has the property that the two contact points $\mathbf{A}$ and $\mathbf{B}$ at the rigid body, shown in Figure 1, have trajectories $\mathbf{A}_1(t)$ and $\mathbf{B}_1(t)$ associated with one manipulation sub-pose and trajectories $\mathbf{A}_2(t)$ and $\mathbf{B}_2(t)$ related to a second manipulation sub-pose. Each of the sub-poses are specified in the vicinity of the fingertip-body contact location and is related to one of the desired manipulation directions. Next, we obtain the velocity and acceleration constraints in each of the two sub-poses using the contact geometry discussed in Sect. 2.1. Thus the required PVA task for the first sub-pose are $P_1 = (-35°, 61, -68)$, $V_1 = (-3°/s, -1, -3)$, $A_1 = (-60°/s^2, 5, 5)$ and for the second sub-pose are $P_2 = (-78°, 66, -73)$, $V_2 = (6°/s, -6, -5)$, $A_2 = (-115°/s^2, -10, 10)$.

The synthesis equations (14) for $k = 1$, $\dot{s} = 0$, $\ddot{s} = 0$ yield 16 real solutions for the designer to choose from. Among these, 8 solutions are with a negative initial length R_0 and were eliminated. This results from the quadratic nature of R in the formulation of the synthesis equations. The two chosen real solutions out of the possible eight for the simultaneous synthesis of the two RPR fingers for roll-slide body manipulation between two sub-poses M_1 and M_2 in the vicinity of the contact are $G_1 = (62.7, -68.3)$, $W_1 = (1.1, 0.02)$, $R_{0_1} = 2.8$, $s_1 = 2.0$ and $G_2 = (56.7, -75.5)$, $W_2 = (-2.3, 8.4)$, $R_{0_2} = 1.5$, $s_2 = 0.6$. The solutions are chosen due to the compactness of the RPR limbs at its initial configuration and its collision free behavior with the manipulated object as it moves between the two sub-poses. Figure 4 shows the resulting simultaneous design and the CAD drawing of the two mechanical fingers, able to roll-slide along the body's curvature between the specified two sub-poses.

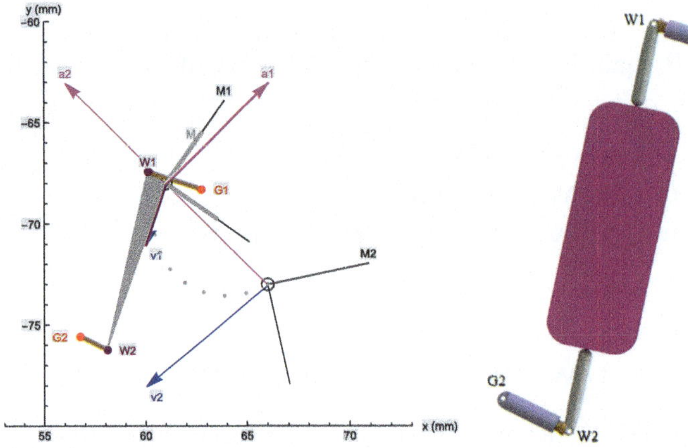

Fig. 4. Left: The two chosen RPR fingers $\mathbf{G}_1\mathbf{W}_1$ and $\mathbf{G}_2\mathbf{W}_2$. The origin of the fixed frame is at (0, 0). The dotted lines denotes the trajectory of the resulting manipulating device as it move between the two sub-poses. Right: CAD drawing of the resulting device.

5 Summary

This paper considers the planar kinematic synthesis with contact and curvature constraints for the simultaneous design of non-anthropomorphic serial mechanical chains for object/environment interaction in multiple directions, a critical focus not fully captured in existing design methodologies. It was shown how the contact direction and curvature constraints can be transformed into conditions on the velocity and acceleration of certain points at the tip of the kinematic serial chain using the task geometry to satisfy a desired bi-directional interaction task in the vicinity of finger-body contact location.

Acknowledgements. We acknowledge the support of NSF Award Id#1404011, Id#1751770 and SUTD-MIT International Design Center. The content is solely the authors' responsibility.

References

1. Boy, G.A.: The Handbook of Human-Machine Interaction: A Human-Centered Design Approach. CRC Press, Taylor & Francis Group, Florida (2017)
2. Zoss, A., Kazerooni, H., Chu, A.: On the mechanical design of the berkeley lower extremity exoskeleton (BLEEX). In: Proceedings of the IEEE/RSJ International Conference on Intelligent Robots and Systems (IROS), pp. 3465–3472 (2005)
3. Chiri, A., Vitiello, N., Giovacchini, F., Roccella, S., Vecchi, F., Carrozza, M.C.: Mechatronic design and characterization of the index finger module of a hand exoskeleton for post-stroke rehabilitation? ASME/IEEE Trans. Mechatron. **17**(5), 884–894 (2012)
4. Grebenstein, M., Chalon, M., Hirzinger, G., Siegwart, R.: A method for hand kinematics design. 7 billion perfect hands? In: Proceedings of the 1st ICABB: International Conference on Applied Bionics and Biomechanics, Venice, Italy (2010)
5. Rimon, E., Burdick, J.W.: A configuration space analysis of bodies in contact - I: 1-st order mobility. Mech. Mach. Theory **30**(6), 897–912 (1995)
6. Rimon, E., Burdick, J.W.: A configuration space analysis of bodies in contact - II: 2-nd order mobility. Mech. Mach. Theory **30**(6), 913–928 (1995)
7. Robson, N., McCarthy, J.M.: Kinematic synthesis with contact direction and curvature constraints on the workpiece. In: Proceedings of the ASME International Design Engineering Technical Conferences, Las Vegas, NV, vol. 8, pp. 581–588 (2007)
8. Robson, N., Ghosh, S., McCarthy, J.M.: Simultaneous geometric design of mechanical fingers for coordinated movement. In: 14th IFToMM World Congress, Taipei, Taiwan (2015)
9. Robson, N., Soh, G.S.: Geometric design of eight-bar wearable devices based on limb physiological contact task. Mech. Mach. Theory **100**, 358–367 (2016)
10. Tan, G.R., Robson, N., Soh, G.S.: Motion generation of passive slider multiloop wearable hand devices. J. Mech. Robot. **9**, 041011–1 (2017)
11. Bottema, O., Roth, B.: Theoretical Kinematics, North-Holland Series in Applied Mathematics and Mechanics, vol. 24. North-Holland Publishing Company, Amsterdam (1979)

A Novel Rotary Positioner with Single Drive: Structural Analysis and Kinematic Design

Alexey Fomin[1(✉)] and Victor Glazunov[2]

[1] Siberian State Industrial University, Novokuznetsk 654007, Russia
alexey-nvkz@mail.ru
[2] Mechanical Engineering Research Institute of the RAS, Moscow 101990, Russia
vaglznv@mail.ru

Abstract. The presented study provides a novel kinematic design of rotary positioner with single drive. The proposed mechanism includes six legs supporting a platform that reproduces different 3D trajectories. Kinematic design of the positioner has been developed as a combination of spatial chain with zero-DoF and a planar linkage with mobility one. Planar linkage is set within a fixed circular guide and serves as a movable base with six kinematic chains ending with carriages that are set under each leg. Motions of the rotary positioner's platform are predefined and controlled by single rotational drive set in the center of the circular guide. Novel kinematic design allows changing trajectories of the platform by variation of only one link length in each kinematic chain of the base linkage. The developed mechanism can be used as a movable platform for spatial orientation of diverse objects, for carrying out cyclic tasks requiring motions on predefined trajectories, in constructions of shaking tables.

Keywords: Mechanism · Parallel structure · Degree-of-freedom
Structural analysis · Kinematic design

1 Introduction

Parallel mechanisms have variety of advantages in comparison with linear structure mechanisms. Such advantages include high rigidity, uniform forces distribution throughout a mechanism's structure, higher positioning precision, increased lift capacity, a possibility of setting actuators on fixed links to reduce inertia forces of movable links, provision of high velocities and accelerations for end-effectors and other important performance characteristics that directly influence on the effective implementation of technological operations [1–6].

Parallel mechanisms can be designed with different number of drives to perform various 3D trajectories of their end-effectors. In the proposed study we address to parallel mechanisms with circular guide. They are widely used for precise positioning, including micropositioning, and allow reaching full rotation of an end-effector around vertical axis [7–12]. In these mechanisms an additional rotary stage has been eliminated to reduce manufacturing charges and also reduce the number of actuators. However, it is still necessary to have up to six actuators that move driving links, carriages, along the

J. Lenarcic and V. Parenti-Castelli (Eds.): ARK 2018, SPAR 8, pp. 364–371, 2019.
https://doi.org/10.1007/978-3-319-93188-3_42

circular guide to orient the end-effector in space. Additional reduction of number of actuators could simplify a control of such mechanisms and lead to further diminution in their value. The number of actuators can be reduced up to one, while various 3D trajectories of end-effectors can be realized in these mechanisms. Mechanisms designed under this idea are shown in [13, 14] as the shaking table, in [15] as the mechanism for passive motion at the human tibiotalar joint, in [16] as the hexapod for micropositioning, in [17, 18] as walking mechanisms with single motor.

The proposed research aims at the development of novel parallel structure mechanism having single drive and circular guide. Such a mechanism can be designing by merging of a group of links with zero degrees-of-freedom (DoF) and a linkage with mobility one. In this case, 3D motions of an end-effector become controllable by a single drive. These properties can be useful for spatial orientation and moving of different objects along with predetermined trajectories, used in designing of shaking tables for earthquake simulations when the earth's surface can perform various motions including transverse and screw motions, used in development of medical devices for posttraumatic rehabilitation. Consider the novel kinematic design of the proposed rotary positioner in Sect. 2.

2 Kinematic Design

The proposed rotary positioner with single drive is shown in Fig. 1. This mechanism includes circular guide set on the fixed link. The platform serves as an end-effector that is supported by six legs.

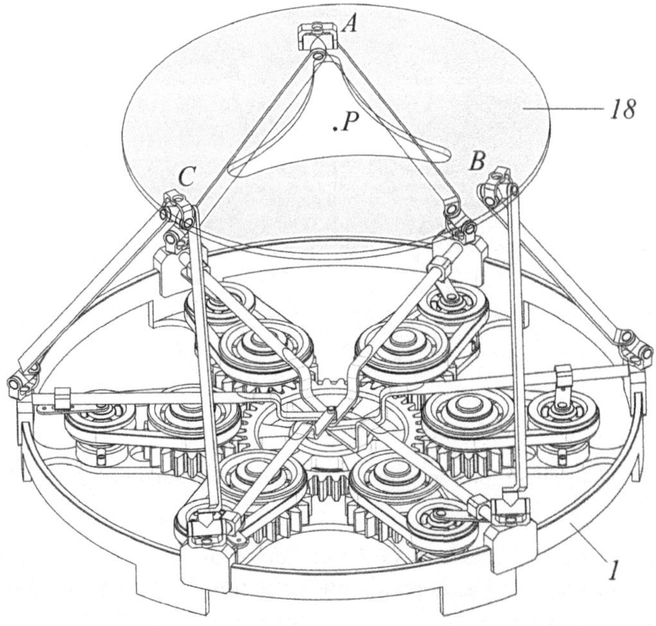

Fig. 1. Kinematic scheme of novel rotary positioner with single drive.

The rotary positioner includes two kinematic chains: one consists of three pairs of chains of links 10–17 and platform 18 separately shown in Fig. 2, another is the planar base linkage that is set within the circular guide separately shown in Fig. 3.

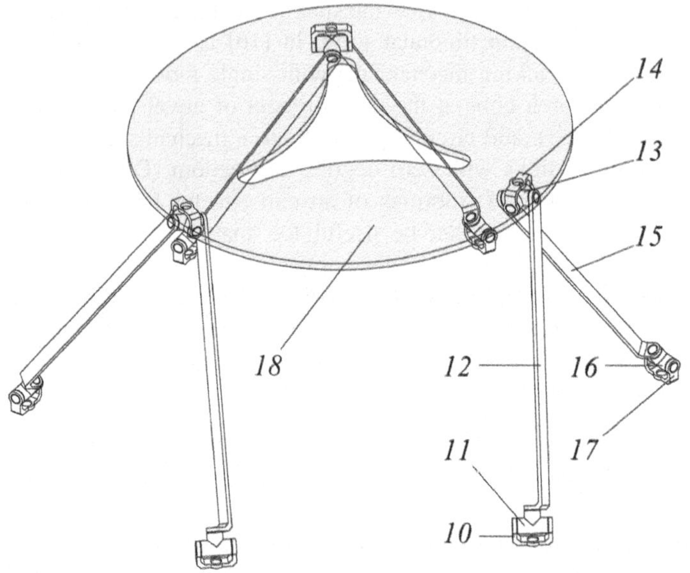

Fig. 2. Spatial zero-DoF kinematic chain with six legs and rotational joints.

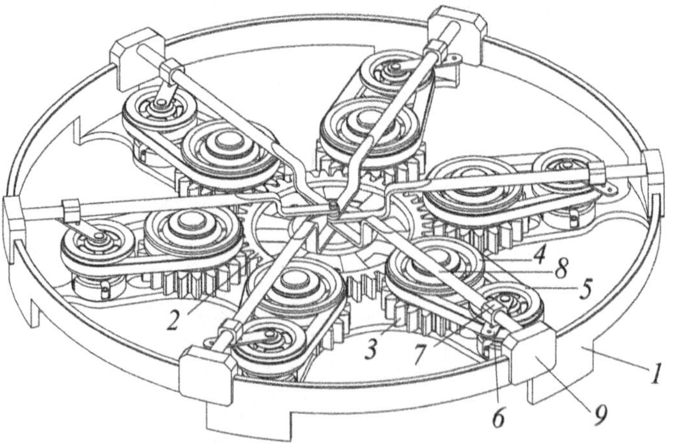

Fig. 3. Planar base linkage with circular guide and single rotational drive.

The planar linkage includes wheel 2 that serves as the driving link and six kinematic chains of links 3–9. The separate chain of links 3–9 is presented in Fig. 4. It includes intermediate wheel 3 that is rigidly connected to intermediate disk 4, and has a gear kinematic pair with driving wheel 2. Intermediate disk 4 is connected with driven disk 5 by means of a flexible coupling. Driven disk 5 is rigidly connected with crank 6 that forms rotational kinematic pair with slide block 7 moving along rocker 8 by means of prismatic kinematic pair. Rocker 8 is rotationally connected with fixed link 1 and is rigidly connected with carriage 9 forming rotational kinematic pair with the circular guide. The axes of rotational kinematic pairs between links 1–2, 1–3 (1–4), 1–5, 1–8 (1–9) are perpendicularly oriented to the circular guide.

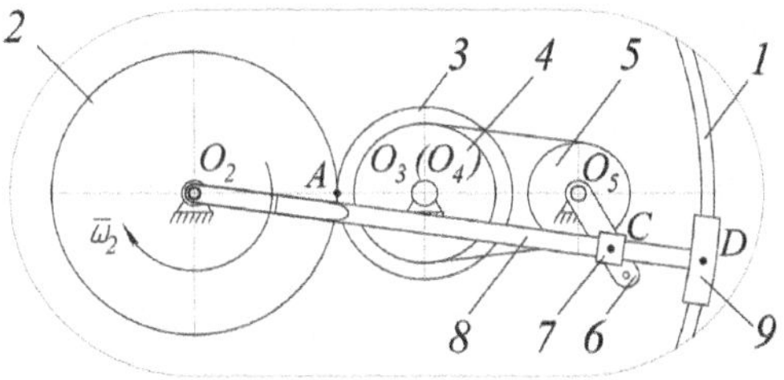

Fig. 4. Connection of driving wheel 2 with one of six kinematic chains of links 3-9.

Carriages 9 in each kinematic chain can have absolutely different displacement amplitudes, which can be varied by changing crank length O_5C. Variable length of crank 6 is achieved by setting holes on it with the opportunity to replace slide block 7 on chosen distances from rotational center O_5. Carriages 9 move each leg on predefined distances orienting platform 18 in space.

3 Mobility Analysis

The mobility of the developed rotary positioner can be calculated as a sum of mobilities of kinematic chains shown in Figs. 2 and 3. The mobility of the spatial kinematic chain can be calculated by Malishev's formula written as

$$W_0 = 6n - 5p_5 - 4p_4 - 3p_3 - 2p_2 - p_1, \tag{1}$$

where W_0 is a mobility, defining number of DoF of a spatial kinematic chain without imposed constraints, n is a number of movable links in a kinematic chain, p_5, p_4, p_3, p_2 and p_1 are numbers of one-, two-, three-, four- and five-DoF kinematic pairs [19, 20]. The kinematic chain from Fig. 2 includes n = 25 and p_5 = 30, so its mobility will be equal to $W_0 = 0$.

Since the base linkage shown in Fig. 3 is planar, its mobility can be calculated by Chebyshev's formula having the following view

$$W_3 = 3n - 2p_5 - p_4, \tag{2}$$

where W_3 is a mobility, defining number of DoF of a planar kinematic chain with three imposed constraints [19, 20]. The planar base linkage includes n = 25, p_5 = 31 and p_4 = 12. According to (2) its mobility is equal to one ($W_3 = 1$). The obtained result means that to have predefined motions of carriages 9, it needs to give one input motion.

Such a motion is the rotational and applied to wheel 2 made as a cylindrical gear with central vertical axis of rotation.

Thus, setting of the spatial kinematic chain with six legs on carriages 9 will provide single active DoF for the rotary positioner. The motion of platform 18 becomes dependent on a rotation of driving wheel 2. Trajectories reproduced by platform 18 may be varied by the change of crank length O_5C (Fig. 4).

4 Kinematic Analysis

In this section we address to the kinematic analysis of the base linkage shown in Fig. 3. The aim here is to analyze movements of carriages 9. It is advisable to investigate only one of the six kinematic chains shown in Fig. 4 as far as five remaining chains can be analyzed in a similar algorithm. Given the fact that the input motion is realized by the rotation of driving wheel 2, we accept its rotation speed as n_2 = 30 rpm. So, angular velocity will be equal to $w_2 = 3.14\ s^{-1}$. The dimensions of all links in this kinematic chain are given parameters. Their values are shown in Table 1.

Table 1. Geometrical dimensions of links of the base linkage.

Parameter	Value, m
Diameter of circular guide 1, D	0.800
Diameter of driving wheel 2, d_2	0.248
Diameter of intermediate wheel 3, d_3	0.130
Diameter of intermediate disk 4, d_4	0.110
Diameter of driven disk 5, d_5	0.090
Length of crank 6, l_6	0.050

Turn angular velocity w_2 along the clockwise direction. Velocity of point A is common for both links 2 and 3 ($V_{A2} = V_{A3}$), it can be calculated as $w_2 \cdot d_2/2$ and is equal to 0.389 m/s. Angular velocity of intermediate wheel 3 (w_3) is equal to angular velocity of intermediate disk 4 because of their rigid connection to each other ($w_3 = w_4$), it can be calculated as $2V_A/d_3$ and is equal to 5.99 s^{-1}. Then it is possible to find angular velocity of driven disk 5 (w_5), it is calculated as $w_4 \cdot d_4/d_5$ and is equal to 7.321 s^{-1}. Angular velocity w_5 is equal to w_6 because links 5 and 6 are rigidly connected to each other. To find velocity of carriage 9, particularly velocity of point D, it needs to find velocity of point C. This point accumulates velocities of links 6, 7 and 8, so there will be velocities V_{C6}, V_{C7} and V_{C8}. Velocity V_{C6} is equal to V_{C7}, because crank 6 and slide block 7 are connected by rotational joint. Velocity V_{C6} can be calculated as $w_6 \cdot l_6$ and is equal to 0.366 *m/s*. Velocity V_{C8} can be found graphically from the following equation system

$$\begin{cases} V_{C8} = V_{C7} + V_{C8C7},\ V_{C8C7} \| CO_2, \\ V_{C8} = V_{O2} + V_{C8O2},\ V_{C8O2} \perp CO2. \end{cases} \quad (3)$$

Figure 5 provides graphical solution of (3) in the form of velocity diagram. Its scale μ_V is calculated as V_{C6}/pc_6 and is equal to 0.001 (m/s)/mm. Velocity V_{C8} is calculated as $\mu_V{\cdot}pc_8$ and equal to 0.225 m/s. After that we can find velocity of point D (V_D) that belongs to carriage 9, as $V_{C8}{\cdot}DO_2/CO_2$ = 0.275 m/s. Angular velocity w_8 of rocker 8 is calculated as V_D/DO_2 = 0.688 s^{-1}.

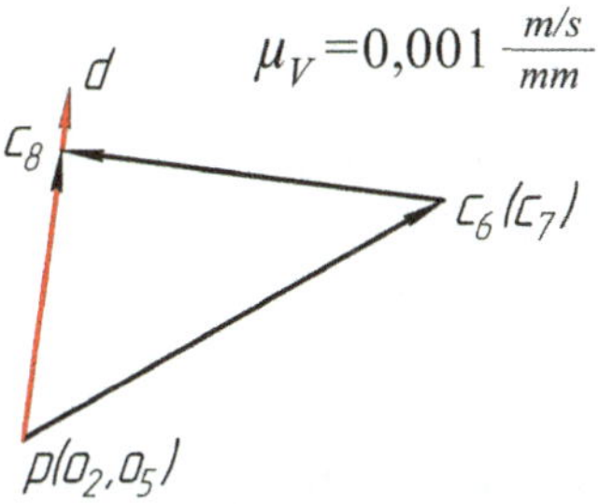

Fig. 5. Graphical determination of velocity of carriage 9.

Kinematic analysis of other five kinematic chains of the base linkage is carried out in the same algorithm. However tilt angles of the links in kinematic chains and their orientation inside of the circular guide should be considered.

Discuss the possibility to reproduce different 3D trajectories by the developed mechanism. Each kinematic chain shown in Fig. 4 includes crank 6 having three holes designed for removals of revolute join 6–7 and changing rotation radius of crank 6. Such removals allow reproducing different 3D trajectories of end-effector 18. The exact number of these trajectories can be calculated through the following formula

$$R = k^m, \tag{4}$$

where R is the number of 3D trajectories reproduced by platform 18 (or the number of possible removals of joints 6–7 at cranks 6), k is the number of holes at each crank, m is the number of cranks. For the developed hexapod k = 3 and m = 6, so parameter R is equal to 729.

Table 2 provides coordinate variation of 3D trajectories of platform's central point P (Fig. 1) in base coordinates for three cases: when joints 6–7 are set in 1st, 2nd and 3rd holes of each crank (1st hole defines shortest length of crank 6, when 3rd hole defines its longest length). Base coordinates are placed in the center of the circular guide. Point P_1 with coordinates X_{P1}, Y_{P1}, Z_{P1} matches with minimal rotation radius of crank 6, when point P_3 with coordinates X_{P3}, Y_{P3}, Z_{P3} matches with maximal rotation radius. For position measurements of points P_1, P_2 and P_3 lengths of legs have been assumed equal to lengths AB, BC and CA forming equiangular triangle on platform 18 (Fig. 1). These lengths are equal to 0.375 m.

Table 2. Coordinate variation of the platform's trajectory with varied lengths of cranks.

X_{P1}, *m*	−0,006	−0,010	−0,014	−0,015	−0,011	−0,002	0,006	0,010	0,006	−0,001
Y_{P1}, *m*	−0,051	−0,042	−0,024	−0,006	0,003	−0.001	−0,012	−0,027	−0,041	−0,051
Z_{P1}, *m*	0,470	0,473	0,480	0,487	0,493	0,494	0,493	0,488	0,480	0,473
X_{P2}, *m*	−0,008	−0,013	−0,017	−0,020	−0,016	−0,003	0,013	0,019	0,011	−0.001
Y_{P2}, *m*	−0,048	−0,042	−0,028	−0,014	−0,006	−0,004	−0,011	−0,023	−0,035	−0,045
Z_{P2}, *m*	0,465	0,470	0,480	0,491	0,498	0,498	0,493	0,485	0,476	0,468
X_{P3}, *m*	−0,008	−0,014	−0,021	−0,027	−0,022	−0,003	0,020	28.79	0,017	0,001
Y_{P3}, *m*	−0,048	−0,042	−0,027	−0,014	−0,007	−0,006	−0,012	−23.79	−0,035	−0,045
Z_{P3}, *m*	0,461	0,466	0,478	0,492	0,500	0,501	0,494	484.82	0,474	0,464

5 Conclusions and Future Works

A new rotary positioner with hexapod's structure and circular guide has been presented in this study. The mechanism allows reproducing various 3D trajectories of its end-effector with the use of single rotational drive placed in the center of the circular guide. The novel kinematic design of this mechanism is based on the combination of the spatial chain and the planar linkage. Such combination provided mobility one for the whole mechanism.

The developed mechanism can be used as a movable platform for a spatial orientation of diverse objects, for performing cyclic operations that require movements on predetermined trajectories. The mechanism can be also used in medical equipment for patient rehabilitation or in shaking tables with one drive for earthquake simulations when the earth's surface can perform various motions including transverse and screw motions. 3D trajectories of the mechanism's end-effector are variable and depend on the length of cranks, rotating within the circular guide.

The future works will include singularity analysis during which singularities of the first type (when the projection of hexapod's leg on horizontal plane will pass through the center of the circular guide) and the second type (when wrenches passing through all hexapod legs are linearly dependent) will be discussed. The future works will also include working zone analysis for the developed hexapod. End positions of the platform will be set and the lengths of cranks will be calculated. Limit values will describe a boundary of the working zone.

Acknowledgments. The reported study has been funded by the Ministry of Education and Science of the Russian Federation according to the research project No. 9.12794.2018/12.2.

References

1. Ceccarelli, M.: Fundamentals of Mechanics of Robotic Manipulation. Springer Science & Business Media, Netherland (2004)
2. Glazunov, V.A., Chunichin, A.: Development of mechanisms of parallel structure. J. Mach. Manuf. Reliab. **3**, 37–43 (2014)

3. Kong, X., Gosselin, C.M.: Type Synthesis of Parallel Mechanisms. Springer, Heidelberg (2007)
4. Carricato, M., Parenti-Castelli, V.: A family of 3-DOF translational parallel manipulators. J. Mech. Des. **125**(2), 302–307 (2014)
5. Kheylo, S.V., Glazunov, V.A., Shirinkin, M.A., Kalendaryov, A.V.: Possible applications of mechanisms of parallel structure. J. Mach. Manuf. Reliab. **5**, 19–24 (2013)
6. Arakelian, V.H., Smith, M.R.: Design of planar 3-DOF 3-RRR reactionless parallel manipulators. Mechatronics **18**(10), 601–606 (2008)
7. Yau, C.L.: (Hexel Corporation, US): Systems and methods employing a rotary track for machining and manufacturing. US Patent No. 6196081 (2001)
8. Janabi-Sharifi, F., Shchokin, B.: A rotary parallel manipulator: modelling and workspace analysis. In: Proceedings of the International Conference on Robotics and Automation, pp. 3671–3677. IEEE, New Orleans (2004)
9. Rashoyan, G.V., Lastochkin, A.B., Glazunov, V.A.: Kinematic analysis of a spatial parallel structure mechanism with a circular guide. J. Mach. Manuf. Reliab. **44**(7), 54–60 (2015)
10. Coulombe, J., Bonev, I.A.: A new rotary hexapod for micropositioning. In: Proceedings of the International Conference on Robotics and Automation, pp. 877–880. IEEE, Karlsruhe (2013)
11. Liu, X.-J., Wang, J., Gao, F., Wang, L.-P.: Mechanism design of a simplified 6-DOF 6-RUS parallel manipulator. Robotica **20**(1), 81–91 (2002)
12. Shchokin, B., Janabi-Sharifi, F.: Design and kinematic analysis of a rotary positioner. Robotica **25**, 75–85 (2007)
13. Borozna, A.G., Glazunov, V.A., Zhuk, V.P., Miodushevsky, P.V., Chutaev, A.A.: 6-coordinate shaking table. RU Patent No. 1753322 (1992)
14. Glazunov, V., Kraynev, A.: Design and singularity criteria of parallel manipulators. In: Proceedings of Romansy 16. CISM Courses and Lectures, vol. 487, pp. 15–22. Springer, Wien (2006)
15. Franci, R., Parenti-Castelli, V., Belvedere, C., Leardini, A.: A new one-DOF fully parallel mechanism for modelling passive motion at the human tibiotalar joint. J. Biomech. **42**(10), 1403–1408 (2009)
16. Fomin, A., Glazunov, V., Didenko, V., Paik J.: The design of a new rotary hexapod with a single active degree of freedom. In: Proceedings of the Fifth International Symposium on Robotics & Mechatronics. IFToMM, Sydney, Australia (2017)
17. Rooney, T., Pearson, M., Welsby, J., Horsfield, I., Sewell, R., Dogramadzi, S.: Artificial active whiskers for guiding underwater autonomous walking robots. In: The 14th International Conference on Climbing and Walking Robots and the Support Technologies for Mobile Machines. WSPC, Paris, France (2011)
18. Nansai, S., Elara, M.R., Iwase, M.: Dynamic analysis and modeling of Jansen mechanism. Procedia Eng. **64**, 1562–1571 (2013)
19. Dvornikov, L.T., Fomin, A.S.: Development of the basic conditions for division of mechanisms into subfamilies. Procedia Eng. **150**, 882–888 (2016)
20. Fomin, A., Dvornikov, L., Paik, J.: Calculation of general number of imposed constraints of kinematic chains. Procedia Eng. **206**, 1309–1315 (2017)

A New Mechanism for the Deployment of Modular Solar Arrays: Kinematic and Static Analysis

Stefano Seriani[1,3], Paolo Gallina[1], Lorenzo Scalera[2], Alessandro Gasparetto[2(✉)], and Armin Wedler[2]

[1] University of Trieste, Trieste, Italy
{sseriani,pgallina}@units.it
[2] University of Udine, Udine, Italy
scalera.lorenzo@spes.uniud.it, gasparetto@uniud.it, armin.wedler@dlr.de
[3] Robotik und Mechatronik Zentrum, Deutsches Zentrum für Luft- und Raumfahrt e.V. (DLR), Oberpfaffenhofen, Germany

Abstract. This paper presents a new solar arrays deployment mechanism for space applications. It consists of a modular kinematic structure, which is operated by a single cable (1 DoF). Compared to traditional methods, this mechanism has the advantage of being reversible in the movement. Kinematic analysis of the mechanism is carried out in this work, as well as static analysis. They allow to define the main actuation parameters such as cable pull tension and spring stiffness. Moreover, suitable values for the mechanism parameters are computed by means of a dedicated algorithm.

Keywords: Solar arrays deployment · Cable-based mechanisms
Foldable mechanisms

1 Introduction

In planetary explorations, the possibility of automating the deployment of solar arrays stored in transportable modules through robotic systems is of primary importance [1]. Space-faring systems have benefited from solar panel arrays since early in the exploration of space, the first notable example being Vanguard 1. Soon after, the first examples of deployable solar panels (Explorer 6 and Pioneer 5) demonstrated the advantages in terms of power output and small footprint [2]. Today, many systems have become available, especially for orbital systems. However, the large majority are non-reversible mechanisms, which means that the deployment operation cannot be undone [3]; reversible deployment systems [4] have been studied but seldom implemented in real missions since there is limited need for reconfiguration in orbit; it is however of paramount importance in mobile or modular planetary architectures. For example, the Exomars rover due to the planet Mars in the early 2020 is equipped with a 4-axis solar panel array [5]; this allows the rover to orient the panels to increase light incidence, thus enhancing power production. This approach, however, is complex as it

J. Lenarcic and V. Parenti-Castelli (Eds.): ARK 2018, SPAR 8, pp. 372–379, 2019.
https://doi.org/10.1007/978-3-319-93188-3_43

involves a large number of actuators and has thus multiple points of failures; we propose a lightweight mechanism which takes advantage of a single actuator, while at the same time providing deployment reversibility. Since the top surface area of the module is comparably small ($\approx 0.1m^2$), and since power production is directly proportional to surface area, an efficient solar power deployment mechanism is essential. Four basic configurations are approachable: (i) Two-panel, storable array; (ii) Four-panel, foldable, storable array; (iii) Sun-tracking, foldable, storable array; (iv) Multi-panel, ground-deployed, foldable, storable array.

Conceptual illustrations are visible in Fig. 1, (a), (b), (c) and (d) respectively; in this paper, focus is set on the fourth configuration. If the area of the solar panels is taken into consideration, with a carrying capacity of the robot of approximately 20 kg, the values shown in Table 1 are obtained. A panel efficiency of 30% was considered [6], as well as a maximum solar irradiance of 1361 W/m^2 on the Moon [7] and around 300 W/m^2 on Mars [8], depending on the conditions of the atmosphere. The mass of the panels was estimated based on the data of NASA's Juno spacecraft [9]. Optimal, as well as average values are given, where the first refers to the sun at the zenith, while the former relates to a 45° solar incidence. The multi-panel configuration ensures the highest power production of the considered arrays, while at the same time being completely passive after deployment, contrary to, for example, the sun-tracking array. Furthermore, the stability of the module is not hindered by the panels, as is the case with the first two configurations.

Table 1. Achievable surface, mass and optimal power output of the four modules configurations.

	Surface	Mass	Power output [W]			
			Moon		Mars	
	[m^2]	[Kg]	Optimal	Average	Optimal	Average
i	0.25 ÷ 0.3	1.5	122	86	27	19
ii	0.5 ÷ 1.2	4.7 ÷ 8	490	346	108	76
iii	0.4 ÷ 0.6	>8	245	245	54	54
iv	1 ÷ 2.7	6 ÷ 12	1102	779	242	171

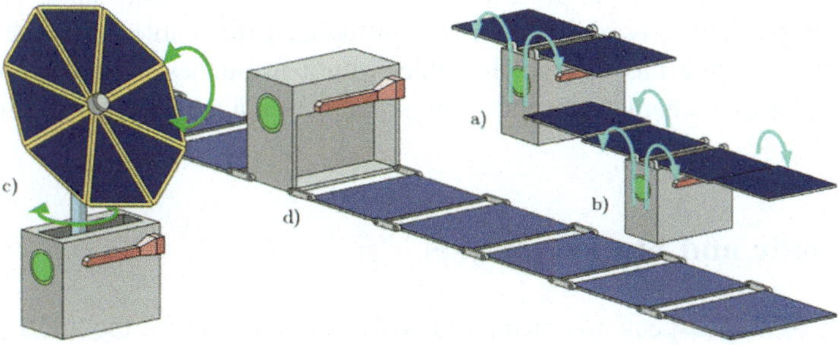

Fig. 1. Solar panel-equipped payload modules configurations.

As a mechanism, the multi-panel solar array shown in Fig. 2, is a planar serial chain which has a set of constraints on the rotations of each link (90°). This chain is loaded with a spring-based mechanism when folded – or stored – inside the module; during the unfolding and folding procedures a cable is used to actuate each link. The cable is wound around the entire array and this allows the complete "roll-up" to a stored configuration.

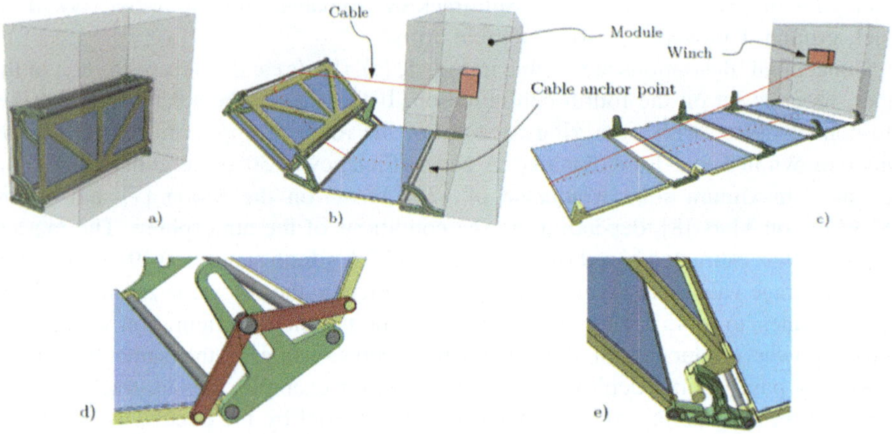

Fig. 2. A 3D model of the multi-panel, ground-deployed, foldable, storable solar array. Green indicates the hinges assemblies. In (a) the array is shown in its stored phase. In (b) the deployment of the second panel is shown, whereas in (c) the fully deployed array is visible. In (d) and (e) the auxiliary mechanisms are shown, respectively the double quadrilateral and the lock guide.

The hinged segments are necessary in order to provide the necessary spacing between the panels, when in the stored position. These segments are deployed by means of the double quadrilateral mechanism highlighted in Fig. 2d. In Fig. 2e, a set of curved guides is shown. The purpose of the guides is to constrain the folded panels in the stored position. Each panel is connected to the next through a mechanical linkage composed of 6 revolution joints, a prismatic joint and a spring. A single cable actuates all the mechanisms. In literature, there are few examples of mechanical linkages operated by cables [10–12].

The paper is structured as follow: in Sect. 2, Kinematic and static analysis are performed. In particular, two conditions are introduced (the cable has to be always in tension and the system has to "remain folded" until it touches the ground during the deployment phase) to characterize the springs; in Sect. 3, a real case has been evaluated and discussed.

2 Kinematic and Static Analysis

There are three main aspects to consider in describing a model for the multi-panel solar array: the interaction between the cable and the segments of the serial chain; the double quadrilateral mechanism that provides the moments for the deployment through a spring-based action; the consequences of friction in the lock guides and the constraint that the guides provide.

Let us consider the deployment of a general section i of a solar panel array formed by n sections (see Fig. 2c). Equilibrium equation on the spring allows to calculate the force acting on point A_i:

$$\boldsymbol{F}_{link,i,1} = \frac{1}{2}\left(\frac{A_i - B_i}{A_i - B_i}\right)\|\boldsymbol{F}_{k,i}\| cos\ \beta_i \tag{1}$$

The same applies for $\boldsymbol{F}_{link,i,2}$, related to β_i. It follows easily that $\boldsymbol{M}_{Q,i} = (\boldsymbol{Q}_i - \boldsymbol{A}_i) \times \boldsymbol{F}_{link,i,1}$ and $\boldsymbol{M}_{P,i} = \left(\boldsymbol{A}_i^{'} - \boldsymbol{P}_i\right) \times \boldsymbol{F}_{link,i,2}$. For sake of simplicity, the equilibrium of moments has been calculated in point $\boldsymbol{Q}_i$. Therefore, an *equivalent deployment moment* $\boldsymbol{M}_{depl,i} = \boldsymbol{M}_{P,i} + \boldsymbol{M}_{Q,i} = f(\vartheta_i)$ can be introduced, according to the principle of virtual works. The tension on the cable $\boldsymbol{F}_{depl,i}$ as a function of ϑ_i can be determined from the following equilibrium equation:

$$\boldsymbol{M}_{depl,i}(\vartheta_i) \cong \overline{\boldsymbol{Q}_i\boldsymbol{P}_{i+1}} \times \boldsymbol{F}_{depl,i} + \overline{\boldsymbol{Q}_i\boldsymbol{G}_i} \times m_i\boldsymbol{g} \tag{2}$$

where $m_i = \sum_{j=i}^{n} m_{p,j}$ is the sum of all the contributions of the folded panels; $m_{p,j}$ is the mass of the jth panel and hinge. Applying the principle of virtual works, the error introduced by the previous equation is given by $\overline{\boldsymbol{Q}_i\boldsymbol{P}_i} \times \boldsymbol{F}_{depl,i}$, which is neglectable for small values of $\overline{\boldsymbol{Q}_i\boldsymbol{P}_i}$.

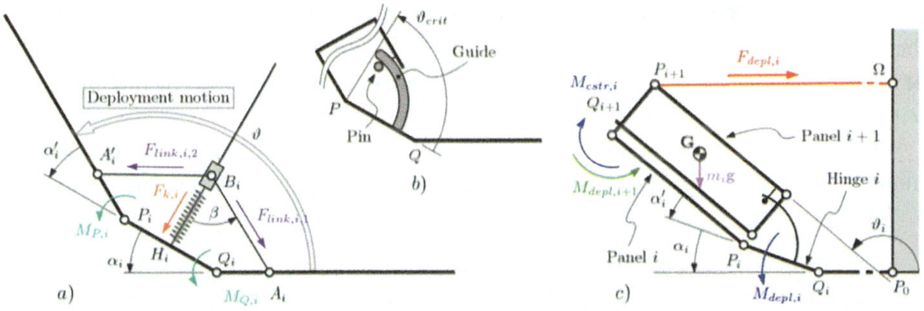

Fig. 3. General definition of the geometrical entities and values of the kinematic chain. In (a) the geometrical and force entities. The subscript i indicates the ith section in the chain. In (b) the constraint guide. In (c) the global configuration of the panels.

In order for the mechanism to operate properly during the deployment phase, the module of the vector $\boldsymbol{F}_{depl,i}$ (cable tension) has to remain always positive (**Condition I**). Naming $\hat{\boldsymbol{x}}$ the unit vector of the reference frame (rightward direction), and assuming that the cable is almost horizontal, the condition becomes:

$$\boldsymbol{F}_{depl,i}.\hat{\boldsymbol{x}} > 0 \tag{3}$$

Let us define $\tilde{\boldsymbol{M}}_{depl,i}$ as the minimum value of torque ($\forall\vartheta_i \in [0, \pi]$) capable to unfold the mechanism. In general, the main condition that allows the mechanism to "remain folded" before $\boldsymbol{Q}_{i+1}$ touches the ground during the deployment is the following (**Condition II**):

$$\boldsymbol{M}_{cstr,i}(\vartheta_i) + \boldsymbol{M}_{g,fold,i}(\vartheta_i) > \boldsymbol{M}_{depl,i+1}(\vartheta_{i+1} = 0)\forall\vartheta_i \in [\vartheta_{crit,i}, \pi] \tag{4}$$

where the first member is the sum of the moment generated by $\boldsymbol{F}_{depl,i}$ around $\boldsymbol{Q}_{i+1}$ and the contribution of the gravity force, $\boldsymbol{M}_{cstr,i} = \overline{\boldsymbol{Q}_{i+1}\boldsymbol{P}_{i+1}} \times \boldsymbol{F}_{depl,i}$ and $\boldsymbol{M}_{g,fold,i}(\vartheta_i) = \overline{\boldsymbol{Q}_{i+1}\boldsymbol{G}_i} \times m_i\boldsymbol{g}$. The second member tends to unfold the mechanism. In other words, during the deployment of the ith segment, the moment generated by the cable and by the gravitational force must be higher than the moment generated by the spring in the $(i+1)^{th}$ hinged segment.

The definition interval of the angle ϑ is limited by the angle ϑ_{crit} and π. The panels are constrained in a closed position by the guide on the bottom right of the sub-figure in the interval $0 \leq \vartheta < \vartheta_{crit}$ (see Fig. 3b).

Let us define $\tilde{\boldsymbol{M}}_{cstr,i}$ as the minimum value of torque that satisfies Eq. (4). This value can be increased by a factor k_{depl} that works as a *factor of safety*. The constraint guides that are responsible for ϑ_{crit} are a source of friction as well. This can be modeled as,

$$\boldsymbol{M}_{\mu,i} \cong \mu r_{g,i}\left(\boldsymbol{M}_{depl,i+1}/\|\boldsymbol{Q}_{i+1} - \boldsymbol{P}\|_i\right) \tag{5}$$

where the factor in parentheses is the radial force caused by the spring mechanism in the $(i+1)^{th}$ hinge and $r_{g,i}$ is the radius of the ith guide. In general it happens that $\boldsymbol{M}_{\mu,i} \ll \boldsymbol{M}_{depl,i}$, since $r_{g,i} \ll \|\boldsymbol{P}_{i+1} - \boldsymbol{Q}_{i+2}\|$, hence, for the purpose of this paper, we will disregard it.

It is possible, from these considerations, to define a procedure to compute the moments and forces that ensure Eqs. 3 and 4 are satisfied for each link in a chain with n elements:

i. Determine $\tilde{\boldsymbol{M}}_{depl,n}$ by using Eq. 1 (the moment generated by the spring system seen in Fig. 3c;
ii. Calculate the value of $\boldsymbol{F}_{depl,n}$ $\forall\vartheta_n \in [\vartheta_{crit,n}, \pi]$ here called $\tilde{\boldsymbol{F}}_{depl,n}$, from Eq. 2 (the cable tension produced by the deployment moment $\tilde{\boldsymbol{M}}_{depl,n}$), while verifying that Eq. 3 is satisfied
iii. Calculate $\tilde{\boldsymbol{M}}_{cstr,n-1}$ from $\tilde{\boldsymbol{F}}_{depl,n}$ while verifying that the condition in Eq. 4 is satisfied (the minimum moment required to keep the next panel in the chain folded);

Steps i–iii can be iterated, for $n-1, n-2, \ldots, 0$ until $\tilde{\boldsymbol{M}}_{depl,0}$ is reached. The input to this procedural computation is the value of the spring force of the n-th panel $\boldsymbol{F}_{link,n,1}$ and the masses m_i of the panels; the outputs are the deployment moments $\tilde{\boldsymbol{M}}_{depl,i}$ of all the n panels. At this point, the procedure allows to derive a direct relation $\boldsymbol{M}_{depl,n-1} \rightarrow \boldsymbol{M}_{depl,0}(n)$, starting from the inputs.

3 Discussion

Having defined a complete model for the multi-panel solar array and at the same time the algorithm for solving it, the main values for the mechanism presented in the previous Section are illustrated. In Table 2 the deployment moments $\tilde{\boldsymbol{M}}_{depl,i}$ for each section i are reported, along with the maximum moments of $\tilde{\boldsymbol{M}}_{\mu,i}$ generated by the friction on the constraint guides and the forces $\boldsymbol{F}_{depl,i}$ acting on the cable. In general, it seems that the deployment moment $\tilde{\boldsymbol{M}}_{depl,i}$ applied to the zeroth hinge grows with n. It is immediately apparent that $\tilde{\boldsymbol{M}}_{\mu,i} \ll \tilde{\boldsymbol{M}}_{depl,i}$ in every occurrence; this is a strong evidence that the guide friction can be safely neglected during the computation, as expected. A friction coefficient of $\mu = 0.5$ was used.

Table 2. Deployment Torques comparison. The values of $\tilde{\boldsymbol{M}}_{depl}$ and $\tilde{\boldsymbol{F}}_{depl}$ are shown for the 4 sections of the multi-panel solar array for gravity values relative to the surfaces of the Earth, Mars and the Moon. $k_{depl} = 1.2$.

Celestial body	Interaction	Links (n = 4)			
		0	1	2	3
Earth	$\tilde{\boldsymbol{M}}_{depl,i}$[Nm]	8.83	2.79	1.33	0.62
	$\tilde{\boldsymbol{M}}_{\mu,i}$[Nm]	0.08	0.03	0.01	–
	$\tilde{\boldsymbol{F}}_{depl,i}$[N]	78.69	26.9	13.7	19.2
Mars	$\tilde{\boldsymbol{M}}_{depl,i}$[Nm]	3.34	1.06	0.51	0.23
	$\tilde{\boldsymbol{M}}_{\mu,i}$[Nm]	0.03	0.01	<0.00	–
	$\tilde{\boldsymbol{F}}_{depl,i}$[N]	29.78	10.18	5.18	7.27
Moon	$\tilde{\boldsymbol{M}}_{depl,i}$[Nm]	1.46	0.46	0.22	0.1
	$\tilde{\boldsymbol{M}}_{\mu,i}$[Nm]	0.01	<0.00	<0.00	–
	$\tilde{\boldsymbol{F}}_{depl,i}$[N]	13.01	4.45	2.27	3.18

The influence of the parameter k_{depl} is paramount; as our research shows, this parameter has a huge impact on the increment speed of $\tilde{\boldsymbol{M}}_{depl,i}$ along the links. Indeed, the exponential relation $\boldsymbol{M}_{depl,0}(n) > f\left(k^n_{depl}\right)$ occurs. Having defined the deployment moments for the links of the chain, the forces $\boldsymbol{F}_{link,i}$ produced in the quadrilateral mechanism illustrated in Fig. 3a can be computed using Eq. 1. This shows comparably large values for the zero-th link; however, it should be noted that these values are strongly dependent on the geometry of the mechanism, with the overall value decreasing when the size of the involved entities (e.g. the links, or the distance $\overline{\boldsymbol{Q}_i\boldsymbol{A}_i}$) increases. In general, this means that the larger the mechanism, the lower the force needed; this ultimately calls for a trade-off, which should be carefully determined.

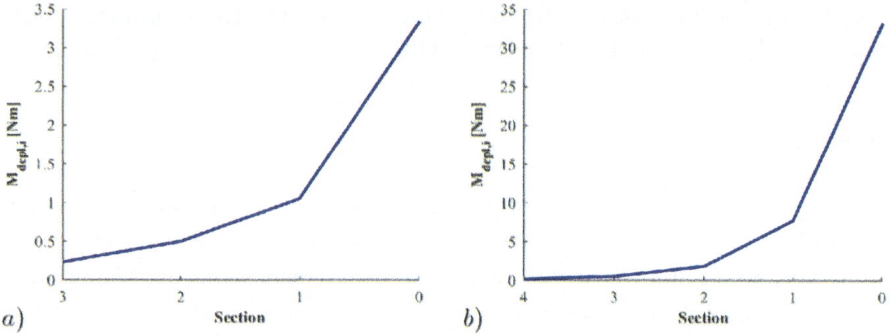

Fig. 4. Results of the algorithm on the surface of Mars ($g = 3.711 ms^{-2}$). In (a), the figure shows $\boldsymbol{M}_{depl,i}$, computed with $k_{depl} = 1.2$. In (b) a comparison of the algorithm is shown, with $k_{depl} = 1.6$.

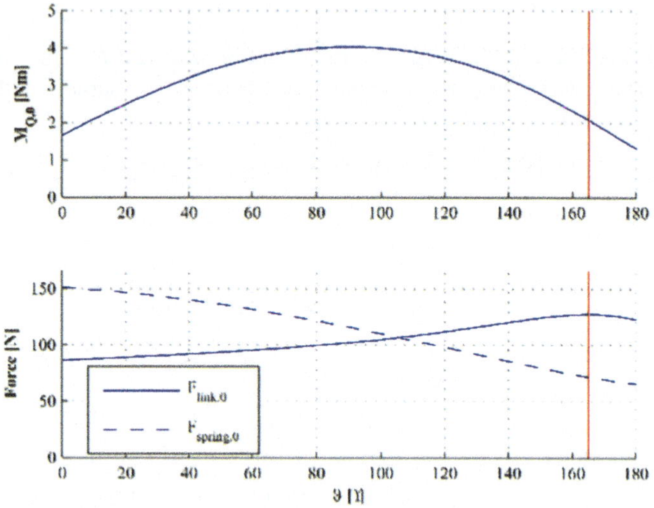

Fig. 5. Forces acting on the double quadrilateral mechanism of the 0th link. In the top plot the values of the moment $\tilde{\boldsymbol{M}}_{depl,0}$ is shown, whereas in the bottom one those of the forces $\boldsymbol{F}_{link,0}$ and $\boldsymbol{F}_{spring,0}$, which act on the links (solid) and on the spring (dashed), are illustrated. The red line marks the location of ϑ_{crit}.

The computation of a 5-links system is shown in Fig. 4 as an extension to the 4-links one. It should be noted that a slightly higher k_{depl} value was used. Indeed, with $k_{depl} = 1.2$ (Fig. 4, a), the 2nd condition cannot be satisfied. In general, we find that values of $k_{depl} > 1.5$ give adequate results with this geometry.

With regards to the spring-loaded actuation of the links, the result of the double quadrilateral described are shown in Fig. 5 for the most loaded link, that is, the zero-th. From Eq. 1 it is possible to compute the forces acting on the mechanism as a function of ϑ. In particular, starting with the $\tilde{\boldsymbol{M}}_{depl,0} = 3.34$ Nm (see Table 2, Mars scenario), it is possible to determine the spring characteristics. In particular, we found a value of $k_{spring} = 1518\ N/m$ to be suitable. The forces illustrated in the bottom plot of the figure show values below 130 N on each link, peaking at 150 N on the spring. A peculiarity of this mechanism is that is causes a beneficial excess of torque at ϑ_{crit} as is visible in the figure.

4 Conclusion

A new solar arrays deployment mechanism for space applications has been conceptualized. It consists of a modular kinematics structure, which is operated by a single cable (1 DoF). This mechanism has the advantage of being reversible in the kinematic movement. The mechanism is made up of a series of panels connected by means of a linkage and a two symmetric springs. The paper presented an algorithm capable to calculate the values of the parameters to be included in the deployment mechanism: the value of the spring stiffness that provides the stored energy to unfold the mechanism and the minimum value of the cable tension needed to fold the mechanism.

Acknowledgments. This work was supported in part by the Helmholtz Association, project alliance ROBEX, under contract number HA-304.

References

1. Wedler, A., et al.: First results of the ROBEX analogue mission campaign: robotic deployment of seismic networks for future lunar missions. In: 68th International Astronautical Congress (IAC) (2017)
2. Flood, D.J.: Space photovoltaics - history, progress and promise. Mod. Phys. Lett. B **15**, 561–570 (2001)
3. Lee, D.Y., et al.: Maximizing photovoltaic power generation of a space-dart configured satellite. Acta Astronaut. **111**, 283–299 (2015). Article no. 5335
4. Kumar, P., Pellegrino, S.: Deployment and retraction of a cable-driven rigid panel solar array. J. Spacecr. Rocket. **33**(6), 836–842 (1996)
5. Ferrando, E. et al.: Photovoltaic assemblies for the power generation of the exomars missions. In: E3S Web of Conferences, vol. 16, Article. no. 04002 (2017)
6. Wanlass, M.: Recent progress in developing ultra-thin, monolithic GaInP/GaAs/GaInAs tandem solar cells. In: SPRAT XX 9/25-9/27 (2007)
7. Kopp, G., Lean, J.: A new, lower value of total solar irradiance: evidence and climate significance. Geophys. Res. Lett. **38**, 1–7 (2011). L01706
8. Spiga, A., Forget, F.: Fast and accurate estimation of solar irradiance on Martian slopes. Geophys. Res. Lett. **35**(15), 1–6 (2008). L15201
9. David, L.: Juno to Jupiter: piercing the veil. Aerosp. Am. **49**(7), 40–45 (2011)
10. Seriani, S., Seriani, M., Gallina, P.: Workspace optimization for a planar cable-suspended direct-driven robot. Robot. Comput. Integr. Manuf. **34**, 1–7 (2015)
11. Pigani, L., Gallina, P.: Cable-direct-driven-robot (CDDR) with a 3-link passive serial support. Robot. Comput. Integr. Manuf. **30**(3), 265–276 (2014)
12. Trevisani, A., Gallina, P., Williams II, R.L.: Cable-direct-driven robot (CDDR) with passive SCARA support: theory and simulation. J. Intell. Rob. Syst. **46**(1), 73–94 (2006)

Cable Driven Parallel Manipulators

Energy Consumption Reduction of a Cable-Driven Storage and Retrieval System

Tobias Bruckmann[1(✉)], Christopher Reichert[2], and Hongqian Ji[1]

[1] Chair of Mechatronics, University of Duisburg-Essen, Duisburg, Germany
tobias.bruckmann@uni-due.de,
hongqian.ji0119@gmail.com

[2] SEW-EURODRIVE GmbH & Co. KG, Bruchsal, Germany
christopher.reichert@sew-eurodrive.de

Abstract. This paper addresses multiple strategies to reduce the energy consumption of a rack feeder that is based on cable-driven robot technology. Here, both the hardware design and the trajectories to be followed are optimized.

Keywords: Cable robot · Wire robot · Rack feeder
Cable-driven storage and retrieval system · Trajectories · Counterweights
Optimization

1 Introduction

Storage-and-retrieval-systems (also known as rack feeders) are fundamental components of any modern automated warehousing using storage principles based on high racks. Currently, a consortium of mechatronics manufacturers, warehouse software experts, logistics end users and the University of Duisburg-Essen is developing, realizing and optimizing a demonstrator that outperforms the conventional serial kinematics used for storage-and-retrieval machines by using a cable-driven parallel robot (see Fig. 1).

This allows to limit the moving mass to a moving platform carrying a load handling device that takes hold of the payload. Similar approaches have been presented e.g. by Hassan [2], and cable-driven parallel robots have been proven to have a potential for outperforming serial systems in terms of energy efficiency and peak power, see [3, 4]. A first feasibility study was done in the previous project "EffizienzCluster LogistikRuhr". Now, the project "LEAN - Entwicklung von Leichtbau-Regalbediengeräten auf Basis von Seilroboter-Technik aus NRW" realizes, tests and optimizes a cable-driven rack feeder within the facilities of an international logistics company. Within the project, the high rack uses a width of ± 6 m on either side of the origin of the inertial coordinate system (see Fig. 1) and a height of 9 m. The optimization of the geometry and stiffness of the cable robot was already carried out within this project and delivered a massive improvement. The resulting geometry of the investigated system can be found in [5].

Still, the potential for energy consumption reduction needs to be fully unlocked: As the cable-driven rack feeder uses flexible cables that can only transmit tensile forces, a relevant share of the torque provided by the motors is needed to always maintain

J. Lenarcic and V. Parenti-Castelli (Eds.): ARK 2018, SPAR 8, pp. 383–391, 2019.
https://doi.org/10.1007/978-3-319-93188-3_44

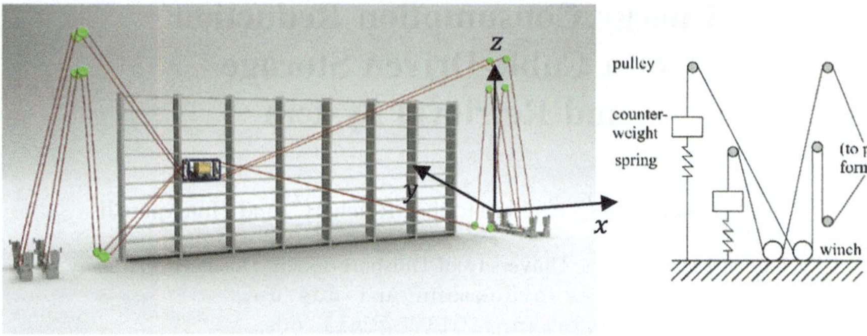

Fig. 1. CAD drawing of current system including coordinate system (image source: [1]) and optional arrangement of passive mechanical energy storage devices

tension in the cables. Here, the integration of passive mechanical energy storage devices (e.g. springs or counterweights) must be investigated. As each trajectory will require a different optimal set of spring or counterweight parameters, respectively, this can be only done efficiently if it includes the planning of optimal trajectories at the same time. In the following sections this task is addressed as follows: In Sect. 2, the applied models for force computations and power consumption are introduced. In Sect. 3, the power requirements as well as the energy consumption are optimized. Section 4 provides an outlook and future steps.

2 System Modeling

2.1 Force Equilibrium

The force equilibrium is computed for a given pose $\boldsymbol{x}$ of the platform that is guided by m cables with corresponding cable lengths $\boldsymbol{l} \in \mathbb{R}^m$. As the cable-driven robot has a parallel topology, each cable applies a wrench onto the platform which summarizes to a resulting cable force wrench of all cables. Per cable i, the applied tension can be decomposed in a unit vector in the direction of the cable and a scalar tension f_i. As the cable force wrench and the external loads $\boldsymbol{w}$ need to be in equilibrium, it follows

$$\boldsymbol{A}^T\boldsymbol{f} + \boldsymbol{w} = 0, \tag{1}$$

where $\boldsymbol{A}^T$ is the transpose of the Jacobian of the system, called structure matrix. Its columns $\boldsymbol{a}^T$ stack the unit vectors in the direction of the cables onto the cross product of the unit vectors and the position vector from the platform center to the cable connection points of the platform [5]. Further details can be found in [6, 7].

2.2 Power Requirements and Energy Consumption

In the next section, so-called double cycles are introduced, where three double cycles $k = A, B, C$ will be employed as benchmarks. For each double cycle, three trajectories $j = 1, \ldots, 3$ are needed. Accordingly, and with respect to the application at hand, so-called double cycles are used as reference trajectories. As a high rack includes dedicated zones for goods with high, medium and low turnover (A, B and C classes, see Fig. 2), per class one reference double cycle is assigned, each of them composed by three trajectories (see Table 1). A double cycle employed in this work starts at the input/output position of the storage-and-retrieval system (pose A1, B1, C1), picks a payload of 50 kg, moves to an empty storage position of interest (pose A2 or B2 or C2), stores the payload, moves empty to a second storage position that is occupied (pose A3 or B3 or C3), picks another payload of 50 kg and moves back to the input/output position.

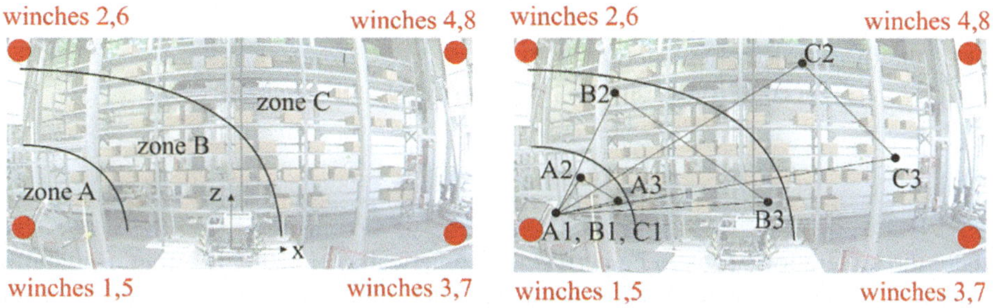

Fig. 2. Zones and double cycles

Within the simulations in this paper, these three double cycles $k = A, B, C$ will be employed as benchmarks. For each double cycle, the three trajectories $j = 1, \ldots, 3$ are needed and the overall power requirement at time t is

$$P_{\text{total}}(t) = \sum_{i=1}^{m} P_i(t) = \begin{cases} \sum_{i=1}^{m} f_{m_i}(t)\dot{l}_i(t) \text{ for } \operatorname{sgn} f_i = \operatorname{sgn} \dot{l}_i \\ \sum_{i=1}^{m} \rho f_{m_i}(t)\dot{l}_i(t) \text{ for } \operatorname{sgn} f_i \neq \operatorname{sgn} \dot{l}_i \end{cases} \tag{2}$$

where f_{m_i} is the wire force corresponding to the torque required from the electric motor after all losses and contributions, details see Eq. (6). As the electrical drives installed are able to recuperate energy in generator mode, and as the drives share an intermediate circuit to shift energy between drives in generator and actuator mode, respectively, a recuperation efficiency factor ρ needs to be included. Here, the electrical model is not detailed further and the energy turnover is approximated as mechanical energy within this paper. In future works this must be completed by detailed models of the electric drives as demonstrated e.g. in [8].

Another characteristic value is the maximum power needed during a motion. For each trajectory j as part of double cycle k, a maximum peak power

$$P_{k_j} = \max(P_{\text{total}}(t)), \text{t} = 0, \ldots, t_{k_j} \tag{3}$$

can be identified. Here, t_{k_j} is the time needed for a trajectory j. Accordingly, for each double cycle k, a maximum peak power

Table 1. Waypoints of double cycles. Unit is meters [m].

	pose 1 (I/O position)		pose 2		pose 3		pose 1 (I/O position)
payload	empty	50 kg		empty		50 kg	empty
class A	$[-5{,}5\ \ 1]^T$		$[-4\ \ 4{,}5]^T$		$[-1\ \ 2]^T$		$[-5{,}5\ \ 1]^T$
class B	$[-5{,}5\ \ 1]^T$		$[-3\ \ 6{,}5]^T$		$[3\ \ 3]^T$		$[-5{,}5\ \ 1]^T$
class C	$[-5{,}5\ \ 1]^T$		$[1\ \ 9]^T$		$[5{,}5\ \ 5{,}5]^T$		$[-5{,}5\ \ 1]^T$

$$P_k = \max\left(P_{k_j}\right), j = 1, \ldots, 3 \tag{4}$$

out of the three trajectories is computed.

The energy needed for a trajectory j as part of double cycle k is

$$E_{k_j} = \int_0^{t_{k_j}} P_{\text{total}}(t)dt \tag{5}$$

where the integration needs to be performed numerically. Note that the trapezoidal rule is applied which might lead to artificial damping effects and thus a decrease of the energy computed.

The commanded forces $\boldsymbol{f}_m$ for the m winches can be computed as follows

$$\boldsymbol{f}_m = \boldsymbol{J}\ddot{\boldsymbol{l}} + \boldsymbol{f} + \boldsymbol{f}_r - \boldsymbol{f}_c - \boldsymbol{f}_s, \tag{6}$$

wherein the term $\boldsymbol{J}\ddot{\boldsymbol{l}}$ defines the inertia forces considering each winch drive. Based on a force distribution algorithm [7], the feedforward term $\boldsymbol{f}$ is computed to move the platform along a desired trajectory. Furthermore, a compensation of all friction forces $\boldsymbol{f}_r$ is performed.

When the winches are be pretensed by counterweights with the masses $\boldsymbol{m}_c$, the following forces are acting the winch drives:

$$\boldsymbol{f}_c = \boldsymbol{m}_c \circ \left(\boldsymbol{g} - \ddot{\boldsymbol{l}}\right), \tag{7}$$

where $\circ$ is the Hadamard product for element-wise multiplication, $\boldsymbol{g} = [g, \cdots, g]^T \in \mathbb{R}^m$ describes the gravity acceleration and $\ddot{\boldsymbol{l}}$ defines the acceleration of the actuated cables. In addition to counterweights, also pretensed springs can be used to reduce the energy consumption of a cable-driven parallel robot. The forces $\boldsymbol{f}_s$ caused by the pretensioned springs are calculated as follows:

$$\boldsymbol{f}_s = \boldsymbol{k} \circ \Delta\boldsymbol{l}(t) + \boldsymbol{f}_p \tag{8}$$

Therein, the term $\boldsymbol{f}_p$ defines the initial pretension forces of the springs (i.e. for deflection of the spring away from its rest length $\Delta\boldsymbol{l}(t = 0) = 0$) and $\boldsymbol{k}$ describes the vector of the spring stiffnesses. Accordingly, the term $\boldsymbol{k} \circ \Delta\boldsymbol{l}(t)$ defines the varying spring forces during the motion of the platform.

3 Optimization of Power Requirements and Energy Consumption

To unlock the potential of the cable-driven robot approach for the application at hand, the power requirements of the drives and the energy consumption of the overall system must be optimized. This requires a dynamical model of the cable robot including masses and inertia of the platform as well as inertia and friction of the winches. Modeling details can be found in [6] where power requirements and energy consumption are discussed in detail in [9]. For simplicity and without loss of generality, in this section all motions of the platform are only planar, and only two translational degrees-of-freedom were used.

Within this paper, two approaches to optimize the energy and power demands are chosen:

- From the application, only point-to-point motions are required. Thus, the motions (trajectories) between two points can be arbitrarily chosen within the workspace. Again, in order to save both drive power and energy, the exact trajectory parameters can be made subject to optimization (typically, the straight line between two points is neither the fastest nor the most energy saving one [10]). Noteworthy, for the practical operation of the cable robot, this needs to be performed for each possible trajectory, i.e. from each storage position to all others. This is part of future work in realizing the control system. Several approaches have been demonstrated in the past [11–14].
- A fully tensed cable robot always needs to create an inner tension. Conventionally, this is realized by the torque of the drives which is associated with increased power demands of each drive. In case of recuperation capabilities, the additional power invested by a winch acting as an actuator is recuperated by other winches that act as generators at the same time. Still, as the recuperation efficiency is limited (here, $\rho = 85\%$ was chosen), this means a loss of energy which can be avoided by using other mechanical elements to create inner tension, see Fig. 1. These elements are called passive elements in the scope of this paper. From a mechanical engineering design point-of-view, counterweights and springs are known to be able to store potential energy and to provide pretension for the system [9, 15]. The counterweights usually reduce the required pretension, but increase the mass of the system which might be an issue for highly dynamical motions. It is now subject to an optimizer to identify the optimal parameters of counterweights and springs. Clearly, this is not very efficient for arbitrary trajectories within the workspace. Thus, within this paper representative trajectories from the high, medium and low turnover classes (A, B and C, respectively) are employed, see Sect. 2.

From a purely mechanical point of view, a cyclic motion only requires energy to compensate for recuperation and friction losses. Accordingly, and depending on the quality of bearings, gears, drive power supply and intermediate power circuit, the saving potential is varying.

On the other hand, the power demands of the drives are of high interest as this directly influences the hardware costs. Note, that usually the maximum peaks of power and torque, respectively, are crucial for selecting the drives.

For the optimization of the system, the following parameters were chosen:

- The minimum required force to avoid slackness in the cables is $f_{\min} = 100$ N.
- The maximum allowable force in the cables is $f_{\max} = 5000$ N. (Also the force f_{m_i} corresponding to the actuator torque might be reasonable instead, but this is not employed here.)
- The mass of the end effector is 50 kg.
- The maximum allowable total power consumption taken from the electric network is $P_{\text{total}} \leq 15$ kW.
- The maximum cable velocity provided by the winches is $|\dot{l}_i| = 7.5\,\frac{\text{m}}{\text{s}}, i = 1, \ldots, m$.
- To reduce the number of optimization parameters, the parameters of springs and counterweights of the winches 1 and 5, 2 and 6, 3 and 7 as well as 4 and 8, respectively, were defined to be identical (see Fig. 2).

As power optimal trajectories and designs may generally differ from energy optimal and designs, both need to be included in the cost function to be applied, using weights as introduced below. A certain optimization sequence was applied. Here, both the pattern search algorithm as well as the particle swarm optimizer available from MATLAB® toolboxes where tested since they make only little requirements to the properties of the cost function. Note, both converged to the same results – also in repeated runs – where the pattern search algorithm was faster. An investigation on alternative optimizers will be part of future work:

1. Initially, optimal trajectories are computed for the trajectories of the three double cycles. Clearly, per double cycle three sets of trajectory parameters are needed as three independent trajectories are required for a complete double cycle, giving a total of nine trajectories. For the optimization, a cost function c_{k_j} is defined below. It includes the energy consumption plus the peak power along a trajectory and the trajectory time t_{k_j} per trajectory. Weights for the cost functions are introduced, reflecting both the different physical units of the quantities to be optimized and the application-specific user demands (i.e. minimum time vs. minimum peak power vs. minimum energy consumption). In the scope of this paper, short cycle times were emphasized. Thus, the time needed for a trajectory uses a factor of $K_t = 15000$, the total energy required uses a factor of $K_{\text{Energy}} = 1$ and the peak power identified along a trajectory is weighted by $K_{\text{Power}} = 5$. Thus for a trajectory j as part of double cycle k the following optimization problem needs to be solved

$$\begin{aligned} \text{minimize} \quad & c_{k_j} = K_t \cdot t_{k_j} + K_{\text{Energy}} \cdot E_{k_j} + K_{\text{Power}} \cdot P_{k_j} \\ \text{s.t.} \quad & f_{\min} \leq f_i \leq f_{\max}, \left|\dot{l}_i\right| \leq 7.5\,\frac{\text{m}}{\text{s}}, i = 1, \ldots, m, \\ & P_{\text{total}} \leq 15 \text{ kW} \end{aligned} \tag{9}$$

2. In this first step, all passive elements are neglected in the simulation. The optimization parameters are only the trajectory parameters, described by the coefficients of a spline of 7th order per degree-of-freedom. This allows to specify position, velocity and acceleration on the start and end points of each segment plus the positions of two intermediate points at $1/3t_{k_j}$ and $2/3t_{k_j}$ per segment. This gives a

wide range of possible shapes of the trajectory, and both horizontal and vertical velocity of the platform can be independently addressed.

3. The simulated system is now completed by counterweights which act directly on the winches. The initial counterweight masses are chosen such that the robot is in equilibrium in the workspace center. In a second run of the system, the optimal mass parameters are to be found while the trajectory parameters from the last step are applied. Within this optimization step, the different turnover shares of the classes A, B and C are reflected by weighing factors $\gamma_A, \gamma_B, \gamma_C$ in the cost function: Here, the double cycle in class A was considered to represent 85% of all cycles (thus $\gamma_A = 0.85$), class B was considered 10% (thus $\gamma_B = 0.1$) and class C was considered 5% (thus $\gamma_C = 0.05$). Note, per double cycle the first and the last trajectory including the payload are weighted by a factor of $\sigma_1 = \sigma_3 = 1.2$ in the cost function while the unloaded intermediate trajectory is weighted by $\sigma_2 = 0.9$. This emphasizes the importance of loaded movements, as in practice also single loaded operations (known as single cycles) are performed. As the trajectory times are fixed by the previous steps, only the weights $K_{\text{Energy}} = 1$ and $K_{\text{Power}} = 3$ are chosen based on requirements of the warehouse at hand:

$$\begin{aligned} &\text{minimize} \quad c = \sum_{k=A,B,C} \gamma_k \left[\sum_{j=1}^{3} \left(\sigma_j \cdot K_{\text{Energy}} \cdot E_{k_j} \right) + K_{\text{Power}} \cdot P_k \right] \\ &\text{s.t.} \quad f_{\min} \leq f_i \leq f_{\max}, \left| \dot{l}_i \right| \leq 7.5\ \tfrac{\text{m}}{\text{s}}, i = 1, \ldots, m, \\ &\qquad P_{\text{total}} \leq 15\,\text{kW} \end{aligned} \tag{10}$$

4. Now again the trajectory parameters are optimized again for the cable robot including counterweights, using the counterweight parameters from the last step.

5. Steps 2 and 3 are repeated, where the system is completed by springs. This allows to separately evaluate the impact of springs.

The resulting optimal parameters of the masses of the counterweights are $\boldsymbol{m}_c = [37, 8\,\text{kg}, 53.3\,\text{kg}, 10.0\,\text{kg}, 41.1\,\text{kg}, 37, 8\,\text{kg}, 53.3\,\text{kg}, 10.0\,\text{kg}, 41.1\,\text{kg}]$. As Table 2 shows, there is a massive improvement of power and energy demands around 20–50% using counterweights. For their optimal efficiency, small extensions of the trajectory time (around 5%) are needed, but this is acceptable compared to the gains in peak power and energy savings.

For the springs, optimal stiffness $\boldsymbol{k} = \left[25\frac{N}{m}, 0, 0, 0, 25\frac{N}{m}, 0, 0, 0\right]$ and optimal pretension of $\boldsymbol{f}_p = [10\,\text{N}, 0, 0, 0, 10\,\text{N}, 0, 0, 0]$ were found. Interestingly, only the lower left winches 1 and 5 (see Fig. 2) are assigned to springs which corresponds to the location of the input/output position of the storage-and-retrieval system (pose A1, B1, C1). Furthermore, according to Table 2, the benefit of using springs in terms of the optimization goals is generally limited. Therefore, for cost reasons it was decided not to use springs in first prototypes.

Table 2. Optimization results. In the sum, no weighting factors are considered.

		No passive elements	Counter	Counter + Spring	Change
Zone A	$t_A = t_{A_1} + t_{A_2} + t_{A_3}[\mathrm{s}]$	4,78	4,9	5,05	–5,6%
	$E_A = E_{A_1} + E_{A_2} + E_{A_3}[\mathrm{kWs}]$	4,94	3,63	3,76	23,9%
	$P_A[\mathrm{kW}]$	4,72	2,96	2,37	49,8%
Zone B	$t_B = t_{B_1} + t_{B_2} + t_{B_3}[\mathrm{s}]$	6,25	6,52	6,27	–0,3%
	$E_B = E_{B_1} + E_{B_2} + E_{B_3}[\mathrm{kWs}]$	8,57	7,29	7,01	18,2%
	$P_B[\mathrm{kW}]$	5,92	4,06	4,45	24,8%
Zone C	$t_C = t_{C_1} + t_{C_2} + t_{C_3}[\mathrm{s}]$	7,16	7,48	7,54	–5,3%
	$E_C = E_{C_1} + E_{C_2} + E_{C_3}[\mathrm{kWs}]$	12,21	7,42	7,71	36,9%
	$P_C[\mathrm{kW}]$	7,22	5,18	4,89	32,3%
Summary	$t_{\mathrm{total}} = t_A + t_B + t_C[\mathrm{s}]$	18,19	18,9	18,86	–3,7%
	$E_{\mathrm{total}} = E_A + E_B + E_C[\mathrm{kWs}]$	25,72	18,34	18,48	28,1%
	$P_{\max} = \max(P_A, P_B, P_C)[\mathrm{kW}]$	7,22	5,18	4,89	32,3%

4 Conclusions and Outlook

This paper demonstrated the efficiency of selecting counterweights and springs for realizing the inner tension of a cable robot used as a storage and retrieval system. This included simplified modelling of the energy and power demands as well as a careful design of the selected trajectories suitable to represent the application-specific motions. Massive improvements of power and energy demands around 20–50% using counterweights are possible and will be validated in endurance tests using a demonstrator to be realized in the next months. Additionally, models of the electric drives as well as alternative optimizers will be investigated.

References

1. Bruckmann, T., et al.: An energy-efficient wire-based storage and retrieval system. In: IEEE/ASME International Conference on Advanced Intelligent Mechatronics (AIM). IEEE, Wollongong (2013)
2. Hassan, M., Khajepour, A.: Analysis of a large-workspace cable-actuated manipulator for warehousing applications. In: International Design Engineering Technical Conferences and Computers and Information in Engineering Conference Volume 7: 33rd Mechanisms and Robotics Conference, Parts A and B, ASME 2009, San Diego, California, USA, August 30–September 2 2009, vol. 49040, pp. 45–53 (2009)
3. Kraus, W., Spiller, A., Pott, A.: Energy efficiency of cable-driven parallel robots. In: IEEE International Conference on Robotics and Automation (ICRA) (2016)
4. Yan, L., Bone, G.M.: Are parallel manipulators more energy efficient? In: Proceedings of the IEEE International Symposium on Computational Intelligence in Robotics and Automation (Cat. No. 01EX515) (2001)

5. Reichert, C., Bruckmann, T., Optimization of the geometry of a cable-driven storage and retrieval system. In: Yang, R., Takeda, Y. (eds.) 5th IFToMM International Symposium on Robotics & Mechatronics (ISRM 2017). Springer, Sydney (2017)
6. Reichert, C., Bruckmann, T.: Unified contact force control approach for cable-driven parallel robots using an impedance/admittance control strategy. In: Proceedings of the 14th World Congress in Mechanism and Machine Science, Taipei (2015)
7. Gouttefarde, M., et al.: A Versatile Tension Distribution Algorithm for n-DOF Parallel Robots Driven by n + 2 Cables. Robotics **31**(6), 1444–1457 (2015)
8. Borgstrom, P.H., et al.: Generation of energy efficient trajectories for NIMS3D, a three-dimensional cabled robot. In: IEEE International Conference on Robotics and Automation (2008)
9. Reichert, C., Unterberg, U., Bruckmann, T.: Energie-optimale Trajektorien für seilbasierte Manipulatoren unter Verwendung von passiven Elementen. In: Proceedings of the 1st IFToMM D-A-CH Conference, Dortmund (2015)
10. Fang, S.: Design, Modeling and Motion Control of Tendon-Based Parallel Manipulators, in Lehrstuhl für Mechatronik. Duisburg-Essen, Duisburg (2004)
11. Gagliardini, L., et al.: Discrete reconfiguration planning for cable-driven parallel robots. Mech. Mach. Theor. **100**, 313–337 (2016)
12. Tempel, P., et al.: Design and programming for cable-driven parallel robots in the German pavilion at the EXPO 2015. Machines **3**(3), 223–241 (2015)
13. Trevisani, A.: Underconstrained planar cable-direct-driven robots: a trajectory planning method ensuring positive and bounded cable tensions. Mechatronics **20**(1), 113–127 (2010)
14. Barbazza, L., et al.: Trajectory planning of a suspended cable driven parallel robot with reconfigurable end effector. Robot. Comput. Integr. Manuf. **48**, 1–11 (2017)
15. von Zitzewitz, J., et al.: Use of passively guided deflection units and energy-storing elements to increase the application range of wire robots. In: Bruckmann, T., Pott, A. (eds.) First International Conference on Cable-driven Parallel Robots (CableCon 2012), pp. 167–184. Springer, Stuttgart (2013)

Computing Cross-Sections of the Workspace of a Cable-Driven Parallel Robot with 6 Sagging Cables Having Limited Lengths

Jean-Pierre Merlet(✉)

HEPHAISTOS Project, Université Côte d'Azur, Inria, Sophia Antipolis, France
Jean-Pierre.Merlet@inria.fr

Abstract. Although workspace is essential for the design and control of cable-driven parallel robots very few works have been devoted to this topic when sagging cables are considered, most probably because of the complexity of the cable model. In this paper we consider a CDPR with 6 sagging cables whose lengths have a limited maximal value. We propose an algorithm to compute the border of horizontal cross-section of the workspace for a given altitude and orientation of the platform. We show that the singularity of the inverse kinematics equations have to be taken into account for a proper determination of the border.

Keywords: Cable-driven parallel robots · Saggging cables · Workspace

1 Introduction

Workspace is an essential element for the design and control of a robot. As far as parallel robots with rigid legs are concerned efficient algorithms have been proposed to compute the border of cross-section of the workspace when the orientation and altitude are fixed, taking into account various limitations such as constraints on the leg lengths, on the amplitude of the passive joint motion or interference between the legs [6]. For cable-driven parallel robots (CDPR) the calculation is somewhat more complex because the signs of the cable tensions have to be taken account. For CDPR having *ideal* cables (no deformation due to the weight of the cable, no elasticity) a cable will have to be taken into account only if its tension is positive. Workspace calculation is therefore no more only a geometrical problem but a kineto-static problem. Several papers have addressed workspace calculation of 6 d.o.f. CDPR with ideal cables but considering only tension limitation [1–5,7,9,11,17,18,21–23] while CDPR with elastic cables have been less addressed [10,13]. The workspace of planar robot with sagging cables has been presented in [19] while, to the best knowledge of the author, 6 d.o.f CDPR has been considered only in [12]. However this paper has only considered as constraint the singularity of the inverse kinematic equations

J. Lenarcic and V. Parenti-Castelli (Eds.): ARK 2018, SPAR 8, pp. 392–400, 2019.
https://doi.org/10.1007/978-3-319-93188-3_45

(IK). Our motivation for this work is to introduce another constraint, namely an upper limit on the cable lengths, although we will see that the singularity of the IK has to be taken into account. In this paper we will consider CDPRs with only six cables, this limitation being explained in the next section.

2 Inverse and Forward Kinematics

In this paper we will use the Irvine sagging cable model that is valid for elastic and deformable cable with mass [8] and that has been shown to be in very good agreement with experimental results [20]. This model is established in the cable plane in which we have $A_i = (0, 0)$ as upper attachment point of the cable and $B_i = (x_b \geq 0, z_b < 0)$ as lower attachment point. Vertical and horizontal forces $F_z, F_x > 0$ are exerted on the cable at point B_i. For a cable with length at rest L_0 the coordinates of B are given by [8]:

$$x_b = F_x(\frac{L_0}{EA_0} + \frac{sinh^{-1}(F_z) - sinh^{-1}(F_z - \frac{\mu g L_0}{F_x})}{\mu g}) \tag{1}$$
$$z_b = \frac{F_z}{EA_0} - \mu g L_0^2/2 + \frac{\sqrt{F_x^2 + F_z^2} - \sqrt{F_x^2 + (F_z - \mu g L_0)^2}}{\mu g}$$

where E is the Young modulus, A_0 the cable cross-section area, L_0 the cable length at rest and μ the cable linear density. We consider a CDPR with $n \geq 6$ cables which is suspended i.e. the winches output point A_i is always higher than the attachment point B_i of the cable on the platform. If the pose of the platform is given, then the IK unknowns are the $3n$ (F_x^i, F_z^i, L_0^i). We consider the vector $\mathbf{A_iB_i}$ with components x_r^i, y_r^i, z_r^i and define the angle α_i such that

$$x_r^i \sin(\alpha_i) + y_r^i \cos(\alpha_i) = 0 \tag{2}$$

If $(F_x^i, 0, F_z^i)$ are the components of the force that is applied by the platform on the cable in its plane, then the force exerted by the cable on the platform at B_i may be written in the reference frame as $\mathbf{F_a^i} = (-F_x^i \cos(\alpha_i), -F_x^i \sin(\alpha_i), -F_z^i)$. We assume that the platform is submitted only to gravity which exert a vertical force $\mathcal{F}$ and no torque around the platform center of mass C. Hence the mechanical equilibrium imposes

$$\sum_{j=1}^{j=n} \mathbf{F_a^j} + \mathcal{F} = \mathbf{0} \qquad \sum_{j=1}^{j=n} \mathbf{CB_i} \times \mathbf{F_a^j} = \mathbf{0} \tag{3}$$

With these 6 constraints and the $2n$ provided by (1) we have $2n + 6$ constraints for $3n$ unknowns. If $n = 6$ we have a square system that will admit in general a finite number of solution(s) while if $n > 6$ there may be an infinite number of solutions. In this paper we will assume that $n = 6$ to deal with a finite number of solution(s). Note that it has been shown that solving the IK for $n = 6$ is a difficult task and may have indeed several solutions [14,15].

For the forward kinematics problem (FK) the length L_0^i are known and the pose of the platform has to be determined. The unknowns are the 6 parameters $\mathbf{X}$ defining the pose of the platform and the $2n$ (F_x^i, F_z^i) for a total of $2n+6$ unknowns. As for the constraints we have the $2n$ constraints (1) and the 6 constraints of the mechanical equilibrium (3). Note that it is useful to add the $n\alpha_i$ as unknowns and the n equations (2) as constraints so that we have $3n+6$ unknowns/constraints. Therefore the forward kinematics is always a square system that is very difficult to solve and admit usually several solutions. It is important to note that both the FK and IK are **not algebraic** as they involve inverse hyperbolic functions.

3 Workspace

We are interested in calculating cross-sections of the workspace of a CDPR with 6 cables for given altitude and orientation of the platform under the constraint that the cable lengths L_0 cannot exceed a limit L_0^m. The position of C will be denoted x, y, z. If we let x, y be free the FK has 26 unknowns and 24 equations which therefore describe a surface. If we set one of the L_0 to L_0^m the FK will describe a 1-dimensional variety, i.e. a curve in the $x-y$ plane that will be the border of the workspace. Unfortunately being given the FK equations the elimination of F_x^i, F_z^i, α_i from the FK in order to get an implicit equation $G_i(x, y) = 0$ seems to be very difficult. We will thus use another approach: basically we will determine poses potentially on the border and we will use a continuation approach to obtain a polygonal approximation of the G_i curve. Hence the first step of the algorithm is to find poses that may be on the border, that will be denoted *extremal poses*.

3.1 Finding Possible Extremal Poses

An extremal poses is characterized by the property that at least one of its IK solution has at least one L_0 being equal to L_0^m. For finding such pose we will consider a pose $\mathbf{X_0}$, called an *initial pose*, that is most probably inside the workspace and compute its n IK solution(s) using the method proposed in [14]. We consider the pose $\mathbf{X_1}$ defined by the coordinates $(x_0 + \epsilon, y_0)$ where ϵ is a small value that is determined with the Kantorovitch theorem so that the IK system is guaranteed to have a single solution around $\mathbf{X_0}$ which is obtained by using the Newton-Raphson scheme using as initial guess the solution obtained for $\mathbf{X_0}$. We then check if for $\mathbf{X_1}$ all the L_0^j satisfies $L_0^j \leq L_0^m$. If one of the L_0^j is close to L_0^m we consider the FK equations with x as unknown and impose $y = y_0, L_0^j = L_0^m$. Solving this system with Newton provides an extremal point for one of the branch of the IK at $\mathbf{X_0}$. This extremal point lies on a potential arc for the border (that will be denoted S_j^k), that is associated to $L_0^j = L_0^m$, where k denotes the solution number of the IK at $\mathbf{X_0}$ that has been used to derive $\mathbf{X_1}$. If the L_0 at $\mathbf{X_1}$ all satisfies $L_0^j \leq L_0^m$, then we repeat a motion along the $x+$ direction to find a new pose $\mathbf{X_2}$ and so on until we find an extremal pose.

This process may be repeated by moving from $\mathbf{X_0}$ in the $x-, y+, y-$ directions. After this step we have a set of extremal poses $\mathbf{X_j^e}, \mathbf{X_k^e}, \ldots$ that belong to different arcs $S_j, S_k, \ldots$. We may now present the continuation algorithm that allow us to obtain a polygonal approximation of the arcs $S_j, S_k, \ldots$.

3.2 The Continuation Algorithm

Let $\mathbf{X_l} = (x_l, y_l)$ be an extremal pose, that has been obtained when moving along the $x+$ direction, lying on the curve S_l, and such that $L_0^l \approx L_0^m$. We then consider a pose $\mathbf{X_k}$ whose y coordinate will be set to $y_l - \gamma$, where $\gamma > 0$ is very small. We set L_0^l to L_0^m and by using Newton we are able to determine the x coordinate of $\mathbf{X_k}$ so that this pose belongs to S_l. With $\mathbf{X_l}, \mathbf{X_k}$ on S_l we are able to get an approximation of the unit tangent vector $\mathbf{T_l^k}$ of S_l at $\mathbf{X_k}$. We consider the largest component u of $\mathbf{T_l^k}$ and define $x(y)_{k+1} = x(y)_k + \epsilon u$. The value of ϵ is determined using Kantorovitch theorem so that Newton is able to calculate $y(x)_{k+1}$ such that $\mathbf{X_{k+1}} = (x_{k+1}, y_{k+1})$ belongs to S_l. We check if $\mathbf{X_{k+1}}$ is such that $L_0^j \leq L_0^m$ for all $j \in [1, 6], j \neq l$. If this is the case we have obtained a new pose on S_l and we repeat the process. The continuation process stops in the following cases:

- if L_0^j, $j \neq l$ is close to L_0^m we consider the IK with $L_0^l = L_0^m$ and $L_0^j = L_0^m$ with free x, y and use Newton with $\mathbf{X_k}$ as solution guess to solve this system. At this pose we have obtained a special extremal pose, called an *end-point* of S_l, which has the property of having 2 cable lengths equal to L_0^m
- we get a pose $\mathbf{X_u}$ that is very close to a previously determined $\mathbf{X_v}$ on S_l. This means that S_l is a closed loop

When the continuation has stopped we will start a new continuation process for S_l using $-\mathbf{T_l^k}$ for fixing the continuation direction. This continuation process is repeated for all starting points of all branches.

3.3 Finding New Starting Points

When the continuation has completed we consider the set of end-points that has been obtained. Each of the end-point $\mathbf{X_l^e} = (x_l^e, y_l^e)$ originates from an arc S_l that meets another arc S_k. We examine if any arc S_k that has been previously determined includes $\mathbf{X_l^e}$. If this not the case we consider the 4 IK problem with $x = x_l^e + \pm\epsilon$ or $y = y_l^e \pm \epsilon$, $x(y)$ being free and L_0^k set to L_0^m. For each of this IK we use Newton with as initial guess the F_x, F_z, L_0 obtained at $\mathbf{X_l^e}$ and $x_l^e(y_l^e)$ as guess for the unknown $x(y)$ As soon as we get one solution that satisfy $L_0^u \leq L_0^m$ for all u in $[1, 6]$, $u \neq k$, then we have obtained a new starting point for the branch S_k and we use the continuation algorithm to determine the arc S_k. This process is repeated until all end-points belongs to one of the determined S_l. This normally ensure that we have determined a closed-region that will represent the external border of the workspace. Our algorithm provides a polygonal approximation of the border whose vertices are guaranteed to belong to the border. A question arise however: between two successive vertices the border is assumed to be close to the segment joining the 2 vertices but may the real border be significantly different from the segment? The answer to this question is no as the Kantorovitch theorem is based on the evaluation of the maximum of the norm of the Hessian matrix all over a ball centered at one vertex that enclose the next vertex and a drastic change of the border will be reflected on the maximum of the norm.

3.4 Example

We consider as example our large scale robot MARIONET-CRANE [16], probably one of the largest CDPR ever deployed. This robot is a suspended CDPR with 6 cables, whose A_i, B_i coordinates are given in Table 1. We set $L_0^m = 25m$ and use $x = 2, y = 8$ as initial point for finding starting points. At this point the IK has a single solution and we use the method described in Sect. 3.1 for finding extremal points on S_5, S_6. These points are used to compute these arcs (Fig. 1) with the continuation algorithm of Sect. 3.2. Using the end-points of these arcs we find new starting points for S_3, S_4 by using the strategy described in Sect. 3.3 and the continuation algorithm allows us to establish these arcs. At this stage there is no new starting point and we get the workspace presented in Fig. 1. Although it seems that the workspace is a closed-region, details of the upper-left corner of the workspace (presented in Fig. 1) show that the curves S_3, S_6 intersect at two poses M, N. The pose M is an end-point of theses curves but the point N is not. Indeed the IK at N has 2 solutions: one with $L_0^3 = L_0^m, L_0^6 < L_0^m$ and another one with $L_0^6 = L_0^m, L_0^3 < L_0^m$. Let us consider the pose $B2$ that is very close of S_3 so that Newton can be used for finding one IK solution for B_2. Starting from B_2 if we move along the $x-$ or $y+$ direction we end up at a pose on S_3 while moving along the $y-$ direction we end up on S_5. But if we move along $x+$ we cross S_6 and end up in a singular configuration. Now consider $B1$ that is close to S_3 but outside the determined workspace. This pose has 2 valid IK solutions which shows that we have missed reachable poses. For one of them we reach S_3 if we move along the $x-, y+$ direction, S_5 for $y-$ and a singularity for $x+$. For the other one when moving along $x-$ we cross S_3 and stop at S_6 while when moving along $y-$ we stop at S_6. When moving along $x+$ we reach a singularity while when moving along $y+$ we cross S_3 once and end up on another part of S_3.

Table 1. Coordinates of the A_i, B_i points on the base and on the platform (in cm)

x	y	z	x	y	z
-325.9	-47.5	882.6	942.1	-348.2	1155.5
953.8	379.7	1153.3	557.0	2041.4	870.4
-250.5	1681.0	864.9	-334.2	942.1	878.8

x	y	z	x	y	z
-10	-93	-3	10	-93	-3
27	50	-7	27	50	-7
-27	50	-7	-27	50	-7

From this analysis it appears first that we have to perform a local analysis if two S_j and S_k curves intersect but not at an end-point. Finding such an intersection point is not difficult as the polygonal approximation provides an

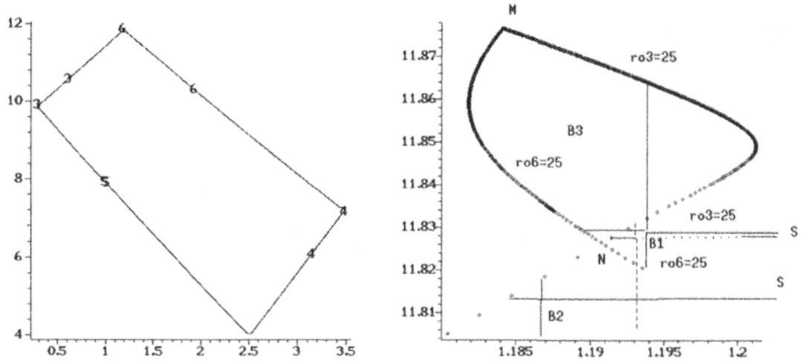

Fig. 1. The workspace border for $L_0^m = 25m$ and a detail of its upper-left corner

estimate of its x, y coordinates. Then we consider two IK systems, one having $L_0^j = L_0^m$, the other one having $L_0^k = L_0^m$. This amounts to 42 equations in 42 unknowns with a very good initial guess for all them so that Newton will provide the solution. Second the analysis shows that we have to consider singularities that may contribute to part of the border of the workspace.

3.5 Managing Singularities

Let us assume that we have found a pose $\mathbf{X_s}$ that has an IK solution but is very close to an IK singularities so that when moving from $\mathbf{X_s}$ in a given direction $\mathbf{v}$ we are no more able to use Newton as the IK become singular at $\mathbf{X_s^1} = \mathbf{X_s} + \epsilon\mathbf{v}$. A continuation scheme may be designed to follow the singularity curve. First we use Newton to determine an IK solution for a pose $\mathbf{X_s'}$ that is close to $\mathbf{X_s}$ but not on the line that includes $\mathbf{X_s}$ with direction $\mathbf{v}$. From $\mathbf{X_s'}$ we move along $\mathbf{v}$ until we reach a pose $\mathbf{X_s^2}$ that is singular. An approximation of the tangent unit vector $\mathbf{T_s}$ of the singularity curve may then be obtained from $\mathbf{X_s^1}, \mathbf{X_s^2}$. Let $\mathbf{N_s}$ be an unit vector that is perpendicular to $\mathbf{T_s}$ and such that $\mathbf{X_s X_s^1}.\mathbf{N_s} < 0$. We consider then a new pose $\mathbf{X_3} = \mathbf{X_s'} + \gamma\mathbf{T_s} + \delta\mathbf{N_s}$ where γ has a small value and δ is a small positive constant. This pose is designed so that $\mathbf{X_3}$ lies on the same side of the line $\mathbf{X_s^1 X_s^2}$ than $\mathbf{X_s}, \mathbf{X_s'}$ and sufficiently away from the singularity curve so that we can compute its IK solution by using Newton with the solution at $\mathbf{X_s'}$ as initial guess. From $\mathbf{X_3}$ we then moves in the direction $-\mathbf{N_s}$ by small increment until we reach a singular pose $\mathbf{X_s^3}$. This process is repeated until we reach a pose $\mathbf{X_s^n}$ that is very close to a S_j curve and is repeated by starting from $\mathbf{X_s}$ but using $-\mathbf{T_s}$ as motion direction.

This strategy has been used in our example using the singular pose that has been found by moving from $B1$ in the $x+$ direction. We find a singularity curve that is displayed in Fig. 2. The analysis of this new region shows that in the region `U2NU1U3` the IK has two solutions meaning that any motion in this zone may be executed by following two different kinematic branches. For one of this branch the motion is restricted to lie in the region `U0NU1U3` while for the other branch it is restricted to `U2NMU1U3`. Any pose in the region `NMU1` is reachable but a specific motion strategy may have to be designed. For example if the CDPR has to move from $B2$ to $B3$ it will have first to move in the region U0NU1U3 e.g. to $B1$) by following the same kinematic branch than at $B2$. Then at $B1$ the cables lengths will have to be adjusted to get to the other available kinematic branch and then the CDPR may move toward $B3$.

The workspace calculation has been implemented in Maple but a rough approximation of the computation time for an algorithm implemented in C is about 5 min for computing a cross-section border if no singularity is encountered and about 10mn if there are singularities.

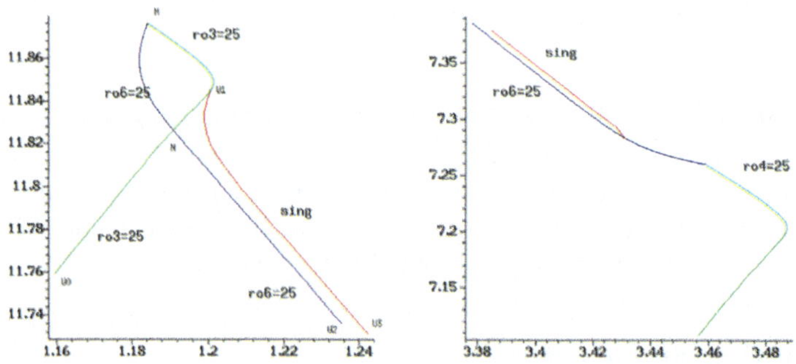

Fig. 2. View of the singularity curve close to N (left) and its connection to S_6 (right)

4 Conclusion

Calculating workspace for CDPR with sagging cables is a difficult task because of the complexity of the cable model and because it induces several solutions to the IK. The algorithm proposed in this paper allows one to calculate the workspace border in a reasonable amount of time. Still there are numerous issues that have to be addressed. First of all the algorithm is able to compute the external border of the workspace but may miss void in the inside of the workspace. Second as the computation time is still relatively high one may wonder if it will not be more efficient to start from the workspace border obtained for ideal cable (that is easy and very fast to determine) and use a continuation method on the Young modulus and linear density to slowly modify the shape of the border for getting the result. Finally the case of CDPR with more than 6 cables has to be investigated: the method is available and will be presented in another paper.

References

1. Albus, J., Bostelman, R., Dagalakis, N.: The NIST ROBOCRANE. J. Robot. Syst. **10**(5), 709–724 (1993)
2. Aref, M., Taghirad, H.: Geometrical workspace analysis of a cable-driven redundant parallel manipulator: KNTU CDRPM. In: IEEE International Conference on Intelligent Robots and Systems (IROS), Nice, France, pp. 1958–1963, 22–26 September 2008
3. Diao, X., Ma, O.: Workspace determination of general 6 d.o.f. cable manipulators. Adv. Robot. **22**(2–3), 261–278 (2008)
4. Duan, Q., Duan, X.: Workspace calculation and quantification calculations of cable-driven parallel robots. Adv. Mech. Eng. **2014**, 9 (2014). Article ID 358727
5. Ferraresi, C., Paoloni, M., Pescarmona, F.: A new methodology for the determination of the workspace of six-DOF redundant parallel structures actuated by nine wires. Robotica **25**(1), 113–120 (2007)
6. Gosselin, C.: Determination of the workspace of 6-DOF parallel manipulators. ASME J. Mech. Des. **112**(3), 331–336 (1990)
7. Gouttefarde, M., Daney, D., Merlet, J.P.: Interval-analysis based determination of the wrench-feasible workspace of parallel cable-driven robots. IEEE Trans. Robot. **27**(1), 1–13 (2011)
8. Irvine, H.M.: Cable Structures. MIT Press, Cambridge (1981)
9. Jeong, J., Kim, S., Kwak, Y.: Kinematics and workspace analysis of a parallel wire mechanism for measuring a robot pose. Mech. Mach. Theor. **34**(6), 825–841 (1999)
10. Korayem, M., Bamdad, M., Saadat, M.: Workspace analysis of cable-suspended robots with elastic cable. In: IEEE International Conference on Robotics and Biomimetics, ROBIO 2007, pp. 1942–1947 (2007)
11. Lim, W., et al.: A generic force closure algorithm for cable-driven parallel manipulators. Mech. Mach. Theor. **46**(9), 1265–1275 (2011)
12. Merlet, J.P.: Computing cross-sections of the workspace of cable-driven parallel robots with 6 sagging cables. In: Computational Kinematics, Poitiers (2017)
13. Merlet, J.P.: On the workspace of suspended cable-driven parallel robots. In: IEEE International Conference on Robotics and Automation, Stockholm, 16–20 May 2016
14. Merlet, J.P.: A new generic approach for the inverse kinematics of cable-driven parallel robot with 6 deformable cables. In: ARK, Grasse, 27–30 June 2016
15. Merlet, J.P.: On the inverse kinematics of cable-driven parallel robots with up to 6 sagging cables. In: IEEE International Conference on Intelligent Robots and Systems (IROS), Hamburg, Germany, pp. 4536–4361, 28 September–2 October 2015
16. Merlet, J.P., Daney, D.: A portable, modular parallel wire crane for rescue operations. In: IEEE International Conference on Robotics and Automation, Anchorage, pp. 2834–2839, 3–8 May 2010
17. Ouyang, B., Shang, W.W.: A new computation method for the force-closure workspace of cable-driven parallel manipulators. Robotica **33**(3), 537–547 (2015)
18. Pusey, J., et al.: Design and workspace analysis of a 6–6 cable-suspended parallel robot. Mech. Mach. Theor. **139**(7), 761–778 (2004)
19. Riehl, N., et al.: On the static workspace of large dimension cable-suspended robots with non negligible cable mass. In: 34th Annual Mechanisms and Robotics Conference, Montréal, 15–18 August 2010

20. Riehl, N., et al.: On the determination of cable characteristics for large dimension cable-driven parallel mechanisms. In: IEEE International Conference on Robotics and Automation, Anchorage, pp. 4709–4714, 3–8 May 2010
21. Stump, E., Kumar, V.: Workspaces of cable-actuated parallel manipulators. ASME J. Mech. Des. **128**(1), 159–167 (2006)
22. Tavolieri, C., Ceccarelli, M., Merlet, J.P.: A workspace analysis of a fully constrained cable-based parallel manipulator by using interval analysis. In: Musme, San Juan, Argentina, 8–12 April 2008
23. Verhoeven, R.: Analysis of the workspace of tendon-based Stewart platforms. Ph.D. thesis, University of Duisburg-Essen, Duisburg (2004)

A Unified Approach to Forward Kinematics for Cable-Driven Parallel Robots Based on Energy

Andreas Pott[1(✉)] and Philipp Tempel[2]

[1] Institute for Control Engineering of Machine Tools and Manufacturing Units, Fraunhofer IPA, University of Stuttgart, Stuttgart, Germany
andreas.pott@isw.uni-stuttgart.de

[2] Institute for Control Engineering of Machine Tools and Manufacturing Units, University of Stuttgart, Stuttgart, Germany
Philipp.Tempel@isw.uni-stuttgart.de

Abstract. This paper deals with a unified approach to forward kinematics for both over-constrained and under-constrained cable robots. Moreover, little assumptions on the cable model are required. This makes the proposed method applicable to a variety of currently discussed cable models including the standard model, an elastic cable model, static cable sagging, or even finite elements model. Solutions to the forward kinematics problem are computed by minimizing potential energy in the cable robot. As shown in this paper, the method unifies forward kinematics of both over-constrained and under-constrained cable robots.

Keywords: Cable-driven parallel robots · Forward kinematics
Cable model

1 Introduction

The computation of the pose of a cable-driven parallel robot from given cable lengths is a challenging yet important task in analysis and control. The difficulty of this problem arises from a number of sources: The coupled nature of the constrained equations, the cable model, and the kinematic classification of the cable robot. As the mobile platform of a cable robot is constrained by a number of cables, the geometric relations are highly coupled and depend on the number of cables m. With the degree-of-freedom n of the mobile platform, a common classification [13] includes *incompletely restrained* (IRPM, $m < n$), *complete restrained* (CRPM, $m = n$), and *redundantly restrained* positioning mechanisms (RRPM, $m > n$) where this classification has relevance to the applicable kinematic methods. The geometric results from conventional parallel robots apply to the CRPM class: Forward kinematics of Steward-Gough platforms of general geometry can have up to 40 real solutions [5,7]. An algorithm to compute the solutions with interval analysis is presented by Merlet [10]. The IRPM class is

J. Lenarcic and V. Parenti-Castelli (Eds.): ARK 2018, SPAR 8, pp. 401–409, 2019.
https://doi.org/10.1007/978-3-319-93188-3_46

addressed by Carricato and Merlet who discuss kinematics of under-constrained cable robots by means of the Lagrangian functions [3] and also express some connections between stability and energy [2].

A detailed analysis of the cables reveal effects such as elastic reactions and sagging of the cables making the estimation of the pose involved. So-called *cable models* account for the non-ideal behavior of the transmission element. A simplified consideration of sagging for inverse kinematics is given by Nguyen [4]. More sophisticated results taking into account sagging cables are presented by Merlet [12] and lately continuation is proposed to deal with the problem [11].

The work presented in this paper, is a consequent advancement of our previous work. A numerical algorithm to estimate the pose of an over-constrained cable robot is presented in [14]. The algorithm is inspired by minimization of the squared constrained error, which is related to potential energy stored in the cables. Criteria for stable convergence are discussed in [15] and this method is also applicable to take into account panning pulleys [16]. Kraus [9] proposes some modification on the functional to be minimized making it a measure of energy with clear physical meaning.

From state of the art, we conclude the following. Many approaches for non-trivial cable models and for under-constrained cable robots are not real-time capable for usage in the controller. The ill-conditioned model leads to poor numerical behavior at certain poses. It remains unclear if this is an artificial result of the assumptions made in modelling or if this is an intrinsic property of the cable robot. The experimental results obtained for cable robots widely show that cable robots behave quite smooth but inaccurate when fed with error-prone control parameters. This observation supports the assumption that the extreme values obtained from some computations are artifacts from modelling and numerical evaluation rather than natural properties of the cable robot.

In this paper, we propose a new approach to consider the problem which is based on energy and optimization. This new perspective allows to model the cable robot in a unified way where unification is achieved in terms of the kinematic classification of the robot bringing together both over- and under-constrained robots. Furthermore, different cable models can be considered in a common framework allowing unification also in terms of the cable model. The rest of the paper is organized as follows: Sect. 2 provides the geometric and energy equations used in this paper. Energy expressions for some cable models are introduced in Sect. 3. In Sect. 4, we present computational results for a simple example and discuss and interpret some well-known properties in the light of the energy formulation.

2 An Energy Formulation for Kinematics

The usage of energy is the key element to understand the behavior of cable robots. The underlying idea is that energy draws the connection between the pose of the platform and the (unstretched) cable length. If the cable lengths are fixed, the pose of the cable robot is expected to settle such that energy is (locally) minimized. Especially, the notion of local minimization is employed to understand why the cable robot moves to a specific solution out of all local minima. The usage of the energy function allows to consistently introduce a holistic robot state in terms of both configuration space and operational space parameters. Thus, the robot's kinematic model becomes *stateful* making future states depending on present states. Therefore, given a state of the robot system in terms of pose and cable length $(\mathbf{y}_0, \mathbf{l}_0)$ along with a change in cable length $\Delta\mathbf{l}$, it is meaningful to compute the resulting pose with forward kinematics. Within the energy space, there must be a trajectory from the state $(\mathbf{y}_0, \mathbf{l}_0)$ to the new state $(\mathbf{y}, \mathbf{l}_0 + \Delta\mathbf{l})$, and the path follows from physical principle of energy minimization (Hamilton's principle). Thus, we extract a criterion to decide which of the multiple solutions of the conventional kinematic solutions is reached.

Current kinematic codes fall short to explain which solution is reached when changing the input variables. As conventional formulations in kinematics are stateless, it is clear that information from a previous state are not conserved and also cannot be taken into account in physically correct way. It is common to use the last values as initial values in numerical schemes and this proved useful in some applications. However, the appropriateness and correctness is not covered by physics.

A general approach to describe the motion of a holonomic mechanical system is based on the Lagrangian formalism, where the energy of a mechanical system is defined by the kinetic energy T and the potential energy U. Then, the Lagrangian function L is defined as

$$L = U - T, \tag{1}$$

where U is configuration dependent potential energy and T represents kinetic energy dependent on pose and velocities. The well-known Lagrangian equations of second kind allow to derive the equations of motion of a mechanical system. The approach is based on general assumptions and it is applicable for cable robots (see e.g. [1,8,9]).

If we consider the system at rest, all kinetic energy T is either converted to potential energy U or vanishes from the system in terms of dissipation. Therefore, the energy of the system at rest will be a (local) minimum of the potential energy U. Let $\mathbf{q} \in \Re^m$ be the vector of the generalized coordinates in configuration space and $\mathbf{y} \in \Re^n$ be a parameterization of the pose of the cable robot's platform. The potential energy is a scalar function of these coordinates $U(\mathbf{y}, \mathbf{q}) : \Re^{m+n} \rightarrow \Re$. To setup the potential energy of the robot, one has to formulate the potential energy of all its components. This includes especially the following aspects: (i): potential energy U_P of the *mobile platform mass* m_P and its height in the gravitational field. (ii): the elongation of the cable and thus the potential energy U_i stored in the i-th cable (may it be a linear or nonlinear spring). (iii): the cable mass in the gravity field to account for sagging. This potential energy contributes to the potential energy of the cable U_i and depends on the curve formed by the cable. (iv): elastic deformations in robot components such as mobile platform, robot frame, winch, gearbox, or motor. (v): pose-dependent potential energy in the drive-trains and cable guiding system, e.g. translating counter-masses, movable parts in drive-trains, panning pulleys, traveling winches.

The latter two effects can basically be captured by the proposed energy considerations as they represent pose-dependent energy but are neglected for the sake of simplicity in this contribution.

The potential energy U of the overall system at rest is received by summing up the energy of its components

$$U = \sum_i U_i \quad \text{where} \quad i \in \{\mathrm{P}, 1, \ldots, m\} \tag{2}$$

In this setting, m denotes the number of cables. Note that this energy expression can be setup independent of the number of cables m. Especially, there is no difference between under-constrained robots with $m < n$ (IRPM) and over-constrained robots with $m \geq n$ (RRPM). To this end, the energy approach can be formulated regardless of the robot's kinematic classification and for all kinds of cable models (see Sect. 3).

The mobile platform is assumed to be a rigid body with its mobile frame $\mathcal{K}_\mathrm{P}$ located at the center of gravity. The mass of the platform is given by m_P and for the sake of completeness the momentum of inertia is given by $\boldsymbol{\Theta}_\mathrm{P}$ where the inertia tensor is given in the components of frame $\mathcal{K}_\mathrm{P}$. The pose of the platform is given by the position vector $\mathbf{r}$ and the rotation matrix $\mathbf{R}$. The distal anchor points located at the platform are given in local coordinate $\mathcal{K}_\mathrm{P}$ as ${}^\mathrm{P}\mathbf{b}_i$. The potential energy of the platform with respect to the world frame $\mathcal{K}0$ becomes

$$U_\mathrm{P} = m_\mathrm{P}\, \mathbf{g} \cdot \mathbf{r} \tag{3}$$

where $\mathbf{g}$ is the vector of gravitational acceleration respective to $\mathcal{K}0$.

Forward kinematics is derived from the energy by minimizing the potential energy as proposed in a special case in [14] and more generalized by Kraus [9]. By doing so, the effect of inertia in the system is neglected. One provides the unstretched lengths l_i^0 as input parameters and searches for the pose $\mathbf{r}, \mathbf{R}$ that minimize the potential energy

$$\phi^{\mathrm{DK}}(\mathbf{l}^0) = (\mathbf{r}, \mathbf{R}) \quad \text{such that} \quad \min_{\mathbf{r},\mathbf{R}} U. \tag{4}$$

The effectiveness of this approach is already demonstrated while neglecting the potential energy of the platform [14] as well as for using a linear elastic model for the cables [9]. Furthermore, Schmidt showed that one can also cope with panning pulleys [16]. Different numerical schemes can be used to solve the optimization problem. The real-time implementations [14,16] are based on a Levenberg-Marquardt scheme. Interval analysis is also efficient in finding all local optima [6] and preliminary results in this paper are determined with a simplex method.

3 Cable Models

As a prerequisite to modeling the cables, it is important to distinguish between geometric relations and their (common) kinematic interpretation. Without loss

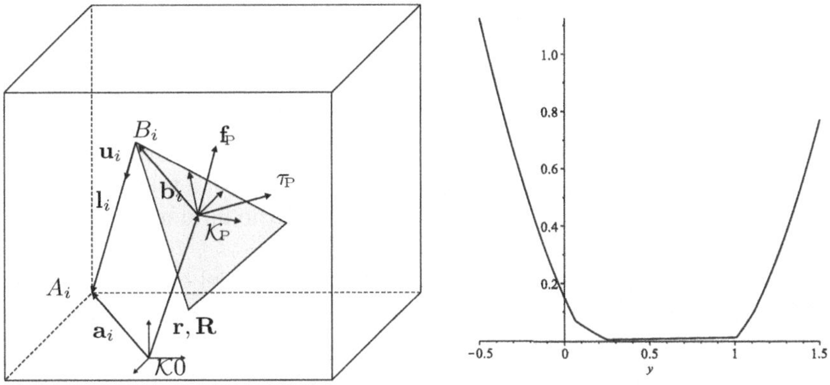

Fig. 1. Left: Geometry and kinematics of a general cable robot. Right: Small section of the energy function for slightly longer cables for $\phi = 0.5$ and $x = 0.2$. The doglegs on the curve result from the unilateral (piece-wise) definition of the energy function for cables.

of generality, the following well-known vector loop $\mathbf{d}_i = \mathbf{a}_i - \mathbf{r}_\text{P} - \mathbf{R}^\text{P}\mathbf{b}_i$ holds true as shown in Fig. 1 where $\mathbf{d}_i$ is the Euclidian displacement between the proximal A_i and distal B_i reference points on robot's base and platform, respectively. Let $d_i = |\mathbf{d}_i|_2$ be a shorthand for the lengths. This distance is a *purely geometric object* and exists whatever is the cable model. It also exists if the cable is slack, broken, or disconnected. In conventional kinematics, the length of this vector $\mathbf{d}_i$ is identified with the effective length l_i of the cable. The following equations are solely derived from the geometric nature.

In the following, some cable models are introduced in terms of energy. These quantities can be identified with the conventional solutions of the inverse kinematics of the standard model. Note that consistent energy formulations depend on the current (unstretched) cable lengths l_i^0, on the pose $(\mathbf{r}, \mathbf{R})$ of the platform, and some physical constants for stiffness.

Inelastic Unilateral Cable: The inelastic (rigid) cable model is rather simple although it presents an ill-conditioned mathematical function in terms of energy. If the distance between cable ends is below l_i^0, energy and force is 0. If it is beyond, the potential energy of the cable is infinitely high. Thus, the cable operates at the step function at length l_i^0 or is ineffective. This energy function shows that the assumption of inelastic cables leads to a rather crude energy model with little physical meaning. In contrast, the derived constraint equation allows the well-known kinematics formulation. The inelastic model needs specific numerical treatment when used in an optimization function.

Strut Model of the Cable: A simple model assumes the cable to be straight slender bar. This assumption disregards the unilateral nature of the cable, however its advantage lies in the simplicity of the model. The potential energy of linear elastic cable can simply be expressed using the proximal and distal anchor points $\mathbf{a}_i$ and $\mathbf{b}_i$, respectively. The energy becomes

$$U_i^{\mathrm{S}} = \frac{1}{2} c_i \left(d_i - l_i^0\right)^2 \tag{5}$$

where $c_i = EA/l_i^0$ is the cable's spring constant composed from the Young's modulus E, the cable's cross-section A and l_i^0 is the unstretched cable length. Note that this differs from the first study [14] where the influence of unstretched cable length on cable energy is neglected.

Unilateral Linear Elastic Cable Model: The strut model captures the elongation of the cable with a high accuracy but it falls short in describing compression as cables are hardly able to transmit pushing forces. As real cables still possess a finite bending stiffness, cable lengths in the magnitude of the diameter are able to transmit also pushing forces but for relevant cable lengths the elastic bending forces are secondary to the gravity of the cable and can be neglected.

The strut model for the energy in the cable seems to be reasonable on first sight but does not accurately reflect the unilateral nature of the cable force transmission. A more accurate model of the cable cancels the potential energy if the distance between the proximal and distal anchor points is smaller than the unstretched cable length l_i^0. Therefore, the corrected function reads

$$U_i^{\mathrm{U}} = \begin{cases} \frac{1}{2} c_i \left(d_i - l_i^0\right)^2 & \text{for} \quad d_i > l_i^0 \\ 0 & \text{otherwise} \end{cases} . \tag{6}$$

4 Results and Discussion

To demonstrate the energy method, a simple example of a planar 1R2T cable robot is selected where a setup with $m = 2$ and $m = 4$ cables, i.e. an under-constrained and an over-constrained model is analyzed. The geometry parameters of the cable robot are as follows $\mathbf{a}_1 = [-1, 1]^{\mathrm{T}}$, $\mathbf{a}_2 = [1, 1]^{\mathrm{T}}$, $\mathbf{a}_3 = [1, 0]^{\mathrm{T}}$, $\mathbf{a}_4 = [-1, 0]^{\mathrm{T}}$, $\mathbf{b}_1 = [0.3, 0.2]^{\mathrm{T}}$, $\mathbf{b}_2 = [0.6, 0.2]^{\mathrm{T}}$, $\mathbf{b}_3 = [0.6, -0.2]^{\mathrm{T}}$, and $\mathbf{b}_4 = [0.3, -0.2]^{\mathrm{T}}$. The mass of the platform is set to $m_{\mathrm{P}} = 1\,\mathrm{kg}$ and the stiffness of cables is assumed to $EA = 1000\,\mathrm{N}$. Using the geometry data in the table and cable length $l_1 = l_4 = 1.334\,\mathrm{m}$ and $l_2 = l_3 = 0.5\,\mathrm{m}$, the energy fields for both the over- and under-constrained cable robots are evaluated with

$$U = U_{\mathrm{P}} + \sum_i U_i, \tag{7}$$

where the only difference between suspended and fully constrained robots is the maximum index of the sum which is $m = 2$ for the IRPM and $m = 4$ for the RRPM. The energy of the system is shown in Fig. 2. Ranges for position x, y and orientation ϕ are chosen such that the pre-computed solution is close by. One can see from the form of the energy function that a rather smooth convergency is expected for optimization of the potential energy U.

In Fig. 1, a specific part of the energy function for the IRPM robot is extracted for $\phi = 0.5$ and $x = 0.2$. The doglegs of the curve at $y = 0.05$, $y = 0.23$, and

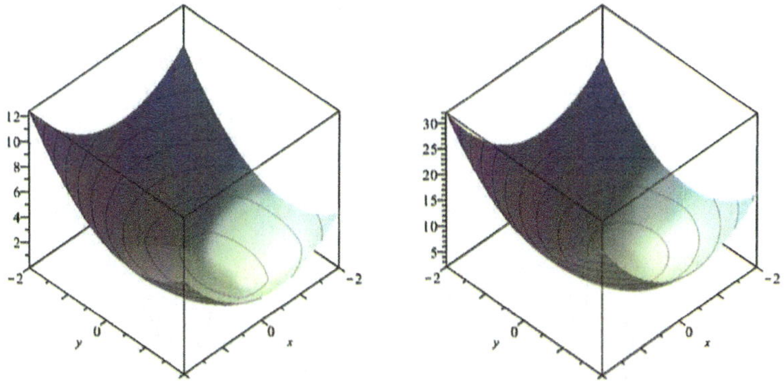

Fig. 2. Energy function for the IRPM version (left) and RRPM (right) version of the cable robot for $\phi = 0.5$.

$y = 1.0$ result from the unilateral (piece-wise) definition of the energy function for cables. The interesting part is the very small gradient on the almost horizontal line between $y = 0.23$ and $y = 1$ which is slight titled towards $y = 0.23$ where the local minimum is found. This gradient is caused by gravity in the region where both cables are slack. Note that real cables are some six orders of magnitude more stiff making the transition from the valley floor to the steep flank a big numerical challenge.

Solutions that cancel the gradient may be stable or instable (see Fig. 3a and b). This can be checked using the Hessian matrix $\mathbf{H}$ with the partial derivatives of second order. If the Hessian matrix is positive definite, the function is convex and one has indeed determined a minimum in the energy functions that is stable under changes in the pose.

An analysis of the Hessian matrix $\mathbf{H}$ for the linear elastic model is given in [15] showing that the forward kinematic solution converges if the unstretched cable lengths are sufficiently close to a solution. This Hessian matrix is related to the stiffness matrix which is the second order derivative of U

$$\mathbf{K} = \frac{\nabla^2 U(\mathbf{l}_0)}{\partial \mathbf{y}^2} \tag{8}$$

Given that the cable robot's potential energy U has multiple local minima, the transition between two minima can be characterized by the amount of energy required to achieve the change (see Fig. 3c).

Any motion of the platform is generated from changes in cable length. This leads to a translation of the minima in the energy function (see Fig. 3d). Thus, the depth ΔU of these local minima is a measure for the stability of the minima and the cable robot's tendency to switch to another minima. Which minimum is reached depends on the previous states of the cable robot thus explaining why physical robots move to a certain minimum out of the set of all local minima. Furthermore, the geometric shape of the energy function in the vicinity of the minimum is a measure for stiffness and stability (see Fig. 3e and f). The Hessian

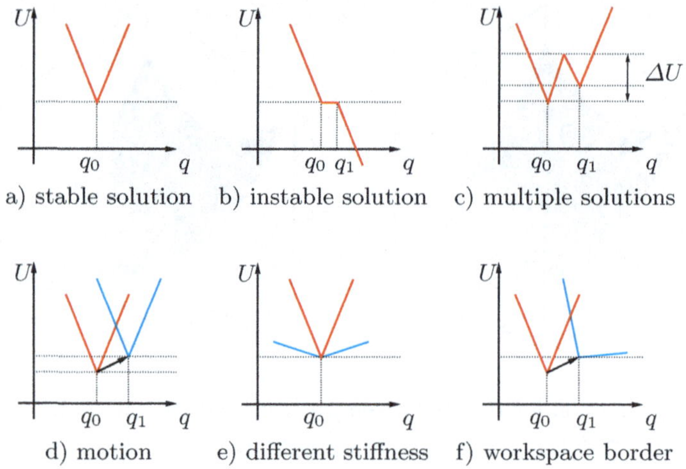

Fig. 3. Characteristic configurations of the energy function in forward kinematics

matrix is a first order estimation of this stiffness. However, also nonlinear changes in the local area around the pose are captured by the energy function and the function can be asymmetric, for example at the border of the workspace.

Identification and consideration of slack cables integrated into the energy formulation of respective cable models and does not require a combinatorial split of the kinematic computation in many special cases.

5 Conclusions

In this paper, an energy-based method is proposed for computing forward kinematics of cable robots. The formulation of the kinematic problem in terms of energy allows to provide a description which is meaningful for over- and under-constrained cable robots. Moreover, the energy method gives a framework to consider different cable models. By embedding the well-known geometric results for forward kinematics into a consistent energy field, one can study the connection between the multiple solutions and derive criteria which solutions are connected in trajectories and how likely changes amongst the so-called solution branches are.

The region around the local minima allows to characterize poses with respect to stiffness and numerical convergency. Moreover, singularities can be understood as the fusion of two local minima. It is conjectured that the analysis of the energy leads to new meaningful metrics to measure the proximity to a singularity as the energy space is a well-defined physical framework.

The results generalize to conventional parallel robots as Stewart-Gough platforms are embedded in the theory when using the strut (cable) model and the number of cables as set to $m = n = 6$.

Acknowledgement. The authors would like to thank the German Research Foundation (DFG) for financial support of the project within the Cluster of Excellence in Simulation Technology (EXC 310/1) at the University of Stuttgart.

References

1. Afshari, A., Meghdari, A.: New Jacobian matrix and equations of motion for a 6 d.o.f cable-driven robot. Int. J. Adv. Robot. Syst. **4**(1), 63–68 (2007)
2. Behzadipour, S., Khajepour, A.: Stiffness of cable-based parallel manipulators with application to stability analysis. J. Mech. Des. **128**(1), 303–310 (2006)
3. Carricato, M., Merlet, J.-P.: Stability analysis of underconstrained cable-driven parallel robots. IEEE Trans. Robot. **29**(1), 289–296 (2013)
4. Nguyen, D.Q., Gouttefarde, M., Company, O., Pierrot, F.: On the simplifications of cable model in static analysis of large-dimension cable-driven parallel robots. In: IEEE International Conference on Intelligent Robots and Systems (IROS) (2013)
5. Dietmaier, P.: The Stewart-Gough platform of general geometry can have 40 real postures. In: Lenarčič, J., Husty, M.L. (eds.) Advances in Robot Kinematics, pp. 7–16. Kluwer Academic Publishers, Salzburg, Austria (1998)
6. Hansen, E., Walster, G.W.: Global Optimization Using Interval Analysis. Marcel Dekker, NEw York (2004)
7. Husty, M.L.: An algorithm for solving the direct kinematic of Stewart-Gough-type platforms. Mech. Mach. Theor. **31**(4), 365–380 (1996)
8. Khosravi, M.A., Taghirad, H.D.: Dynamic modeling and control of parallel robots with elastic cables: singular perturbation approach. IEEE Trans. Robot. **30**(3), 694–704 (2014)
9. Kraus, W.: Force control of cable-driven parallel robots. Ph.D. thesis. Universität Stuttgart, Germany (2015)
10. Merlet, J.-P.: Solving the forward kinematics of a gough-type parallel manipulator with interval analysis. Int. J. Robot. Res. **23**(3), 221–235 (2004)
11. Merlet, J.-P.: A generic numerical continuation scheme for solving the direct kinematics of cable-driven parallel robot with deformable cables. In: IEEE International Conference on Intelligent Robots and Systems, pp. 4337–4343 (2016)
12. Merlet, J.-P., Alexandre dit Sandretto, J.: The forward kinematics of cable-driven parallel robots with sagging cables. In: Pott A., Bruckmann T. (eds) Cable-Driven Parallel Robots, pp. 3–15. Springer, Cham (2014)
13. Ming, A., Higuchi, T.: Study on multiple degree-of-freedom positioning mechanism using wires (part 2) - development of a planar completely restrained positioning Mechanism. Int. J. Japan Soc. Precis. Eng. **28**(3), 235–242 (1994)
14. Pott, A.: An algorithm for real-time forward kinematics of cable-driven parallel robots. In: Advances in Robot Kinematics, pp. 529–538. Springer, Dordrecht (2010)
15. Pott, A., Schmidt, V.: On the forward kinematics of cable-driven parallel robots. In: IEEE/RSJ International Conference on Intelligent Robots and Systems, pp. 3182–3187 (2015)
16. Schmidt, V., Pott, A.: Implementing extended kinematics of a cable-driven parallel robot in real-time. In: Cable-Driven Parallel Robots. Mechanisms and Machine Science, pp. 287–298. Springer, Heidelberg (2013)

Kinematic Modeling and Twist Feasibility of Mobile Cable-Driven Parallel Robots

Tahir Rasheed[1], Philip Long[2], David Marquez-Gamez[3], and Stéphane Caro[4(✉)]

[1] École Centrale de Nantes, Laboratoire des Sciences du Numérique de Nantes, UMR CNRS 6004, 1, rue de la Noë, 44321 Nantes, France
Tahir.Rasheed@ls2n.fr

[2] RIVeR Lab, Department of Electrical and Computing Engineering, Northeastern University, Boston, USA
p.long@northeastern.edu

[3] IRT Jules Verne, Chemin du Chaffault, 44340 Bouguenais, France
david.marquez-gamez@irt-jules-verne.fr

[4] CNRS, Laboratoire des Sciences du Numérique de Nantes, UMR CNRS 6004, 1, rue de la Noë, 44321 Nantes, France
stephane.caro@ls2n.fr

Abstract. The kinematic performance of any robotic system can be analyzed from its first order kinematic model. Recently a novel concept of Mobile Cable-Driven Parallel Robots (MCDPRs) has been introduced to achieve an autonomous reconfiguration of Reconfigurable Cable-Driven Parallel Robots (RCDPRs). To analyze the kinematic performance of MCDPRs, this paper presents the first order kinematic model of the latter. Using the proposed kinematic model, the Available Twist Set (ATS) is determined that is required to analyze the twist capacities of MCDPRs. In contrast to ATS of classical CDPRs we show that the ATS of MCDPRs depends, not only on the cables velocities, but on the velocities of Mobile Bases (MBs) as well. As illustrative examples, the twist capacities of two and six Degree of Freedom (DoF) MCDPR is presented.

Keywords: Cable-driven parallel robots · Mobile bases
Kinematic modeling · Available twist set

1 Introduction

A Cable-Driven Parallel Robot (CDPR) is a type of parallel manipulator with limbs as cables, connecting the moving-platform with a fixed base frame. The platform is moved by appropriately controlling the cable lengths or tensions. CDPRs contains numerous advantages over conventional robots, *e.g*, high accelerations [6], large payload capabilities [1], and large workspace [7].

However, a major drawback in classical CDPRs having fixed cable layout, *i.e*, fixed exit points and cable configuration, is the potential collisions between the cables and the surrounding environment which can significantly reduce the robot

J. Lenarcic and V. Parenti-Castelli (Eds.): ARK 2018, SPAR 8, pp. 410–418, 2019.
https://doi.org/10.1007/978-3-319-93188-3_47

workspace. Better performances can be achieved with an appropriate CDPR architecture. Cable robots with a possibility of undergoing a change in their geometric structure are known as Reconfigurable Cable-Driven Parallel Robots (RCDPRs). Different strategies have been proposed for maximizing the robot workspace or increasing platform stiffness in the recent work on RCDPRs [3]. However, reconfigurability is typically performed manually for most existing RCDPRs.

To achieve autonomous reconfigurability of RCDPRs, a novel concept of Mobile Cable-Driven Parallel Robots (MCDPRs) was introduced in [10]. The first MCDPR prototype has been designed and built in the context of Echord++ "FASTKIT" project[1]. The targeted application for such MCDPR prototype is logistics.

Some papers deals with the velocity analysis of parallel manipulators [9]. However, few focus on the twist analysis of CDPRs [8]. This paper deals with the kinematic modeling of MCDPRs that is required to analyze the kinematic performance of the robot. The paper is organized as follows. Section 2 presents the kinematic model of MCDPRs. Section 3 deals with the determination of the Available Twist Set (ATS) for MCDPRs using the kinematic modeling of the latter. The ATS can be used to obtain the twist capacities of the moving-platform. Section 4 presents the twist capacities of the moving-platform for the MCDPRs under study. Finally, conclusions are drawn and future work is presented in Sect. 5.

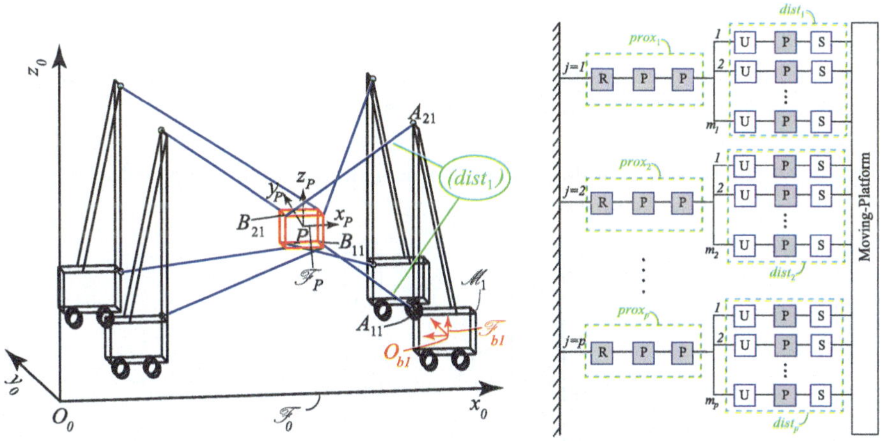

Fig. 1. (a) MCDPR parameterization (b) Kinematic architecture of MCDPRs, active joints are highlighted in gray, passive joints are highlighted in white

[1] https://www.fastkit-project.eu/.

2 Kinematic Modeling

A MCDPR is composed of a classical CDPR with m cables and a n degree-of-freedom (DoF) moving-platform mounted on p Mobile Bases (MBs). The jth mobile base is denoted as $\mathscr{M}_j$, $j = 1, \ldots, p$. The ith cable mounted onto $\mathscr{M}_j$ is named as $\mathscr{C}_{ij}$, $i = 1, \ldots, m_j$, where m_j denotes the number of cables carried by $\mathscr{M}_j$. $\mathbf{u}_{ij}$ denotes the unit vector along the cable $\mathscr{C}_{ij}$. Each jth mobile base along with its m_j number of cables is denoted as jth PD (pd_j) module. Each pd_j consists of a proximal ($prox_j$) and a distal ($dist_j$) module. $dist_j$ consists of m_j cables between $\mathscr{M}_j$ and the moving-platform. In this paper, cables are assumed to be straight and massless, thus can be modeled as a Universal-Prismatic-Spherical (UPS) kinematic chain. Generally MBs are four-wheeled planar robots with two-DoF translational motions and one-DoF rotational motion, thus, $prox_j$ can be modeled as a virtual RPP kinematic chain between the base frame $\mathscr{F}_0$ and the frame $\mathscr{F}_{bj}$ attached to $\mathscr{M}_j$. An illustrative example with $p = 4$ MBs and $m = 8$ cables is shown in Fig. 1a. A general kinematic architecture of a MCDPR is shown in Fig. 1b.

2.1 Kinematics of the Distal Module

A classical CDPR is referred as distal modules in MCDPR. The twist ${}^0\mathbf{t}_P^{dist}$ of the moving-platform due to the latter is expressed as [4,11]:

$$\mathbf{A}\ {}^0\mathbf{t}_P^{dist} = \dot{\mathbf{l}}, \tag{1}$$

where $\mathbf{A}$ is the $(m \times n)$ parallel Jacobian matrix, containing the actuation wrenches due to the cables on the mobile platform. The twist ${}^0\mathbf{t}_P = [\boldsymbol{\omega}, \dot{\mathbf{p}}]^T$ is composed of the platform angular velocity vector $\boldsymbol{\omega} = [\omega_x,\ \omega_y,\ \omega_z]^T$ and linear velocity vector $\dot{\mathbf{p}} = [\dot{p}_x,\ \dot{p}_y,\ \dot{p}_z]^T$, expressed in $\mathscr{F}_0$. ${}^0\mathbf{t}_P^{dist}$ denotes the platform twist due to the distal module motion. $\dot{\mathbf{l}}$ is a m-dimensional cable velocity vector. Here, Eq. (1) can be expressed as:

$$\begin{bmatrix} \mathbf{A}_1 \\ \mathbf{A}_2 \\ \vdots \\ \mathbf{A}_j \\ \vdots \\ \mathbf{A}_p \end{bmatrix} {}^0\mathbf{t}_P^{dist} = \begin{bmatrix} \dot{\mathbf{l}}_1 \\ \dot{\mathbf{l}}_2 \\ \vdots \\ \dot{\mathbf{l}}_j \\ \vdots \\ \dot{\mathbf{l}}_p \end{bmatrix}, \tag{2}$$

where $\dot{\mathbf{l}}_j = [\dot{l}_{1j}, \dot{l}_{2j}, \ldots, \dot{l}_{m_j j}]^T$. $\mathbf{A}_j$ is expressed as:

$$\mathbf{A}_j = \begin{bmatrix} [({}^0\mathbf{b}_{1j} - {}^0\mathbf{p}) \times \mathbf{u}_{ij}]^T & \mathbf{u}_{1j}^T \\ [({}^0\mathbf{b}_{2j} - {}^0\mathbf{p}) \times \mathbf{u}_{2j}]^T & \mathbf{u}_{2j}^T \\ \vdots & \\ [({}^0\mathbf{b}_{m_j j} - {}^0\mathbf{p}) \times \mathbf{u}_{m_j j}]^T & \mathbf{u}_{m_j j}^T \end{bmatrix}, \tag{3}$$

where ith row of $\mathbf{A}_j$ is associated with the actuation wrench of the ith cable mounted onto $\mathscr{M}_j$. ${}^0\mathbf{b}_{ij}$ denotes the Cartesian coordinate vector of the anchor points B_{ij} in $\mathscr{F}_0$. ${}^0\mathbf{p}$ denotes the platform position in $\mathscr{F}_0$.

2.2 Kinematic Modeling of a MCDPR

The twist ${}^0\mathbf{t}_P^j$ of the moving-platform due to pd_j can be expressed in $\mathscr{F}_0$ as:

$$ {}^0\mathbf{t}_P^j = {}^0\mathbf{t}_P^{prox_j} + {}^0\mathbf{t}_P^{dist_j} \tag{4} $$

where ${}^0\mathbf{t}_P^{prox_j}$ (${}^0\mathbf{t}_P^{dist_j}$, resp.) is the twist of the moving-platform due to the motion of the proximal (distal, resp.) module of pd_j expressed in $\mathscr{F}_0$. Equation (4) take the form:

$$ {}^0\mathbf{t}_P^j = {}^{bj}\mathbf{A}d_P \; {}^0\mathbf{t}_{bj}^{prox_j} + {}^0\overline{\mathbf{R}}_{bj} \; {}^{bj}\mathbf{t}_P^{dist_j} \tag{5} $$

where ${}^{bj}\mathbf{A}d_P$ is called the adjoint matrix, which represents the transformation matrix between twists expressed in $\mathscr{F}_{bj}$ and twist expressed in $\mathscr{F}_P$,

$$ {}^{bj}\mathbf{A}d_P = \begin{bmatrix} \mathbf{I}_3 & 0_3 \\ -{}^{bj}\hat{\mathbf{r}}_P & \mathbf{I}_3 \end{bmatrix} \tag{6} $$

where ${}^{bj}\hat{\mathbf{r}}_P$ is the cross-product matrix of vector $\overrightarrow{O_{bj}P}$ expressed in $\mathscr{F}_0$. ${}^{bj}\mathbf{t}_P^{dist_j}$ is the moving-platform twist due to $dist_j$ expressed in $\mathscr{F}_{bj}$. The augmented rotation matrix (${}^0\overline{\mathbf{R}}_{bj}$) is used to express ${}^{bj}\mathbf{t}_P^{dist_j}$ in $\mathscr{F}_0$:

$$ {}^0\overline{\mathbf{R}}_{bj} = \begin{bmatrix} {}^0\mathbf{R}_{bj} & 0_3 \\ 0_3 & {}^0\mathbf{R}_{bj} \end{bmatrix} \tag{7} $$

where ${}^0\mathbf{R}_{bj}$ is the rotation matrix between frames $\mathscr{F}_{bj}$ and $\mathscr{F}_0$. As the Proximal module is being modeled as a virtual RPP limb, ${}^0\mathbf{t}_{bj}^{prox_j}$ from Eq. (4) can be expressed as:

$$ {}^0\mathbf{t}_{bj}^{prox_j} = \mathbf{J}_{bj} \; \dot{\mathbf{q}}_{bj} \tag{8} $$

where $\mathbf{J}_{bj}$ is a (6×3) serial Jacobian matrix of $prox_j$ and $\dot{\mathbf{q}}_{bj}$ is the virtual joint velocities of the latter, namely,

$$ {}^0\mathbf{t}_{bj}^{prox_j} = \begin{bmatrix} \mathbf{k}_0 & \mathbf{0}_3 & \mathbf{0}_3 \\ \mathbf{k}_0 \times {}^0\mathbf{p} & {}^0\mathbf{R}_{bj}\mathbf{i}_0 & {}^0\mathbf{R}_{bj}\mathbf{j}_0 \end{bmatrix} \begin{bmatrix} \dot{\theta}_j \\ \dot{\rho}_{1j} \\ \dot{\rho}_{2j} \end{bmatrix} \tag{9} $$

where $\mathbf{i}_0$, $\mathbf{j}_0$ and $\mathbf{k}_0$ denotes the unit vector along x_0, y_0 and z_0 respectively. Upon multiplication of Eq. (5) with $\mathbf{A}_j$:

$$ \mathbf{A}_j {}^0\mathbf{t}_P^j = \mathbf{A}_j {}^{bj}\mathbf{A}d_P \; \mathbf{J}_{bj} \; \dot{\mathbf{q}}_{bj} + \mathbf{A}_j {}^0\overline{\mathbf{R}}_{bj} \; {}^{bj}\mathbf{t}_P^{dist_j}. \tag{10} $$

As $\mathbf{A}_j {}^0\overline{\mathbf{R}}_{bj}\ {}^{bj}\mathbf{t}_P^{dist_j}$ represents the cable velocities of the $dist_j$ (see Eq. (2)), Eq. (10) can also be expressed as:

$$\mathbf{A}_j {}^0\mathbf{t}_P^j \ = \ \mathbf{A}_j {}^{bj}\mathbf{Ad}_P\ \mathbf{J}_{bj}\ \dot{\mathbf{q}}_{bj} + \dot{\mathbf{l}}_j. \tag{11}$$

The twist of the moving-platform $\mathbf{t}_P$ and the twists generated by the limbs are the same, namely,

$${}^0\mathbf{t}_P^1 = {}^0\mathbf{t}_P^2 = {}^0\mathbf{t}_P^3 \ldots = {}^0\mathbf{t}_P^j \ldots = {}^0\mathbf{t}_P^p = \mathbf{t}_P \tag{12}$$

Thus, the twist of the moving-platform in terms of all the p number of limbs can be expressed as:

$$\begin{bmatrix}\mathbf{A}_1\\ \mathbf{A}_2\\ \vdots\\ \mathbf{A}_p\end{bmatrix}\mathbf{t}_P = \begin{bmatrix}\mathbf{A}_1{}^{b1}\mathbf{Ad}_P\ \mathbf{J}_{b1} & \mathbf{0} & \mathbf{0}\cdots & \mathbf{0}\\ \mathbf{0} & \mathbf{A}_2{}^{b2}\mathbf{Ad}_P\ \mathbf{J}_{b2} & \mathbf{0}\cdots & \mathbf{0}\\ \vdots & \vdots & \vdots & \vdots\\ \mathbf{0} & \mathbf{0} & \mathbf{0}\cdots & \mathbf{A}_p{}^{bp}\mathbf{Ad}_P\ \mathbf{J}_{bp}\end{bmatrix}\dot{\mathbf{q}}_b + \dot{\mathbf{l}} \tag{13}$$

where $\dot{\mathbf{q}}_b = [\dot{\mathbf{q}}_{b1}, \dot{\mathbf{q}}_{b2}, \ldots, \dot{\mathbf{q}}_{bp}]^T$ and $\dot{\mathbf{l}} = [\dot{\mathbf{l}}_1, \dot{\mathbf{l}}_2, \ldots, \dot{\mathbf{l}}_p]^T$. Equation (13) can be expressed in the matrix form as:

$$\mathbf{A}\mathbf{t}_P = \mathbf{B}_b\dot{\mathbf{q}}_b + \dot{\mathbf{l}} \tag{14}$$

$$\mathbf{A}\mathbf{t}_P = \mathbf{B}\dot{\mathbf{q}} \tag{15}$$

where $\mathbf{B} = [\mathbf{B}_b\ \ \mathbf{I}_m]$ is a $(m \times (3p+m))$-matrix while $\dot{\mathbf{q}} = [\dot{\mathbf{q}}_b\ \ \dot{\mathbf{l}}]^T$ is a $(3p+m)$-dimensional vector containing all joint velocities. Equation (15) represents the first order kinematic model of MCDPRs.

3 Available Twist Set of MCDPRs

This section aims at determining the set of available twists for MCDPRs that can be generated by the platform. For a classical CDPR, the set of twist feasible poses of its moving platform are known as Available Twist Set (ATS). A CDPR posture is called twist-feasible if all the twists within a given set, can be produced at the platform, within given joint velocity limits [8].

According to [8], ATS of a CDPR corresponds to a convex polytope that can be represented as the intersection of the half-spaces bounded by its hyperplanes known as *Hyperplane-Shifting Method* (HSM) [2,5]. Although HSM can be utilized to determine the ATS of MCDPRs, the approach in [8] is not directly applicable due to the difference in the kinematic models of CDPR (Eq. 1) and MCDPR (Eq. 15) as matrix $\mathbf{B} \neq \mathbf{I}$.

The kinematic model of the MCDPRs is used to determine the ATS of the moving-platform. In case $m = n$, $\mathbf{A}$ is square, Eq. (15) can be expressed as:

$$\mathbf{t}_P = \mathbf{A}^{-1}\mathbf{B}\,\dot{\mathbf{q}} \implies \mathbf{t}_P = \mathbf{J}\,\dot{\mathbf{q}} \tag{16}$$

where $\mathbf{J}$ is a Jacobian matrix mapping the joint velocities onto the platform twist. The ATS will correspond to a single convex polytope, constructed under the mapping of Jacobian $\mathbf{J}$.

In case $m \neq n$, matrix $\mathbf{A}$ is not square, however there exist in total C_m^n $(n \times n)$ square sub-matrices of matrix $\mathbf{A}$, denoted by $\mathbf{A}^k$, $k = 1, \ldots, C_m^n$, obtained by removing $m - n$ rows from $\mathbf{A}$. For each sub-matrix we can write:

$$\hat{\mathbf{t}}_p^k = {\mathbf{A}^k}^{-1}\mathbf{B}^k \implies \hat{\mathbf{t}}_p^k = \mathbf{J}^k\;\dot{\mathbf{q}}, \qquad k = \{1, \ldots, C_m^n\} \tag{17}$$

where $\hat{\mathbf{t}}_p^k$ is the twist generated by the kth sub-matrix $\mathbf{A}^k$ out of C_m^n $(n \times n)$ square sub-matrices of matrix $\mathbf{A}$. $\mathbf{B}^k$ is a sub matrix of $\mathbf{B}$ using corresponding rows that are chosen in $\mathbf{A}^k$ from $\mathbf{A}$. HSM in [2,5] is directly applicable to compute all the hyperplanes for C_m^n convex polytopes knowing the minimum and maximum joint velocity limits. Thus, the ATS of MCDPRs is the region bounded by all of the foregoing hyperplanes.

4 Results

This section deals with the twist feasibility analysis of two different case studies. From the ATS acquired using the kinematic model of a given MCDPR configuration, we aim to study the difference in the moving platform twist considering fixed and moving MBs.

4.1 Case Study: $p = 2$, $m = 4$ and $n = 2$ DoF MCDPR

The first case study is a planar MCDPR with a point mass end-effector shown in Fig. 2a. The MBs have only one degree of freedom along $\mathbf{i}_0$. The joint velocity limits are defined as:

$$-0.8\ \text{m.s}^{-1} \leq \dot{\rho}_{1j} \leq 0.8\text{m.s}^{-1}, -2\text{m.s}^{-1} \leq \dot{l}_{ij} \leq 2\text{m.s}^{-1}, i = \{1, 2\},\ j = \{1, 2\}, \tag{18}$$

Matrix $\mathbf{A}$ has six 2×2 sub-matrices Thus, ATS is the region bounded by the hyperplanes formed by these six convex polytopes. The difference of ATS between fixed (correspond to a classical CDPR) and moving MBs can be observed in Fig. 2b and c. To illustrate the difference, a Required Twist Set (RTS) equal to $[1.15\,\text{m.s}^{-1}, 1.675\,\text{m.s}^{-1}]^T$ is considered, depicted by a red point in Fig. 2b and c. For fixed MBs, it should be noted that RTS is outside the ATS. By considering moving MBs, RTS is within the ATS.

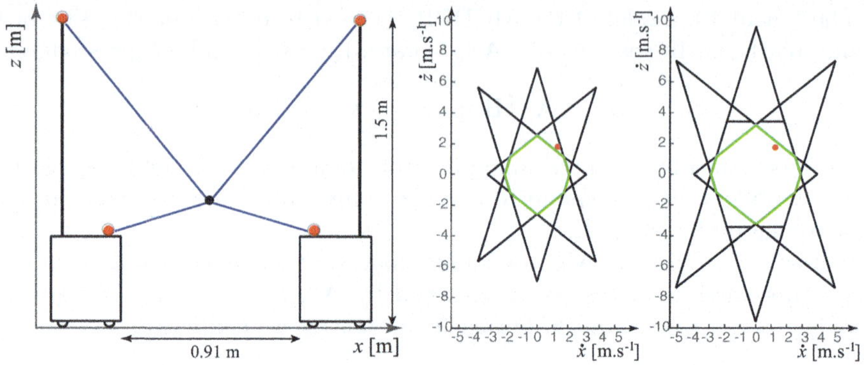

Fig. 2. (a) Configuration under study of $p = 2$, $m = 4$ and $n = 2$ MCDPR (b) ATS in green for fixed MBs (c) ATS in green for moving MBs

4.2 Case Study: $p = 2$, $m = 8$ and $n = 6$ DoF MCDPR

The same approach is adopted to determine the ATS for a given FASTKIT configuration in Fig. 3a. The joint velocity limits are defined as:

$$-0.2\ \text{m.s}^{-1} \leq \dot{\theta}_j,\ \dot{\rho}_{1j},\ \dot{\rho}_{2j} \leq 0.2\ \text{m.s}^{-1}, \quad j = \{1, 2\}, \tag{19}$$

$$-2\ \text{m.s}^{-1} \leq \dot{l}_{ij} \leq 2\ \text{m.s}^{-1}, \quad j = \{1, 2\},\ i = \{1, \ldots, 4\}, \tag{20}$$

The maximum absolute twist that the platform can achieve in each Cartesian direction by considering fixed and moving MBs are illustrated in Fig. 3b and c. The maximum absolute wrench of the moving platform is illustrated in red where f_x, f_y, f_z and m_x, m_y, m_z represent the forces and the moments that can be generated by the cables onto the moving platform. For the analysis, the cable tensions are bounded between 0 and 20 N respectively. It can be observed the

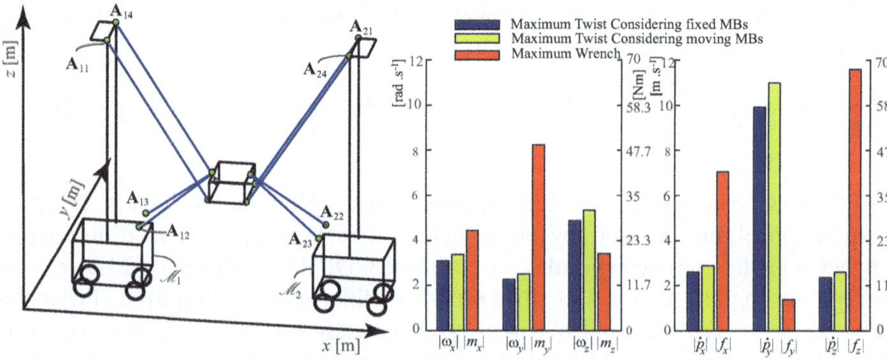

Fig. 3. (a) A FASTKIT configuration (b, c) Maximum absolute twist and wrenches that FASTKIT platform can generate about each Cartesian direction

twist capacity of the moving-platform is increased when MBs are moving. On the contrary, high velocity capability of the moving-platform in certain directions also results in very less wrench capability in the respective directions. Thus, this velocity is unattainable outside certain dynamic conditions.

5 Conclusion

This paper dealt with the kinematic modeling of Mobile Cable-Driven Parallel Robots (MCDPRs) that can be used to analyze its kinematic performance. The developed kinematic model was utilized to determine the Available Twist Set (ATS) of MCDPRs. It considers the joint velocity limits for cables and the Mobile Bases (MBs). Using ATS, the twist capacities of the moving-platform was determined. Two case studies have been used in order to illustrate the effect of the moving MBs onto the platform twist. Future work will focus the trajectory planning of MCDPRs and experimental validations with FASTKIT prototype.

Acknowledgements. This research work is part of the European Project ECHORD++ "FASTKIT" dealing with the development of collaborative and mobile cable-driven parallel robots for logistics.

References

1. Albus, J., Bostelman, R., Dagalakis, N.: The NIST robocrane. J. Field Robot. **10**(5), 709–724 (1993)
2. Bouchard, S., Gosselin, C., Moore, B.: On the ability of a cable-driven robot to generate a prescribed set of wrenches. J. Mech. Robot. **2**(1), 011010 (2010)
3. Gagliardini, L., Caro, S., Gouttefarde, M., Girin, A.: Discrete reconfiguration planning for cable-driven parallel robots. Mech. Mach. Theory **100**, 313–337 (2016)
4. Gouttefarde, M., Daney, D., Merlet, J.-P.: Interval-analysis-based determination of the wrench-feasible workspace of parallel cable-driven robots. In: 2011 IEEE Trans. Robot. **27**(1), 1–13 (2011)
5. Gouttefarde, M., Krut, S.: Characterization of parallel manipulator available wrench set facets. In: Advances in Robot Kinematics: Motion in Man and Machine, pp. 475–482 (2010)
6. Kawamura, S., Kino, H., Won, C.: High-speed manipulation by using parallel wire-driven robots. Robotica **18**(1), 13–21 (2000)
7. Lambert, C., Nahon, M., Chalmers, D.: Implementation of an aerostat positioning system with cable control. IEEE/ASME Trans. Mechatron. **12**(1), 32–40 (2007)
8. Lessanibahri, S., Gouttefarde, M., Caro, S., Cardou, P.: Twist feasibility analysis of cable-driven parallel robots. In: Cable-Driven Parallel Robots, pp. 128–139. Springer, Heidelberg (2018)
9. Merlet, J.-P.: Efficient computation of the extremum of the articular velocities of a parallel manipulator in a translation workspace. In: 1998 IEEE International Conference on Robotics and Automation (ICRA), vol. 3, pp. 1976–1981. IEEE (1998)

10. Rasheed, T., Long, P., Marquez-Gamez, D., Caro, S.: Tension distribution algorithm for planar mobile cable-driven parallel robots. In: The Third International Conference on Cable-Driven Parallel Robots (CableCon 2017), pp. 268–279. Springer, Heidelberg (2018)
11. Roberts, R.G., Graham, T., Lippitt, T.: On the inverse kinematics, statics, and fault tolerance of cable-suspended robots. J. Field Robot. **15**(10), 581–597 (1998)

Kinetostatic Characterization of a Loading System Based on a Partially-Decoupled Parallel Manipulator

Giulio Spagnuolo(✉), Nicola Sancisi, Rocco Vertechy, and Vincenzo Parenti-Castelli

DIN-Department of Industrial Engineering, University of Bologna, Viale del Risorgimento 2, 40136 Bologna, Italy
{giulio.spagnuolo,nicola.sancisi,rocco.vertechy, vincenzo.parenti}@unibo.it

Abstract. The paper deals with the kinetostatic characterization of a knee loading test-rig with cable-driven parallel mechanism architecture. The system is supposed to work close to a nominal pose where it shows a partially-decoupled behavior. The scope of the study is the determination of the maximum deviations between the wrench components at the nominal and actual platform pose for a generic loading condition. Moreover, the peculiar cable-based actuation system leads to non-linearity issues: a specific procedure is then presented to perform the kinetostatic analysis and to handle such non-linearities.

Keywords: Parallel manipulator · Kinetostatic analysis
Wrench deviation analysis · Geometrical non-linearities

1 Introduction

Biomechanical models have been proved to be fundamental in the prosthesis design and for outlining surgical treatments and rehabilitation strategies [1, 2]. A key point for the definition and validation of such models is the characterization of the kinetostatic and dynamic behaviour of human joints. In the last thirty years a number of in vitro tests has been promoted in order to reproduce the in vivo loading condition. Several tests have been focused on the knee joint, which is subjected to a statistically significant number of injuries and, therefore, surgical interventions. There have been identified two main groups of knee test rigs, based on their architecture, namely the Oxford Knee Rig (OKR) types [3] and the robot-based knee testing systems [4]. The OKR consents six degrees of freedom (dof) to the knee joint, but it cannot provide general loading conditions. Notwithstanding the capability to apply general loading conditions, the robot-based knee systems are complex, costly and less performant, and this prevents their wider adoption. In order to overcome these limitations, a new test rig was developed [5]. It is able to analyse the knee behaviour in a wide range of flexion movements, while the applied loads can be modified in real time, to simulate generical tasks. Despite its simplicity and low cost, the rig was demonstrated to have good accuracy, repeatability and versatility. The core of the rig is a loading system with a

J. Lenarcic and V. Parenti-Castelli (Eds.): ARK 2018, SPAR 8, pp. 419–427, 2019.
https://doi.org/10.1007/978-3-319-93188-3_48

cable-driven parallel mechanism architecture that is partially decoupled for platform motions about a given nominal pose (position and orientation). This simplifies the force control of the rig, improves its dynamic performance, and maximizes the loads applied by actuators to the platform in all directions.

Despite its simplicity, the considered cable-driven loading system leads to face issues in terms of assessment, control and accuracy. Avoiding an in-depth description of the test rig, which is deeply discussed in [5], the present paper is thus focused on the kinetostatic characterization of the loading system. In particular, it assesses the extent to which the chosen parallel mechanism is actually decoupled for platform poses different from the nominal one. Definition of the maximum absolute deviation between the nominal and actual wrench (force and torque) components provides a measure for this decoupling: as long as this deviation remains low, the mechanism can be considered decoupled and the force control of the manipulator does not require measurements of the platform configurations to impose a given load. In addition, the particular actuation system based on cables leads to an interesting issue: for a given platform pose, the direction of forces applied by each actuator changes with the wrench, introducing non-linearities in the control of the system. A method to deal with these non-linearities with respect to (wrt) the kinetostatic problem is proposed.

2 Loading System Characterization

2.1 Modelling and Notation

The aforementioned loading system is equivalent to a 6-6 Stewart platform mechanism [6], with some fundamental differences. With respect to a standard fully-parallel manipulator, the linear actuator of each leg is substituted by a cable closed loop (realized by means of pulleys) which includes a double-stroke pneumatic actuator with double-ended piston rod on the base (Fig. 1a). The use of cables reduces inertias on the platform, while the loop makes each actuator able to apply both positive and negative forces on the platform. Cable arrangement is shown in Fig. 1b: the two ends of each cable loop are connected at two opposite points on the movable platform, in order to apply forces along opposite directions (positive and negative forces). Introducing a notation taken from cable-driven parallel manipulator theory [7], the loading system can be represented as a 6-dof parallel mechanism whose mobile platform is driven by 12 cables, coupled in six pairs $\{i, i'\}$. Each pair of cables belongs to a cable closed loop (Figs. 1a and 2): i and i' represent the line of action respectively of a positive or a negative force exerted by an actuator. Thus, at each instant only six cables are actually active, the others being slack, depending on the signs of the actuator forces. Moreover, the line of action of i and i' are coincident at the nominal pose, but in general are different at other poses.

Let us consider a fixed frame $\mathbf{S}_O = (O, \mathrm{x}, \mathrm{y}, \mathrm{z})$ and a mobile frame $\mathbf{S}_P = (P, \mathrm{x_p}, \mathrm{y_p}, \mathrm{z_p})$, the latter being attached at the ring-platform and centered at P, as shown in Fig. 2. $\mathbf{S}_P$ is coincident with $\mathbf{S}_O$ at the nominal pose. Referring to a cable-pair closed loop, the cable i [i'] is taut between the points A_i and B_i [A_i' and B_i'], which are modeled as spherical joints and assumed to be fixed wrt the base and the mobile platform, respectively.

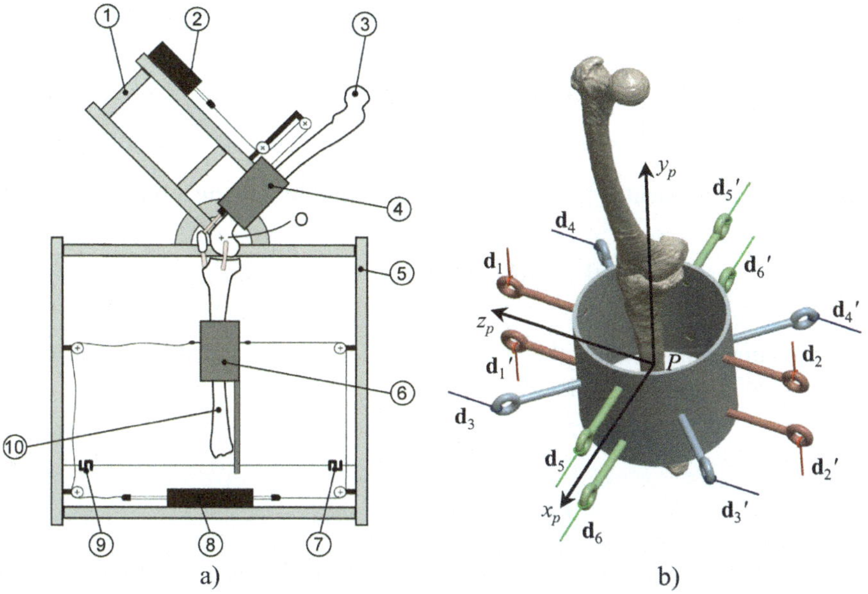

Fig. 1. (a) Test rig structure: portal (1); single-acting pneumatic actuator (2); femur (3); femur fixation system (4); base (5); tibial ring (6); load cells (7) and (9); double-acting pneumatic actuator (8); tibia (10). (b) Cable arrangement on the mechanism platform [5].

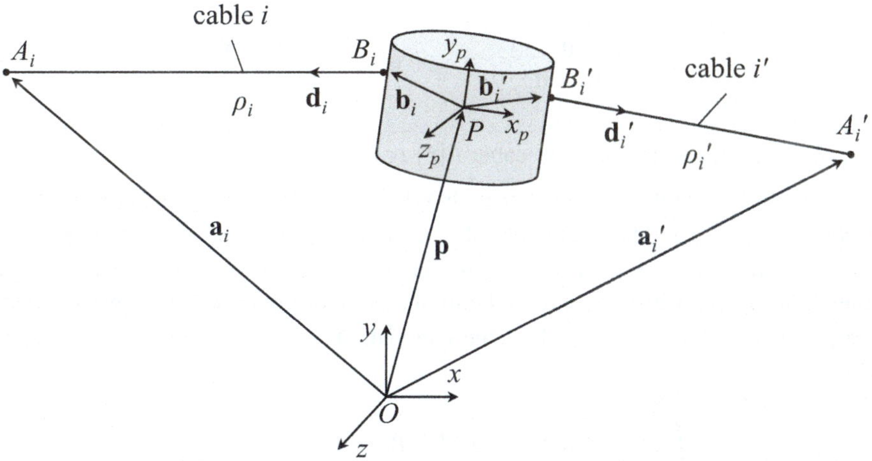

Fig. 2. Cable-pair closed loop notation

The pose of the mobile platform wrt $\mathbf{S}_O$ is defined by the position vector $\mathbf{p}$, of point P ($\vec{OP}$), and by the orientation of the mobile frame wrt the base frame, given by a rotation matrix $\mathbf{R}$. The taut length of the cable i [i'] is denoted $\rho_i\left[\rho'_i\right]$ and the representation of $\vec{OA_i}\left[\vec{OA'_i}\right]$ and $\vec{PB_i}\left[\vec{PB'_i}\right]$ in $\mathbf{S}_O$ are denoted as $\mathbf{a}_i\left[\mathbf{a}'_i\right]$ and $\mathbf{b}_i\left[\mathbf{b}'_i\right]$, respectively. Let us identify each cable i [i'] direction by its unit vector

$$\mathbf{d}_i = \vec{A_iB_i}/\rho_i \quad \left[\mathbf{d}'_i = \vec{A'_iB'_i}/\rho'_i\right] \tag{1}$$

The relationship between the twelve cable tensions $\boldsymbol{\tau}$ and the wrench $\mathbf{f}$ at the platform reference point P is [8, 9]:

$$\mathbf{W}\boldsymbol{\tau} = \mathbf{f} \tag{2}$$

$$\mathbf{f} = \left(f_x, f_y, f_z, m_x, m_y, m_z\right)^T \tag{3}$$

where $\mathbf{f}$ is a 6×1 vector (in order, force through P and moment wrt P), $\boldsymbol{\tau}$ is a 12×1 vector, and $\mathbf{W}$ is the 6×12 pose-dependent *wrench matrix* whose columns are denoted by $\mathbf{w}_i$ [$\mathbf{w}'_i$]

$$\mathbf{W} = \left(\mathbf{w}_1, \ldots, \mathbf{w}_6, \mathbf{w}'_1, \ldots, \mathbf{w}'_6\right) \tag{4}$$

where

$$\mathbf{w}_i = \begin{pmatrix} \mathbf{d}_i \\ \mathbf{b}_i \times \mathbf{d}_i \end{pmatrix}_{6\times 1}, \quad \mathbf{w}'_i = \begin{pmatrix} \mathbf{d}'_i \\ \mathbf{b}'_i \times \mathbf{d}'_i \end{pmatrix}_{6\times 1} \tag{5}$$

Thus, the wrench applied at P by cable i [i'] is $\mathbf{f}_i = \mathbf{w}_i\tau_i \left[\mathbf{f}'_i = \mathbf{w}'_i\tau_i\right]$.

As mentioned above, the system is driven by twelve cables, coupled in six pairs $\left\{\mathbf{d}_i, \mathbf{d}'_i\right\}(i = 1, \ldots, 6)$. The vector of cable tensions can be written as $\boldsymbol{\tau} = \left(\boldsymbol{\tau}_+^T, \boldsymbol{\tau}_-^T\right)^T$, where $\boldsymbol{\tau}_+$ and $\boldsymbol{\tau}_-$ are 6×1 vectors, representative of the cable tensions induced by the actuators during a positive ($\delta_i > 0$, δ_i being the actuator pressure) and negative ($\delta_i < 0$) actuation, respectively. For each cable-pair $\left\{\mathbf{d}_i, \mathbf{d}'_i\right\}$, the following relationship is verified:

$$\begin{cases} \left\{\tau_{+,i}, \tau_{-,i}\right\} = \left\{\tau_{a,i}, 0\right\}, \delta_i > 0 \\ \left\{\tau_{+,i}, \tau_{-,i}\right\} = \left\{0, \tau_{a,i}\right\}, \delta_i < 0 \end{cases} \quad i = 1, \ldots, 6 \tag{6}$$

where $\tau_{a,i}$ represents the magnitude of the force exerted by the i-th actuator for a positive or negative pressure $\left(\tau_{a,i} \geq 0\right)$. Thus, the wrench matrix (3) can be also written as

$$\mathbf{W} = \left(\mathbf{w}_1, \ldots, \mathbf{w}_6, \mathbf{w}_1^{'}, \ldots, \mathbf{w}_6^{'}\right) = (\mathbf{W}_+, \mathbf{W}_-) \tag{7}$$

where $\mathbf{W}_+$ and $\mathbf{W}_-$ are 6×6 pose-dependent *wrench sub-matrices*. Thus, Eq. (2) becomes

$$(\mathbf{W}_+, \mathbf{W}_-)\left(\boldsymbol{\tau}_+^T, \boldsymbol{\tau}_-^T\right)^T = \mathbf{f} \tag{8}$$

For the peculiar cable arrangement (Fig. 1b), at the nominal pose $\mathbf{W}_+ = -\mathbf{W}_-$, and wrench force and moment components are decoupled. For instance, a vertical force is obtained by acting on the cable pair $\{\mathbf{d}_1, \mathbf{d}_2\}$ with the same load, while a moment about the anterior axis x_p is obtained by acting on $\left\{\mathbf{d}_1, \mathbf{d}_2^{'}\right\}$ with the same load.

2.2 Kinetostatic Characterization

Let us define:
(n) *Theoretical (nominal) configuration*, the nominal pose of the platform.
(r) *Actual (real) configuration*, the actual pose of the platform.

According to [10], the loading system kinetostatic characterization refers to the comparison between the wrench obtained at the (n)- and (r)-configurations of the loading system for the same actuator forces $\boldsymbol{\tau}$. Since the wrench matrix $\mathbf{W}$ is pose-dependent, the deviation of the wrench $\mathbf{f}^{(\mathrm{n})}$ wrt $\mathbf{f}^{(\mathrm{r})}$ is

$$\begin{cases} \mathbf{W}^{(\mathrm{n})}\boldsymbol{\tau} = \mathbf{f}^{(\mathrm{n})} \\ \mathbf{W}^{(\mathrm{r})}\boldsymbol{\tau} = \mathbf{f}^{(\mathrm{r})} \end{cases} \Rightarrow \left(\mathbf{W}^{(\mathrm{r})} - \mathbf{W}^{(\mathrm{n})}\right)\boldsymbol{\tau} = \left(\mathbf{f}^{(\mathrm{r})} - \mathbf{f}^{(\mathrm{n})}\right) \Rightarrow \Delta\mathbf{W}\boldsymbol{\tau} = \Delta\mathbf{f} \tag{9}$$

where $\Delta\mathbf{f}$ and $\Delta\mathbf{W} = (\Delta\mathbf{W}_+, \Delta\mathbf{W}_-)$ are the variations of the wrench and of the wrench matrix. The columns of the latter are

$$\begin{cases} \Delta\mathbf{w}_{+,i} = \mathbf{w}_{+,i}^{(\mathrm{r})} - \mathbf{w}_{+,i}^{(\mathrm{n})} \\ \Delta\mathbf{w}_{-,i} = \mathbf{w}_{-,i}^{(\mathrm{r})} - \mathbf{w}_{-,i}^{(\mathrm{n})} \end{cases} \tag{10}$$

Since Eq. (8) is linear, the maximum variation of the k-th wrench component Δf_k $(k = 1, \ldots, 6)$ is obtained when all actuators exert the maximum actuation force in either positive or negative direction. Each Δf_k can assume both positive and negative values. However, due to the non-linearity introduced by Eq. (5), the superior and inferior limit values can be different. If the maximum actuation force is the same for all actuators, the superior and inferior limit values for Δf_k, wrt the maximum theoretical value of the k-th wrench component (namely, the full-scale reading) are

$$\begin{cases} e_{sup,k} = (\max_\tau \Delta f_k)/\left(\max_\tau \left|f_k^{(n)}\right|\right) = \sum_i \max\left\{\Delta \mathrm{w}_{+,(k,i)}, \Delta \mathrm{w}_{-,(k,i)}\right\}/\sum_i \left|\mathrm{w}_{+,(k,i)}^{(\mathrm{n})}\right| \\ e_{inf,k} = (\min_\tau \Delta f_k)/\left(\max_\tau \left|f_k^{(n)}\right|\right) = \sum_i \min\left\{\Delta \mathrm{w}_{+,(k,i)}, \Delta \mathrm{w}_{-,(k,i)}\right\}/\sum_i \left|\mathrm{w}_{+,(k,i)}^{(\mathrm{n})}\right| \end{cases} \tag{11}$$

The maximum absolute error value of the k-th wrench component wrt the full-scale reading is

$$e_k = \max\left(\left|e_{sup,k}\right|, \left|e_{inf,k}\right|\right) \tag{12}$$

Thus, $\mathbf{e} = (e_k), k = 1, \ldots, 6$, is a 6×1 vector, that represents the pose-dependent maximum deviation of the wrench $\mathbf{f}^{(\mathrm{n})}$ wrt $\mathbf{f}^{(\mathrm{r})}$.

3 Numerical Results

The kinetostatic performance of the loading system was evaluated with the procedure described above, by computing the pose-dependent maximum deviation $\mathbf{e}$ of the wrench $\mathbf{f}^{(\mathrm{n})}$ wrt $\mathbf{f}^{(\mathrm{r})}$. In order to perform an overall analysis of the system, a generic reference geometry was chosen, considering the cable arrangement represented in Fig. 1b. The projection of B_i and B_i' on the $\mathrm{x_p z_p}$-plane, denoted π_{xz}^{p}, lies on a circumference centered at P and with radius $r_B = 100$ mm. Moreover, the distance between B_i [B_i'] (with i = 1, 2, 5, 6) and π_{xz}^{p} is $l_B = 100$ mm. Finally, the taut length of the cable i and i' at the nominal pose is $\rho_i^{(\mathrm{n})} = \rho_i'^{(\mathrm{n})} = 500$ mm.

Since the error vector $\mathbf{e}$ is pose-dependent, the (r)-configurations analyzed here were chosen considering a spherical workspace (SW), described by its spherical coordinates SW $= (r_{SW}, \theta, \varphi)$, where: r_{SW} is the sphere radius; θ is the azimuth angle, measured counterclockwise in the xy-plane wrt $\mathbf{S}_o$, from the positive x-axis; φ is the elevation angle from the xy-plane wrt $\mathbf{S}_o$. It is worth mentioning that, due to the symmetry of the reference geometry, only the following spherical sector was analyzed:

$$\mathrm{SW} = (r_{SW}, \theta, \varphi)|\{\theta \in [0, 90°], \varphi \in [0, 90°]\} \tag{13}$$

The maximum deviations due to platform translation and rotation were analyzed separately. In particular, the following platform motion conditions were taken into consideration:

- a translation of the platform center P along the SW surface, considering a radius value of $r_{SW} = 10$ mm;
- a rotation by 5° about a vector lying on a generic SW radius, for several azimuth and elevation angles θ, φ.

Results are reported in Figs. 3 and 4. In particular, Fig. 3 shows the maximum deviation due to the platform translation, while Fig. 4 shows the maximum deviation due to the platform rotation. Results show that the loading system performs well wrt the

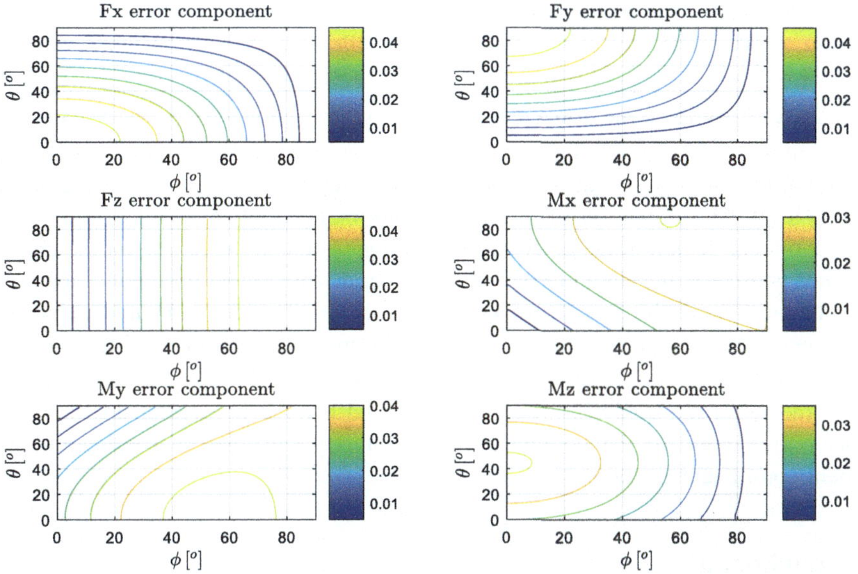

Fig. 3. Maximum deviations vs platform translation: the color bar refers to the maximum absolute error value of each wrench component wrt the full-scale, due to platform translation.

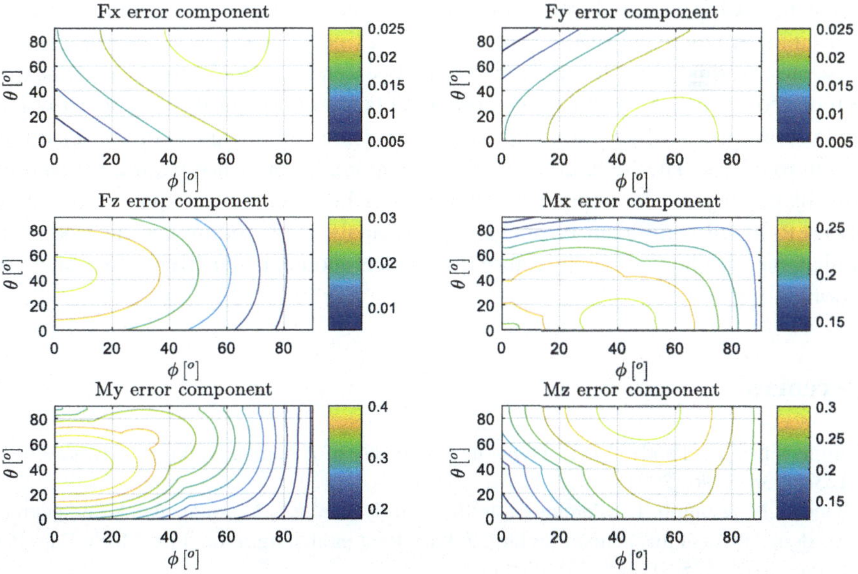

Fig. 4. Maximum deviations vs platform rotation: the color bar refers to the maximum absolute error value of each wrench component wrt the full-scale, due to platform rotation.

exerted forces, i.e., force components are almost insensitive to small variations of the platform pose: force deviation due to both platform translation and rotation is $e_i < 0.05$ $(i = 1, \ldots, 3)$. As a consequence of the mechanism architecture, since the ratio $r_{SW}/\rho_i^{(n)}$ remains low, the force components of the wrench $\mathbf{f}^{(n)}$ can be considered equivalent to the ones of $\mathbf{f}^{(r)}$. On the contrary, a platform rotation produces a significant increase of the moment deviation error (Fig. 4): a relatively low rotation of the platform leads to a 20–40% deviation of the moment components at some rotation directions. This significant deviation depends on the loading system architecture, on the analyzed reference geometry and, in addition, on the general wrench and pose conditions of the present analysis: considering all rotations and load combinations in cables, changes in the moment arm are significant even for low rotations of the platform, which is reflected on the maximum deviation of the moment components of $\mathbf{f}^{(n)}$ wrt to those of $\mathbf{f}^{(r)}$.

By means of the analysis procedure proposed here, it is possible to show that a proper optimization of the loading system geometry can lead to a reduction of the moment component deviation.

4 Conclusions

A preliminary kinetostatic characterization of a loading system with a cable-driven parallel mechanism architecture has been performed. The system is supposed to work close to a nominal pose where it shows a partially-decoupled behavior. The scope of this analysis is the determination of the maximum deviations between the wrench components at the nominal and actual platform pose for generic loading conditions. The peculiar arrangement of the cables introduces non-linearities: a specific procedure is presented in this paper to perform the analysis and to handle these non-linearities. The reported results show that force components are almost insensitive to variation of the platform pose, while moment components are significantly more sensitive to platform rotations. The force control of the manipulator does not require measurement of the platform pose to impose given forces, while moments require more precise analyses to avoid significant errors. The same analysis procedure can also be used to optimize the mechanism architecture to reduce moment sensitivity to real system positioning.

References

1. Anderson, F.C., Pandy, M.G.: Dynamic optimization of human walking. J. Biomech. Eng. **123**(5), 381–390 (2001)
2. Sancisi, N., Zannoli, D., Parenti-Castelli, V., Belvedere, C., Leardini, A.: A one-degree-of-freedom spherical mechanism for human knee joint modelling. Proc. Inst. Mech. Eng. Part H J. Eng. Med. **225**(8), 725–735 (2011)
3. Zavatsky, A.B.: A kinematic-freedom analysis of a flexed-knee-stance testing rig. J. Biomech. **30**(3), 277–280 (1997)

4. Varadarajan, K.M., Harry, R.E., Johnson, T., Li, G.: Can in vitro systems capture the characteristic differences between the flexion–extension kinematics of the healthy and TKA knee? Med. Eng. Phys. **31**(8), 899–906 (2009)
5. Forlani, M., Sancisi, N., Conconi, M., Parenti-Castelli, V.: A new test rig for static and dynamic evaluation of knee motion based on a cable-driven parallel manipulator loading system. Meccanica **51**(7), 1571–1581 (2016)
6. Innocenti, C., Parenti-Castelli, V.: Direct position analysis of the Stewart platform mechanism. Mech. Mach. Theory **25**(6), 611–621 (1990)
7. Gouttefarde, M., Daney, D., Merlet, J.P.: Interval-analysis-based determination of the wrench-feasible workspace of parallel cable-driven robots. IEEE Trans. Rob. **27**(1), 1–13 (2011)
8. Roberts, R.G., Graham, T., Lippitt, T.: On the inverse kinematics, statics, and fault tolerance of cable-suspended robots. J. Field Robot. **15**(10), 581–597 (1998)
9. Hiller, M., Fang, S., Mielczarek, S., Verhoeven, R., Franitza, D.: Design, analysis and realization of tendon-based parallel manipulators. Mech. Mach. Theory **40**(4), 429–445 (2005)
10. Sancisi, N., Zannoli, D., Parenti-Castelli, V.: A procedure to analyse and compare the sensitivity to geometrical parameter variations of one-DOF mechanisms. In: ASME 2011 International Design Engineering Technical Conferences and Computers and Information In Engineering Conference, pp. 453–460. American Society of Mechanical Engineers (2011)

Biorobotics

Kinematic Analysis of a Novel Parallel 2SPRR+1U Ankle Mechanism in Humanoid Robot

Shivesh Kumar[1(✉)], Abhilash Nayak[3], Heiner Peters[1], Christopher Schulz[1], Andreas Müller[4], and Frank Kirchner[1,2]

[1] Robotics Innovation Center, German Research Center for Artificial Intelligence (DFKI GmbH), 28359 Bremen, Germany
{shivesh.kumar,heiner.peters,christopher.schulz,frank.kirchner}@dfki.de
[2] Fachbereich Mathematik und Informatik, Arbeitsgruppe Robotik, Universität Bremen, 28359 Bremen, Germany
[3] LS2N, Ecole Centrale de Nantes, 44321 Nantes, France
abhilash.nayak@ls2n.fr
[4] Institute of Robotics, Johannes Kepler University, Linz 4040, Austria
a.mueller@jku.at

Abstract. Parallel mechanisms are increasingly being used as modular subsystems in various robots and man-machine interfaces for their good stiffness, payload to weight ratio and dynamic properties. This paper presents the kinematic analysis of a novel parallel mechanism of type 2SPRR+1U for application as a humanoid ankle joint with two degrees of freedom. Tools from computational algebraic geometry are used to provide solutions to the forward and inverse kinematics problems. These are further used to characterize the workspace of this mechanism and provide description of its singularity curves. The kinematic analysis demonstrates that the chosen design can provide human ankle like workspace and good torque transmission capability without suffering from any singularities which makes it an ideal candidate for ankle joint module in humanoid robots.

Keywords: Kinematic analysis · Parallel mechanisms · Ankle joint Humanoids

1 Introduction

The growing popularity of parallel mechanisms includes their new applications as modular subsystems in humanoids, animaloids and man-machine interfaces due to their good stiffness, payload to weight ratio and dynamic properties. Indeed the lower amount of moving mass and freedom to place actuators away from the moving platform motivates researchers to develop new mechanisms which can better mimic the complex anatomy of biological systems. In recent years, closed loop sub-mechanisms such as Active Ankle [4], Stewart platform, 2SPU+1U [7],

J. Lenarcic and V. Parenti-Castelli (Eds.): ARK 2018, SPAR 8, pp. 431–439, 2019.
https://doi.org/10.1007/978-3-319-93188-3_49

double parallelogram linkage [5] etc. have been used in various hybrid robotic systems like hominid CHARLIE [3], multi-legged robot MANTIS [1], RECUPERA full body exoskeleton [2] and humanoid robot LOLA [6]. Most of these hybrid robots utilize parallel submechanisms as two or three DOF orientational units.

In this paper, we present a novel two degrees of freedom orientational parallel mechanism of type 2SPRR+1U which is used as an ankle joint in a humanoid robot developed at DFKI-RIC (see Fig. 1). The kinematic actuation principle of this mechanism comprises of a motion constraint generator leg with a universal joint and two auxiliary actuation legs of type SPRR as shown in Fig. 2. It is well known that during walking, the torque required for the pitch movement is larger than the torque required for the roll movements [6]. When the two motors are actuated in the same direction, the mechanism produces a pitch only movement demonstrating good torque transmission characteristics. It has been shown in biomechanics studies that during the ankle pitch movement of human gait, a peak torque between 105 Nm and 120 Nm is required when flexion/extension angle is between $-6°$ and $-12°$ [8]. To reflect this in the ankle design, the foot attachment points of the two linear actuators may be displaced along the z-axis by 30 mm. Utilizing a common universal joint at the offset points would reduce the workspace of the roll movement. Instead, two skew revolute joints, with axes parallel to axes of universal joint on constraint generator leg, connected by an intermediate offset link are used to provide the desired torque characteristics in the pitch movement with minimal influences on the motion range of roll movement. The main contribution of this paper is a comprehensive kinematic analysis of this mechanism.

The paper is organized as follows: Sect. 2 presents the manipulator's architecture and constraint equations. Section 3 presents the solutions to the direct and inverse kinematic problems by utilizing tools from computational algebraic

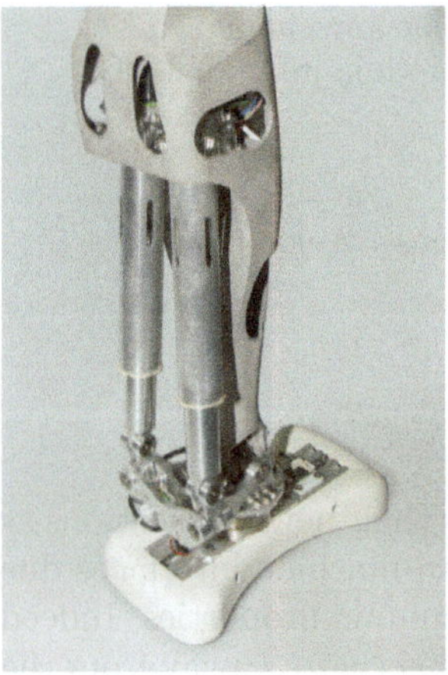

Fig. 1. Built-up prototype of Ankle joint

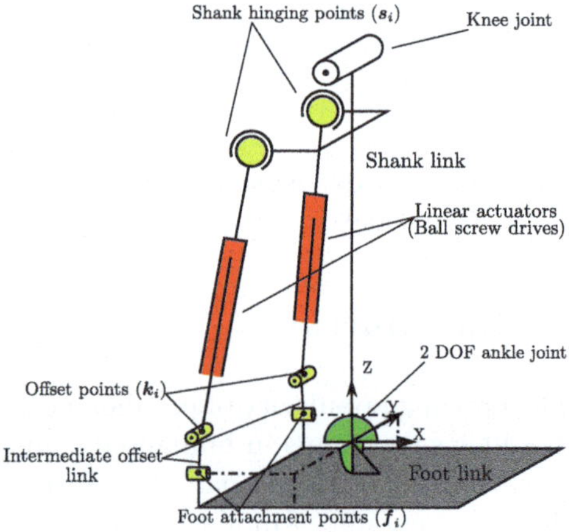

Fig. 2. Scheme of the mechanism

geometry. Section 4 presents the workspace characterisation, description of its singularity curves and performance analysis and Sect. 5 concludes the paper.

2 Architecture and Constraint Equations

The mobility of a mechanism ($\mathcal{M}$) can be calculated with the help of Kutzbach-Grübler criteria as follows:

$$d_s(\mathcal{M}) = s \cdot (n - m - 1) + f = s \cdot (-c) + f, \tag{1}$$

where n is number of links in the mechanism $3 + 3 + 2 = 8$, m is total number of joints $4+4+1 = 9$, f is total dof of joints $2+6+6 = 14$, and s is the motion parameter. Since, it is a spatial mechanism, $s = 6$. Hence, the mobility can be calculated as $d_s(\mathcal{M}) = 6 \cdot (8 - 9 - 1) + 14 = 2$.

The manipulator architecture and geometry is shown in Fig. 3. Let us define a set of three points: shank point ($\mathbf{s}_i$), foot attachment point ($\mathbf{f}_i$) and the offset point ($\mathbf{k}_i$) on the two auxiliary actuation legs of the mechanism. The base frame O is attached to the shank link and is coincident with the end effector (EE) frame E attached to the foot link in zero configuration. The intermediate offset link $\mathbf{f}_i\mathbf{k}_i$ rotates about the x-axis (denoted as $\hat{\mathbf{n}}$) of the frame defined at $\mathbf{f}_i$, thus point $\mathbf{k}_i$ moves on a circle of radius r equal to the length of the link, $\|\mathbf{f}_i - \mathbf{k}_i\|$. The length of the linear actuators (d_i) is the norm of the vector ($\mathbf{k}_i - \mathbf{s}_i$). We also define a vector $\boldsymbol{\delta}_i := (\mathbf{s}_i - \mathbf{f}_i)$.

The constraint equations of the manipulator are the following:

$$d_i^2 = \|\mathbf{k}_i - \mathbf{s}_i\|^2 = \|\mathbf{p}_i - \mathbf{s}_i\|^2 + \|\mathbf{k}_i - \mathbf{p}_i\|^2, i \in \{1, 2\} \tag{2}$$

We can rewrite (2) purely as a function of ($\mathbf{s}_i, \hat{\mathbf{n}}, \mathbf{f}_i$).

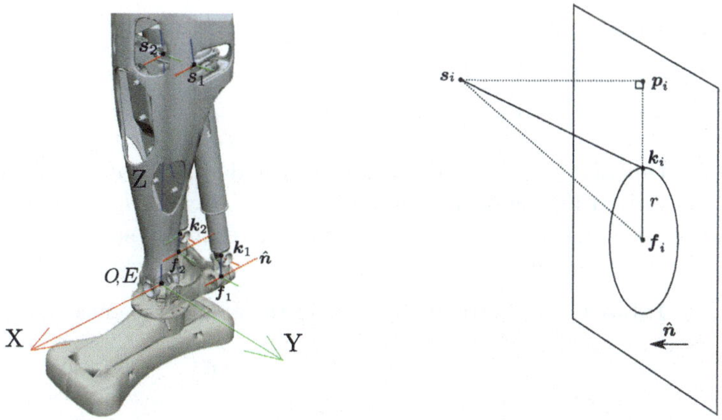

Fig. 3. Geometry of the 2 dof ankle mechanism (RGB colours denote XYZ axes)

$$d_i^2 = ||\hat{\mathbf{n}} \cdot \boldsymbol{\delta}_i||^2 + (||\mathbf{f}_i - \mathbf{p}_i|| - r)^2$$
$$d_i^2 = ||\hat{\mathbf{n}} \cdot \boldsymbol{\delta}_i||^2 + (||\hat{\mathbf{n}} \times \boldsymbol{\delta}_i|| - r)^2 \tag{3}$$

For the purpose of visualization or computing passive joint angles, it is necessary to compute the point $\mathbf{k}_i$ which is given by (4).

$$\mathbf{k}_i = \mathbf{f}_i + r \frac{\boldsymbol{\delta}_i - (\hat{\mathbf{n}} \cdot \boldsymbol{\delta}_i)\hat{\mathbf{n}}}{||\boldsymbol{\delta}_i - (\hat{\mathbf{n}} \cdot \boldsymbol{\delta}_i)\hat{\mathbf{n}}||} \tag{4}$$

The orientation of the moving platform is parameterized by roll (θ, around X axis) and pitch (ϕ, around Y axis) angles such that ${}^O\mathbf{R}_E = \mathrm{Rot}(X, \theta) \cdot \mathrm{Rot}(Y, \phi)$. The revolute joint axis vector ($\hat{\mathbf{n}}$) and the foot attachment point ($\mathbf{f}_i$) are expressed in global coordinate frame using $\hat{\mathbf{n}} = {}^O\mathbf{R}_E \cdot \hat{\mathbf{n}}_E$ and $\mathbf{f}_i = {}^O\mathbf{R}_E \cdot \mathbf{f}_i^E$ respectively where $\hat{\mathbf{n}}_E$ and $\mathbf{f}_i^E$ denote the revolute joint axis and foot attachment point vector in EE frame.

3 Solving Forward and Inverse Kinematics

Algebraic geometry techniques have proven to be useful in solving the forward kinematics of parallel manipulators but they require the constraint equations to be algebraic. Tangent half angle substitutions might leave the constraint equations undefined for π orientations. Hence, in order to have an algebraic description of the mechanism's constraint equations, cosines and sines are replaced by $\cos(\theta) = x$, $\sin(\theta) = y$, $\cos(\phi) = u$ and $\sin(\phi) = v$ in ${}^O\mathbf{R}_E$ though it comes at a cost of adding two more equations to the ideal set. To this end, rearranging Eq. (3) and squaring to avoid the square root term $||\hat{\mathbf{n}} \times \boldsymbol{\delta}_i||$ leads to four algebraic constraint equations:

$$g_1 := (d_1^2 - ||\hat{\mathbf{n}} \cdot \boldsymbol{\delta}_1||^2 - ||\hat{\mathbf{n}} \times \boldsymbol{\delta}_1||^2 - r^2)^2 - 4||\hat{\mathbf{n}} \times \boldsymbol{\delta}_1||^2 r^2 = 0 \tag{5a}$$
$$g_2 := (d_2^2 - ||\hat{\mathbf{n}} \cdot \boldsymbol{\delta}_2||^2 - ||\hat{\mathbf{n}} \times \boldsymbol{\delta}_2||^2 - r^2)^2 - 4||\hat{\mathbf{n}} \times \boldsymbol{\delta}_2||^2 r^2 = 0 \tag{5b}$$
$$g_3 := x^2 + y^2 - 1 = 0 \tag{5c}$$
$$g_4 := u^2 + v^2 - 1 = 0 \tag{5d}$$

After substituting the geometric dimensions provided in Table 1, the constraint equations are only a function of variables x, y, u, v, d_1 and d_2. g_1 and g_2 are 16 degree polynomials and are quite long to show here due to space constraints.

Table 1. Geometric dimensions of the ankle mechanism

i	$\mathbf{s}_i$	$\mathbf{f}_i^E$	$\hat{\mathbf{n}}_E$	$\|\|\mathbf{f}_i\mathbf{k}_i\|\|$
1	$(-22.30, 25, 291.27)^T$	$(-70, 40, 0)^T$	$(1, 0, 0)^T$	30
2	$(-22.30, -25, 291.27)^T$	$(-70, -40, 0)^T$	$(1, 0, 0)^T$	30

The solution to inverse kinematics problem (IKP) of the manipulator is straightforward and unique for a given orientation of the moving platform as the joint variables d_i can be easily calculated from Eq. (3). It is noteworthy that when the roll angle is zero, Eq. (3) yields $d_1 = d_2$.

The direct kinematics problem (DKP) aims to find the variables x, y, u and v when the prismatic joint lengths are specified. In search of maximum number of solutions to DKP (assembly modes), an ideal of the constraint polynomials g_i is defined: $\mathscr{I} = \langle g_1, g_2, g_3, g_4 \rangle \mid \mathscr{I} \subseteq k[u, v, x, y]$. Finding the Groebner basis with a pure lexicographic ordering of the orientation parameters in any order leads to a univariate polynomial of degree 32. Since squaring two of the four constraint equations quadruples the number of solutions, the number of solutions must be quartered. Hence, the upper limit to DKP solutions of the manipulator under study is eight. To investigate the number of real solutions, **RootFinding[Isolate]** function of Maple^{TM} is used. The algorithm behind this function finds out the rational univariate representation of the set of polynomials and isolates the real roots of these univariate polynomials based on Descartes' rule of sign and the bisection strategy in a unified framework. The variables d_1 and d_2 are varied from 221 mm to 331 mm (physical motion range of linear actua-

Table 2. Percentage of real solutions to direct kinematics

Number of solutions	0	2	4	6	8
Number of poses (/2601)	124	268	2146	63	0
Percentage	4.77	10.30	82.51	2.42	0

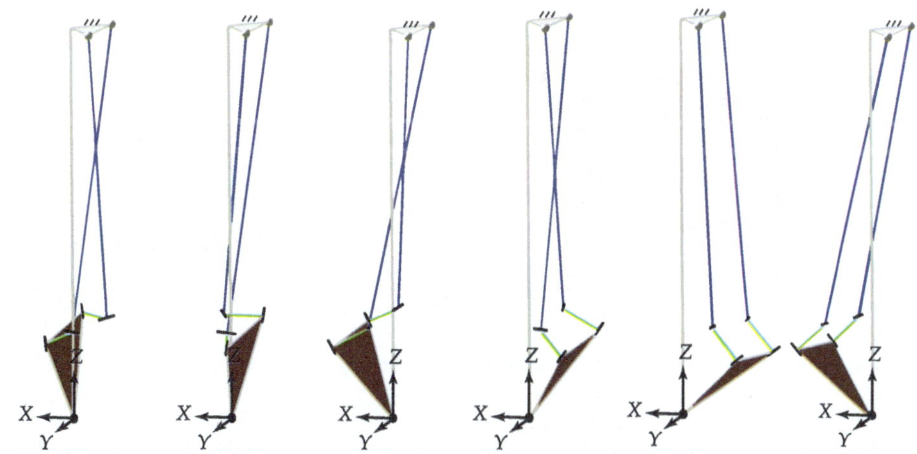

(a) $\theta = 172.32°$ $\phi = -92.94°$ (b) $\theta = 172.33°$ $\phi = -78.30°$ (c) $\theta = 175.32°$ $\phi = -116.30°$ (d) $\theta = 175.30°$ $\phi = -54.91°$ (e) $\theta = 5.03°$ $\phi = 39.28°$ (f) $\theta = 5.09°$ $\phi = 131.92°$

Fig. 4. Assembly modes for d_1 =221 mm and d_2 =228.3 mm tors) with an increment of 6 mm and the percentage of the number of real DKP solutions is listed in Table 2. It is evident that the maximum number of real solutions for the considered set of prismatic joint lengths is six. Figure 4 shows six such assembly modes when $d_1 = 221$ mm and $d_2 = 228.3$ mm. It is speculated that a different choice of design parameters might lead to eight real solutions to DKP. In the physical construction of the ankle joint, passive joint limits are chosen such that there exists a unique solution to forward kinematics for a given input of actuator lengths in their feasible motion range (for instance Fig. 4e).

4 Workspace, Singularity and Performance Analysis

To demonstrate the suitability of the novel 2-SPRR+1U mechanism as a humanoid ankle joint, it is important to compute and characterize its feasible workspace in orientation and configuration domains. The feasible configuration space is calculated by varying the orientation variables describing foot rotation, roll (θ) and pitch (ϕ) angles, in the range $[-\pi, \pi]$. Then the physical limits of the linear actuators ($d_i \in [221, 331]$ mm) are imposed to compute the workspace of the mechanism under actuator constraints. The resulting configuration space and orientation workspace are shown in Fig. 5. It is possible to take into account physical limits of passive joints in the mechanism to further compute the physically realizable workspace which is indicated with a closed curve in the figure. The final range of motion (ROM) of the proposed ankle mechanism is more than that of an average human and is presented in Table 3 (compare with [8]).

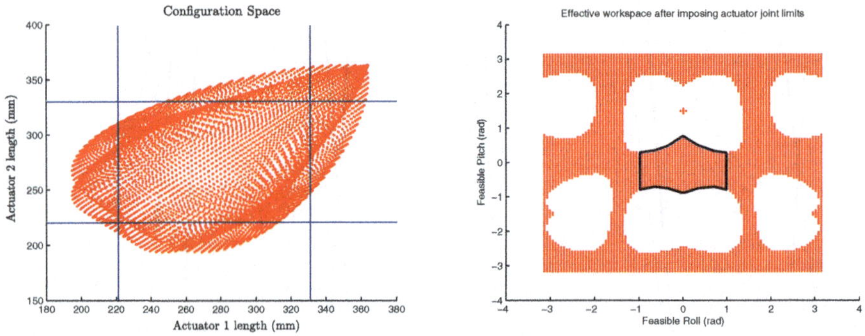

Fig. 5. Configuration space and orientation workspace under actuator physical limits

The ankle mechanism under study does not have any limb singularities since the auxiliary actuation legs do not generate any constraints on the moving platform. Nonetheless, the actuation scheme results in the so called actuation singularities that can be determined through the kinematic Jacobian matrix of the manipulator obtained by the partial differentiation of the constraint polynomials in Eq. (5) with respect to the orientation parameters:

$$\mathbf{J} = \begin{bmatrix} \frac{\partial g_1}{\partial \theta} & \frac{\partial g_1}{\partial \phi} \\ \frac{\partial g_2}{\partial \theta} & \frac{\partial g_2}{\partial \phi} \end{bmatrix} \tag{6}$$

The configurations for which the determinant of the Jacobian matrix $\mathbf{J}$ vanishes are called actuation singularities. The determinant of $\mathbf{J}$ depends only on θ and ϕ. An implicit plot of the equation $\det(\mathbf{J}) = 0$ in terms of the orientation variables θ and ϕ is shown in Fig. 6 which shows the singularity curves in the mechanism's workspace. Also, it can be observed that there exist four singularities each for pure roll ($\phi = 0$) and pure pitch ($\theta = 0$) movements. Figure 7 shows

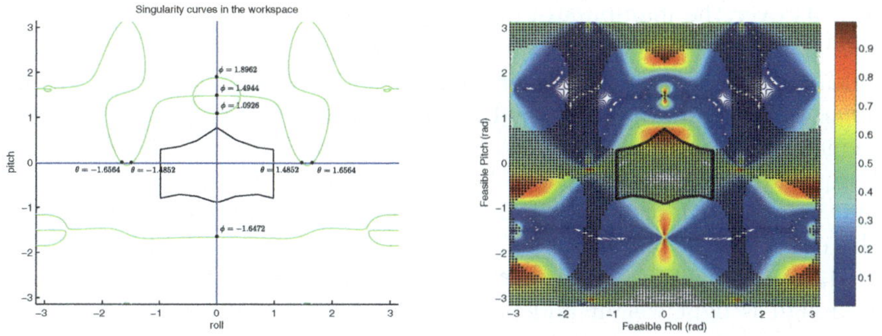

Fig. 6. Singularity curve and inverse of condition number over workspace

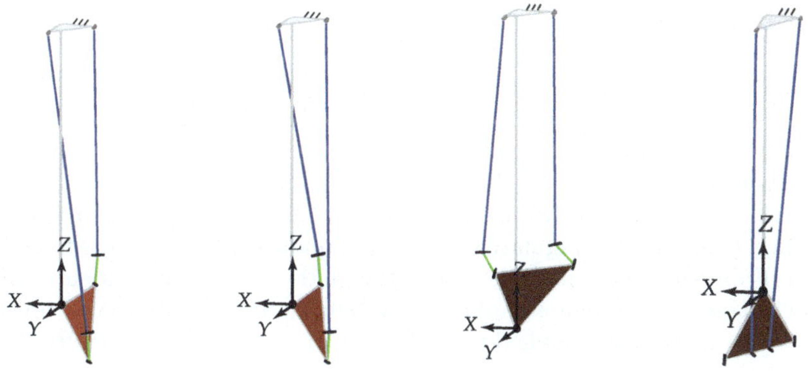

(a)	(b)	(c)	(d)
$\theta = -1.49$ rad	$\theta = 1.49$ rad	$\phi = 1.09$ rad	$\phi = -1.65$ rad
$d_1 = 306$ mm	$d_1 = 228$ mm	$d_1 = 216$ mm	$d_1 = 362$ mm
$d_2 = 228$ mm	$d_2 = 306$ mm	$d_2 = 216$ mm	$d_2 = 362$ mm

Fig. 7. Singularity configurations for pure roll (θ) and pure pitch (ϕ) movements

Table 3. Ankle joint specification (total weight of lower leg = 3.2 kg, weight of one actuator = 0.44 kg)

Range (min. to max.)	Position	Max. abs. force	Max. abs. velocity
Ankle pitch	$-51.5°$ to $45°$	43.8 N m to 110.1 N m	$61° s^{-1}$ to $154° s^{-1}$
Ankle roll	$-57°$ to $57°$	30.6 N m to 57 N m	$118° s^{-1}$ to $222° s^{-1}$
Linear actuator	221 mm to 331 mm	754 N	$81 mm s^{-1}$

the singular poses for the pure roll and pure pitch movements which are closest to the zero configuration of the mechanism.

The quality of velocity or force transmission of a parallel manipulator can be measured by plotting the inverse of condition number of the kinematic Jacobian matrix (**J**) over the manipulator's workspace. The inverse of condition number of the Jacobian is calculated with $c(\mathbf{J}) = \frac{1}{\|\mathbf{J}\| \cdot \|\mathbf{J}^{-1}\|}$ where $\| \cdot \|$ represents the Euclidean norm of the matrix. From Fig. 6, it is evident that the kinematic Jacobian matrix is well-conditioned in the feasible orientation workspace of the ankle mechanism.

For practical purposes, it is crucial to calculate the maximum absolute velocity and torque available at the EE, from the maximum force and velocity that can be delivered by the actuators. These are computed with the help of kinematic Jacobian matrix and actuator specification (see Table 3). The proposed ankle design is light weight and provides good force and velocity transmission characteristics along pure pitch and roll movements which are highly desired in modern humanoid robots.

5 Conclusion

This paper presents a comprehensive kinematic analysis of a novel 2SPRR+1U parallel mechanism for application as a humanoid ankle joint with two degrees of freedom. Using tools from computational algebraic geometry, an upper bound to the number of solutions to the direct kinematics problem and the real assembly modes have been studied. Inverse kinematics is used to study the mechanism's workspace, compute the singularity curves and quality of velocity and force transmission. From the study, it is clear that the ankle mechanism is highly suitable for application as a humanoid ankle subsystem module. The analysis presented in this work can be easily applied to 2SPU+1U mechanism by substituting $r = 0$.

Acknowledgement. The work presented in this paper was performed within the project TransFIT, funded by the German Aerospace Center (DLR) with federal funds from the Federal Ministry for Economic Affairs and Energy (BMWi) (Grant Nos. 50RA1701, 50RA1702 and 50RA1703). The fifth author acknowledges that this work has been partially supported by the Austrian COMET-K2 program of the Linz Center of Mechatronics.

References

1. Bartsch, S., Manz, M., Kampmann, P., Dettmann, A., Hanff, H., Langosz, M., von Szadkowski, K., Hilljegerdes, J., Simnofske, M., Kloss, P., Meder, M., Kirchner, F.: Development and control of the multi-legged robot mantis. In: International Symposium on Robotics (2016)
2. Kirchner, E.A., Will, N., Simnofske, M., Benitez, L.M.V., Bongardt, B., Krell, M.M., Kumar, S., Mallwitz, M., Seeland, A., Tabie, M., Wöhrle, H., Yüksel, M., Heß, A., Buschfort, R., Kirchner, F.: Recupera-reha: exoskeleton technology with integrated biosignal analysis for sensorimotor rehabilitation. In: Transdisziplinäre Konferenz SmartASSIST, pp. 504–517 (2016)
3. Kuehn, D., Bernhard, F., Burchardt, A., Schilling, M., Stark, T., Zenzes, M., Kirchner, F.: Distributed computation in a quadrupedal robotic system. Int. J. Adv. Robot. Syst. **11**(7), 110 (2014)
4. Kumar, S., Bongardt, B., Simnofske, M., Kirchner, F.: Design and kinematic analysis of the novel almost spherical parallel mechanism active ankle. J. Intell. Robot. Syst. 1–23 (2018). https://doi.org/10.1007/s10846-018-0792-x
5. Kumar, S., Simnofske, M., Bongardt, B., Mueller, A., Kirchner, F.: Integrating mimic joints into dynamics algorithms – exemplified by the hybrid recupera exoskeleton. In: Advances in Robotics (AIR 2017), 28 June–2 July, New Delhi, India. ACM-ICPS (2017)
6. Lohmeier, S., Buschmann, T., Schwienbacher, M., Ulbrich, H., Pfeiffer, F.: Leg design for a humanoid walking robot. In: 2006 6th IEEE-RAS International Conference on Humanoid Robots, pp. 536–541 (2006)
7. Serracín, J., Puglisi, L., Saltaren, R., Ejarque, G., Sabater-Navarro, J., Aracil, R.: Kinematic analysis of a novel 2-d.o.f. orientation device. Robot. Auton. Syst. **60**(6), 852–861 (2012)
8. Zoss, A.B., Kazerooni, H., Chu, A.: Biomechanical design of the Berkeley lower extremity exoskeleton (BLEEX). IEEE/ASME Trans. Mech. **11**(2), 128–138 (2006)

Online Calibration Procedure for Motion Tracking with Wearable Sensors Using Kalman Filtering

Alessandro Filippeschi(✉), Emanuele Ruffaldi, Lorenzo Peppoloni, and Carlo Alberto Avizzano

Scuola Superiore Sant'Anna, 56124 Pisa, Italy
{a.filippeschi,e.ruffaldi,l.peppoloni,c.avizzano}@santannapisa.it

Abstract. Several methods have been developed to estimate human limbs poses through inertial measurement units (IMUs). Although a big effort has been dedicated to the selection of the sensor fusion algorithm, less attention has been paid to some aspects that may cause considerable estimation errors. This paper presents a novel method for human upper limb motion tracking that accounts for the motion of the IMUs with respect to the attached limb and a calibration procedure for the estimation of limbs lengths. Three Unscented Kalman Filters are proposed to estimate joint angles, IMUs poses, and the limbs lengths based on the IMUs measurements. We validate our method by means of an optical motion tracking system that we used to calculate wrists position. This approach shows to be able to estimate unknown link lengths, to update correctly IMUs position and to improve the wrists position estimation.

Keywords: Unscented kalman filtering · Sensor fusion
Kinematic chains

1 Introduction

Human motion tracking based on wearable sensors has received a considerable attention in the past years. The availability of inertial measurement units (IMUs) makes tracking of human motion both affordable and unobtrusive, thus being used in several applications which include sports [9], medicine [5], and teleoperation [7].

Many methods focus on sensor fusion algorithms and the kinematic model of the human limbs. However, all these methods require some parameters, which often include the body limbs lengths and the pose of the IMU with respect to (w.r.t.) its parent limb (i.e. the limb to which it is attached). These quantities are not always directly measurable, e.g. in the human's limbs case. This problem has been deeply studied for marker-based optical motion tracking systems, for which specific protocols were developed to precisely track human motion [3]. The same attention has not been given when dealing with IMU-based motion tracking. Considering the human example, several algorithms (e.g. [2,10]) take

J. Lenarcic and V. Parenti-Castelli (Eds.): ARK 2018, SPAR 8, pp. 440–448, 2019.
https://doi.org/10.1007/978-3-319-93188-3_50

into account human kinematics, but limbs lengths are either measured out or estimated from anthropometric tables. This approximation is a source of errors [6], and methods at the state of the art do not deal with this problem specifically. Therefore, we propose to use IMUs data to better estimate unknown lengths by means of a novel calibration procedure, and a filter that we call K3.

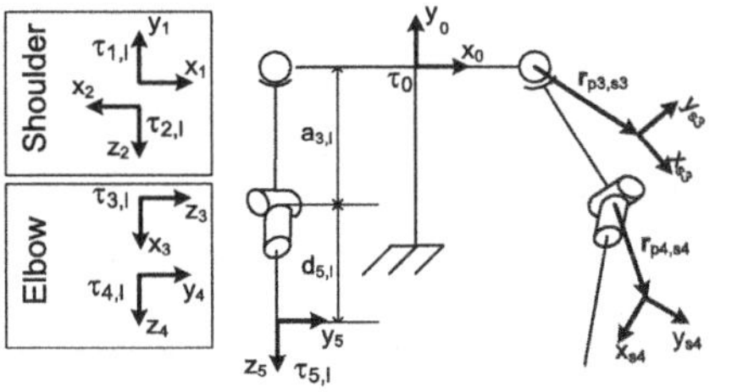

Fig. 1. Kinematic model of the upper limbs shown from behind. The model shows the DH reference frames as well as sensors ones.

In addition to limbs' length, IMUs' pose requires being estimated as it is difficult to be measured. This problem has not been studied in depth: in most of the works (see [4] for a review), IMUs are considered to be aligned with and to "sit" on one of the axes of its parent limb frame, and their positions are measured out before tracking trials. This solution is a twofold source of error: first, the error on the position of the IMU, that affects linear acceleration estimation. Second, fixed values of IMUs positions cannot account for eventual IMUs motion with respect to their parent limbs. To limit this error, the position of the sensors must be well known and constant, thus requiring a tight mount of the sensors on the limbs. In this chapter, we propose to estimate IMUs position in the filter that we call K2. Based on our previous work [8] (filter K1), we evaluate the influence of limbs lengths and sensor pose estimation on the accuracy of motion reconstruction.

2 Motion Reconstruction Method

The proposed motion reconstruction method is based on the signals of four IMUs, namely $s_1 \dots s_4$, attached to the upper arms and to the forearms of left and right arms respectively. In the following, $\mathbf{r}^c_{(.)}$ is a free vector written in the frame c, whereas $\mathbf{r}^c_{a,b}$ is a vector connecting frames a and b origins, represented in the frame c, R^b_a is the rotation matrix from b to a frame, and $c_\kappa = \cos\kappa$ and $s_\kappa = \sin\kappa$.

2.1 Upper Limbs Kinematic Model

The upper limbs were modeled as two kinematic chains rooted in the chest, that we consider as a fixed frame. Each chain is composed of two rigid bodies representing the upper arm and the forearm. The shoulder is modeled as a spherical joint, whereas the forearm has two rotational DoFs with respect to the upper arm, i.e. elbow flexion and forearm pronation. According to the adopted Denavit-Hartenberg (DH) convention, 6 frames are defined for each kinematic chain. The root frame, called τ_0, is attached to the chest. The others are called $\tau_{i,j}$ where i represents the DoF between $i-1$-th and i-th frame, and j represents the side (i.e. left l or right r). We then define $\tau_{i-1,j}$ as parent frame of $\tau_{i,j}$. Frame $\tau_{i,j}$ can then be obtained from its parent frame by defining four parameters, namely $a_{i,j}$, $d_{i,j}$, $\alpha_{i,j}$ and $\theta_{i,j}$, through the homogeneous matrix

$$T_{i,j}^{i-1,j} = \begin{bmatrix} c_{\theta_{i,j}} & -s_{\theta_{i,j}} c_{\alpha_{i,j}} & s_{\theta_{i,j}} s_{\alpha_{i,j}} & a_{i,j} c_{\theta_{i,j}} \\ s_{\theta_{i,j}} & c_{\theta_{i,j}} c_{\alpha_{i,j}} & -c_{\theta_{i,j}} s_{\alpha_{i,j}} & a_{i,j} s_{\theta_{i,j}} \\ 0 & s_{\alpha_{i,j}} & c_{\alpha_{i,j}} & d_{i,j} \\ 0 & 0 & 0 & 1 \end{bmatrix} = \begin{bmatrix} R_{i,j}^{i-1,j} & \mathbf{r}_{i,j}^{i-1,j} \\ \mathbf{0} & 1 \end{bmatrix} \quad (1)$$

where Fig. 1 shows the kinematic model, the notation, and the six $\tau_{i,l}$ frames.

2.2 Motion Reconstruction Approach

Given this kinematic model, motion reconstruction consists of estimating **joint variables** $\theta_{i,j}$. However, this task involves other variables belonging to different categories. **Sensor variables** include all the $\mathbf{r}_{p,s}^{s}$ and R_p^s independent variables: defined τ_s as the frame attached to the s-th sensor, three variables of R_s^p (p denotes the frame attached to the s-th sensor parent limb) and the three variables of $\mathbf{r}_p^{s,p}$ are required to define the pose of each sensor w.r.t τ_0. **Anatomical variables** include all the $a_{i,j}$, $d_{i,j}$ and $\alpha_{i,j}$ required for the DH representation of arms kinematics. Among anatomical variables, $a_{1,j}$, $a_{3,j}$, $\alpha_{4,j}$, $a_{4,j}$, $d_{5,j}$ are not known, depend on anatomic features of the user, and are constant. In this work, we use one calibration procedure and three filters to evaluate the effect of sensor and anatomical variables estimation. In the first (K1) joint variables are estimated by means of an Unscented Kalman Filter (UKF), whereas both sensor and anatomical variables are treated as static parameters which are initially provided, as it occurs in most of the methods in the literature [4]. The second filter K2, extends K1 by estimating sensor positions $\mathbf{r}_{p,s}^{s}$. The third algorithm K3 is used only during the third calibration step to estimate limbs lengths, whose final estimate is then used in either K1 or K2. To this aim, only variables in $\mathbf{\Pi}_j = [a_{3,j}, d_{5,j}], j \in l, r$ are estimated. If K3 is not used, $\mathbf{\Pi}_j$ are measured out before starting the trial. The remaining anatomic variables are either measured out before the trial or they are determined according to the DH convention. They are hence parameters regardless the setup and they are estimated at the beginning of the trial. Rotation matrices R_p^s are considered to be constant as we suppose only small displacement to occur during motion.

The calibration procedure three steps. In the first step, IMUs are rotated to gather data from as many different orientations as possible. Data are then used for hard calibration of magnetometers. In the second step, the data are recorded while the user keeps statically first the N-pose (arms neutral besides the body) and then the T-pose (arms abducted by 90° from the N-pose). During these phases we assume $\theta_{i,j} = 0 \quad \forall i, j$ and $\theta_{i,j} = 0 \quad \forall i, j\, i \neq 1$, whereas $\theta_{1,l} = \theta_{1,r} = \pi/2$ respectively. Under these assumptions we calculate R_s^p as the best fit of IMUs' accelerometer data in terms of least square error. Once obtained all R_s^p, we calculate $\mathbf{h}^0$ as the vector that best fits IMUs' magnetometer data. In the third calibration step the user is asked to move his/her arms spanning his/her available workspace always keeping one hand palm fixed with respect to the other. We found that 40 s suffice to have a stable estimation of Π_j parameters.

2.3 Kalman Filters for Motion Reconstruction

This section presents the aforementioned K1-K3 filters. Each K-th filter is an Unscented Kalman Filter, for which a state vector $\mathbf{x}_m$ is defined along with process and measurement models as follows:

$$\mathbf{x}_m(k+1) = f_m(\mathbf{x}_m(k)) + \nu_m(k) \quad (2)$$

$$\mathbf{y}_m(k) = h_m(\mathbf{x}_m(k)) + \epsilon_m(k), \quad m = 1 \ldots 3 \quad (3)$$

where k is the time sample, $\nu_m(k) \sim \mathcal{N}(0, Q_m)$ is the white Gaussian process noise and $\epsilon_m(k) \sim \mathcal{N}(0, P_m)$ is the white Gaussian measurements noise.

The K1 filter estimates, for each arm, joint variables of the j-th arm (we neglect j index referred to the arm). The state vector of K1 is $\mathbf{x}_1 = [\theta_1, \dot{\theta}_1, \ddot{\theta}_1, \ldots \ddot{\theta}_5]^T$, whereas process and measurement models are

$$\begin{bmatrix} \theta_i(k+1) \\ \dot{\theta}_i(k+1) \\ \ddot{\theta}_i(k+1) \end{bmatrix} = \begin{bmatrix} 1 & T_s & \frac{1}{2}T_s^2 \\ 0 & 1 & T_s \\ 0 & 0 & 1 \end{bmatrix} \begin{bmatrix} \theta_i(k) \\ \dot{\theta}_i(k) \\ \ddot{\theta}_i(k) \end{bmatrix} + \nu_i(k), \quad (4)$$

and

$$\ddot{\mathbf{r}}_{0,s}^s = R_s^p \ddot{\mathbf{r}}_{0,p}^p + S(\dot{\omega}_s^s)\mathbf{r}_{p,s}^s + S(\omega_s^s)^2 \mathbf{r}_{p,s}^s + R_s^0 \mathbf{g}^0 + \epsilon_{r_i} \quad (5)$$

$$\omega_s^s = R_s^p(\omega_p^p + \dot{\vartheta}_{p+1}\mathbf{z}_0) + \epsilon_{\omega_i} \quad (6)$$

$$\mathbf{h}_s^s = R_s^0 \mathbf{h}^0 + \epsilon_{h_i} \quad (7)$$

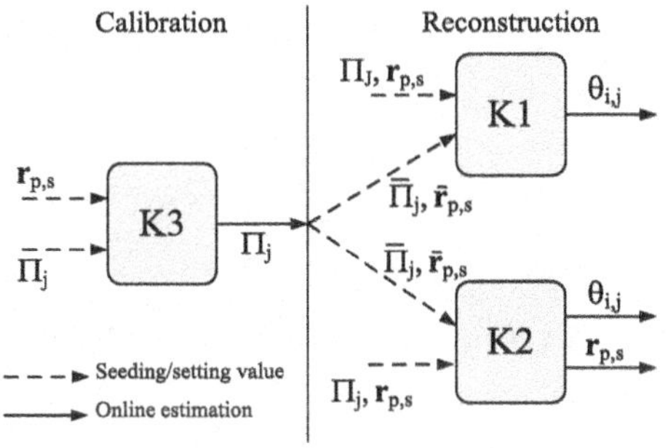

Parameters	K1 active	K2 active	K1/K2 + K3 active
R_s^p		calibration step 2	
$\mathbf{r}_{p,s}^s$	measured	estimated online	estimated online
$\mathbf{\Pi}_j$	measured	measured	set in K1/K2 according to K3 final outcome (calibration step 3)
$\mathbf{h}_0$		calibration step 2	

Fig. 2. K1-K3 filters architecture and parameters required for the motion reconstruction.

where T_s is the sample time and, for the s-th sensor, $\mathbf{a}_s$ is the measured acceleration, $\mathbf{o}_s$ the angular velocity, and $\mathbf{m}_s$ the Earth's magnetic field, which are estimated respectively by $\ddot{\mathbf{r}}^s_{0,s}$, ω^s_s, and $\mathbf{h}^s_s$. IMUs data are all available in the s-th IMU's frame

The K2 filter is an extension of the K1 filter in which the state includes also the sensor positions. These are estimated along with the joint variables, thus allowing us both to refine the initial estimation of sensors positions and to detect their changes over time. For each arm, the state vector is $\mathbf{x}_2 = [\mathbf{x}_1 \mathbf{r}^{s1}_{s1,p1} \mathbf{r}^{s2}_{s2,p2}]^T$, where $s1$ and $s2$ are the sensors attached to the upper arm and the forearm of the selected arm. Positions $\mathbf{r}^{s1}_{s1,p1}$ and $\mathbf{r}^{s2}_{s2,p2}$ models are random walks:

$$\mathbf{r}^s_{s,p}(k+1) = \mathbf{r}^s_{s,p}(k) + \nu_s \tag{8}$$

where ν_r $\mathcal{N}(0, Q_s)$. The measurement model is inherited from the K1 filter and extended to relate $\mathbf{r}^s_{s,p}$ to accelerometer measurements. It is worth noting that Eq. 5 already relates $\mathbf{r}^s_{s,p}$ to $\mathbf{a}_s$. However, in order to have more flexibility in allowing measurements to influence $\mathbf{r}^s_{s,p}$ estimation, we added two measurements equations that relate linear acceleration of sensor s to its $\mathbf{r}^s_{s,p}$ variables. By writing linear acceleration for each sensor s we complete the measurements model:

$$\mathbf{a}_s - R^0_s \mathbf{g}^0 = \ddot{\mathbf{r}}^s_{s,0} = R^p_s \ddot{\mathbf{r}}^p_{p,0} + [S(\dot{\omega}^s_s) + S^2(\omega^s_s)]\mathbf{r}^s_{s,p}. \tag{9}$$

K3 filter targets the estimation of $\mathbf{\Pi}_j$ parameters. This filter relies on the assumption that there exists a point E that belongs to both left and right forearms. This assumption is acceptable if wrists are kept fixed with respect to each other. Under this condition, it is possible to add the constraint:

$$\mathbf{r}^0_{E_l,0} = \mathbf{r}^0_{E_r,0}, \quad \dot{\mathbf{r}}^0_{E_l,0} = \dot{\mathbf{r}}^0_{E_r,0}, \quad \ddot{\mathbf{r}}^0_{E_l,0} = \ddot{\mathbf{r}}^0_{E_r,0} \tag{10}$$

where E_l and E_r are the E point when though as belonging to respectively the left and the right arm. Similarly to the K2 filter, the process for $\mathbf{x}_1$ variables is represented by Eq. 4, whereas $\mathbf{\Pi}_l$ and $\mathbf{\Pi}_r$ model is a random walk:

$$\mathbf{\Pi}_j(k+1) = \mathbf{\Pi}_j(k) + \nu_{\pi_j}, \tag{11}$$

and $\nu_{\pi_j} \sim \mathcal{N}(0, Q_{\pi_j})$, and Q_{π_j} reflects the belief on the initial estimation of π_j. The measurements model is inherited from K1 and it is completed by Eq. 10 by writing E_l and E_r positions, velocities and accelerations as functions of $\mathbf{\Pi}_{l/r}$ parameters:

$$\begin{bmatrix} \mathbf{r}^{i-1,j}_{E_j,i-1} \\ 1 \end{bmatrix} = T^{i-1,j}_{i,j} \begin{bmatrix} \mathbf{r}^{i,j}_{E_j,i} \\ 1 \end{bmatrix} = \begin{bmatrix} R_{\theta_{i,j}} \lambda_{i,j} \\ 1 \end{bmatrix}, \text{ with } R_{\theta_{i,j}} = \begin{bmatrix} c_{\theta_{i,j}} & 0 \\ s_{\theta_{i,j}} & 0 \\ 0 & 1 \end{bmatrix}, \lambda_{i,j} = \begin{bmatrix} a_{i,j} \\ d_{i,j} \end{bmatrix}$$

$$\dot{\mathbf{r}}^0_{E_j,0} = \sum_{l=0}^{4} S(\omega^0_{l+1,j})\mathbf{r}^0_{l-1,l} = \sum_{l=0}^{4} S(\omega^0_{l+1,j}) R^0_{l,j} R_{\theta_{l+1,j}} \lambda_{l+1,j} = \sum_{l=0}^{4} \Phi_{l,j} \lambda_{l+1,j}$$

$$\ddot{\mathbf{r}}^0_{E_j,0} = \sum_{l=0}^{4} [S(\dot{\omega}^0_{l+1,j}) + S^2(\omega^0_{l+1,j})]\mathbf{r}^0_{l-1,l} = \sum_{l=0}^{4} \Psi_{l,j} \lambda_{l+1,j}.$$

Defined $\lambda_j = [\lambda_{1,j}^T \dots \lambda_{5,j}^T]^T$ we have

$$\mathbf{r}^0_{E_j,i-k} = \left[\Gamma_{i-1,j} \middle| \Gamma_{i-2,j} \middle| \cdots \middle| \Gamma_{0,j} \right] \lambda_j = \tilde{\Gamma} \lambda_j \tag{12}$$

$$\dot{\mathbf{r}}^0_{E_j,0} = \left[\Phi_{i-1,j} \middle| \Phi_{i-2,j} \middle| \cdots \middle| \Phi_{0,j} \right] \lambda_j = \tilde{\Phi} \lambda_j \tag{13}$$

$$\ddot{\mathbf{r}}^0_{E_j,0} = \left[\Psi_{i-1,j} \middle| \Psi_{i-2,j} \middle| \cdots \middle| \Psi_{0,j} \right] \lambda_j = \tilde{\Psi} \lambda_j. \tag{14}$$

Since $\mathbf{\Pi}_j \subset \lambda_j$, Eqs. 12, 13, and 14 show that $\mathbf{r}^0_{E_j,0}$ linearly depends on $\mathbf{\Pi}_j$. By selecting rows of Eqs. 12, 13, and 14 that involve $\mathbf{\Pi}_j$ we obtain

$$\left[\mathbf{r}^{0^T}_{E_j,0} \; \dot{\mathbf{r}}^{0^T}_{E_j,0} \; \ddot{\mathbf{r}}^{0^T}_{E_j,0} \right]^T = \left[\hat{\Gamma}^T_j \; \hat{\Phi}^T_j \; \hat{\psi}^T_j \right]^T \mathbf{\Pi}_j = C_j \mathbf{\Pi}_j. \tag{15}$$

Equation 15 allows us to write constraints Eq. 10 as

$$\mathbf{0}_{9,1} = \left[C_l \; - C_r \right] \left[\mathbf{\Pi}_l^T \; \mathbf{\Pi}_r^T \right], \tag{16}$$

which completes the measurement model of K3.

3 Experimental Evaluation

3.1 Materials and Method

An experimental validation was set to compare K1-K3 effects on motion reconstruction. IMU based reconstruction was compared against ground truth data obtained from the optical motion capture system Vicon MX2+ with 7 cameras (OMG, Oxford, UK). Four IMUs Invensense MPU9150 (Borregas Ave Sunnyvale, CA, USA) running at 100 Hz were attached to upper arms and forearms. Markers were placed on the manubrium sternal, acromions, and on the wrists to reconstruct the scaphoid-pisiform axes. The volunteer who participated in the experiment was asked to move while keeping his thorax as steady as possible. After the three calibration steps described in Sect. 2.2, he performed these functional movements: elbow flexion/extension; shoulder abduction/adduction and shoulder flexion/extension. The three filters and the calibration procedure were tested in the following setups. **Traditional**: reconstruction is made by K1 filter. This setup is used as a baseline, that is other setups will be compared to this one to check improvements. **Sensors**: reconstruction is made by K2 filter without using the K3 filter. **Full**: the K3 filter is run before online motion tracking, the output of K3 are $\bar{\mathbf{\Pi}}_j$ parameters, i.e. the final estimations of Π_j, that are used as input for either the K1 or the K2 filter. In this case, $\mathbf{r}^s_{s,p}$ are constant for the K3 filter and variables for the K1/K2 filters. Figure 2 shows the components of the three setups along with the role of the variables.

Variables $\mathbf{r}^s_{s,p}$ and $\mathbf{\Pi}_j$ can be either set or seeded and then estimated online. To set or seed $\mathbf{r}^s_{s,p}$ when the optical motion capture is not used, we assume $\mathbf{r}^s_{s,p} = [0 \, l_s \, 0]^T$. In this case, the quantities to be measured are l_s, $a_{3,l}$, $d_{5,l}$, $a_{3,r}$, and $d_{5,r}$. Otherwise, these quantities are obtained from the optical tracking. We adopted three ways of setting and seeding such quantities to test the algorithm. **Vicon**: this is based on optical capture data. It is the most precise way we used to measure selected quantities and it is used as a reference. **Hand**: the aforementioned quantities are measured by means of a tape measure. **Calibrated**: the K3 filter is used to estimate $\mathbf{\Pi}_j$, obtaining $\bar{\mathbf{\Pi}}_j$. $\mathbf{r}^s_{s,p}$ is then estimated by keeping the hand length ratio between $\bar{\mathbf{\Pi}}_j$ parameters and l_s.

3.2 Results and Discussion

Experimental results and their discussion are reported in this section. We considered the wrists positions in the $\tau_{1,j}$ frames to reduce the uncertainty in locating the Vicon frame with respect to the τ_0 frame. Figure 3(a) shows the estimations of anatomic variables $\mathbf{\Pi}_j$ made by the K3 filter with *hand*, and *Vicon* settings of the $\mathbf{r}^s_{s,p}$ parameters. The last values of $\mathbf{\Pi}_j$ are $\bar{\mathbf{\Pi}}_j$ in K1/K2 filters. Figure 3(b) presents the estimation of sensor positions obtained with either *hand* settings or *hand* updated by K3 filter. Figure 3(c) wrists position estimation with *hand* settings and either K1 or K2+K3 filters. Finally, we report the average root mean square error (ARMSE) of wrists position estimation during the whole trial. In the *traditional* setup the ARMSE was 103.9 mm and 123.9 mm for *Vicon* and *hand* settings respectively. In *sensor* setup with *hand* setting the ARMSE was 123.0 mm. The *full* setup was tested with *hand* settings, the ARMSE was 105.0 mm and 103.1 mm for K1 and K2 respectively.

The results allow for the following observations. Regardless the seeding, anatomic variables estimation of K3 filter takes about 5 s to get close to the Vicon estimate and less than 40 s to stabilize. However, it is to be investigated why a different initial seeding lead to even slightly different results. Filter K2 allows for refining sensors placement and to detect their displacements during the trial. We note that different seeding of K2 leads to very similar evolution of the sensor positions. Wrist ARMSE show that applying limb length estimation improves the estimation of the wrist position, whereas IMU position estimation has a smaller effect. However, whereas the limb lengths affect the estimation along the whole trial, sensors position variations are significant only in the last part, thus reducing their average effect on position estimation. Using the K3 filter improves the position estimation of 15–16% (see K1-*hand* against K1+K3-*hand*) whereas K2 causes an improvement of 0.7–1.8%. This result is remarkable as adding K3 allow for obtaining a good position estimation without the need of accurately measuring out limbs lengths. However, a larger error in the $\mathbf{y}_0$ component of the left wrist position requires further investigation. Finally, results show that our filters do not suffer from drift. Data and code used for this test are available online [1].

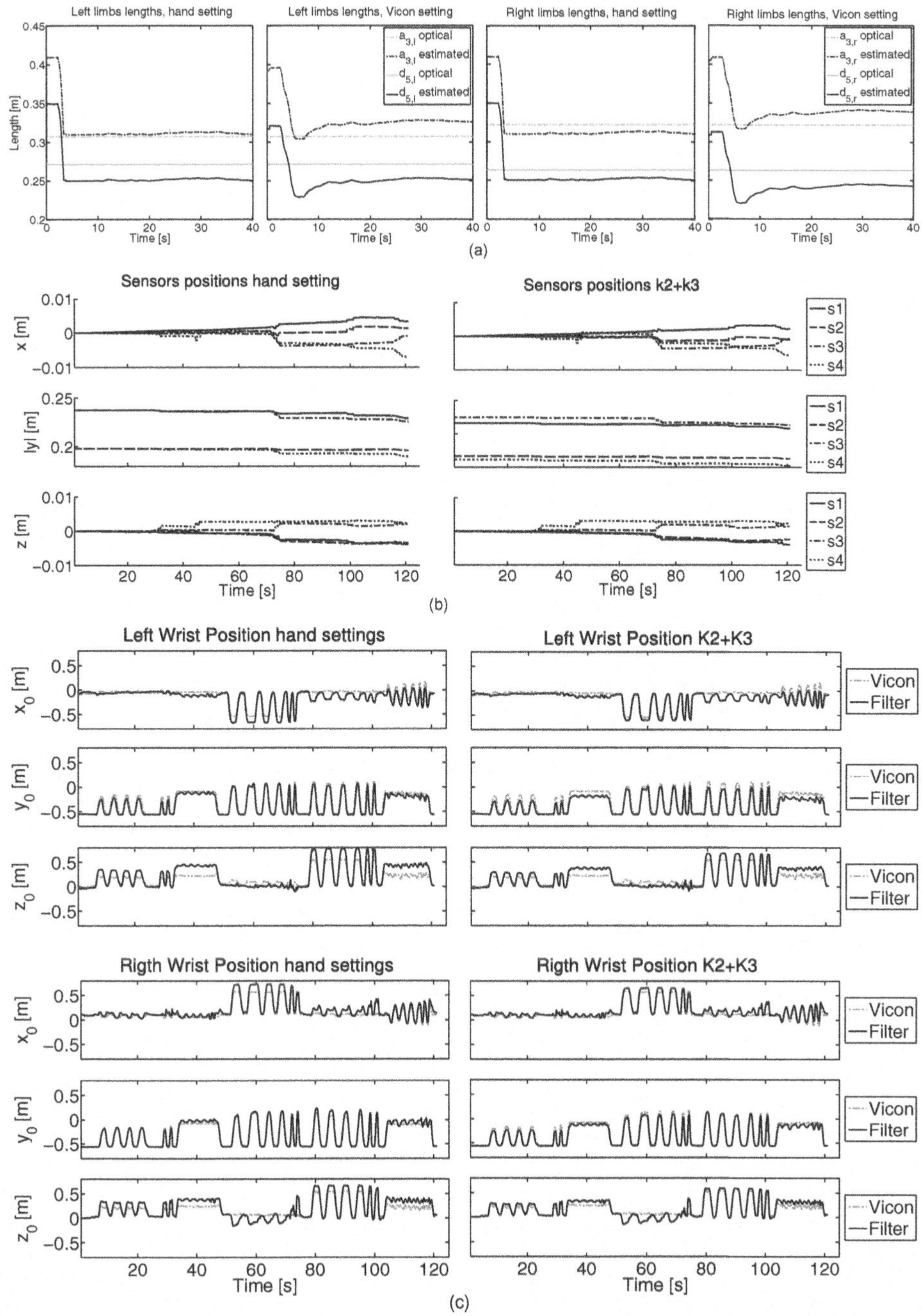

Fig. 3. (a) $\mathbf{\Pi}_l$ and $\mathbf{\Pi}_r$ estimation with either *hand* or *Vicon* settings of the $\mathbf{r}^s_{s,p}$ parameters. (b) Estimation $\mathbf{r}^s_{s,p}$ with *hand* (plots on the left), and *hand* updated by K3 filter and *Vicon* settings of the $\mathbf{\Pi}_j$ parameters and seeding of $\mathbf{r}^{s1}_{s1,p}$ in the plots on the right. (c) Wrist position reconstruction with the K1 and the K2+K3 setup and *hand* seeding of variables and setting of parameters.

4 Conclusions

The validation of the proposed calibration procedure showed that our method provides a stable and consistent estimation of both the anatomic variables and wrist positions, whose estimation is improved by using the K3 filter. Further research is aimed at evaluating effectiveness of K2 with larger sensor displacements.

References

1. https://doi.org/10.5281/zenodo.1206985
2. Bleser et al, G.: Using egocentric vision to achieve robust inertial body tracking under magnetic disturbances. In: ISMAR, pp. 103–109. IEEE (2011)
3. Davis, R.B., et al.: A gait analysis data collection and reduction technique. Hum. Mov. Sci. **10**(5), 575–587 (1991)
4. Filippeschi, A., et al.: Survey of motion tracking methods based on inertial sensors: a focus on upper limb human motion. Sensors **17**(6), 1257 (2017)
5. Lee, G.X., Low, K.: A factorized quaternion approach to determine the arm motions using triaxial accelerometers with anatomical and sensor constraints. IEEE T. Instrum. Meas. **61**(6), 1793–1802 (2012)
6. Miezal, M., Taetz, B., Bleser, G.: On inertial body tracking in the presence of model calibration errors. Sensors **16**(7), 1132 (2016)
7. Miller, N., et al.: Motion capture from inertial sensing for untethered humanoid teleoperation. In: Humanoid Robots, vol. 2, pp. 547–565. IEEE (2004)
8. Peppoloni, L., et al.: A novel 7 degrees of freedom model for upper limb kinematic reconstruction based on wearable sensors. In: SISY, pp. 105–110. IEEE (2013)
9. Ruffaldi, E., Peppoloni, L., Filippeschi, A.: Sensor fusion for complex articulated body tracking applied in rowing. Proc. Inst. Mech. Eng. Part P J. Sports Eng. Tech. **229**(2), 92–102 (2015)
10. Yun, X., Bachmann, E.R.: Design, implementation, and experimental results of a quaternion-based Kalman filter for human body motion tracking. IEEE Trans. Robot **22**(6), 1216–1227 (2006)

Sensitivity Analysis and Identification of Human Parameters for an Adaptive, Underactuated Hand Exoskeleton

Antonio Di Guardo(✉), Mine Sarac, Massimiliano Gabardi, Daniele Leonardis, Massimiliano Solazzi, and Antonio Frisoli

PERCRO Laboratory, TeCIP Institute, Scuola Superiore Sant'Anna of Pisa, Pisa, Italy
{antonio.diguardo,mine.sarac,massimiliano.gabardi}@santannapisa.it, {d.leonardis,solazzi,a.frisoli}@sssup.it

Abstract. Designing an underactuated hand exoskeleton system that contains user's finger in the kinematics loop improves the adjustability for different hand sizes. On the downside, the configuration and motion level kinematics become highly responsive to the finger properties and how the device is worn. In this paper, first we investigate the sensitivity of these variable parameters on the finger pose and force transmission analysis. Considering that the manual measurements of these parameters might be inaccurate and time consuming, we present an adaptive calibration algorithm to estimate them when a user is asked to wear the device. Finally, comparing the results of the calibration algorithm to the readings of an optical tracking system shows that the proposed solution is capable of estimating the variable parameters well enough to improve the kinematics analysis.

1 Introduction

Hand exoskeletons exert forces to user's fingers either to assist them during rehabilitative therapy exercises or to render realistic force feedback based on virtual or teleoperated interactions. The exceptional number of Degrees of Freedom (DoFs) of the human hand increases the complexity of the design of a hand exoskeleton, which yields the designers to search alternative approaches, such as soft exoskeletons [1] or fingertip devices [2].

The intrinsic adaptation of the exoskeleton to user's hand is a recent and promising approach. In particular, intrinsic adaptation to different hand sizes was achieved previously by closing the rigid linkage mechanism on the finger, resulting in a whole parallel kinematic system including both the exoskeleton and finger joints [3,4]. The device was effectively implemented in neurorehabilitation [5,6], adapting to the hands of different patients using the same device in a tight sequence of rehabilitation sessions. The intrinsically adaptive kinematic design was previously merged with the underactuation concept [7,8] to be used for both rehabilitation and teleoperation tasks. The pose of user's finger can be estimated through kinematics analysis using the actuator sensor and

J. Lenarcic and V. Parenti-Castelli (Eds.): ARK 2018, SPAR 8, pp. 449–457, 2019.
https://doi.org/10.1007/978-3-319-93188-3_51

an additional sensor utilized along a passive joint of the underactuated device. The kinematics of the device also depends on the properties of the user's finger and how the device is worn: since these parameters are important for an accurate estimation and some of them are challenging to be measured manually, a calibration process might be useful for their prediction. One calibration approach [9,10] is to track the fingertip using external sensors and obtain hand parameters recursively. Alternatively, the distance between the thumb and the index fingertips were calibrated without additional sensors [11,12].

In this paper, we analyze the sensitivity of hand parameters for the pose estimation and force transmission of an adaptive, underactuated exoskeleton [7]. Such analysis provides useful information about the most critical parameters for the kinematics to be measured before operating the device. Once the parameters to be estimated are highlighted, we developed a novel calibration algorithm to be executed once the exoskeleton is fastened to the user hand. The proposed solution is advantageous for the adaptive, underactuated exoskeleton during application scenarios, where the finger pose estimation and force transmission is essential to be computed.

2 Materials and Methods

5 fingered adaptive, underactuated hand exoskeleton shown in Fig. 1a can assist 2 *DoFs* of user's each finger through a single actuator and a planar kinematics independently. This paper will focus on a single finger component of the device for simplicity. The mechanical modelling of the structure and the deterministic model of user's finger are detailed in Fig. 1b.

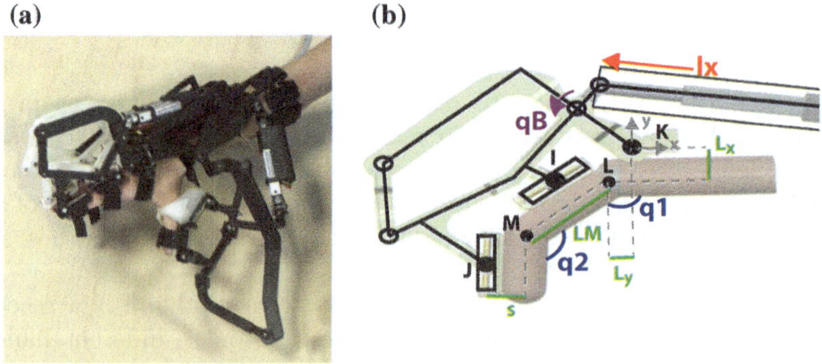

Fig. 1. (a) Overview of the 5-fingered exoskeleton. (b) Kinematics model of the underactuated hand exoskeleton with useful definitions depicted on it.

The instant finger pose (q_{o1}, q_{o2}) and transmitted forces to the finger phalanges can be estimated through kinematics analysis [7,8] using the sensory measurements (l_x, q_B), the finger articulation parameters and the position of the metacarpophalangeal joint with respect to the exoskeleton. In particular, $\overline{LM}$ is the length of the proximal phalange, s is the distance between the passive slider axes and the phalange axes along the proximal and middle phalanges (see Fig. 2). The parameter s is supposed to be constant and equal for both phalanges. Finally, Lx and Ly are horizontal and vertical distances between the points K and L. The last two parameters drastically depend by the way the user is wearing the device and cannot be directly measured.

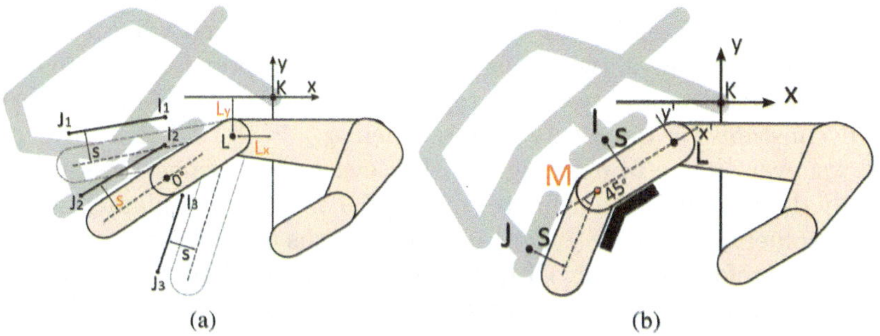

Fig. 2. The calibration notations during (a) the first and (b) the second phase.

2.1 Analysis of Sensitivity to Different Hand Parameters

Even though the aforementioned parameters need to be known for the kinematics analysis of the exoskeleton, how much they affect the output of these analyses is not clear. In order to understand how accurate these parameters should be measured when the user wears the device, a sensitivity analysis should be performed.

The one-at-a-time (OAT) sensitivity analysis can be used to evaluate the effect of each parameter by means of a cost function in which other parameters are kept fixed. To study these changes, a known hand model was used as a base and each parameters are changed with 1 μm difference while others are kept the same to highlight how much impact this little variation causes on the analysis outcome through different hand poses.

Given a parameter E to be analyzed, the pose estimation simulation is run for the focused parameter using $E_1 = E_{av} - 0.001$ and $E_2 = E_{av} + 0.001$, where E_{av} is the measurement of the base model to compute a corresponding P_1 and P_2 as $P_i = sqrt((q^i_{o1})^2 + (q^i_{o2})^2)$. Finally, the impact of the parameter can be calculated as:

$$err = (P_2 - P_1)/(E_2 - E_1). \tag{1}$$

Similarly, the determinant of the Jacobian matrix S_1 and S_2 were calculated as $S_i = det(J_i)$ for the parameters E_1 and E_2 correspondingly. The sensitivity of the current parameter was achieved as:

$$SI = \frac{(S_2 - S_1)/S_{av}}{(E_2 - E_1)/E_{av}} \tag{2}$$

where S_{av} is found when the cost function is computed using the parameter E_{av}. Whenever SI results greater than 1 the parameter can be claimed to be efficient.

2.2 Adaptive Algorithm for Estimation of Hand Parameters on the Fly

Since it is difficult to precisely determine the centre of an anatomical joint, the four parameters s, L_x, L_y, $\overline{LM}$ cannot be easily measured when the device is worn. For this reason, a two steps algorithm is proposed for estimating the unknown parameters. In the first step s, L_x, L_y are determined. Since s is assumed constant for the entire finger, when $q_{o2} = 0°$, $\overline{IJ}$ is always parallel to $\overline{ML}$ for every value of q_{o1} (Fig. 2a). For every system pose it is possible to univocally determine the coordinates of I and J in the $x - y$ plane just knowing the potentiometers readings (l_x, q_B) and the lengths of both finger and exoskeleton links. Therefore $\forall(q^i_{o1}, q^i_{o2})$ Eq. 3 identifies the straight line parallel to $\overline{IJ}$ and passing through the finger joint L as function of x, y and the offset s in the $x - y$ reference frame centered in K with the x-axis parallel to the metacarpal direction.

$$\begin{bmatrix} 1 & -m_i & \frac{-1}{\cos(\tan^{-1}(m_i))} \end{bmatrix} \begin{bmatrix} L_y \\ L_x \\ s \end{bmatrix} - \frac{y_{I_i}(x_{I_i} - x_{J_i}) - x_{J_i}(y_{I_i} - y_{J_i})}{x_{I_i} - x_{J_i}} = 0 \tag{3}$$

where $m_i = \frac{y_{I_i} - y_{J_i}}{x_{I_i} - x_{J_i}}$. As shown in Fig. 2a three poses $[(q^1_{o1}, 0\ deg), (q^2_{o1}, 0\ deg), (q^3_{o1}, 0\ deg)]$ are sufficient to determine the coordinates of L in the $x - y$ plane and the offset s. It should be noted that it is not necessary to know the three q^i_{o1} joint coordinates to perform this identification step. Since the elasticity and some mechanical backlash in the attachments allow little relative movements between the exoskeleton and the finger during the flexion/extension motion, in order to find out the closest solution for L_x, L_y and s to all the possible solutions obtained along the entire opening and closure path, thirty different poses equally spaced on the entire stroke of joint M have been take into account. The minimum square method has been used to solve the linear optimization problem defined by thirty equations of the type of Eq. 3, each one derived from a different pose. Please note that L_x, L_y and s are sufficient to univocally determine q_{o1} independently on q_{o2} value. The second estimation procedure step allows defining the phalanx length $\overline{LM}$. Referring to Fig. 2b, known q_{o1} and the coordinates of L, the straight line passing through $\overline{LM}$ is defined. Since, as previously mentioned, J and s are also known, both M and $\overline{LM}$ can be determined by knowing q_{o2}. In the experimental procedure q_{o2} has been fixed to 45° by locking the user's finger joint with a 3D-printed piece properly shaped. Equation 4 shows the mathematical expression for $\overline{LM}$ in the reference frame $L_{x'y'}$ (Fig. 2b).

$$\overline{LM} = \left| J_{x'} - \frac{J_{y'}}{\tan q_{o2}} - \frac{s}{\cos q_{o2}} \right| \tag{4}$$

where (J_x', J_y') are the coordinates of J expressed in $L_{x'y'}$. In order to increase the algorithm robustness in the finger workspace, the $\overline{LM}$ value has been evaluated as the mean among thirty $\overline{LM^i}$ values obtained for different q_{o1}^i values equally spaced in the stroke of joint L.

3 Results and Discussion

3.1 Results of the Analysis of Sensitivity to Different Hand Parameters

The impact of the finger length, finger width and horizontal and vertical offsets of joint K with respect to joint M was first analyzed based on the pose estimation. For almost 100 possible pose combinations, the error in pose estimation was calculated using Eq. 1 for an intentional change of 0.001 mm for each parameter. Figure 3 shows that the finger width and the vertical offset causes a higher error in degrees per mm compared to others (two central plots).

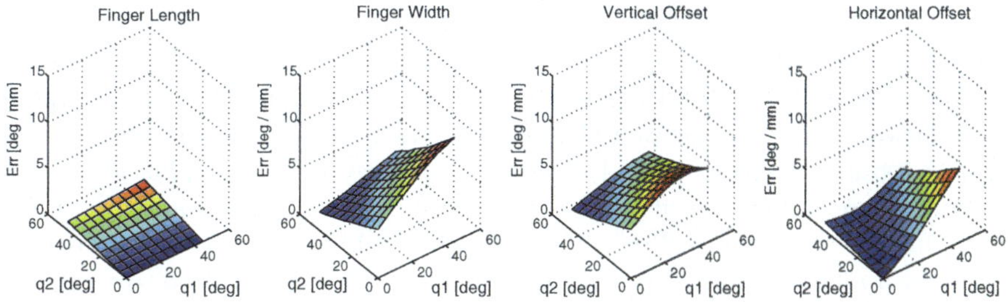

Fig. 3. Parameter error estimation according to pose

The same simulation was extended to the motion level kinematics analysis by computing the difference of the determinant for the Jacobian matrix. Please note that the sensitivity index in Eq. 2 shows a dramatic impact of a parameter if the index is higher than 1. Figure 4 shows the output of SI for Jacobian matrix for each finger pose, and the most crucial parameters can be found as finger width and vertical offset.

3.2 Validation of the Adaptive Algorithm for Estimation of Hand Parameters

In order to experimentally evaluate the estimation of hand parameters through the presented algorithm, an optical tracking system (Optitrack V120 Trio) has been used to measure phalanxes position and orientation. The precision of the tracking system, which depends on the distance of markers from the infrared cameras, has been experimentally evaluated to be below 0.1 mm (estimated as repeatability of distance between two markers fixed to a solid body, which

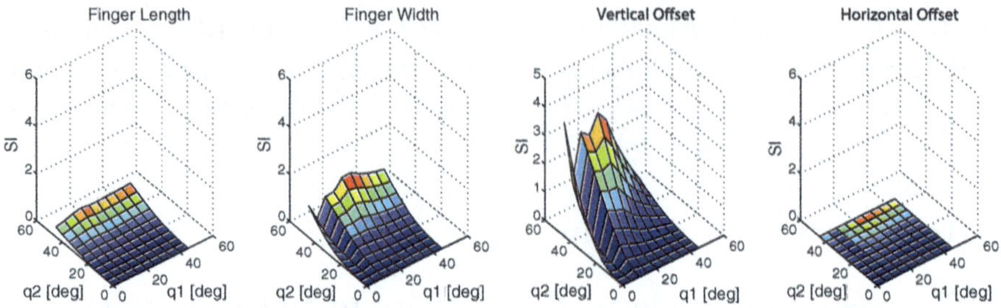

Fig. 4. Sensitivity analysis output for single pose

is moved and rotated across the workspace). Two markers were fixed to the exoskeleton base (one corresponding to the K point in Fig. 2, two to the link fixed to the first phalanx, and two to the link fixed with the second phalanx. Through positions of these markers, the joint angles q_{o1} and q_{o2} could be computed. Three subjects participated in the experimental session (male, aged 25–31 years old) with the length of the middle finger proximal phalanx measuring respectively 53 mm, 54 mm and 66 mm. Each experimental session consisted in wearing the exoskeleton, calibrating the algorithm presented above through two opening-closing movements, and finally in accomplishing a free hand closing and opening for validation. The two opening-closing movements for system calibration were executed fixing q_{o2} (to 0° and 45° respectively) by means of plastic supports. During such movements, sensors measurements were recorded and provided to the estimation algorithm offline. The following table presents measured (pedix m) and estimated (pedix e) parameters of the subjects. All data are expressed in mm.

Subject	$\overline{LM_m}$	$\overline{LM_e}$	s_m	s_e	L_x	L_y
1	56 ± 1	54.98	8 ± 1	10.81	1.96	22.50
2	53 ± 1	51.77	9 ± 1	9.22	9.52	24.61
3	54 ± 1	49.98	8 ± 1	8.50	0.19	23.76

During the free hand validation phase, the finger trajectory was computed by means of both the optical tracking system and the direct kinematic (Fig. 5). The absolute error of q_{o1} is comparable with the angular backlash between the phalanx and the exoskeleton. The error on q_{o2} estimation results higher, because of the second phases of the algorithm not only propagates the error of the first phases but suffers the difficulty of keeping fixed the angle between proximal and median phalanxes due to the human skin softness.

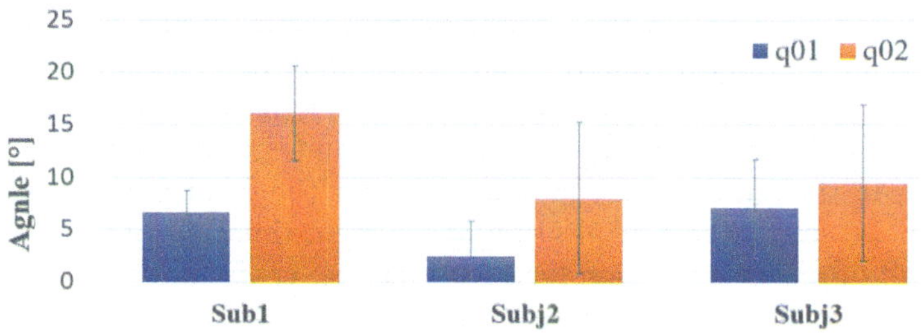

Fig. 5. Bar plot representing errors along test trajectory

4 Discussion and Conclusions

In this paper we investigate kinematics of an underactuated hand exoskeleton, that is intrinsically adaptive to different hand sizes due to the inclusion of user's hand joints in the parallel kinematics of the system. We analyzed the sensitivity of the system to variations of hand parameters, depending on wearability and hand size of different users. The analysis of sensitivity evaluated how estimation of finger pose and force transmission of the adaptive mechanism were affected by a variation of the above parameters. The analysis highlighted that certain parameters, in particular the offset of the metacarpal finger joint with respect to the exoskeleton base, become critical in the estimation of finger joint angular position and in the estimation of force transmission between actuators and finger links. Moreover, such parameters are related to the wearability of the exoskeleton to specific hand dimensions and shape, thus they are difficult to be estimated before wearing the exoskeleton and definitely challenging to be measured once the exoskeleton is worn. Due to the above reasons, a calibration algorithm, to be run once the exoskeleton is already worn at the user's hand, has been developed and presented in the paper. The algorithm made use of data acquired during a constrained calibration motion to geometrically compute the unknown parameters. The algorithm has been experimentally evaluated by means of an optical tracking system that measured the real position and orientation of finger phalanxes for three subjects with different hand sizes. The finger trajectory measured by the tracking system has been then compared to the trajectory computed by the exoskeleton direct kinematic, using embedded sensors values and parameters estimated by the algorithm. Results showed good estimation results, although the not perfectly rigid fixing of the phalanxes to the exoskeleton introduces errors in the pose estimation (mean errors: $e_{q_{o1}} = 5.40°$ and $e_{q_{o2}} = 11.18°$). In addition, the softness of human tissues and complexity of articulations makes very difficult to build an accurate model in which position and orientation of the axes of any human joint are assumed to be fixed. [13,14]. On this regard it has to be noted that the adopted model considers a rigid link between the finger and the exoskeleton. However, the peculiar exoskeleton kinematics is designed to apply only normal forces, and not torques, to finger phalanxes. Such design minimizes displacement between the exoskeleton links and finger phalanxes, and makes more effective the use of fastened soft belts for fixing the hand to the exoskeleton. Sensitivity analysis and the developed algorithm investigates critical aspects the new exoskeleton designs try to overcome: wearability and adaptability to different user hands. Design solutions including hand joints in the parallel kinematic design offer the great potential of intrinsic adaptation, yet at the cost of position and force estimation problems. The presented work proposes a solution for such systems, envisaging useful application in the teleoperation and neurorehabilitation scenarios, where both finger pose and force interaction are relevant information to be estimated.

References

1. Lee, S., Landers, K.A., Park, H.-S.: Development of a biomimetic hand exotendon device (biomhed) for restoration of functional hand movement post-stroke. IEEE Trans. Neural Syst. Rehabil. Eng. **22**(4), 886–898 (2014)
2. Iqbal, J., Tsagarakis, N., Caldwell, D.: Human hand compatible underactuated exoskeleton robotic system. Electr. Lett. **50**(7), 494–496 (2014)
3. Troncossi, M., Foumashi, M.M., Carricato, M., Castelli, V.P.: Feasibility study of hand exoskeleton for rehabilitation of post-stroke patients. In: Proceedings of the ASME 11th Biennial Conference on Engineering Systems Design and Analysis, ESDA2012, Nantes, France (2012)
4. Foumashi, M.M., Troncossi, M., Castelli, V.P.: Design of a new hand exoskeleton for rehabilitation of post-stroke patients. In: Romansy 19–Robot Design, Dynamics and Control, pp. 159–166. Springer (2013)
5. Leonardis, D., Barsotti, M., Loconsole, C., Solazzi, M., Troncossi, M., Mazzotti, C., Castelli, V.P., Procopio, C., Lamola, G., Chisari, C., et al.: An EMG-controlled robotic hand exoskeleton for bilateral rehabilitation. IEEE Trans. Haptics **8**(2), 140–151 (2015)
6. Barsotti, M., Sotgiu, E., Leonardis, D., Sarac, M., Sgherri, G., Lamola, G., Chiara, F., Procopio, C., Chisari, C., Frisoli, A.: A novel approach for upper limb robotic rehabilitation for stroke patients. In: International Conference on Human Haptic Sensing and Touch Enabled Computer Applications, pp. 459–469. Springer (2016)
7. Sarac, M., Solazzi, M., Sotgiu, E., Bergamasco, M., Frisoli, A.: Design and kinematic optimization of a novel underactuated robotic hand exoskeleton. Meccanica **52**(3), 749–761 (2017)
8. Sarac, M., Solazzi, M., Leonardis, D., Sotgiu, E., Bergamasco, M., Frisoli, A.: Design of an underactuated hand exoskeleton with joint estimation. In: Advances in Italian Mechanism Science, pp. 97–105. Springer (2017)
9. Griffin, W.B., Findley, R.P., Turner, M.L., Cutkosky, M.R.: Calibration and mapping of a human hand for dexterous telemanipulation. In: ASME IMECE 2000 Symposium on Haptic Interfaces for Virtual Environments and Teleoperator Systems, pp. 1–8 (2000)
10. Fischer, M., van der Smagt, P., Hirzinger, G.: Learning techniques in a dataglove based telemanipulation system for the DLR hand. In: 1998 IEEE International Conference on Robotics and Automation, 1998, Proceedings, vol. 2, pp. 1603–1608. IEEE (1998)
11. Bennett, D.J., Hollerbach, J.M.: Closed-loop kinematic calibration of the Utah-MIT hand. In: Experimental Robotics I, pp. 539–552. Springer (1990)

12. Bennett, D.J., Hollerbach, J.M.: Self-calibration of single-loop, closed kinematic chains formed by dual or redundant manipulators. In: Proceedings of the 27th IEEE Conference on Decision and Control, 1988, pp. 627–629. IEEE (1988)
13. Cempini, M., De Rossi, S.M.M., Lenzi, T., Vitiello, N., Carrozza, M.C.: Self-alignment mechanisms for assistive wearable robots: a kinetostatic compatibility method. IEEE Trans. Robot. **29**(1), 236–250 (2013)
14. Bottlang, M., Madey, S., Steyers, C., Marsh, J., Brown, T.D.: Assessment of elbow joint kinematics in passive motion by electromagnetic motion tracking. J. Orthop. Res. **18**(2), 195–202 (2000)

Gait Phases Detection Using a 6 d.o.f. Ankle Joint Electro-Goniometer

Viet Anh Dung Cai[1(✉)], Philippe Bidaud[2], and Long Triet Giang Huynh[3]

[1] Eastern International University, Thu Dau Mot, Binh Duong, Viet Nam
dung.cai@eiu.edu.vn
[2] University of Pierre et Marie Curie, Paris, France
bidaud@isir.upmc.fr
[3] Saigon Technology University, Ho Chi Minh, Viet Nam
huynhlong.cdt@gmail.com

Abstract. This paper proposes a method for the determination of the kinematics of the human ankle joint using a 6 d.o.f. electro-goniometer. The mechanical design and the kinematic model of the device are presented. The authors performed the Instantaneous Helical Axis (IHA) measurement during walking experiment and proposed a kinematics descriptor for the identification of gait phases and sub-phases, based on the measure of the direction of the IHA and the magnitude of the rotational vector. First results reveal manifest potential and perspective of development of this method, as good repeatability is obtained for the measurement of the kinematics descriptor during different gait cycles.

1 Introduction

The ankle-foot complex plays a major role in the stabilization of walking and more generally in the dynamic performance of the locomotor system. It has been subject of research for several authors in the last decades (Brockett and Chapman 2016; Rodgers 1988). It is one of the most complex articular system of human anatomy comprising several anatomical joints in series. This articular complex ensures the "fine" adaptation of the locomotion system to variations in geometry of the support with the ground. It also adapts through all its mobility to the gait of walking and more broadly to all the needs of locomotor activity. Beyond its mobility, this complex like all the other joints of the locomotion system sees its mechanical impedance varied during locomotor activity (Lee and Hogan 2015).

One can find in the literature several works on the analysis of the ankle joint kinematics and dynamics during walking (Rodgers 1988; McPoil and Knecht 1985), or in-vivo studies (Isman and Isman 1969). However, to our knowledge, detailed studies on the kinematics analysis of the ankle joint during sub-phases of the gait cycle have not been realized yet. It is maybe due to the difficulty to measure the joint during this activity. According to the definition proposed by (Torricelli et al. 2016), the gait cycle can be divided into 2 phases and 7 sub-phases that are:

J. Lenarcic and V. Parenti-Castelli (Eds.): ARK 2018, SPAR 8, pp. 458–464, 2019.
https://doi.org/10.1007/978-3-319-93188-3_52

- Swing phase: 40% of the gait cycle, comprising 3 sub-phases: Initial swing (12%), Mid-Swing (16%) and Terminal Swing (12%).
- Stance phase: 60% of the gait cycle, comprising 4 sub-phases: Loading Response (12%), Mid-Stance (18%), Terminal Stance (20%) and Pre-Swing (10%).

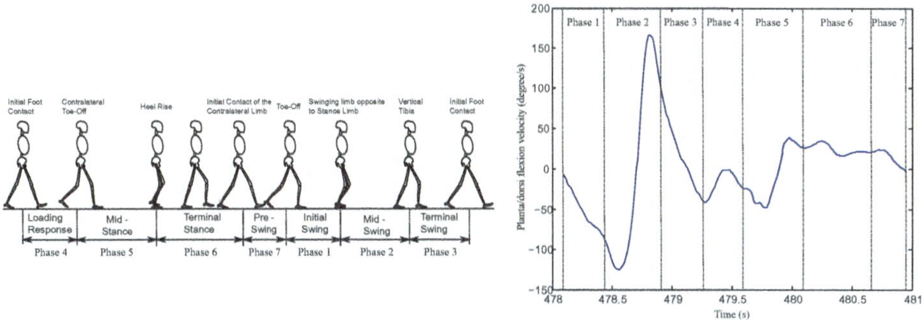

Fig. 1. Definition of the gait phases and associated rotational velocity of the planta/dorsi flexion.

One of the hypotheses that we formulate and which motivates this work is that the ankle-foot complex, because of its function and the "richness" of its activity, can constitute a particularly valuable source of information for various purposes, such as the detection and analysis of abnormal/pathological gait. It can also improve diagnosis and help surgical decision when necessary. In another perspective, the activity of the foot can be a mean of detecting the intentions of locomotor activities and in this sense to control the proper coordination movements of an exoskeleton in particular. This information is indeed essential to ensure the order transparency of the system and provide the necessary mechanical actions for assistance to locomotion (Cai et al. 2015; Ferris and Lewis 2009). The understanding of this activity can also help to design the locomotive systems of humanoid robots (Alfayat et al. 2009), especially when great agility is sought or the ability to change pace.

In order the evaluate the activity of ankle foot complex, a method for the determination of the Instantaneous Helical Axis (IHA) of the human ankle joint using a 6 d.o.f. electro-goniometer is proposed. We show, through measurements realized on an individual, how this IHA evolves during locomotion. We are interested afterward in the problem of the fine detection of different phases observed during human locomotion. For this purpose, a descriptor of the gait phases that should allow a robust detection of the changes on all 7 sub-phases (Harris and Wertsch 1994) is proposed. The mechanical design and the kinematic model of the device are described in Sects. 2 and 3. The result of the descriptor measurement during walking experiment, which is presented in Sect. 4, shows a manifest potential and perspective of development of the method, as we obtained good repeatability for the measure.

2 Design of the Ankle Joint Electro-Goniometer

2.1 Mechanical Design

The mechanical design of the ankle joint measurement device must comply with the ankle joint kinematics and allows the measurement of all its 6 d.o.f. Here the serial architecture is chosen for the sake of simplicity. Each joint is equipped with optical encoder with a resolution of 400 pulses per revolution. The transmission between the measurement device and the computer is assured by Bluetooth, at a sampling frequency of 140 Hz.

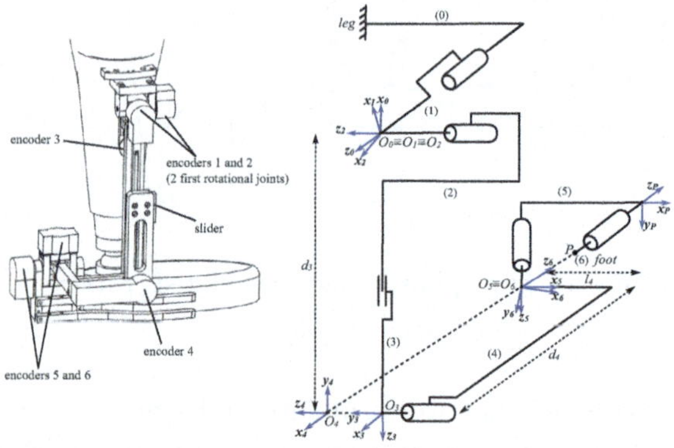

Fig. 2. Design of the ankle joint electro-goniometer and its frame assignment, using modified D.H. notation (Nguyen et al. 2017).

The design and the kinematic scheme of the device are shown in Fig. 2. The mechanism is attached to the user's calf at the surface of the tibia bones. It is attached to his/her foot at the surface of the tarsals, metatarsals, and calcaneus bones. When the system is secured onto the user's limbs, we make the assumption that the frames $(\boldsymbol{R_0})$ and $(\boldsymbol{R_6})$ are attached to the calf and the foot of the subject and their kinematic parameters represent that of these limbs. The 4th and the 6th joints, whose rotation axes are located near the anatomical axes, allow the mechanism to follow the ankle joint's principal movements, which are dorsi/planta flexion and pronation/supination, in their range of motion (ROM). The 5th joint assures the measurement of the internal/external rotation. These 3 rotational joints are connected to joints 1 and 2 through a prismatic joint. The latter allow an adaptation of the mechanism to different limb sizes. The whole structure have six degrees of freedom, all equipped with optical encoders, allowing the determination of all the 6 spatial parameters of the ankle joint.

3 Determination of the Instantaneous Helical Axis

The instantaneous helical axis (IHA) allows the representation of all the ankle joint's kinematic parameters in space. Its variation is regardless of the coordinate system being used. Several authors (Kinzel et al. 1972; Cai et al. 2010) have described a method for the determination of FHA (Finite Helical Axis) and IHA by using an electro-goniometer to measure the 6 d.o.f. of the mechanism. In this paper, our method consists of writing the Jacobian matrix at 2 points $\boldsymbol{P_1} = (a_1, b_1, c_1)'$ and $\boldsymbol{P_2} = (a_2, b_2, c_2)'$ located on the IHA. The resulting system of equations can then be solved in order to determine theses 2 points' coordinates, which also give the equation of the IHA. A similar method has been proposed for the knee joint in (Cai et al. 2010).

By writing the loop-closure equation, we obtain the expression of the velocity vector at point P as follow

$$\left(\omega_P \; v_P\right)_{/B_3} = \left(\omega_x \; \omega_y \; \omega_z \; v_x \; v_y \; v_z\right)^{\top}_{/B_3} = \boldsymbol{J(P)}_{/B_3}\dot{\boldsymbol{q}} \tag{1}$$

with $\boldsymbol{J(P)}_{/B_3}$ is the Jacobian matrix of the mechanism, written at point P in the base B_3, $\dot{\boldsymbol{q}}$ is the vector of mechanical joint velocities. This leads to the system of equations below:

$$\begin{cases} \omega_x = K_4 = -K_8; \\ \omega_y = -K_2 = K_7; \\ \omega_z = K_1 = -K_5; \\ v_x = K_1.b + K_2.c + K_3; \\ v_y = K_4.c + K_5.a + K_6; \\ v_z = K_7.a + K_8.b + K_9; \end{cases} \tag{2}$$

Here $P_{/R_3} = (a, b, c)^{\top})$ is a point located on the IHA. $K_i s$ are expressions of the system of equations issued from Eq. 1. They are independent of P and were formed by regrouping these equations in function of a, b and c.

When the ankle joint rotates, the twist component ω_y (measured in the base $\boldsymbol{B_3}$) is different to zero. Therefore, the coordinate b of point P can be chosen arbitrarily in order to compute the 2 other components a and c, as shown in Eq. 3. Here P is a point

$$c = (\omega_z b + K_3 - v_x)/\omega_y, \qquad a = (\omega_x b - K_9 + v_z)/\omega_y. \tag{3}$$

Given $P' = (a', b', c')$ another point located on the IHA. The linear velocity vectors of points P and P' are the same and can be written as follow

$$\boldsymbol{v_P} = \boldsymbol{v'_P} = \left(v_x \; v_y \; v_z\right)^{\top} = \epsilon \boldsymbol{PP'} = \epsilon \left(a' - a \; b' - b \; c' - c\right)^{\top} \tag{4}$$

From Eq. 2, we obtain

$$\begin{cases} K_1(b - b') + K_2(c - c') = 0 \\ K_4(c - c') + K_5(a - a') = 0 \\ K_7(a - a') + K_8(b - b') = 0 \end{cases} \Leftrightarrow \begin{cases} K_1 v_y + K_2 v_z = 0 \\ K_4 v_z + K_5 v_x = 0 \\ K_7 v_x + K_8 v_y = 0 \end{cases} \tag{5}$$

From Eqs. 2, 3 and 5, the values of v_y, v_x and v_z can be computed

$$\begin{aligned} v_y &= (\omega_x \omega_y K_3 + {\omega_y}^2 K_6 + \omega_y \omega_z K_9)/({\omega_x}^2 + {\omega_y}^2 + {\omega_z}^2) \\ v_x &= -K_8 v_y / K_7 = \omega_x v_y / \omega_y, v_z = -K_1 v_y / K_2 = \omega_z v_y / \omega_y \end{aligned} \tag{6}$$

The IHA coordinates and velocity vectors are then expressed in the frame $\boldsymbol{R_6}$, which is attached to the user's foot by using the homogenous transform matrix $\boldsymbol{T_{63}}$.

4 Results and Discussion

This section presents the result of a 5-steps walking experiment. The IHA is computed for each of the 7 gait sub-phases. As it was noticed in the Sect. 3, the IHA can only be determined provided that the principal rotational velocity is not zero (here the condition for HA computation is: $|\omega| > 5°/s$). Figures 3 present the 3D view of the IHA of 7 different gait phases during 1 gait cycle.

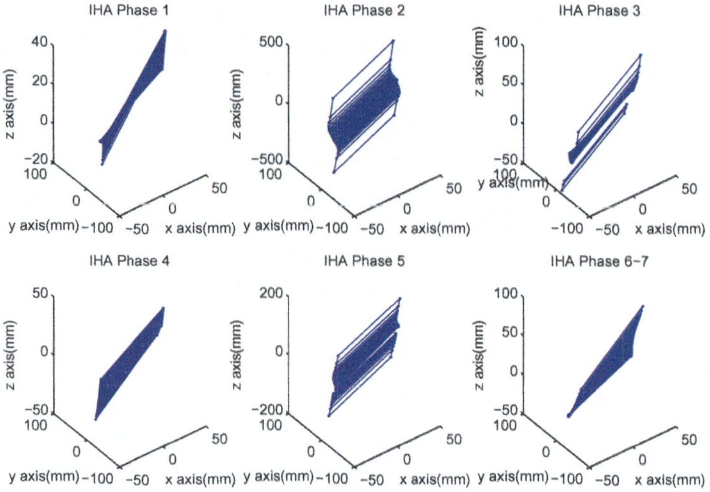

Fig. 3. Instantaneous helical axis of the 7 gait phases during one gait cycle.

Figure 4 presents the evolution of the principal velocities as well as the IHA's orientation of the 2 ankle joints during one gait cycle. At time t, the orientation determined by the angle between the 2 vectors: the actual and the first IHA of the gait cycle. Even if a light dis-symmetry can be observed on the curves, the curves' shapes obtained from the 2 ankles remain similar.

Figure 5a and b presents the cyclogram of the ankle joint kinematic descriptor, measured during 1 gait cycle. The kinematic descriptor is determined by the IHA orientation $(\boldsymbol{u_x}, \boldsymbol{u_y}, \boldsymbol{u_z})$ and the magnitude of the rotational velocity vector $|\omega| = norm(\omega_x, \omega_y, \omega_z)$. Here the gait phases were determined in accordance with the definition given in Fig. 1. Despite some dis-symmetry for the left/right ankle cyclograms, the results obtained from the experiments, however, highlight repeatability. This information provided by the measurement of the IHA and the angular velocity vector can be exploited to develop a robust gait phase detection algorithm. Figure 5c illustrates the result obtained on the same subject for different gait cycles. The repeatability of the measure is verified here, as the points measured during different gait cycles are located on the same regions, especially during the swing-phase. One can also resort to trajectory recognition methods (Black and Jepson 1998) in order to obtain a better discrimination of the gait phases in case of points overlapping.

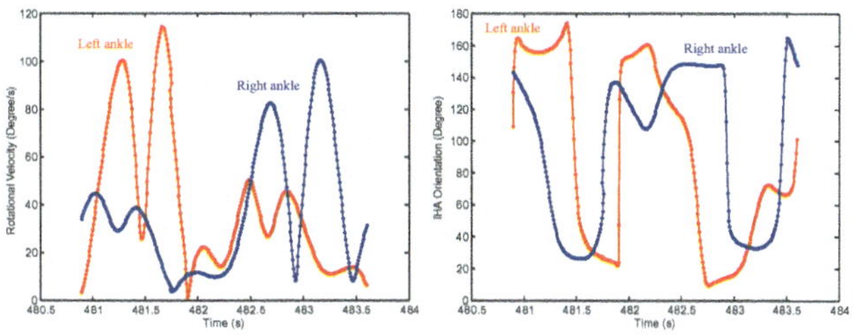

Fig. 4. Comparison of the magnitude of the rotational velocity vector and the IHA orientation between the left and right ankles.

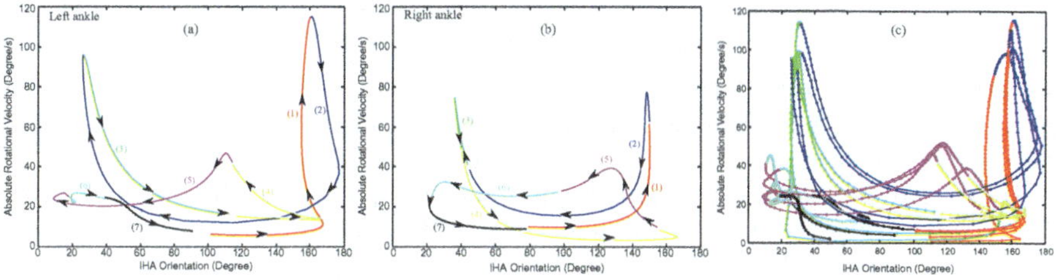

Fig. 5. (a) and (b) The cyclogram of the ankle joint kinematic descriptor, measured during 1 gait cycle. (c) Measures obtained during 4 consecutive gait cycles. The 7 gait sub-phases are determined following the rule proposed by (Toricelli 2016).

5 Conclusions

In this paper, a method for the measurement of the human ankle joint kinematics during sub-phases of the gait cycle, using a 6 d.o.f. electro-goniometer, is presented. The design and the kinematic model of the system is described in details and first experimental results are also given. The measurement, realized simultaneously onto the 2 ankles during walking, highlights an apparent repeatability as well as a dis-symmetry in range of motion of the 2 ankles. An ankle joint kinematics descriptor is also defined, comprising the rotational velocity magnitude and the direction of the IHA (which is here represented in function of the orientation between actual IHA and an axis defined on the attach of the mechanism to the user foot). Base on this device, robust gait phases detection algorithms and methods for the determination of ankle joint anomalies can be developed.

References

Alfayat, S., Ouezdou, F.B., Namoun, F.: New three DOF ankle mechanism for humanoid robotic application: modeling, design and realization. In: IEEE/RSJ International Conference on Intelligent Robots and Systems, pp. 4969–4976 (2009)

Black, M.J., Jepson, A.D.: A probabilistic framework for matching temporal trajectories: condensation-based recognition of gestures and expressions. In: Computer Vision - ECCV 1998. Lecture Notes in Computer Science, vol 1406. Springer, Heidelberg (1998)

Brockett, C.L., Chapman, G.: Biomechanics of the ankle. Orthop. Trauma **30**(3), 232–238 (2016)

Cai, D., Bidaud, P., Hayward, V., Gosselin, F.: Estimation of complex anatomical joint motions using a spatial goniometer. In: ROMANSY 18, pp. 399–406 (2010)

Cai, D., Bidaud, P., et al.: In search of transparency for lower limb exoskeleton devices with a new mechanical design and a robust gait phases detection method. In: IEEE ROBIO 2015, pp. 1151–1156 (2015)

Ferris, D.P., Lewis, C.L.: Robotic lower limb exoskeletons using proportional myoelectric control. In: IEEE Conference on Engineering in Medicine and Biology Society, EMBS (2009)

Harris, G.F., Wertsch, J.J.: Procedures for gait analysis. Arch. Phys. Med. Rehabil. **75**(2), 216–225 (1994)

Isman, R.E., Isman, V.T.: Anthropometric studies of the human foot and ankle. Bull. Prosthet. Res. 97–129 (1969)

Kinzel, G.L., Hall, A.S., Hillberry, B.M.: Measurement of the total motion between two body segments-I. Analytical development. J. Biomech. **5**, 93–105 (1972)

Lee, H., Hogan, N.: Time-varying ankle mechanical impedance during human locomotion. IEEE Trans. Neural Syst. Rehabil. Eng. **23**(5), 755–764 (2015)

McPoil, T.G., Knecht, H.G.: Biomechanics of the foot in walking: a function approach. J. Orthop. Sport Phys. Ther. **7**, 69–72 (1985)

Nguyen, V.T., et al.: Kinematic measurement of the ankle joint using a six d.o.f. electrogoniometer. In: IEEE International Conference on System Science and Engineering (ICSSE), pp. 24–28 (2017)

Rodgers, M.M.: Dynamic biomechanics of the normal foot and ankle during walking and running. Phys. Ther. **68**(12), 1822–1830 (1988)

Torricelli, D., et al.: Human-like compliant locomotion: state of the art of robotic implementations. Bioinspir. Biomim. **11**, 1–20 (2016). paper 051002

Fluoroscopy Validation of Noninvasive 3D Bone-Pose Tracking via External Pressure-Foils

Nikolas Bufe[1(✉)], Gregor Kuntze[2,3], Janet L. Ronsky[2,3], and Andrés Kecskeméthy[1]

[1] Chair of Mechanics and Robotics, University of Duisburg-Essen, Duisburg, NRW, Germany
{Nikolas.Bufe,andres.kecskemethy}@uni-due.de

[2] Department of Mechanical and Manufacturing Engineering, University of Calgary, Calgary, AB, Canada
{gkuntze,jlronsky}@ucalgary.ca

[3] McCaig Institute for Bone and Joint Health, University of Calgary, Calgary, AB, Canada

Abstract. Precise noninvasive bone motion tracking is a topic of increasing interest, as it provides medical practitioners with important information about musculoskeletal disorders and/or their treatment. Currently, the only non-invasive method that achieves 3D tracking precision below 1 mm is Dual Fluoroscopy (DF). However, this technique is expensive, restricted to only small measuring volumes and time periods, and can only be applied sporadically due to X-ray radiation limits. In a series of previous papers, the authors presented a new concept for bone motion tracking based on external virtual palpation of typical bone protuberances via motion-tracked pressure foils. This paper presents a DF validation of the method for three basic movements of the shank: flexion/extension, abduction/adduction and internal rotation. It is shown that by simple pressure-foil palpation, bone-tracking precisions of 0.5 to 1.0 mm and 0.3° to 0.6° can be attained with respect to manually-registered DF, reaching the same order of magnitude as state-of-the-art model-based tracking algorithms for DF using CT volumes.

Keywords: In-vivo bone motion tracking · Pressure-foil palpation
Three-ellipsoid registration

1 Introduction

Accurate determination of bone motion via noninvasive external tracking devices is crucial for many biomechanical tasks and patient-specific medical objective diagnosis and treatment plans of the musculoskeletal system. The most popular approach in this respect consists in tracking bone motion by retroreflective markers placed on the skin, and computing from these the sought bone motion, see

J. Lenarcic and V. Parenti-Castelli (Eds.): ARK 2018, SPAR 8, pp. 465–473, 2019.
https://doi.org/10.1007/978-3-319-93188-3_53

e.g. [6]. However, it is known that these methods suffer from significant soft tissue artifacts (STA) between retroreflective markers and bones, leading to up to several centimeters of tracking errors, depending on the marker location and the bone motion of interest [11]. Advanced approaches seek to minimize STAs by using point clusters [4], but there still remains a systematic error [11]. To circumvent STAs, currently, the only existing approaches are invasive methods such as intracortical pins used during surgeries [5], or noninvasive procedures with radiation exposure such as dual fluoroscopy [7,12]. An accurate noninvasive and radiation-free method for bone motion tracking is thus a still open issue.

A novel approach for noninvasive radiation-free bone motion estimation was proposed in [8]. The approach consists in palpating prominent bone protuberances with three planar pressure sensor foils. In a first step, the geometrical problem was analyzed using Gröbner bases [3], showing that the planar case yields in general 64 complex solutions, of which 48 real solutions could be found. Of these, one physically meaningful solution could be identified by considering that palpated bone protuberances should be as close as possible to the pressure sensor foils. The feasibility of the 3D case was analyzed in [2] by comparing in vivo shank motion to a reference motion tracked by retroreflective markers, however not being able to verify precision due to marker-based STAs. This paper continues this research by validating the pressure foil (PF) with a Dual Fluoroscopy (DF) system for the three basic motions of the shank flexion/extension, abduction/adduction, and internal rotation. The results show that pressure-foil tracking can render comparable accuracy to gold standard bone motion estimation by dual fluoroscopy [11].

2 Brief Review of Method

The basic concept of 3D bone pose estimation using pressure foils is briefly reviewed here (see [2] for more details). The pressure foils are devices that can measure pressure distribution with a small grid size and are fixed to planar rigid plates which are tracked in space by marker-based cameras (see concept sketch on Fig. 1 left). For the target bone, three protuberances at different, well-separated bone locations that are clearly palpable from the skin are identified and their outer interface (facing the skin) is approximated by corresponding ellipsoids (Fig. 1 right)

$$\underline{\mathcal{E}}_i(u_i, v_i) = \begin{bmatrix} a_i \cos(u_i)\cos(v_i) \\ b_i \cos(u_i)\sin(v_i) \\ c_i \sin(u_i) \end{bmatrix} \tag{1}$$

with $-\pi/2 \leq u_i \leq \pi/2$ and $-\pi \leq v_i \leq \pi$, and where a_i, b_i, c_i are the corresponding half axes and u_i, v_i are the corresponding surface coordinates (here angles). By slight pressure to the skin above the protuberances, the PFs will render "pressure points" as the extremes of pressure distribution. These pressure points will arise at the minimal distance d_i between the PF and the protuberance outer face.

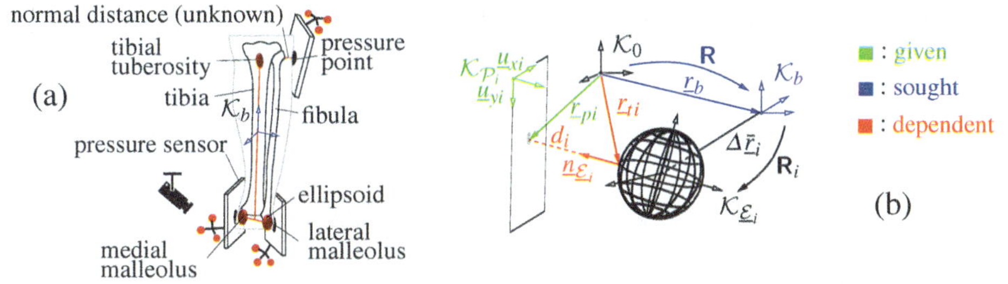

Fig. 1. Concept sketch and ellipsoid parameters for 3D bone pose detection by pressure points.

Assuming a constant relative pose (radius vector $\Delta\underline{\bar{r}}_i$, orientation matrix $\mathbf{R}_i$) of the ellipsoids with respect to the bone reference frame $\mathcal{K}_b$, and with given pressure points (as measurements), the corresponding points of the ellipsoid fulfill two sets of orthogonality constraints:

(1) [Gradient orthogonality] The gradient of $\underline{\mathcal{E}}_i$ at point $\underline{r}_{ti}$,

$$\underline{n}_{\mathcal{E}_i} = \mathbf{R}\mathbf{R}_i \left(\frac{\partial \mathcal{E}_i}{\partial u_i}(u_i, v_i) \times \frac{\partial \mathcal{E}_i}{\partial v_i}(u_i, v_i) \right) . \tag{2}$$

must be orthogonal to the PF plane:

$$g_{i,x} : \quad \underline{n}_{\mathcal{E}_i}^{\mathrm{T}} \, \underline{u}_{xi} = 0 \tag{3}$$

$$g_{i,y} : \quad \underline{n}_{\mathcal{E}_i}^{\mathrm{T}} \, \underline{u}_{yi} = 0 . \tag{4}$$

(2) [Distance orthogonality] The distance vector from the measured pressure point $\underline{r}_{pi}$ on the PF plane to the closest ellipsoid point $\underline{r}_{ti}$ must be normal to the PF plane:

$$d_{i,x} : \quad (\underline{r}_{ti} - \underline{r}_{pi})^{\mathrm{T}} \, \underline{u}_{xi} = 0 \tag{5}$$

$$d_{i,y} : \quad (\underline{r}_{ti} - \underline{r}_{pi})^{\mathrm{T}} \, \underline{u}_{yi} = 0 . \tag{6}$$

By choosing three ellipsoids rigidly attached to the bone, twelve constraint equations are obtained, which in principle suffice to determine the twelve unknowns

$$\underline{q} = [\, \underline{r}_b \quad \underline{\Phi} \quad u_1 \quad v_1 \quad u_2 \quad v_2 \quad u_3 \quad v_3 \,]^{\mathrm{T}} , \tag{7}$$

where $\underline{r}_b$ and $\underline{\Phi}$ define the position and orientation matrix $\mathbf{R}$ of the bone reference frame $\mathcal{K}_b$ with respect to the world frame $\mathcal{K}_0$, and u_i and v_i are the internal surface parameters. However, the geometric solution of this problem can lead to a large variety of real and/or complex solutions [3]. Hence the problem is numerically restructured by requiring only a minimum for the composed cost function

$$F(\underline{q}) = c_1 \sum_{i=1}^{3} \{\, g_{i,x}^2 + g_{i,y}^2 \,\} + c_2 \sum_{i=1}^{3} \{\, d_{i,x}^2 + d_{i,y}^2 \,\} + c_3 \sum_{i=1}^{3} {d_i}^2 , \tag{8}$$

with weights c_i defined such as to balance between preciseness $\{g_{i,x}^2+g_{i,y}^2\}$, $\{d_{i,x}^2+d_{i,y}^2\}$ and feasibility (d_i), and also to homogenize units among terms. Note that from this formulation the bone pose can be determined without knowing the distance d_i between PF and bone surface, which is eliminated and can be arbitrary.

3 Validation Using Dual Fluoroscopy

One participant (sex: male, age: 34, height: 173 cm, weight: 82 kg, no injuries) volunteered to perform three different movement patterns of the right shank: (1) flexion/extension, (2) abduction/adduction and (3) internal rotation of the knee, each of which was repeated approximately three times back and forth (limited by the allowed radiation duration). The Conjoint Health Research Ethics Board approved this study, and informed consent was obtained from the participant before the experiment. Three Tekscan 5027 PF foils (0.6 mm grid size) attached to planar plates equipped with retroreflective marker tetrads were placed on the right shank of the participant at (1) the tibial tuberosity, (2) the medial and (3) lateral malleolus (Fig. 2a). Absolute 3D sensor poses were tracked by using the 8 camera motion capture system Motion Analysis (Santa Rosa, CA) with Cortex version 2.1 software. Bone motion was tracked simultaneously with the PF and DF systems at sampling frequencies $\approx$ 20 Hz and 30 Hz, respectively. Data from both systems was post-synchronized by linearly interpolating raw data. Figure 2 (right) schematically shows the setup of the DF system: both image intensifiers were placed parallel to the ground with an interbeam angle of 60° and a distance of approximately 56 cm and 61 cm away from the shank at the point of intersection of both beams. The three-dimensional in vivo motion was determined with the radiographic model-based tracking technique using a computed tomography (CT) scan of the examined bone [10]. Manual 2D-3D registration was performed using Joint Track Biplane (University of Florida, Gainesville, Florida, USA) for all three movement patterns. For the PF, ellipsoids were visually fitted to the CT scan with the program library "Mobile-C++" [9], obtaining the parameters shown in Fig. 3, where the local y axis points along the tibial shaft in the proximal, the x axis in the lateral, and the z axis in the

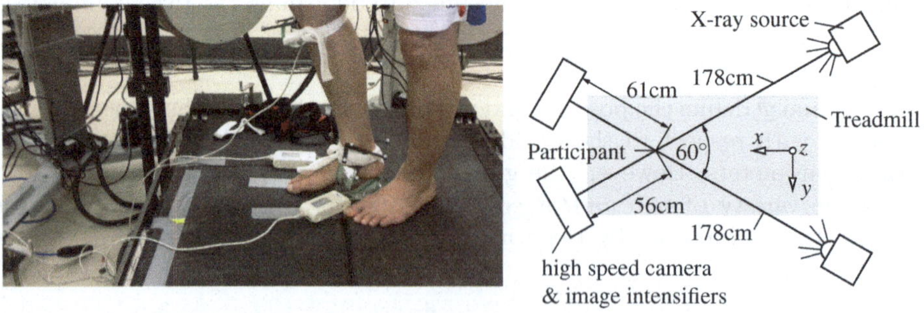

Fig. 2. Sensor placement on the right shank (left) and setup of the DF system (right).

posterior direction, respectively. Weighting factors of the cost function (8) where chosen as $c_i = 1, i = 1, \ldots, 3$. The comparison was evaluated as relative pose difference from PF to DF (in mm/degrees), where rotational deviations were assumed as very small and thus as vectors $\Delta\underline{\varphi} = [\Delta\varphi_x, \Delta\varphi_y, \Delta\varphi_z]^{\mathrm{T}}$ according to EULER's Theorem for small rotations.

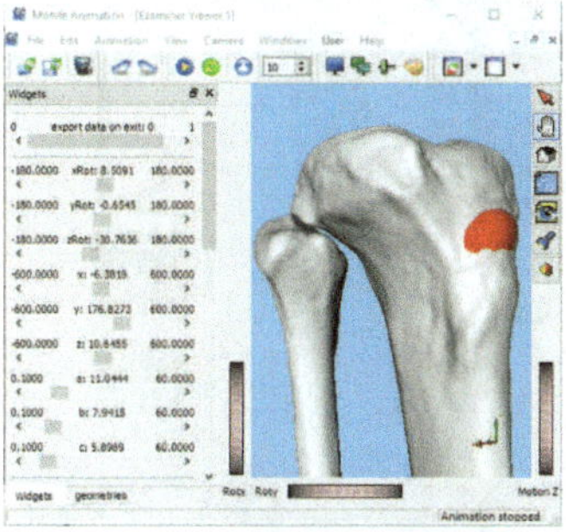

Ellipsoid	Semi-axes [mm]	Cardan xyz [°]	Offsets [mm]
	6.1	-59.8	-8.7
1	9.5	-56.7	177.1
	12.0	68.6	11.3
	3.1	4.6	35.0
2	8.6	25.4	-183.4
	6.0	6.8	4.5
	4.1	19.8	-27.3
3	3.7	-17.3	-167.6
	5.6	5.6	-3.9

Fig. 3. Screenshot of ellipsoid fitting for tibial tuberosity and table of ellipsoid parameters

The results are shown in Figs. 4 and 5. During the experiments, some time intervals did not yield correct values due to dereliction of medial ankle PF marker positions from the camera's field of view. Also, during internal rotation, the tibial tuberosity lost contact with the pressure sensing area for a small period of time. These periods are highlighted with light gray bars and were excluded from the evaluation.

Initially, large discrepancies between the manually registered DF poses and the PF results were identified (around 6 mm and 4°, see Fig. 4 upper left). However, analysis of the manual DF registration showed that this was not sufficiently precise (see difference between dashed yellow and dot-dashed orange contour of the projected CT scan and solid blue contours of the DF X-ray images in Fig. 4a). This is due to the fact that manual registration requires rotations of the CT volume in 3D using Euler angles to match the DF projections, which is very unintuitive to accomplish. Thus, in a next round, the PF results were used to find better initial-guess poses for manual registration (Fig. 4b). This resulted in near-to-perfect results for the ensuing fine-fitting by manual registration (see Fig. 4c). The PF tracking thus can be seen as a workable method for automatically finding suitable start poses for manual DF 2D-3D registration, in particular for translations parallel to the X-ray directions and rotations about longitudinal bone axes, which are less visible on the 2D fluoroscopy images.

By analyzing the plots of Fig. 4 bottom left and Fig. 5, one can see that the PF tracking differs from the DF, apart from outliers, in the order of magnitude of 0.6 mm and 0.2° for flexion/extension, 0.5 mm and 0.3° for abduction/adduction, and 1.0 mm and 0.6° for internal rotation. Some spiky outliers were verified indicating noise effects which could be filtered by future post-processing.

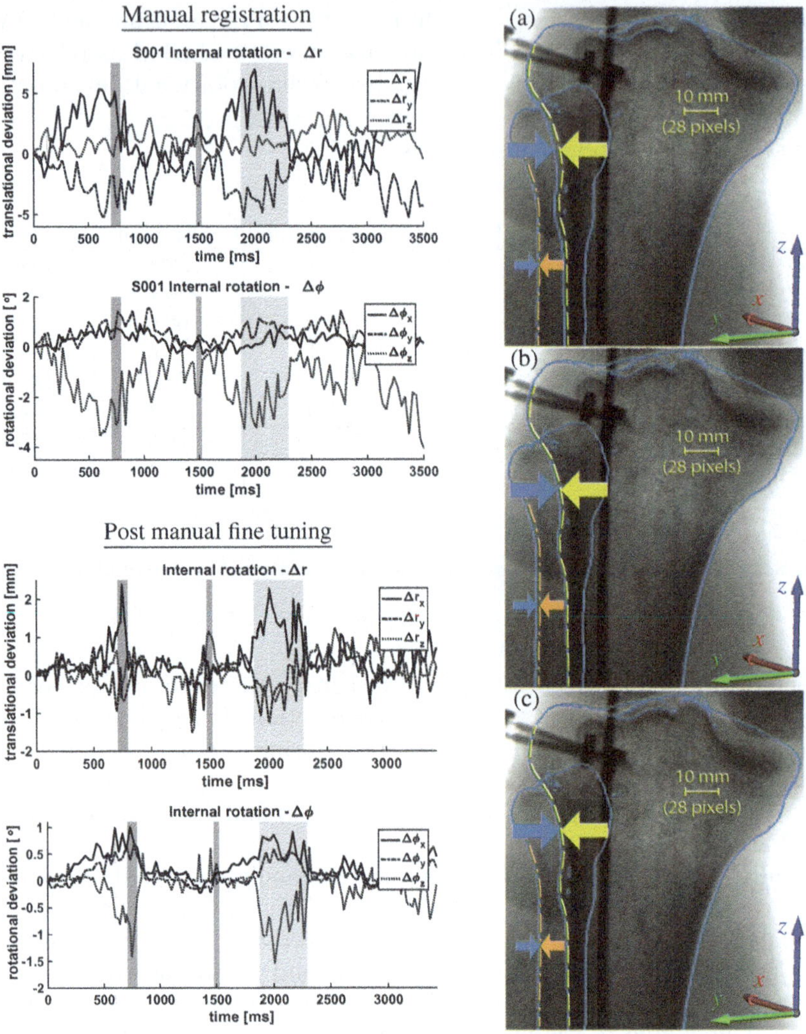

Fig. 4. Translational and rotational deviations between PF and DF system before and after manual fine tuning (left) and 2D-3D registration comparison of (a) initial manual registration, (b) pressure foil registration, (c) and pressure foil registration with post manual fine tuning.

In comparison, in [1] it was shown that the accuracy of dynamic model-based Dual Fluoroscopy tracking with respect to bone-embedded Tantalium beads was 0.3° to 0.9° in rotation and 0.3 mm to 0.7 mm in translation (with one outlier reaching 1.0 mm). This seems to indicate that the presented pressure-foil tracking method is of comparable precision to current DF bone tracking.

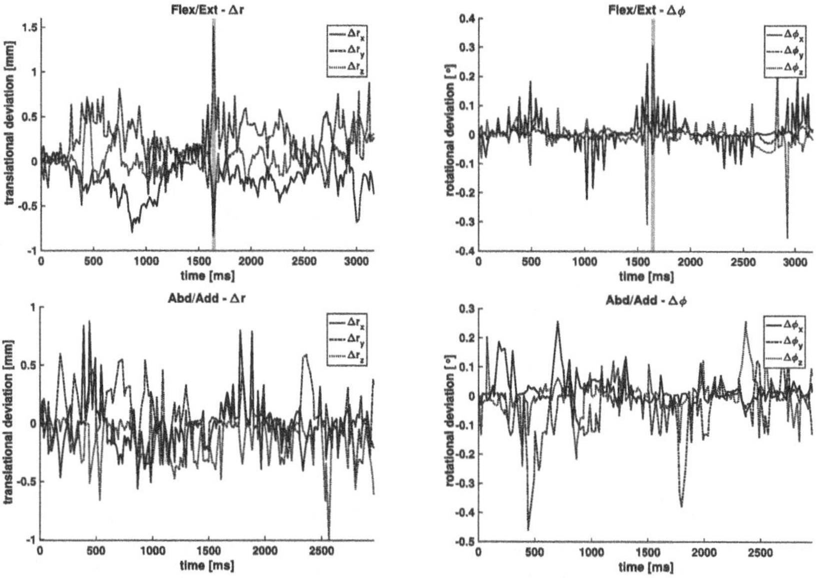

Fig. 5. Translational (left) and rotational (right) deviations betwen PF and DF after fine tuning.

4 Sensitivity Analysis

The influence of ellipsoid bone geometry fitting was tested for the example of abduction/adduction motion by (a) independently varying the centers as well as the semi-axes of all three ellipsoids using Gaussian noise with a mean value $\mu = 0.0$ mm and a standard deviation $\sigma = 0.5$ mm and (b) by proportionally increasing the initial semi-axes up to a factor of 1000. Hereby, due to the tangent conditions (3) to (6), the size of the ellipsoid is immaterial. Starting from the known solution shown in Fig. 5, maximal deviations of relative bone positions Δr and rotations $\Delta\phi$ with respect to the DF motion were evaluated using the concept of the characteristic length κ for coupling translations and rotations to the equivalent maximal translational change $\Delta s = \max(\|\Delta r\| + \kappa\|\Delta\phi\|)$, with

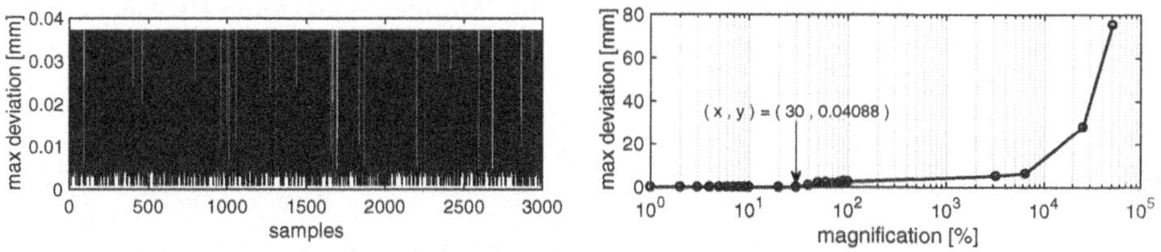

Fig. 6. Maximum translational deviation after (a) adding Gaussian noise to geometry dimensions (left) and (b) increasing proportionally all semi-axes (right).

$\kappa = 2$ mm/°, according to the experimental dimensions of translational and rotational motion (see Fig. 5).

The results of 3000 bone motion computations using (a) Gaussian noise are displayed in Fig. 6 left, showing a maximum translational deviation of $\Delta s = 0.037$ mm. Also, from Fig. 6 right, it can be seen that (b) the maximum translational deviation for the abduction/adduction data set remains almost constant and below 0.04 mm when the semi-axes of all ellipsoids are simultaneously increased by up to 30%. Thus, it can be concluded that the influence of ellipsoid fitting to bone geometry is rather insensitive for 3D bone-tracking using pressure foils, thus rendering this a robust method.

5 Conclusions and Outlook

The present paper shows that non-invasive 3D bone tracking by external pressure-foil palpation using a three-ellipsoid model of the bone renders precisions in the order of magnitude of dual fluoroscopy for the example of flexion/extension, abduction/adduction and internal rotation of the shank The feasibility and robustness of the approach is shown by a sensitivity analysis using small variations of the bone-fixed ellipsoid parameters. While the method is very promising and validated, there are a number of issues to be tackled in the future. On the one hand, a technique to define ellipsoid models that are independent of CT scans would be valuable for reducing the amount of data to be surveyed for the bone tracking and making the PF approach more universally applicable. Also, the attachment of the pressure foil sensors on the skin can be optimized such that retroreflective markers remain visible all the time for the marker cameras, and no lift-off of the PF with respect to the protuberance occurs. Moreover, the external pressure-foil approach could be further developed as a tool for determining suitable initial poses for dual-fluoroscopy 2D-3D registration, which is likely one of the most significant sources of errors introduced with manual registration. On the other hand, further numerical tests would be valuable to determine optimal weighting factors for the combination of precision and feasibility of the optimization cost function. Furthermore, tests for application to other bony structures would be applicable.

Acknowledgements. The authors would like to thank Jessica Küpper and Payam Zandiyeh for their valuable contributions and technical assistance. The authors gratefully acknowledge the support of the Centre for Mobility and Joint Health, McCaig Institute for providing us a CT image of the participant using a GE Revolution GSI, GE Healthcare, Milwaukee, USA.

References

1. Anderst, W., Zauel, R., Bishop, J., Demps, E., Tashman, S.: Validation of three-dimensional model-based tibio-femoral tracking during running. Med. Eng. Phys. **31**(1), 10–16 (2009). https://doi.org/10.1016/j.medengphy.2008.03.003. http://www.sciencedirect.com/science/article/pii/S1350453308000441

2. Bufe, N., Heinemann, A., Köhler, Kecskeméthy, A.: An approach for bone pose estimation via three external ellipsoid pressure points. In: Proceedings of the 15th Conference on Advances in Robot Kinematics (ARK), Grasse, France (2016)
3. Bufe, N., Kecskeméthy, A.: Position analysis of a planar rigid-body tracked by three ellipse pressure points along straight lines. In: Proceedings of the 14th World Congress in Mechanism and Machine Science, Taipei, Taiwan (2015)
4. Cappello, A., Stagni, R., Fantozzi, S., Leardini, A.: Soft tissue artifact compensation in knee kinematics by double anatomical landmark calibration: performance of a novel method during selected motor tasks. IEEE Trans. Biomed. Eng. **52**(6), 992–998 (2005)
5. Cappozzo, A., Catani, F., Leardini, A., Benedetti, M., Croce, U.D.: Position and orientation in space of bones during movement: experimental artefacts. Clin. Biomech. **11**(2), 90–100 (1996)
6. Cappozzo, A., Croce, U.D., Leardini, A., Chiari, L.: Human movement analysis using stereophotogrammetry: part 1: theoretical background. Gait Posture **21**(2), 186–196 (2005)
7. Sharma, Gulshan B., Kuntze, G., Kukulski, D., Ronsky, J.L.: Validating dual fluoroscopy system capabilities for determining invivo knee joint soft tissue deformation: a strategy for registration error management. J. Biomech. **48**(10), 2181–2185 (2015)
8. Kecskeméthy, A.: Non-invasive sensor for the examination of the human or animal locomotor system (2010). http://www.directorypatent.com/EP/2186477-a1.html
9. Kecskeméthy, A., Hiller, M.: An object-oriented approach for an effective formulation of multibody dynamics. Comput. Methods Appl. Mech. Eng. **115**, 287–314 (1994)
10. Lemieux, L., Jagoe, R., Fish, D.R., Kitchen, N.D., Thomas, D.G.T.: A patient-to-computed-tomography image registration method based on digitally reconstructed radiographs. Med. Phys. **21**(11), 1749–1760 (1994). http://dx.doi.org/10.1118/1.597276
11. Peters, A., Galna, B., Sangeux, M., Morris, M., Baker, R.: Quantification of soft tissue artifact in lower limb human motion analysis: a systematic review. Gait Posture **31**(1), 1–8 (2010)
12. Sati, M., de Guise, J., Larouche, S., Drouin, G.: Quantitative assessment of skin-bone movement at the knee. Knee **3**(3), 121–138 (1996)

Author Index

J. Lenarcic and V. Parenti-Castelli (Eds.): ARK 2018, SPAR 8, pp. 475–477, 2019.
https://doi.org/10.1007/978-3-319-93188-3

The manufacturer's authorised representative in the EU is Springer Nature Customer Service Centre GmbH, Europaplatz 3, 69115 Heidelberg, Germany. If you have any concerns regarding our products, please contact ProductSafety@springernature.com

Printed and bound by CPI Group (UK) Ltd, Croydon, CR0 4YY
30/09/2026
02225230-0017